ANALYSIS ON GAUSSIAN SPACES

ANALYSIS ON GAUSSIAN SPACES

Yaozhong Hu
University of Kansas , USA

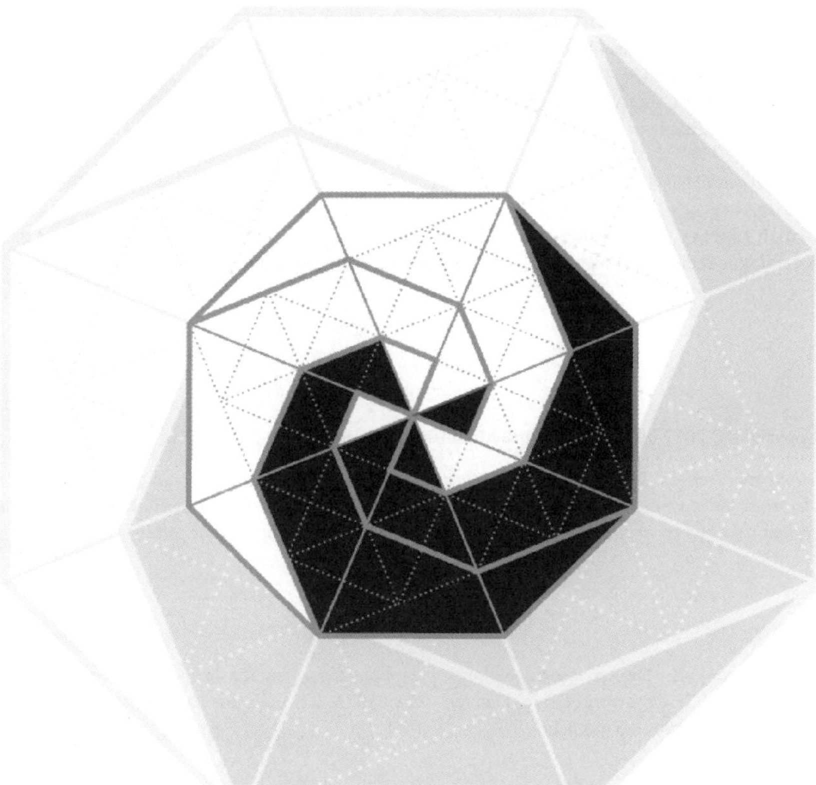

World Scientific

NEW JERSEY • LONDON • SINGAPORE • BEIJING • SHANGHAI • HONG KONG • TAIPEI • CHENNAI • TOKYO

Published by

World Scientific Publishing Co. Pte. Ltd.
5 Toh Tuck Link, Singapore 596224
USA office: 27 Warren Street, Suite 401-402, Hackensack, NJ 07601
UK office: 57 Shelton Street, Covent Garden, London WC2H 9HE

Library of Congress Cataloging-in-Publication Data
Names: Hu, Yaozhong, 1961–
Title: Analysis on Gaussian spaces / by Yaozhong Hu (University of Kansas, USA).
Description: New Jersey : World Scientific, 2016. |
Includes bibliographical references and index.
Identifiers: LCCN 2016025210 | ISBN 9789813142176 (hardcover : alk. paper)
Subjects: LCSH: Spaces of measures. | Gaussian measures. | Measure theory. |
Gaussian distribution. | Distribution (Probability theory)
Classification: LCC QA312 .H8 2016 | DDC 515/.42--dc23
LC record available at https://lccn.loc.gov/2016025210

British Library Cataloguing-in-Publication Data
A catalogue record for this book is available from the British Library.

Printed in Singapore by B & Jo Enterprise Pte Ltd

To Jun Fu and To Ruilong Hu

Preface

To study the classical analysis, one needs to understand well the Euclidean space $\mathbb{R}^d$ and the Lebesgue measure. One expects the same when one studies the nonlinear Gaussian functionals.

The study of infinite dimensional Banach space with a Gaussian measure (Gaussian space) and analysis of nonlinear functionals on Gaussian space have been studied since long and have found many applications. However, they are usually in different books aiming at different communities. One of the goals of this book is to put them together. It intends to present some basic materials on Gaussian space and on analysis of nonlinear Gaussian functionals. It differs from some existing books on Malliavin calculus not only because it contains some material on Gaussian space but also because it contains some material on analysis of nonlinear Gaussian functionals which are absent in popular Malliavin calculus books. For example, this book includes some important topics such as Littlewood-Paley-Stein-Meyer theory, interpolation inequality, complex hypercontractive inequality, correlation inequality, polarization technique, Wick product and so on. Effort has been made to allow this book accessible to broader audience. For example, some probabilistic proofs of (complex) hypercontractivity, correlation inequality and so on are now given in purely analytic ways. To facilitate the reading of the book, an appendix is included to present some basic results and their proofs from analysis and probability.

Due to page limitation the book excludes many other important applications such as applications to fractional Brownian motions, stochastic partial differential equations, and mathematical finance.

I take this opportunity to thank my advisor, P. A. Meyer, who taught me probability from the beginning. Although I had had a master degree in system science before I went to study under his guidance, I had very limited knowledge on probability theory back then. I am so grateful for his endless patience and generous encouragement.

Many mathematicians have helped me through my career of study and research. I very much appreciate their open-handed help and constant support. Here, let me mention particularly my postdoctoral advisors: Sergio Albeverio, Gopinath Kallianpur, Bernt Øksendal, and Weian Zheng. I also thank my colleagues, Tyrone Duncan, David Nualart, and Bozenna Pasik-Duncan. Special thanks go to David Elworthy and Jiaan Yan.

Finally, I would like express my sincere appreciation to Ms Rok Ting Tan and other staff from World Scientific for their generous support in the publication of this book.

Yaozhong Hu
Lawrence, Kansas
and Oslo, Norway

Contents

Chapter 1

Introduction

Analysis of functions on finite dimensional Euclidean space $\mathbb{R}^d$ with Lebesgue measure dx is one of the most important mathematical subjects. Researchers have been working to extend it to infinite dimension. In the absence of measure, there have been a great amount of works on nonlinear functional analysis, see for example, [Zeidler (1986, 1990a,b, 1985, 1988)].

Infinite dimensional Lebesgue measure has been (formally) used in physics such as in the famous Feynman path integral, which is one of the backbones of quantum physics. However, it is well-known (see Proposition 4.1) that in infinite dimensional space there is no Lebesgue measure. The next natural choice for infinite dimensional measure is the Gaussian measure, which is a family of measures depending on the mean and covariance. In fact, there have been large amount of works to justify the Feynman path integral by using analytic extension of integrals with respect to Gaussian measures. Another important application to infinite dimensional Gaussian measure is the Euclidean quantum field theory (see [Glimm and Jaffe (1987); Simon (1974, 2005)]) and stochastic quantization (see [Hu and Kallianpur (1998)] and the references therein).

However, even in finite dimensional case, analysis with respect to Gaussian measure is not a trivial analogy of the classical analysis with respect to Lebesgue measure. Many most frequently used properties with respect to Lebesgue measure is no longer true for Gaussian measure. For example, Lebesgue measure is translation invariant. Namely, if B is a Borel measurable subset of $\mathbb{R}^d$, then B and $a + B = \{a + b, b \in B\}$ have the same Lebesgue measure. But this is no longer true for Gaussian measure, which concentrates around the mean. However, it is quasi-invariant in the sense that Girsanov theorems hold (see Sections 6.4 and 6.5). Thus, to analyze function with respect to Gaussian measures, one needs to study in more details the Gaussian measure itself. There have been several books on Gaussian measure ([Bogachev (1998); Kuo (1975); Ledoux and Talagrand (2011); Lifshits (2012)]).

In this book, we shall first present some properties of Gaussian measures. In particular, we shall present the Brunn-Minkowski inequality, spectral gap and logarithmic Sobolev inequality (and their unification), hypercontractive inequality, variance inequality, correlation inequality (see Section 3). Usually, if an inequality holds true in finite dimension and if the constants appeared in the inequality do not depend on dimension, then this inequality can be extended to infinite dimension (via finite dimensional approximation). Their forms are straightforward and we shall not repeat them.

In probability theory, a Gaussian measure is usually given by a Gaussian stochastic process. The Gaussian measure can also be characterized by the corresponding process. As for general stochastic process a basic property is the sample path Hölder continuity. This amounts to say that the Gaussian measure is supported by the space of Hölder continuous functions. In this direction, the Garsia-Rodemich-Rumsey inequality seems to be quite powerful. We present this inequality and its extension from metric to volumetric case in Section 2.

In the infinite dimension, we shall deal with some inequalities (in Section 7) such as complex hypercontractive inequality, multiplier inequality, Littlewood-Paley-Stein-Meyer theory, Meyer's inequality, and interpolation inequality. The Grothendieck inequality is not for Gaussian measure. But an elegant proof uses Gaussian measure. Then, we also include it in this book.

One of the first things to do for a nonlinear Gaussian functional is to define the derivative and integration. In nonlinear functional analysis the Fréchet and Gâteaux (directional) derivatives are well-studied. However, in probability theory, many interesting functionals of Brownian motion that we encounter is not even continuous with respect to the sup norm. Here, we give a concrete example of Lévy area. Let $\mathbb{B} = \left\{\omega : [0,1] \to \mathbb{R}^d, \omega(0) = 0\right\}$ be the space of continuous functions starting with 0. It is a Banach space with respect to the sup norm: $\|\omega\| = \sup_{0\le t\le 1} |\omega(t)|$. The classical Wiener measure $\mathbb{P}$ is a probability measure on the Borel σ-algebra of $\mathbb{B}$ such that the coordinate mapping $\{W_t = W_t(\omega) = \omega(t)\,, 0 \le t \le 1\}$ is a d-dimensional Brownian motion (in probability theory, we quite often omit the dependence on sample random element ω). Consider the case $d = 2$. Let $\left\{(W_t^1, W_t^2), 0 \le t \le 1\right\}$ be a two dimensional Brownian motion. Then, the

Lévy area is defined by

$$L(W^1, W^2) = \int_0^1 W_t^1 dW_t^2 - \int_0^1 W_t^2 dW_t^1$$

which is in the second chaos (as a nonlinear functional on $\mathbb{B}$). When restricted to the space of continuously differentiable functions,

$$L(f,g) = \int_0^1 f_t dg_t - \int_0^1 g_t df_t \,.$$

In particular, $L(0,0) = 0$. Now, let $f_n(t) = \frac{1}{n}\sin(n^3 t)$ and $g_n(t) = -\frac{1}{n}\cos(n^3 t)$. We have

$$\lim_{n\to\infty} \sup_{0\le t\le 1} |f_n(t)| = \lim_{n\to\infty} \sup_{0\le t\le 1} |g_n(t)| = 0\,.$$

On the other hand, $L(f_n, g_n) = n \to \infty$. This means that the nonlinear functional $L(W^1, W^2)$ defined on $\mathbb{B}$ is not continuous with respect to the sup norm. If a nonlinear functional is not continuous, then its Fréchet derivative does not exist.

In infinite dimensional case, we also encounter another serious problem which we will now explain. When we apply the infinite dimensional analysis to probability theory, a nonlinear functional is usually given by a random variable $F : \mathbb{B} \to \mathbb{R}$. This random variable is defined only almost surely with respect to the Wiener measure $\mathbb{P}$. However, there is subset $\mathbb{B}_0 \subseteq \mathbb{B}$ such that $\mathbb{P}(\mathbb{B}_0) = 1$ and such that for any $x \in \mathbb{B}_0$, $\mathbb{P}(x_0 + \mathbb{B}) = 0$. Thus, $F(x + \varepsilon x_0)$ is not even well-defined and the Gâteaux derivative $D_h F(x) = \frac{d}{d\varepsilon}|_{\varepsilon=0} F(x + \varepsilon h)$ cannot be defined for all $h \in \mathbb{B}$. On the other hand, the Cameron-Martin space

$$\mathbb{H} = \left\{ h \in \mathbb{B}\,,\ h \text{ is absolutely continuous and } \int_0^1 |\dot{h}(t)|^2 dt < \infty \right\}$$

is a dense subset of $\mathbb{B}$ and for any $h \in \mathbb{H}$, $\mathbb{P}(\mathbb{B}+h) = 1$. Thus, one may define the derivative $D_h F(x) = \lim_{\varepsilon\to 0} \frac{F(x+\varepsilon h) - F(x)}{\varepsilon}$ for all $h \in \mathbb{H}$ (the limit can be taken in L^p or in probability). Gross developed a general framework for this triple $(\mathbb{B}, \mathbb{H}, \mathbb{P})$ which is called abstract Wiener space ([Gross (1967)]. See also [Kuo (1975)]). In the probability theory this derivative is called Gross-Sobolev derivative or more commonly Malliavin derivative.

Although the Gross-Sobolev derivative can be defined only along the directions $h \in \mathbb{H}$ and $\mathbb{P}(\mathbb{H}) = 0$ this concept and corresponding analysis have found many applications in analysis, geometry (see [Bismut (1984a,b, 2013); Elworthy, Le Jan and Li (1999)]), probability theory, and mathematical finance. One of the most famous examples is the application of the representation of the density of a random variable to the proof of Hörmander's theorem on hypoellipticity. Let us briefly explain this representation formula for one dimensional random variable. If $F : \mathbb{B} \to \mathbb{R}$ is a random variable and if there is a probability density function $\rho(x) : \mathbb{R} \to \mathbb{R}_+$ such that $\mathbb{P}(F \le a) = \int_{-\infty}^a \rho(x)dx$ for all $a \in \mathbb{R}$, then we say that the law of F is absolutely continuous (with respect to the Lebesgue measure). The representation formula for $\rho(x)$ can be stated as follows. If $DF : \mathbb{B} \to \mathbb{H}$ exists and if there is a $u : \mathbb{B} \to \mathbb{H}$ such that $\langle DF, u\rangle_{\mathbb{H}}^{-1} u$ is in the domain the adjoint operator δ of D and $\delta\left(\langle DF, u\rangle_{\mathbb{H}}^{-1} u\right) \in L^p$ for a certain $p \ge 1$, then $\rho(x)$ exists and can be represented by

$$\rho(x) = \mathbb{E}\left[I_{\{F\ge x\}}\delta\left(\frac{u}{\langle DF, u\rangle_{\mathbb{H}}}\right)\right].$$

The proof of the above identity is also easy. Denote the right hand side by $\tilde\rho(x)$. Then, for any smooth function $g : \mathbb{R} \to \mathbb{R}$ with compact support, we have (denoting $G(x) = \int_{-\infty}^x g(u)du$)

$$\begin{aligned}
\int_{\mathbb{R}} g(x)\tilde\rho(x)dx &= \mathbb{E}\left[\int_{\mathbb{R}} g(x)I_{\{F\ge x\}}dx\delta\left(\frac{u}{\langle DF, u\rangle_{\mathbb{H}}}\right)\right] \\
&= \mathbb{E}\left[G(F)\delta\left(\frac{u}{\langle DF, u\rangle_{\mathbb{H}}}\right)\right] \\
&= \mathbb{E}\left[\left\langle DG(F), \frac{u}{\langle DF, u\rangle_{\mathbb{H}}}\right\rangle\right] \\
&= \mathbb{E}\left[g(F)\left\langle DF, \frac{u}{\langle DF, u\rangle_{\mathbb{H}}}\right\rangle\right] = \mathbb{E}(g(F)).
\end{aligned}$$

This means that $\tilde\rho(x)$ is indeed the probability density function of F. It is interesting to note that in the above argument, the only serious property we have used is the chain rule for D ($DG(F) = g(F)DF$, namely, D is a derivative operator).

In probabilistic language the Hörmander theorem states that if F is the solution of a stochastic differential equation satisfying the hypoelliptic condition, then F has smooth density. The probabilistic approach to the Hörmander theory was first proposed in [Malliavin (1978)]. The first rigorous proof of the fact that $\langle DF, DF\rangle^{-1} \in L^p$ (in many cases one take $u = DF$) for the solution of stochastic differential equation satisfying the hypoelliptic condition can be found in [Watanabe (1984)]. See Section 10.2 for details. Those readers who are only interested in the probabilistic proof of Hörmander's theorem refer to [Hairer (2011)].

The first six chapters of the book will be devoted to general theory of Gaussian measure and analysis of nonlinear Gaussian functionals. Most of these materials may be found in a variety of sources. We put them together in an organic way so that it conforms with the theme of analysis of nonlinear Gaussian functionals in infinite dimensional Gaussian space.

After general theory of analysis of nonlinear Gaussian functionals, we dedicate ourselves to analysis of some particular classes of Gaussian functionals. We pay attention to three classes of functionals: Wiener chaos (multiple Wiener-Itô and Stratonovich integrals); local time or self-intersection local time of Brownian motion; and solution of stochastic differential equations. The focus will be different from the existing books and will focus on author and collaborators' work.

Recently, there has been a great amount of work on analysis of fractional Brownian motions (see [Biagini, Hu, Øksendal and Zhang (2008); Hu (2005)]) and stochastic partial differential equations. Due to limitation of the page we choose to completely omit these topics.

To the best of author's knowledge, some proofs of inequalities appear the first time in this book. For example, it is agreed by some researchers that one of the most elegant proofs of Nelson's hypercontractivity is Neveu's proof by using Brownian motion and Itô formula ([Neveu (1976)], see also [Dellacherie, Maisonneuve and Meyer (1987)]). In this book, we construct a function by using heat semigroup and the positivity of the derivative implies the hypercontractive inequality (see Theorem 3.19 for a generalization of the hypercontractivity). The Janson's original proof of complex hypercontractive inequality uses also Brownian motion and martingales (see [Janson (1997)]). We present also a new analytic proof in the book. See Section 7.1.

To facilitate the reading of the book, we include an appendix where one may find some results and their proofs from classical analysis and probability theory used in this book.

Chapter 2

Garsia-Rodemich-Rumsey Inequality

Given a function f (of one, finite, or infinitely many variables), the Garsia-Rodemich-Rumsey inequality can be used to bound the increment $|f(t) - f(s)|$ or rectangular increment of f by an integral with respect to the modulus of continuity of f. This type of inequality is powerful in establishing the sample path continuity and Hölder continuity of a stochastic process. It is also useful in dealing with the uniform convergence in the absence of martingale structure. The first section concerns with the distance $|f(t) - f(s)|$. The second section deals with volume-metric like increment (rectangular increment) of the form $|f(s_2, t_2) - f(s_1, t_2) - f(s_2, t_1) + f(s_1, t_2)|$. The third section contains a simple application.

2.1 Metric Garsia-Rodemich-Rumsey Inequality

The following inequality plays an important role in proving the Hölder continuity of a stochastic process.

Theorem 2.1 (Garsia-Rodemich-Rumsey inequality). *Let p and Ψ be continuous, strictly increasing functions on $[0, \infty)$ such that*

$$p(0) = \Psi(0) = 0 \quad \text{and} \quad \lim_{t\to\infty} \Psi(t) = \infty\,.$$

Given $T > 0$ and $\phi \in C([0,T], \mathbb{R}^d)$, if there is a constant B such that

$$\int_0^T \int_0^T \Psi\left(\frac{|\phi(t) - \phi(s)|}{p(|t-s|)}\right) dsdt \le B\,, \tag{2.1.1}$$

then for all $0 \le s \le t \le T$,

$$|\phi(t) - \phi(s)| \le 8 \int_0^{|t-s|} \Psi^{-1}\left(\frac{4B}{u^2}\right) p(du)\,. \tag{2.1.2}$$

Proof. We follow the argument of [Garsia, Rodemich and Rumsey (1970)] and divide the proof into several steps.

Step 1. Define

$$d_{-1} = T \quad \text{and} \quad I(t) = \int_0^T \Psi\left(\frac{|\phi(t) - \phi(s)|}{p(|t-s|)}\right) ds\,.$$

Since $\int_0^T I(t)dt \le B$ there is a $t_0 \in (0, d_{-1})$ such that $I(t_0) \le B/T$.

Step 2. First, we prove the following statement. If $d > 0$ is a given number and if two continuous and positive functions f and g satisfy $\int_0^d f(s) \le A$ and $\int_0^d g(s)ds \le B$ for two positive constants A and B, then there is a point $t_0 \in (0, d)$ such that $g(t_0) \le 2B/d$ and $f(t_0) \le 2A/d$. In fact, the conditions $\int_0^d f(s) \le A$ and $\int_0^d g(s)ds \le B$ imply $\int_0^d \left[\frac{f(s)}{A} + \frac{g(s)}{B}\right] ds \le 2$. Thus there is a point $t_0 \in (0, d)$ such that $\frac{f(t_0)}{A} + \frac{g(t_0)}{B} \le 2/d$ which implies that $f(t_0) \le 2A/d$ and $g(t_0) \le 2B/d$.

We define t_n and d_n by the following induction. d_{-1} and t_0 are defined as in above step 1. Given t_{n-1}, define d_{n-1} by $p(d_{n-1}) = \frac{1}{2}p(t_{n-1})$. Then, $d_{n-1} \le t_{n-1}$. Using the above statement and noticing that

$$\int_0^{d_{n-1}} I(t)dt \le B\,,$$

and

$$\int_0^{d_{n-1}} \Psi\left(\frac{|\phi(t) - \phi(t_{n-1})|}{p(|t - t_{n-1}|)}\right) dt = I(d_{n-1}) \le I(t_{n-1})\,,$$

we can choose $t_n \in (0, d_{n-1})$ such that

$$I(t_n) \le 2B/d_{n-1} \tag{2.1.3}$$

and

$$\Psi\left(\frac{|\phi(t_n) - \phi(t_{n-1})|}{p(|t_n - t_{n-1}|)}\right) \le 2I(t_{n-1})/d_{n-1}\,. \tag{2.1.4}$$

This way, we define t_n and d_n recursively.

Step 3. From the above construction we see

$$2p(d_{n+1}) = p(t_{n+1}) \le p(d_n) = \frac{1}{2}p(t_n)\,.$$

This implies that $d_{n+1} \le t_{n+1} \le d_n \le t_n$ and $t_n \downarrow 0$. Moreover, we have

$$\begin{aligned} p(t_n - t_{n+1}) \le p(t_n) = 2p(d_n) &= 4\left(p(d_n) - \frac{p(d_n)}{2}\right) \\ &\le 4\left(p(d_n) - \frac{p(t_{n+1})}{2}\right) \le 4(p(d_n) - p(d_{n+1}))\,. \end{aligned} \tag{2.1.5}$$

Step 4. Now we combine (2.1.3)-(2.1.5) to obtain

$$\begin{aligned}|\phi(t_n)-\phi(t_{n+1})| &\le \Psi^{-1}\left(2I(t_d)/d_d\right)p(t_n-t_{n+1})\\ &\le \Psi^{-1}\left(4B/d_{n-1}d_d\right)4(p(d_n)-p(d_{n+1}))\\ &\le 4\Psi^{-1}\left(4B/d_d^2\right)(p(d_n)-p(d_{n+1}))\\ &\le 4\int_{d_{n+1}}^{d_n}\Psi^{-1}\left(4B/u^2\right)p(du)\,.\end{aligned}$$

Sum over $n \ge 0$ to obtain (since $t_n \to 0$)

$$|\phi(t_0)-\phi(0)| \le 4\int_0^T \Psi^{-1}(4B/u^2)p(du)\,. \tag{2.1.6}$$

Replacing $\phi(t)$ by $\phi(T-t)$ we have

$$|\phi(T)-\phi(t_0)| \le 4\int_0^T \Psi^{-1}(4B/u^2)p(du)\,. \tag{2.1.7}$$

The above two inequalities yield

$$|\phi(T)-\phi(0)| \le 8\int_0^T \Psi^{-1}(4B/u^2)p(du)\,. \tag{2.1.8}$$

Step 5. For any $0 < s < t < T$, apply the above inequality (2.1.8) to the function

$$\tilde\phi(u) = \phi\left(s+\frac{t-s}{T}u\right), \quad u\in[0,T] \tag{2.1.9}$$

and

$$\tilde p(u) = p\left(\frac{t-s}{T}u\right), \quad u\in[0,T]\,. \tag{2.1.10}$$

It is easy to verify

$$\int_0^T\int_0^T \Psi\left(\frac{|\tilde\phi(v)-\tilde\phi(u)|}{\tilde p(|v-u|)}\right)dudv \le \frac{T^2}{(t-s)^2}B = \tilde B\,. \tag{2.1.11}$$

Thus, the equality (2.1.8) yields

$$|\tilde\phi(T)-\tilde\phi(0)| \le 8\int_0^T \Psi^{-1}(4\tilde B/u^2)\tilde p(du)\,. \tag{2.1.12}$$

This inequality can be simplified to (2.1.2). □

One immediate consequence of the above theorem is the following Kolmogorov lemma.

Corollary 2.1. *Let $\{X_t, 0 \le t \le T\}$ be a stochastic process satisfying that all the moments of X exist and there are constants $\alpha > 0$ and $p_0 > 1$ such that*

$$\mathbb{E}\left[|X_t - X_s|^p\right] \le C_p |t-s|^{\alpha p}, \quad \forall\, 0 \le s,t \le T \quad \text{and} \quad \forall\, p \ge p_0. \tag{2.1.13}$$

Then, for any $\beta \in (0,\alpha)$, the process X has a version which is Hölder continuous with exponent β.

Proof. Fix an arbitrary $p \ge p_0$. Take $\Psi(x) = x^p$ and $p(x) = x^\alpha$. Then $\Psi^{-1}(x) = x^{1/p}$. It is easy to see from (2.1.13) that

$$\mathbb{E}\left[\int_0^T \int_0^T \Psi\left(\frac{|X_t - X_s|}{p(|t-s|)}\right) dsdt\right] \le C_p\,.$$

Thus, the random variable $B = \int_0^T \int_0^T \Psi\left(\frac{|X_t - X_s|}{p(|t-s|)}\right) dsdt$ is almost surely finite. The Garsia-Rodemich-Rumsey inequality implies that

$$|X_t - X_s| \le 8\int_0^{|t-s|} \Psi^{-1}\left(\frac{4B}{u^2}\right) p'(u)du \le C|t-s|^{\alpha - 2/p} \le C|t-s|^{\beta}\,,$$

where we choose a p such that $p \ge p_0 \vee [2/(\alpha-\beta)]$. □

Corollary 2.2. *Let $\{X_t, 0 \le t \le T\}$ be a stochastic process satisfying that there are positive constants $\alpha, p > 0$ such that*

$$\mathbb{E}\left[|X_t - X_s|^p\right] \le C|t-s|^{1+\alpha}, \quad \forall\, 0 \le s,t \le T\,. \tag{2.1.14}$$

Then, X has a continuous version on $[0,T]$.

Proof. *Step 1.* First, we claim that for any fixed $\beta \in (0, \alpha/p)$ the following inequality holds

$$\mathbb{P}\left(\sup_{0\le s<t\le T} \frac{|X_t - X_s|}{|t-s|^\beta} > \lambda\right) \le C/\lambda^p\,. \tag{2.1.15}$$

Take $\Psi(x) = x^p$ and $p(x) = x^{\frac{p\beta+2}{p}}$. Denote $\Theta_T = \int_0^T \int_0^T \Psi\left(\frac{|X_t - X_s|}{p(|t-s|)}\right) dsdt$. From (2.1.14) it is easy to verify that

$$\mathbb{E}\left[\Theta_T\right] \le C\int_0^T \int_0^T |t-s|^{1+\alpha-p\beta-2} dsdt = CT^{1+\alpha-p\beta} < \infty\,.$$

By Chebyshev inequality, there is a constant C_T such that

$$\mathbb{P}(\Theta_T > \lambda) \leq \frac{C_T}{\lambda}, \quad \forall\, \lambda > 0\,. \tag{2.1.16}$$

If $\Theta_T \leq \lambda$, then by the Garsia-Rodemich-Rumsey inequality, we have

$$|X_t - X_s| \leq 8 \int_0^t \Psi^{-1}\left(\frac{4\lambda}{u^2}\right) p'(u)du = A\lambda^{1/p}|t-s|^\beta$$

for some constant A. This means that the event $\left\{\sup_{0\leq s<t\leq T} \frac{|X_t - X_s|}{|t-s|^\beta} > \lambda\right\}$ is contained in the event $\{\Theta_T > (\lambda/A)^p\}$. This combined with (2.1.16) proves (2.1.15).

Step 2. To prove the corollary, we needs to show that X is uniformly continuous on any countable subset S of $[0,T]$. For any such set S we can write $S = \cup_{n=1}^\infty S_n$, where $S_n := \{t_{0,n}\,, t_{1,n}\,, \cdots, t_{n,n}\}$ with $0 = t_{0,n} < t_{1,n} < \cdots < t_{n,n} = T$. We have

$$\begin{aligned} &\mathbb{P}\,(X \text{ is uniformly continuous on } S \text{ a.s. })\\ &\quad\geq 1 - \lim_{\lambda\uparrow\infty}\lim_{n\uparrow\infty} \mathbb{P}\left(\sup_{s,t\in S_n, s<t} \frac{|X_t - X_s|}{|t-s|^\beta} > \lambda\right). \end{aligned} \tag{2.1.17}$$

To use the result obtained in step 1 to bound the above probability, we define

$$X^{(n)}(t) = X_{t_{i,n}} + \frac{t - t_{i,n}}{t_{i+1,n} - t_{i,n}}\left(X_{t_{i+1,n}} - X_{t_{i,n}}\right).$$

Then, it is easy to check from (2.1.14) that

$$\mathbb{E}\left[|X_t^{(n)} - X_s^{(n)}|^p\right] \leq 3C|t-s|^{1+\alpha}, \quad \forall\, 0 \leq s, t \leq T\,.$$

Applying (2.1.15) to $X^{(n)}$ we see that the probability on the right hand side of (2.1.17) goes to 0 as $\lambda \to \infty$. Thus, $\mathbb{P}\,(X \text{ is uniformly continuous on } S \text{ a.s. }) = 1$, proving the corollary. □

2.2 Volume Metric Garsia-Rodemich-Rumsey Inequality

If $\{X_t, t = (t_1, \cdots, t_d) \in T\}$ is a multi-parameter process, where T is a (possibly an infinite dimensional) metric space, then one can extend the Garsia-Rodemich-Rumsey inequality to obtain $|X_t - X_s| \leq Cd(t,s)$, where d is a distance between t and s. This generalization is straightforward.

To explain the idea of this section, let us consider a function of two variables $f : [0,T]^2 \to \mathbb{R}$. We want to obtain some conditions such that

$$\begin{aligned} &|f(s_2,t_2) - f(s_1,t_2) - f(s_2,t_1) + f(s_1,t_1)|\\ &\qquad\leq C|s_2 - s_1|^{\alpha_1}|t_2 - t_1|^{\alpha_2}, \quad \forall\, s_1, t_1, s_2, t_2 \in [0,T]\,. \end{aligned}$$

The main difference between this type of inequality and the inequality (2.1.2) is that $d((s_1,t_1),(s_2,t_2)) = |s_2 - s_1||t_2 - t_1|$ is no longer a distance function. It is a volume-like metric.

In this section, we shall present a discussion of these types of inequalities and their applications to the joint Hölder continuity of a stochastic process. The results of this section are from [Hu and Le (2013)]. We also mention some further study on unbounded domain cases in [Hu and Le (2016)]. First, we state the following technical lemma which generalizes a crucial argument used in the proof of Theorem 2.1.

Lemma 2.1. *Let $(\Omega\,,\mathcal{F})$ be a measurable space and let μ be a positive measure on $(\Omega\,,\mathcal{F})$. Let $g:\Omega\times[0,T]\to\mathbb{R}^m$ be a measurable function such that*

$$\int_0^T\int_0^T\int_\Omega \Psi\left(\frac{|g(z,t)-g(z,s)|}{p(|t-s|)}\right)\mu(dz)dsdt \le B < \infty.$$

Then, there exist two decreasing sequences $\{t_k\,, k=0,1,\cdots\}$ and $\{d_k\,, k=0,1,\cdots\}$ with

$$t_k \le d_{k-1} = p^{-1}\left(\frac{1}{2}p(t_{k-1})\right), \quad k=1,2,\cdots \tag{2.2.1}$$

such that the following inequality holds

$$\int_\Omega \Psi\left(\frac{|g(z,t_k)-g(z,t_{k-1})|}{p(|s-t|)}\right)\mu(dz) \le \frac{4B}{d_{k-1}^2}. \tag{2.2.2}$$

Proof. We follow the proof of Theorem 2.1. Let

$$I(t) = \int_0^T\int_\Omega \Psi\left(\frac{|g(z,t)-g(z,s)|}{p(|t_k-t_{k-1}|)}\right)\mu(dz)ds.$$

From the assumption $\int_0^T I(t)dt \le B$ it follows that there is some $t_0\in(0,1)$ such that

$$I(t_0)\le B.$$

Now we describe how to obtain the sequences d_k and t_k recursively for $k=1,2,\cdots$. Given t_{k-1}, define

$$d_{k-1} = p^{-1}\left(\frac{1}{2}p(t_{k-1})\right).$$

Then, by the statement proved in Step 2 of the proof of Theorem 2.1 we can choose $t_k\le d_{k-1}$ such that

$$I(t_k) \le \frac{2B}{d_{k-1}} \tag{2.2.3}$$

and

$$\int_\Omega \Psi\left(\frac{|g(z,t_k)-g(z,t_{k-1})}{p(|t_k-t_{k-1}|)}\right)\mu(dz) \le \frac{2I(t_{k-1})}{d_{k-1}}\,. \tag{2.2.4}$$

Now (2.2.3) and (2.2.4) give

$$\int_\Omega \Psi\left(\frac{|g(z,t_k)-g(z,t_{k-1})|}{p(|t_k-t_{k-1}|)}\right)\mu(dz) \le \frac{2I(t_{k-1})}{d_{k-1}} \le \frac{4B}{d_{k-1}d_{k-2}} \le \frac{4B}{d_{k-1}^2}\,.$$

This is (2.2.2). □

Let $x=(x_1,\ldots,x_d)$ and $y=(y_1,\ldots,y_d)$ be two elements in $\mathbb{R}^d$. We denote $x'=(x_1,\ldots,x_{d-1})$ and $y'=(y_1,\ldots,y_{d-1})$. For each integer $k=1,2,\cdots,n$, we define

$$V_{k,y}x=(x_1,\ldots,x_{k-1},y_k,x_{k+1},\ldots,x_d).$$

Let f be a function from $\mathbb{R}^d$ to $\mathbb{R}^m$. We define the operator $V_{k,y}$ acting on f in the following way:

$$V_{k,y}f(x)=f(V_{k,y}x)$$

or $V_{k,y}f=f\circ V_{k,y}$ in short. It is straight forward to verify that

$$V_{k,y}V_{k,y}=V_{k,y}$$

and

$$V_{k,y}V_{l,y}=V_{l,y}V_{k,y}$$

for $k\neq l$. Next, we define the joint (rectangular) increment of a function f on an n-dimensional rectangle, namely

$$\Box_y^d f(x)=\prod_{k=1}^d (I-V_{k,y})f(x)\,,$$

where I denotes the identity operator.

Example 2.1. If $d=2$, then it is easy to see that

$$\Box_y^2 f(x)=f(y_1,y_2)-f(x_1,y_2)-f(y_1,x_2)+f(x_1,x_2)\,,$$

which is the increment of f over the rectangle containing the two points x and y with all sides parallel to the axis. In particular, if $f(x_1,x_2)=x_1x_2$, then $\Box_y^2 f(x)=(x_1-y_1)(x_2-y_2)$, which is the area of the rectangle. In a more general case, when f has the form $f(x)=\prod_{j=1}^d f_j(x_j)$, then

$$\Box_y^d f(x)=\prod_{j=1}^d [f_j(x_j)-f_j(y_j)]\,.$$

The following simple identity enables us to show our theorem by induction and it plays an essential role in our approach:

$$\square_y^d f(x) = \prod_{k=1}^{d-1}(I - V_{k,y})f(x) - V_{n,y}\prod_{k=1}^{d-1}(I - V_{k,y})f(x)$$
$$= \square_{y'}^{d-1} f(x', x_d) - \square_{y'}^{d-1} f(x', y_d). \tag{2.2.5}$$

We are now in the position to state our volumetric version of Theorem 2.1.

Theorem 2.2. *Suppose Ψ, p_k, $k = 1, \cdots, d$, are continuous, strictly increasing functions on $[0, \infty)$ such that*

$$p_k(0) = \Psi(0) = 0 \quad \text{and} \quad \lim_{t\to\infty} \Psi(t) = \infty\,.$$

Let $f(x)$ be a continuous function on $[0, T]^d$ and suppose that

$$\int_{[0,T]^{2d}} \Psi\left(\frac{|\square_y^d f(x)|}{\prod_{k=1}^d p_k(|x_k - y_k|)}\right) dxdy \le B < \infty\,. \tag{2.2.6}$$

Then, for all $s, t \in [0, T]^d$ we have

$$|\square_s^d f(t)| \le 8^d \int_0^{|s_1-t_1|} \cdots \int_0^{|s_d-t_d|} \Psi^{-1}\left(\frac{4^d B}{u_1^2 \cdots u_d^2}\right) dp_1(u_1)\cdots dp_d(u_d). \tag{2.2.7}$$

Proof. We proceed by induction on d. For $d = 1$, it coincides with the original Garsia-Rodemich-Rumsey inequality (2.1.2). Suppose (2.2.7) holds for $d - 1$. Let f be a continuous function on $[0, T]^d$. For any $x', y' \in \mathbb{R}^{d-1}$ and any $s \in [0, T]$, put

$$g(x', y', s) = \frac{\square_{y'}^{d-1} f(x', s)}{\prod_{k=1}^{d-1} p_k(|x'_k - y'_k|)}.$$

Let $\Omega = [0, T]^{d-1} \times [0, T]^{d-1}$, $z = (x', y')$. By (2.2.5) we can rewrite (2.2.6) as

$$\int_0^T \int_0^T \int_\Omega \Psi\left(\frac{|g(z, s) - g(z, t)|}{p_d(|s - t|)}\right) dzdsdt \le B < \infty.$$

Applying Lemma 2.1, we can find sequences $\{t_k\}$ and $\{d_k\}$ such that

$$t_k \le d_{k-1} = p_d^{-1}\left(\frac{1}{2} p_d(t_{k-1})\right) \tag{2.2.8}$$

and

$$\int_\Omega \Psi\left(\frac{|g(z, t_k) - g(z, t_{k-1})|}{p_d(t_k - t_{k-1})}\right) dz \le \frac{4B}{d_{k-1}^2}. \tag{2.2.9}$$

For each $k \in \mathbb{N}$ and $x' \in [0,T]^{d-1}$, let

$$h_k(x') = \frac{f(x',t_k) - f(x',t_{k-1})}{p_d(|t_k - t_{k-1}|)}.$$

Again from (2.2.5) it follows

$$\frac{\Box_{y'}^{d-1} h_k(x')}{\prod_{i=1}^{d-1} p_i(|x_i' - y_i'|)} = \frac{\Box_{y'}^{d-1} f(x',t_k) - \Box_{y'}^{d-1} f(x',t_{k-1})}{\prod_{i=1}^{d-1} p_i(|x_i' - y_i'|) p_d(|t_k - t_{k-1}|)}$$
$$= \frac{g(x',y',t_k) - g(x',y',t_{k-1})}{p_d(|t_k - t_{k-1}|)}.$$

Thus, the inequality (2.2.9) becomes

$$\int_{[0,T]^{d-1}} \int_{[0,T]^{d-1}} \Psi\left(\frac{\Box_{y'}^{d-1} h_k(x')}{\prod_{i=1}^{d-1} p_i(|x_i' - y_i'|)}\right) dx'dy' \le \frac{4B}{d_{k-1}^2}.$$

Now, by induction hypothesis, for every $k \ge 1$, $a, b \in [0,T]^{d-1}$, $a = (a_1, \ldots, a_{d-1})$ and $b = (b_1, \ldots, b_{d-1})$,

$$|\Box_a^{d-1} h_k(b)| \le 8^{d-1} \int_0^{|a_1 - b_1|} \cdots \int_0^{|a_{d-1} - b_{d-1}|} \Psi^{-1}\left(\frac{4^d B}{u_1^2 \cdots u_{d-1}^2 d_{k-1}^2}\right)$$
$$dp_1(u_1) \cdots dp_{d-1}(u_{d-1}).$$

Denoting $A = [0, |a_1 - b_1|] \times \cdots \times [0, |a_{d-1} - b_{d-1}|]$ and $dp(u_1, \cdots, u_{d-1}) = dp_1(u_1) \cdots dp_{d-1}(u_{d-1})$, the above inequality can be rewritten as

$$|\Box_a^{d-1} f(b,t_k) - \Box_a^{d-1} f(b,t_{k-1})|$$
$$\le 8^{d-1} \int_A \Psi^{-1}\left(\frac{4^d B}{u_1^2 \cdots u_{d-1}^2 d_{k-1}^2}\right) dp(u_1, \cdots, u_{d-1}) p_d(t_{k-1} - t_k). \tag{2.2.10}$$

On the other hand, by (2.2.8), we have

$$p_d(t_{k-1} - t_k) \le p_d(t_{k-1})$$
$$= 2p(d_{k-1}) \le 4\left[p(d_{k-1}) - p(d_k)\right].$$

Combining this inequality with (2.2.10) yields

$$|\Box_a^{d-1} f(b,t_0) - \Box_a^{d-1} f(b,0)| \le \sum_{k=1}^{\infty} |\Box_a^{d-1} f(b,t_k) - \Box_a^{d-1} f(b,t_{k-1})|$$
$$\le 8^{d-1} \sum_{k=1}^{\infty} 4\left[p(d_{k-1}) - p(d_k)\right] \int_A \Psi^{-1}\left(\frac{4^d B}{u_1^2 \cdots u_{d-1}^2 d_{k-1}^2}\right) dp(u_1, \cdots, u_{d-1})$$
$$\le 8^{d-1} \sum_{k=1}^{\infty} 4 \int_{d_k}^{d_{k-1}} \int_A \Psi^{-1}\left(\frac{4^d B}{u_1^2 \cdots u_{d-1}^2 u_d^2}\right) dp(u_1, \cdots, u_{d-1}) dp_d(u_d)$$
$$\le 8^{d-1} 4 \int_0^T \int_A \Psi^{-1}\left(\frac{4^d B}{u_1^2 \cdots u_d^2}\right) dp(u_1, \cdots, u_{d-1}) dp_d(u_d).$$

With $f(x', 1-x_d)$ replaced $f(x', x_d)$ we can obtain the same bound for

$$|\square_a^{d-1} f(b, t_0) - \square_a^{d-1} f(b, 1)|.$$

Hence, for every $a, b \in [0,T]^{d-1}$,

$$\begin{aligned}&|\square_a^{d-1} f(b,1) - \square_a^{d-1} f(b,0)|\\&\le 8^d \int_0^T \int_A \Psi^{-1}\left(\frac{4^d B}{u_1^2 \cdots u_d^2}\right) dp(u_1, \cdots, u_{d-1}) dp_d(u_d). \quad (2.2.11)\end{aligned}$$

To obtain (2.2.7) for general s, t in $[0,T]^d$, we set

$$\bar{f}(t', \tau) = f(t', s_d + \tau(t_d - s_d)) \text{ for } \tau \in [0,T]$$

and

$$\bar{p}_d(u) = p_d(u|s_d - t_d|).$$

Restricting the range of the integration in (2.2.6) and carrying out a change of variables we obtain

$$\int_{[0,T]^d} \int_{[0,T]^d} \Psi\left(\frac{|\square_y^d \bar{f}(x)|}{\prod_{k=1}^{d-1} p_k(|x_k - y_k|) \bar{p}_d(|x_d - y_d|)}\right) dx dy \le \frac{B}{|s_d - t_d|^2}.$$

Thus, by (2.2.11) we deduce

$$\begin{aligned}|\square_s^d f(t)| &= |\square_{s'}^{d-1} \bar{f}(t', 1) - \square_{s'}^{d-1} \bar{f}(t', 0)|\\&\le 8^d \int_0^T \int_0^{|s_1 - t_1|} \cdots \int_0^{|s_{d-1} - t_{d-1}|} \Psi^{-1}\left(\frac{4^d B}{u_1^2 \cdots u_d^2 |s_d - t_d|^2}\right)\\&\qquad dp(u_1, \cdots, u_{d-1}) dp_d(u_d |s_d - t_d|).\end{aligned}$$

Another change of variables yields (2.2.7). □

2.3 Sample Path Hölder Continuity of Random Fields

In this section, we study sample path joint Hölder continuity property of a given stochastic process by using Theorem 2.2. For any random variable X we shall use $\mathbb{E}(X)$ to denote the expectation of X.

Theorem 2.3. *Let W be a random field on $[0,T]^d$. Suppose there exist positive constants α, β_k $(1 \le k \le d)$ and K such that for every x, y in $[0,T]^d$,*

$$\mathbb{E}\left[\left|\square_y^d W(x)\right|^\alpha\right] \le K \prod_{k=1}^d |x_k - y_k|^{1+\beta_k}.$$

Then, for every $\varepsilon = (\varepsilon_1, \ldots, \varepsilon_d)$ with $0 < \varepsilon_k \alpha < \beta_k$ $(1 \le k \le d)$, there exists a random variable η with $\mathbb{E}\eta^\alpha \le K$, such that the following inequality holds almost surely

$$|\Box_t^d W(s)| \le C\eta(\omega) \prod_{k=1}^d |t_k - s_k|^{\beta_k \alpha^{-1} - \varepsilon_k}$$

for all s, t in $[0,T]^d$, where C is a constant defined by

$$C = 8^d 4^{d/\alpha} \prod_{k=1}^d \left(1 + \frac{2}{\beta_k - \alpha\varepsilon_k}\right).$$

Proof. Let $\Psi(u) = |u|^\alpha$, $p_k(u) = |u|^{\gamma_k}$, where $\gamma_k \in (\frac{2}{\alpha}, \frac{2+\beta_k}{\alpha})$, $1 \le k \le d$. A direct application of Theorem 2.2 gives that for all s, t in $[0,T]^d$

$$|\Box_s^d W(t)| \le 8^d \prod_{k=1}^d \frac{\gamma_k |t_k - s_k|^{\gamma_k - \frac{2}{\alpha}}}{\gamma_k - \frac{2}{\alpha}} \left(4^d \iint_{[0,T]^{2d}} \frac{|\Box_y^d W(x)|^\alpha}{\prod_{k=1}^d |x_k - y_k|^{\alpha\gamma_k}} dxdy\right)^{\frac{1}{\alpha}}. \tag{2.3.1}$$

Let

$$B(\omega) = \iint_{[0,T]^{2d}} \Psi\left(\frac{|\Box_y^d W(x)|}{\prod_{k=1}^d p_k(x_k - y_k)}\right) dxdy.$$

From our assumption and from Fubini-Tonelli's theorem, it follows

$$\begin{aligned}\mathbb{E}B &= \iint_{[0,T]^{2d}} \frac{\mathbb{E}|\Box_y^d W(x)|^\alpha}{\prod_{k=1}^d |x_k - y_k|^{\alpha\gamma_k}} dxdy \\ &\le K \iint_{[0,T]^{2d}} \prod_{k=1}^d |x_k - y_k|^{1+\beta_k - \alpha\gamma_k} dxdy < \infty.\end{aligned}$$

Hence, the event $\Omega^* = \{\omega : B(\omega) < \infty\}$ has probability one. Therefore, for each ω in Ω^*, the inequality (2.3.1) gives

$$|\Box_s^d W(t,\omega)| \le 8^d \prod_{k=1}^d \frac{\gamma_k |t_k - s_k|^{\gamma_k - \frac{2}{\alpha}}}{\gamma_k - \frac{2}{\alpha}} \left(4^d B(\omega)\right)^{\frac{1}{\alpha}}$$

for every s, t in $[0,T]^d$. For each k, the power $\gamma_k - \frac{2}{\alpha}$ can be made arbitrarily close to $\frac{\beta_k}{\alpha}$. This completes the proof with $\eta = B^{1/\alpha}$. □

Definition 2.1. A stochastic process $\{W_x^H, x \in [0,1]^d\}$ is called a fractional Brownian field with Hurst parameter $H = (H_1, \cdots, H_d)$ if it is a mean 0 Gaussian field with the following covariance

$$\mathbb{E}(W_x^H W_y^H) = \frac{1}{2^d} \prod_{i=1}^d \left(|x_i|^{2H_i} + |y_i|^{2H_i} - |x_i - y_i|^{2H_i}\right), \tag{2.3.2}$$

where $x = (x_1, \cdots, x_d)$, $y = (y_1, \cdots, y_d) \in [0,1]^d$.

Proposition 2.1. *Let W^H be a fractional Brownian field on $[0,1]^d$ with Hurst parameter $H=(H_1,\dots,H_d)$. Then, there exist a random constant $C_{d,H_1,\cdots,H_d}$ satisfying $\mathbb{E}\exp\left\{\kappa C^2_{d,H_1,\cdots,H_d}\right\}<\infty$ for some $\kappa>0$ and such that for every x,y in $[0,1]^d$*

$$|\Box_y^d W^H(x)|\le C_{d,H_1,\cdots,H_d}\sqrt{\left|\log\left(\prod_{k=1}^d|x_k-y_k|\right)\right|}\prod_{k=1}^d|x_k-y_k|^{H_k}. \quad (2.3.3)$$

Proof. From (2.3.2), it is easy to see that

$$\mathbb{E}|\Box_y^d W^H(x)|^2=\prod_{k=1}^d|x_k-y_k|^{2H_k}.$$

Thus, we have (see Chapter 3, (3.1.8))

$$\begin{aligned}\mathbb{E}|\Box_y^d W^H(x)|^{2n}&=\frac{(2n)!}{2^n n!}\prod_{k=1}^d|x_k-y_k|^{2nH_k}\\&\le C^n\Gamma(n+1)\prod_{k=1}^d|x_k-y_k|^{2nH_k}\end{aligned}$$

for some constant $C>0$. Fix a constant $c>0$ to be specified later and consider $\Psi(u)=e^{cu^2}-1$. Let $p_k(u)=u^{H_k}$ and denote

$$B=\int_{[0,1]^{2d}}\Psi\left(\frac{\Box_y^d W^H(x)}{\prod_{k=1}^d p_k(|x_k-y_k|)}\right)dxdy\,.$$

Then, we have

$$\begin{aligned}\mathbb{E}(B)&=\sum_{n=1}^\infty\frac{c^n}{n!}\frac{\mathbb{E}\left|\Box_y^d W^H(x)\right|^{2n}}{\prod_{k=1}^d p_k^{2n}(|x_k-y_k|)}\\&\le\sum_{n=1}^\infty\frac{c^n}{n!}C^n\Gamma(n+1)<\infty\end{aligned}$$

for some constant $c>0$. This means that $B<\infty$ a.s. By Theorem 2.2

$$\begin{aligned}\left|\Box_y^d W(x)\right|&\le 8^d\int_0^{|x_1-y_1|}\cdots\int_0^{|x_d-y_d|}\Psi^{-1}\left(\frac{4^dB}{u_1^2\cdots u_d^2}\right)dp_1(u_1)\cdots dp_d(u_d)\\&\le C_{d,H_1,\cdots,H_d}\int_0^{|x_1-y_1|}\cdots\int_0^{|x_d-y_d|}\sqrt{|\log(u_1\cdots u_d)|}\\&\qquad u_1^{H_1-1}\cdots u_d^{H_d-1}du_1\cdots du_d\,.\end{aligned}$$

Inequality (2.3.3) follows then from a simple estimation of the above integral. It is easy to see that $C_{d,H_1,\cdots,H_d}\le C\sqrt{\log B}$ for some constant $C>0$ and the proposition is proved. □

Chapter 3

Analysis With Respect to Gaussian Measure in $\mathbb{R}^d$

3.1 Gaussian Measure in $\mathbb{R}^d$

Let $\mathbb{R}^d$ be the d-dimensional Euclidean space with the scalar product $\langle x, y\rangle = \sum_{i=1}^{d} x_i y_i$ and the corresponding Euclidean norm $|x| = \sqrt{\sum_{i=1}^d x_i^2}$. The Borel σ-algebra $\mathcal{B}(\mathbb{R}^d)$ of $\mathbb{R}^d$ is the σ-algebra generated by all open sets of $\mathbb{R}^d$. Let Σ be any positive definite (non degenerate) symmetric $d \times d$ matrix and let $\xi \in \mathbb{R}^d$ be any element in $\mathbb{R}^d$. Denote

$$\phi_{\xi,\Sigma}(x) = (2\pi)^{-d/2} \det(\Sigma)^{-1/2} \exp\left\{-\frac{(x-\xi)^T\Sigma^{-1}(x-\xi)}{2}\right\}, \quad x \in \mathbb{R}^d, \tag{3.1.1}$$

where A^T denotes the transpose of a matrix A or a vector A (we always take column form for a vector in $\mathbb{R}^d$). The Gaussian measure $\mu_d(dx; \xi, \Sigma)$ on $\mathbb{R}^d$ is the probability measure given by

$$\mu_d(B; \xi, \Sigma) = \int_B \phi_{\xi,\Sigma}(x)dx, \quad \forall\, B \in \mathcal{B}(\mathbb{R}^d).$$

A random variable $X : (\Omega, \mathcal{F}, P) \to \mathbb{R}^d$ is called a Gaussian (or normal) (random) vector (variable) with mean ξ and variance Σ if the probability law of this random variable X is $\mu_d(dx; \xi, \Sigma)$: Namely,

$$P(\omega \in \Omega;\ X(\omega) \in B) = \mu_d(B; \xi, \Sigma), \quad \forall\, B \in \mathcal{B}(\mathbb{R}^d). \tag{3.1.2}$$

We also use the notation $X \sim N(\xi, \Sigma)$.

The characteristic function of this measure $\mu_d(dx; \xi, \Sigma)$ (or this random variable X) is

$$\begin{aligned}\Psi_{\xi,\Sigma}(z) &:= \mathbb{E}e^{\iota\langle z, X\rangle} = \int_{\mathbb{R}^d} e^{\iota\langle z,x\rangle}\mu_d(dx;\xi,\Sigma)\\ &= \exp\left\{\iota\langle z,\xi\rangle - \frac{z^T\Sigma z}{2}\right\}, \quad \forall\, z \in \mathbb{R}^d, \end{aligned}\tag{3.1.3}$$

where and throughout the book $\iota = \sqrt{-1}$ denotes the imaginary number.

The moment generating function of this measure is

$$M_{\xi,\Sigma}(z) := \mathbb{E}e^{\langle z, X\rangle} = \int_{\mathbb{R}^d} e^{\langle z,x\rangle} \mu_d(dx;\xi,\Sigma)$$
$$= \exp\left\{\langle z,\xi\rangle + \frac{z^T\Sigma z}{2}\right\}, \quad \forall\, z\in\mathbb{R}^d. \tag{3.1.4}$$

Taking the inverse Fourier transform of (3.1.3), we have

$$\phi_{\xi,\Sigma}(x) = \frac{1}{(2\pi)^d}\int_{\mathbb{R}^d} \exp\left\{\iota\langle z, \xi - x\rangle - \frac{z^T\Sigma z}{2}\right\} dz\,, \quad \forall\, z\in\mathbb{R}^d. \tag{3.1.5}$$

We are going to deduce some properties associated with this Gaussian measure $\mu_d(dx;\xi,\Sigma)$.

Proposition 3.1. *Let $X \sim N(\xi,\Sigma)$ and let $f:\mathbb{R}^d\to\mathbb{R}$ be continuously differentiable. Then*

$$\mathbb{E}\left[\left(\Sigma^{-1}(X-\xi)\right) f(X)\right] = -\mathbb{E}(\nabla f(X))\,, \tag{3.1.6}$$

where $\nabla f = (\partial_1 f, \cdots, \partial_d f)^T$.

The above identity is understood component-wise.

Proof. The left hand side of (3.1.6) is

$$\text{LHS of (3.1.6)} = \int_{\mathbb{R}^d} \left(\Sigma^{-1}(x-\xi)\right) f(x)\phi_{\xi,\Sigma}(x)dx$$
$$= \int_{\mathbb{R}^d} f(x)\nabla\phi_{\xi,\Sigma}(x)dx$$
$$= -\int_{\mathbb{R}^d} \nabla f(x)\phi_{\xi,\Sigma}(x)dx\,,$$

proving (3.1.6). □

Similarly, we have

Proposition 3.2. *Let $X \sim N(\xi,\Sigma)$ and let $f:\mathbb{R}^d\to\mathbb{R}^d$ be continuously differentiable. Then*

$$\mathbb{E}\left[\left\langle \Sigma^{-1}(X-\xi)\,, f(X)\right\rangle\right] = -\mathbb{E}(\text{Tr}(\nabla f))\,, \tag{3.1.7}$$

where $\text{Tr}(\nabla f)(x) = \sum_{i=1}^d \partial_i f_i(x)$ is the trace of the matrix $\nabla f(x) = (\partial_j f_i(x))_{1\le i,j\le d}$.

If $\xi = 0$ and $\Sigma = I$ (the identity matrix in $\mathbb{R}^d$), then

$$\mu(dx) = \mu_d(dx) = \mu_d(dx; \xi, \Sigma) = \phi(x)dx\,,$$

where

$$\phi(x) = (2\pi)^{-d/2} e^{-\frac{|x|^2}{2}}$$

is called the standard Gaussian measure on $\mathbb{R}^d$.

Example 3.1. If $X \sim N(0, \sigma^2)$ is a one dimensional mean zero Gaussian with variance σ^2, then

$$\sum_{k=0}^{\infty} \frac{z^k \mathbb{E}(X^k)}{k!} = \mathbb{E}e^{zX} = e^{z^2\sigma^2/2} = \sum_{n=0}^{\infty} \frac{z^{2n}\sigma^{2n}}{2^n n!}\,.$$

Comparing the coefficients of z^k, we obtain

$$\mathbb{E}\left(X^k\right) = \begin{cases} \frac{(2n)!\sigma^{2n}}{2^n n!} & \text{if } k = 2n \text{ is even} \\ 0 & \text{if } k = 2n \text{ is odd}\,. \end{cases} \tag{3.1.8}$$

The Gaussian measure is one of the most important measures in infinite dimensional analysis since the Lebesgue measure does not exists in infinite dimensional space (as we shall see in Section 4).

If $X\,, Y\,:\ (\Omega, \mathcal{F}, \mathbb{P}) \to \mathbb{R}^2$ are two random variables, their covariance, variance, and correlation are defined respectively as

$$\begin{aligned} \mathrm{Cov}(X, Y) &= \mathbb{E}(X - \mathbb{E}(X))(Y - \mathbb{E}(Y)) = \mathbb{E}(XY) - \mathbb{E}(X)\mathbb{E}(Y)\,; \\ \mathrm{Var}(X) &= \mathrm{Cov}(X, X)\,; \\ \mathrm{Corr}(X, Y) &= \frac{\mathrm{Cov}(X, Y)}{\sqrt{\mathrm{Var}(X)\mathrm{Var}(Y)}}\,. \end{aligned}$$

We have easily

Proposition 3.3. *If $X = (X_1, \cdots, X_d)$ is a Gaussian random vector with mean ξ and variance matrix $\Sigma = (\sigma_{ij})_{1\le i,j\le d}$, then ξ is the mean of X and Σ is the covariance matrix of X.*

$$\xi = \mathbb{E}(X) \quad \text{and} \quad \Sigma = \mathbb{E}\left[(X - \mathbb{E}(X))(X - \mathbb{E}(X))^T\right]\,.$$

3.2 Some Inequalities Related to Gaussian Measure

Definition 3.1. A function $f : \mathbb{R}^d \to \mathbb{R}$ is called component-wise nondecreasing if for any $j = 1, 2, \cdots, d$, the function $\mathbb{R} \ni x_j \to f(\cdots, x_j, \cdots)$ is nondecreasing when other variables are kept fixed.

The random variables $X_1, \cdots, X_d$ are said to be (positively) associated if they satisfy the FKG (Fortuin-Kasteleyn-Ginibre) type inequality:

$$\mathbb{E}\left[f(X_1, \cdots, X_d)g(X_1, \cdots, X_d)\right] \geq \mathbb{E}\left[f(X_1, \cdots, X_d)\right]\mathbb{E}\left[g(X_1, \cdots, X_d)\right] \tag{3.2.1}$$

for any two component-wise nondecreasing functions f and g.

The above inequality means that f and g are positively correlated:

$$\mathrm{Corr}(f(X_1, \cdots, X_d), g(X_1, \cdots, X_d)) \geq 0 .$$

Definition 3.2. An open set A in $\mathbb{R}^d$ is called an increasing set if $x = (x_1, \cdots, x^d)^T \in A$ and if $y = (y_1, \cdots, y_d)$ satisfies $y_i \geq x_i$ for all $i = 1, 2, \cdots, d$, then $y \in A$.

If A is an increasing set, then the indicate function $f(x) = I_A(x)$ is a component-wise nondecreasing function.

Example 3.2. If $a = (a_1, \cdots, a_d)^T \in \mathbb{R}^d$, then $\prod_{i=1}^d [a_i, \infty)$ is an increasing set.

Theorem 3.1. *Let $X = (X_1, \cdots, X_d)$ be a Gaussian vector with mean ξ and covariance matrix $\Sigma = (\sigma_{ij})$. Then, X is associated if and only if*

$$\sigma_{ij} \geq 0 \quad \forall\, i \neq j . \tag{3.2.2}$$

Proof. If f is component wise nondecreasing, then $f(x - \xi)$ is also component wise nondecreasing. Thus, we can assume that $\xi = 0$. By a limiting argument, one may also assume that Σ is strictly positive definite.

First, we show the sufficiency. Assume (3.2.2). By a limiting argument, we can assume that the functions f and g in the definition (3.2.1) are continuously differentiable.

Let Z be an independent copy of X. For any $\lambda \in 0, 1]$ we define $Y_\lambda = \lambda X + \sqrt{1 - \lambda^2} Z$. Let us consider

$$\psi(\lambda) = \mathbb{E}\left[f(X)g(Y_\lambda)\right] = \mathbb{E}\left[f(X)g(\lambda X + \sqrt{1 - \lambda^2} Z)\right] .$$

Differentiating the above function $\psi(\lambda)$ with respect to λ, we obtain

$$\begin{aligned}\frac{d}{d\lambda}\psi(\lambda) &= \mathbb{E}\left[f(X)g(Y_\lambda)\right] \\ &= \mathbb{E}\left\langle f(X)\nabla g(\lambda X + \sqrt{1 - \lambda^2} Z), \left[X - \frac{\lambda}{\sqrt{1 - \lambda^2}} Z\right]\right\rangle .\end{aligned}$$

The integration by parts identity (3.1.7) yields

$$\begin{aligned}
\frac{d}{d\lambda}\psi(\lambda) &= \mathbb{E}\left[f(X)g(Y_\lambda)\right] \\
&= \mathbb{E}\left\langle f(X)\Sigma\nabla g(\lambda X+\sqrt{1-\lambda^2}Z)\,,\left[\Sigma^{-1}X-\frac{\lambda}{\sqrt{1-\lambda^2}}\Sigma^{-1}Z\right]\right\rangle \\
&= -\mathbb{E}\left\langle \nabla f(X)\,,\Sigma\nabla g(\lambda X+\sqrt{1-\lambda^2}Z)\right\rangle \\
&\qquad -\lambda\mathbb{E}\left[f(X)\mathrm{Tr}(\Sigma\nabla^2 g(\lambda X+\sqrt{1-\lambda^2}Z))\right] \\
&\qquad +\frac{\lambda}{\sqrt{1-\lambda^2}}\sqrt{1-\lambda^2}\mathbb{E}\left[f(X)\mathrm{Tr}(\Sigma\nabla^2 g(\lambda X+\sqrt{1-\lambda^2}Z))\right] \\
&= -\mathbb{E}\left\langle \nabla f(X)\,,\Sigma\nabla g(\lambda X+\sqrt{1-\lambda^2}Z)\right\rangle \le 0
\end{aligned}$$

from the condition that $\nabla f \ge 0$, $\nabla g \ge 0$ and $\Sigma \ge 0$. This means that $\psi(\lambda)$ is a decreasing function of λ. Namely, $\psi(0) \ge \psi(1)$. But $\psi(0)$ and $\psi(1)$ are the left hand side and the right hand side of (3.2.1), respectively. This proves the sufficiency of (3.2.2).

The necessity is easy. For any $i, j = 1, \cdots, d$, we take $f(x) = x_i$ and $g(x) = x_j$. It is obvious that $\mathbb{E}(f(X)g(Y)) = \sigma_{ij}$ and $\mathbb{E}(f(X))\mathbb{E}(g(Y)) = 0$. Thus, (3.2.1) implies (3.2.2). □

Remark 3.1. Since Σ is positive definite, $\sigma_{ii} \ge 0$ always hold. Thus, condition (3.2.2) is equivalent to $\sigma_{ij} \ge 0$ for all i, j.

The following Prékopa-Leindler inequality ([Prékopa (1971, 1973)]) is a very important inequality and is related to a number of other important inequalities in analysis such as the reverse Hölder inequality, the reverse Young's inequality, and the Brunn-Minkowski inequality.

Theorem 3.2. *Let $0 < \lambda < 1$ and let the measurable functions $f, g, h : \mathbb{R}^d \to \mathbb{R}_+$ satisfy the following log concave relation:*

$$h\left((1-\lambda)x+\lambda y\right) \ge f(x)^{1-\lambda}g(y)^{\lambda} \quad \textit{for all } x \textit{ and } y \textit{ in } \mathbb{R}^d\,. \tag{3.2.3}$$

Then

$$\|h\|_1 := \int_{\mathbb{R}^d} h(x)\,dx \ge \left(\int_{\mathbb{R}^d} f(x)\,dx\right)^{1-\lambda}\left(\int_{\mathbb{R}^d} g(x)\,dx\right)^{\lambda} =: \|f\|_1^{1-\lambda}\|g\|_1^{\lambda}\,. \tag{3.2.4}$$

Proof. We follow the proof given in [Gardner (2002)]. The proof will be done through induction on the dimension d.

First, we consider the case when $d = 1$. Without loss of generality, we assume that there are F and G such that

$$\int_{\mathbb{R}} f(x) = F > 0\,, \quad \text{and} \quad \int_{\mathbb{R}} g(x) = G > 0\,.$$

Define $u; v : [0; 1] \to \mathbb{R}$ as the two smallest numbers such that

$$\frac{1}{F}\int_{-\infty}^{u(t)} f(x)dx = t\,, \quad \text{and} \quad \frac{1}{F}\int_{-\infty}^{v(t)} g(x)dx = t\,, \tag{3.2.5}$$

where $0 \le t \le 1$. The functions u and v may be discontinuous, but they are strictly increasing functions and so they are differentiable almost everywhere. Let

$$w(t) = (1-\lambda)u(t) + \lambda v(t)\,.$$

Differentiating (3.2.5) with respect to t yields

$$\frac{f(u(t))u'(t)}{F} = \frac{g(u(t))u'(t)}{G} = 1\,. \tag{3.2.6}$$

Using the above identities and the arithmetic-geometric mean inequality, we obtain (when $f(u(t)) \neq 0$ and $g(u(t)) \neq 0$)

$$\begin{aligned} w'(t) &= (1-\lambda)u'(t) + \lambda v'(t) \\ &\ge u'(t)^{1-\lambda} v'(t)^{\lambda} \\ &= \left(\frac{F}{f(u(t))}\right)^{1-\lambda} \left(\frac{G}{g(u(t))}\right)^{\lambda}. \end{aligned}$$

Therefore, from the above inequality and the assumption on f, g, h we obtain

$$\begin{aligned} \int_{\mathbb{R}} h(x)dx &\ge \int_0^1 h(w(t))w'(t)dt \\ &\ge \int_0^1 f(u(t))^{1-\lambda} g(v(t))^{\lambda} \left(\frac{F}{f(u(t))}\right)^{1-\lambda} \left(\frac{G}{g(u(t))}\right)^{\lambda} dt \\ &= F^{1-\lambda} G^{\lambda}\,. \end{aligned}$$

This is (3.2.4) when $d = 1$.

Now assume that the inequality (3.2.4) holds true for any dimension less than or equal to $d-1$. We want to show that the inequality also holds true for the dimension d.

For any $s \in \mathbb{R}$, define a function of $d-1$ variables by $h_s(z) = h(s, z)$, $z \in \mathbb{R}^{d-1}$ and same notations for f_s and g_s. For any $x, y \in \mathbb{R}^{d-1}$, $a, b \in \mathbb{R}$, and $\lambda \in [0, 1]$, we let $c = (1-\lambda)a + \lambda b$. Then

$$\begin{aligned} h_c((1-\lambda)x + \lambda y) &\ge h((1-\lambda)a + \lambda b, (1-\lambda)x + \lambda y) \\ &\ge f(a, x)^{1-\lambda} g(b, y)^{\lambda} \\ &= f_a(x)^{1-\lambda} g_b(y)^{\lambda}\,. \end{aligned}$$

From the induction hypothesis it follows that

$$\int_{\mathbb{R}^{d-1}} h_c(x)dx \geq \left(\int_{\mathbb{R}^{d-1}} f_a(x)dx\right)^{1-\lambda} \left(\int_{\mathbb{R}^{d-1}} g_b(x)dx\right)^{\lambda}. \quad (3.2.7)$$

Now define the following functions of one variable:

$$H(c) = \int_{\mathbb{R}^{d-1}} h_c(x)dx, \quad F(a) = \int_{\mathbb{R}^{d-1}} f_a(x)dx\,, \quad G(b) = \int_{\mathbb{R}^{d-1}} g_b(x)dx\,.$$

The inequality (3.2.7) means that the condition (3.2.3) is satisfied for the above three function. Thus, by the inequality (3.2.4) in the case $d = 1$, we have

$$\begin{aligned}\int_{\mathbb{R}^d} h(x)dx &= \int_{\mathbb{R}} \int_{\mathbb{R}^{d-1}} h(c,z)dzdc = \int_{\mathbb{R}} h(c)dc \\ &\geq \left(\int_{\mathbb{R}} F(a)da\right)^{1-\lambda} \left(\int_{\mathbb{R}} G(b)db\right)^{\lambda} \\ &= \left(\int_{\mathbb{R}^d} f(x)dx\right)^{1-\lambda} \left(\int_{\mathbb{R}^d} g(x)dx\right)^{\lambda},\end{aligned}$$

proving the theorem. □

Theorem 3.3. *Let $f(x,y)$ be a (jointly) logarithmic concave function of $(x,y) \in \mathbb{R}^d \times \mathbb{R}^m$. Let A be a convex subset of $\mathbb{R}^m$. Then, the function $\mathbb{R}^d \ni x \to \int_A f(x,y)dy$ is a logarithmic concave function on $\mathbb{R}^d$.*

Proof. Let $x_1, x_2 \in \mathbb{R}^d$ and $\lambda \in (0,1)$. Define

$$f_1(y) = \begin{cases} f(x_1, y) & \text{if } y \in A \\ 0 & \text{if } y \notin A\,; \end{cases}$$

$$f_2(y) = \begin{cases} f(x_2, y) & \text{if } y \in A \\ 0 & \text{if } y \notin A\,; \end{cases}$$

$$f_3(y) = \begin{cases} f(\lambda x_1 + (1-\lambda)x_2, y) & \text{if } y \in A \\ 0 & \text{if } y \notin A\,. \end{cases}$$

Since f is logarithmic concave, we have

$$f_3(y) \geq \sup_{\lambda u + (1-\lambda)v = y} (f_1(u))^{\lambda} (f_2(v))^{1-\lambda}\,.$$

Thus, by Theorem 3.2 we have

$$\begin{aligned}\int_A f(\lambda x_1 + (1-\lambda)x_2, y)dy &= \int_{\mathbb{R}^m} f_3(y)dy \\ &\geq \left(\int_{\mathbb{R}^m} f_1(y)dy\right)^\lambda \left(\int_{\mathbb{R}^m} f_2(y)dy\right)^{1-\lambda} \\ &= \left(\int_A f(x_1,y)dy\right)^\lambda \left(\int_A f(x_2,y)dy\right)^{1-\lambda},\end{aligned}$$

proving the theorem. □

Slepian's inequality plays important role in small ball problems and other fields. Here, we present this inequality following an idea of [Joag-Dev, Perlman and Pitt (1983)]. See also [Adler (1990); Adler and Taylor (2007)]. There are a number of recent developments (see for example, [Eisenbaum (2014); Fang and Zhang (2011)]).

Theorem 3.4. *Let $X_1, \cdots, X_d$ be centered Gaussian with covariance matrix $\Sigma = (\sigma_{ij})_{1\leq i,j\leq d}$. Let $h : \mathbb{R}^d \to \mathbb{R}$ be a twice continuously differentiable function. Assume that for some (fixed) $1 \leq i, j \leq d$ and $i \neq j$,*

$$\partial_i \partial_j h(x) \geq 0 \quad \forall\, x \in \mathbb{R}^d . \tag{3.2.8}$$

Then, $h(\sigma_{ij}) := \mathbb{E}h(X_1, \cdots, X_d)$ is an increasing function of σ_{ij}.

Proof. Denote $\phi(x, \sigma_{ij}) = \phi_{0,\Sigma}(x)$ the density given by (3.1.1) with $\xi = 0$.

Differentiating (3.1.5) with respect to σ_{ij} and with respect to x_i and x_j we obtain

$$\begin{aligned}\frac{\partial \phi}{\partial \sigma_{ij}} &= -\frac{1}{(2\pi)^d}\int_{\mathbb{R}^d} z_i z_j \exp\left\{-i\langle z, x\rangle - \frac{z^T\Sigma z}{2}\right\} dz\,; \\ \frac{\partial^2 \phi}{\partial x_i \partial x_j} &= -\frac{1}{(2\pi)^d}\int_{\mathbb{R}^d} z_i z_j \exp\left\{-i\langle z, x\rangle - \frac{z^T\Sigma z}{2}\right\} dz\,.\end{aligned}$$

These identities yield $\frac{\partial \phi}{\partial \sigma_{ij}} = \frac{\partial^2 \phi}{\partial x_i \partial x_j}$. Therefore, we have

$$\begin{aligned}\frac{\partial h(\sigma_{ij})}{\partial \sigma_{ij}} &= \int_{\mathbb{R}^d} h(x) \frac{\partial}{\partial \sigma_{ij}} \phi(x, \sigma_{ij}) dx = \int_{\mathbb{R}^d} h(x) \frac{\partial^2}{\partial x_i \partial x_j} \phi(x, \sigma_{ij}) dx \\ &= \int_{\mathbb{R}^d} \left(\frac{\partial^2}{\partial x_i \partial x_j} h(x)\right) \phi(x, \sigma_{ij}) dx \geq 0\,.\end{aligned}$$

This shows the theorem. □

An immediate consequence of the above equality is the following:

Theorem 3.5. *Let $X = (X_1, \cdots, X_d)^T$ and $Y = (Y_1, \cdots, Y_d)^T$ be two centered Gaussian vectors in $\mathbb{R}^d$ such that*

$$\mathbb{E}(X_i^2) = \mathbb{E}(Y_i^2) \quad \text{and} \quad \mathbb{E}(X_i X_j) \le \mathbb{E}(Y_i Y_j)\,, \quad \forall\, 1 \le i, j \le d\,,\ i \ne j\,. \tag{3.2.9}$$

Then, for any $a_1, \cdots, a_d \in \mathbb{R}$, we have

$$P\left(\cap_{i=1}^d \{X_i \ge a_i\}\right) \le P\left(\cap_{i=1}^d \{Y_i \ge a_i\}\right)\,. \tag{3.2.10}$$

Proof. We may take $h(x) = \prod_{i=1}^d I_{[a_i,\infty)}(x_i)$ in the previous theorem (we may approximate this h by smooth functions). □

3.3 Brunn-Minkowski Inequality

Let λ denote the Lebesgue measure on $\mathbb{R}^d$. The classical Brunn-Minkowski inequality states that for any two nonempty compact subsets A and B of $\mathbb{R}^d$,

$$[\lambda(A+B)]^{1/d} \ge [\lambda(A)]^{1/d} + [\lambda(B)]^{1/d}\,, \tag{3.3.1}$$

where

$$A + B := \{a + b \in \mathbb{R}^d \mid a \in A,\ b \in B\}$$

is the Minkowski sum of A and B.

Another related result is that for the Lebesgue measure λ, the mapping $A \to [\lambda(A)]^{1/d}$ is concave in the sense that

$$[\lambda(tA + (1-t)B)]^{1/d} \ge t\,[\lambda(A)]^{1/d} + (1-t)\,[\lambda(B)]^{1/d}\,,$$
$$\forall \text{ nonempty compact sets } A, B \subseteq \mathbb{R}^d \text{ and } t \in [0,1]\,. \tag{3.3.2}$$

These inequalities are also related to isoperimetric inequality, Prékopa-Leindler inequality and so on. We refer to [Schneider (1993)].

Attempts have been devoted to extend the above results to Gaussian measures. We follow [Barthe and Huet (2009)], [Borell (2003)] to present some of these results.

Lemma 3.1. *Let $\alpha_1, \cdots, \alpha_m$ be positive numbers such that*

$$\sum_{j=1}^m \alpha_j \ge 1\,. \tag{3.3.3}$$

Then, there are vectors $v_1, \cdots, v_m \in \mathbb{R}^n$ *such that*

$$|v_i| \le 1\,, \quad \forall i = 1, \cdots, m \quad \text{and} \quad |\sum_{j=1}^{m} \alpha_j v_j| = 1\,. \tag{3.3.4}$$

Moreover, if

$$\alpha_j - \sum_{i \ne j} \alpha_i \le 1 \quad \forall\, j = 1, 2, \cdots, m\,, \tag{3.3.5}$$

then there are vectors $v_1, \cdots, v_m \in \mathbb{R}^n$ *such that*

$$|v_i| = 1\,, \quad \forall i = 1, \cdots, m \quad \text{and} \quad |\sum_{j=1}^{m} \alpha_j v_j| = 1\,. \tag{3.3.6}$$

Proof. Taking $v_1 = \cdots = v_m = \frac{1}{\alpha_1 + \cdots + \alpha_m} e_1$, where $e_1 = (1, 0, \cdots, 0)$, we see (3.3.4) holds.

Denote by S^{n-1} the unit sphere in $\mathbb{R}^n$ and consider the non-negative function $h(v_1, \cdots, v_m) = |\sum_{j=1}^m \alpha_j v_j|$ on $\left(S^{n-1}\right)^{\otimes m}$. Taking $v_1 = \cdots = v_m = e_1$ for example we see that one possible value of h is $h(v_1, \cdots, v_m) = \sum_{j=1}^m a_j \ge 1$. If we can show the minimum value $\underline{\text{h}}$ of h on $\left(S^{n-1}\right)^{\otimes m}$ is less than or equal to 1, then since h is a continuous function on $\left(S^{n-1}\right)^{\otimes m}$ it takes all values between $\underline{\text{h}}$ and $\sum_{j=1}^m a_j$. Thus, there are vectors $v_1, \cdots, v_m \in S^{n-1}$ such that $h(v_1, \cdots, v_m) = 1$. The theorem is proved. Our remaining objective is to show $\underline{\text{h}} \le 1$. The case $\underline{\text{h}} = 0$ is obvious. Since $\underline{\text{h}} \ge 0$, we can now assume that $\underline{\text{h}} > 0$. Let $(v_1^0, \cdots, v_m^0)$ be the minimizer. By the Lagrange multipliers theorem, there are real numbers λ_j, $j = 1, \cdots, m$ such that

$$\alpha_j \sum_{i=1}^{m} \alpha_i v_i^0 = \lambda_j v_j^0\,, \quad j = 1, \cdots, m\,. \tag{3.3.7}$$

Define $\bar{v} := \sum_{i=1}^m \alpha_i v_i^0$. If $\bar{v} = 0$, then $h(v_1^0, \cdots, v_m^0) = 0$. This is contradictory to $\underline{\text{h}} > 0$. So, we have $\bar{v} \ne 0$. Then, (3.3.7) means that for all $j = 1, \cdots, m$, $v_j^0 = \beta_j \bar{v}$ for some $\beta_j \in \mathbb{R}$ and this also means that $\beta_j = \pm 1$. Therefore, we have

$$\underline{\text{h}} = h(v_1^0, \cdots, v_m^0) = \sum_{j \in \mathcal{I}_+} \alpha_j - \sum_{j \in \mathcal{I}_-} \alpha_j > 0\,,$$

where $\mathcal{I}_+$ and $\mathcal{I}_-$ are partition of the index set $\{1, 2, \cdots, n\}$. If $\mathcal{I}_+$ contains only one element, then from the assumption (3.3.5), we see that $h(v_1^0, \cdots, v_n^0) \le 1$. Now it remains to show that it is impossible that $\mathcal{I}_+$

contains two or more elements. We use contraction argument to complete this. In fact, suppose there are two or more elements in $\mathcal{I}_+$. We write

$$\gamma_1 = \alpha_{j_1}, \quad \gamma_2 = \sum_{j\in\mathcal{I}_+\backslash\{j_1\}} \alpha_j \quad \text{and} \quad \gamma_3 = \sum_{j\in\mathcal{I}_-} \alpha_j .$$

Namely, $\underline{\mathrm{h}} = \gamma_1 + \gamma_2 - \gamma_3 > 0$. If we choose $v_j = \bar{v}$ for $j \in \{j_1\} \cup \mathcal{I}_-$ and $v_j = -\bar{v}$ for $j \in \mathcal{I}_+\backslash\{j_1\}$, then $h(v_1, \cdots, v_n) = \gamma_1 + \gamma_3 - \gamma_2$. The fact that $\underline{\mathrm{h}}$ is the minimum implies $\gamma_1 + \gamma_2 - \gamma_3 \le \gamma_1 + \gamma_3 - \gamma_2$, which yields $\gamma_2 \le \gamma_3$. Similarly, we have $\gamma_1 \le \gamma_3$. This implies $(\gamma_2 - \gamma_1)^2 \le \gamma_3^2 \le (\gamma_1 + \gamma_2)^2$, or $-1 \le \dfrac{\gamma_3^2 - \gamma_1^2 - \gamma_2^2}{2\gamma_1\gamma_2} \le 1$. Let $\bar{v}_1$ and $\bar{v}_2$ be two unit vectors in $\mathbb{R}^n$ such that $\langle \bar{v}_1, \bar{v}_2\rangle = \dfrac{\gamma_3^2 - \gamma_1^2 - \gamma_2^2}{2\gamma_1\gamma_2}$. Define $\bar{v}_3 = -\frac{\gamma_1}{\gamma_3}\bar{v}_1 - \frac{\gamma_2}{\gamma_3}\bar{v}_2$. Then, it is easy to see that $\bar{v}_3 \in S^{n-1}$ and $\gamma_1\bar{v}_1 + \gamma_2\bar{v}_2 + \gamma_3\bar{v}_3 = 0$. Now let $v_j = \bar{v}_1 I_{\{j=j_1\}} + \bar{v}_2 I_{\{j\in\mathcal{I}_+\backslash\{j_1\}\}} + \bar{v}_3 I_{\{j\in\mathcal{I}_-\}}$. Then, $h(v_1, \cdots, v_n) = |\gamma_1\bar{v}_1 + \gamma_2\bar{v}_2 + \gamma_3\bar{v}_3| = 0$, which is contradictory to $\underline{\mathrm{h}} > 0$. $\square$

Lemma 3.2. *Assume that $C : \mathbb{R}_+ \times \mathbb{R}^d \to \mathbb{R}$ is twice differentiable. Suppose that the following three conditions*

$$\mathrm{Hess}\,(C) \ge 0\,, \ \nabla C = 0\,, \text{and } C \le 0 \tag{3.3.8}$$

imply $\partial_t C \ge 0$. Suppose also that for some $T > 0$

$$\liminf_{|x|\to\infty} \inf_{0\le t\le T} C(t, x) \ge 0\,. \tag{3.3.9}$$

Then

$$C(0, \cdot) \ge 0 \quad \text{implies} \ \ \forall\, t \in [0, T] \ \ C(t, \cdot) \ge 0\,. \tag{3.3.10}$$

Proof. For any $\varepsilon > 0$ denote $C_\varepsilon(t, x) = C(t, x) + \varepsilon t$. If $C_\varepsilon(t, x) < 0$ for some t and x, then by the assumption (3.3.9), C_ε attains its minimum at some point (t_0, x_0). Thus, $\nabla C(t_0, x_0) = 0$ and $\mathrm{Hess}\,(C)(t_0, x_0) \ge 0$, $C(t_0, x_0) < 0$, and $\partial_t C(t_0, x_0) \le -\varepsilon < 0$. But from (3.3.8), we also have $\partial_t C(t_0, x_0) \ge 0$ which is contradictory to $\partial_t C(t_0, x_0) < 0$. This implies $C_\varepsilon \ge 0$. The lemma is proved since $\varepsilon > 0$ is arbitrary. $\square$

We still denote

$$P_t f(x) = \int_{\mathbb{R}^d} p_t(x, y) f(y)\,, \qquad \text{where} \quad p_t(x, y) = (2\pi t)^{-d/2} \exp\left\{ -\frac{|y - x|^2}{2t} \right\}.$$

Denote

$$\Phi(x) = \frac{1}{\sqrt{2\pi}} \int_{-\infty}^{x} e^{-\frac{u^2}{2}} du\,.$$

Proposition 3.4. *Let h and $f_i, i = 1, \cdots, m$ be twice continuously differentiable functions from $\mathbb{R}^d$ to $[0,1]$. Assume that there exist $a_1, \cdots, a_m \in \mathbb{R}$ such that for all $t > 0, x \in \mathbb{R}^d, f = f_1, \cdots, f_m$ and h*

$$|\nabla f(x+ty)|e^{-\frac{y^2}{2}} \to 0 \text{ as } |y| \to \infty \quad \text{and} \tag{3.3.11}$$

$$\limsup_{|x|\to\infty} f_i(x) \le \Phi(a_i)\,, \quad h \ge \Phi(\sum_{i=1}^m \alpha_i a_i)\,. \tag{3.3.12}$$

Let $\alpha_1, \cdots, \alpha_m$ be positive real numbers such that

$$\alpha_j - \sum_{i\neq j} \alpha_i \le 1 \quad \forall\, j = 1, 2, \cdots, m \quad \text{and} \quad \sum_{j=1}^m \alpha_j \ge 1\,. \tag{3.3.13}$$

If for all $x_1, \cdots, x_m \in \mathbb{R}^d$,

$$\Phi^{-1} \circ h \left(\sum_{j=1}^m \alpha_j x_j \right) \ge \sum_{j=1}^m \alpha_j \Phi^{-1} \circ f_j\,(x_j)\,, \tag{3.3.14}$$

then

$$\Phi^{-1} \circ P_t h \left(\sum_{j=1}^m \alpha_j x_j \right) \ge \sum_{j=1}^m \alpha_j \Phi^{-1} \circ P_t f_j\,(x_j)\,. \tag{3.3.15}$$

Proof. Denote

$$H = H(t,x) = \Phi^{-1} \circ P_t h(x) \quad \text{and} \quad F_j = F_j(t,x) = \Phi^{-1} \circ P_t f_j(x)\,.$$

Set

$$C = C(t, x_1, \cdots, x_m) = H\left(t, \sum_{j=1}^m \alpha_j x_j\right) - \sum_{j=1}^m \alpha_j F_j\,(x_j) = H - \sum_{j=1}^m F_j\,.$$

It is straightforward to check

$$\nabla_{x_i} C = \alpha_i (\nabla H - \nabla F_i)$$
$$\nabla_{x_i} \nabla^*_{x_j} C = \alpha_i \alpha_j \mathrm{Hess}(H) = \delta_{ij} \alpha_i \mathrm{Hess}(F_i)\,,$$

where $\delta_{ij} = 1$ when $i = j$ and 0 when $i \neq j$ is the Kronecker symbol. For any f satisfying (3.3.11) denote $F = \Phi^{-1} \circ P_t f$. Then, $\partial_t P_t f = \frac{1}{2} \Delta P_t f$. Using the fact that $(1/\Phi'(x))' = x/\Phi'(x)$ we have easily that for all nice f

$$\partial_t F = \frac{\partial_t P_t f}{\Phi'(F)} = \frac{\Delta P_t f}{\Phi'(F)}\,,$$

$$\nabla F = \frac{\nabla P_t f}{\Phi'(F)}\,,$$

$$\Delta F = \frac{\Delta P_t f}{\Phi'(F)} + F \frac{|\nabla P_t f|^2}{(\Phi'(F))^2}\,.$$

The above three identities yield

$$\partial_t F = \frac{1}{2}\left(\Delta F - F|\nabla F|^2\right).$$

Applying this identity to $f = f_1, \cdots, f_m, h$, we obtain

$$\partial_t C = \frac{1}{2}(I_1 + I_2), \tag{3.3.16}$$

where

$$I_1 = \Delta H - \sum_{j=1}^m \lambda_j \Delta F_j \quad \text{and} \quad I_2 = -H|\nabla H|^2 + \sum_{j=1}^m \alpha_j F_j |\nabla F_j|^2.$$

We are going to use Lemma 3.2 to prove the theorem. First, we check the condition (3.3.9). For any $\delta > 0$, there is $\varepsilon > 0$ such that

$$\Phi^{-1}(\Phi(a_i) + 2\varepsilon) \le a_i + \frac{\delta}{\sum_{j=1}^m \alpha_j}.$$

Take $r_\varepsilon > 0$ such that $\mu_d(r_\varepsilon B_d(1)) = 1 - \varepsilon$. Then, for any $t \in [0, T]$

$$\begin{aligned} P_t f_i(x_i) &= \int_{r_\varepsilon B_d(1)} f_i(x_i + \sqrt{t}y)\mu_d(dy) + \int_{(r_\varepsilon B_d(1))^c} f_i(x_i + \sqrt{t}y)\mu_d(dy) \\ &\le (1-\varepsilon) \sup_{x_i + r_\varepsilon \sqrt{t} B_d(1)} f_i + \varepsilon \sup f_i \\ &\le \sup_{x_i + r_\varepsilon \sqrt{T} B_d(1)} f_i + \varepsilon \sup f_i \le \Phi(a_i) + 2\varepsilon \end{aligned}$$

for $|x_i|$ sufficiently large by the first condition in (3.3.12). Or

$$\Phi^{-1} \circ P_t f_i(x_i) \le \Phi^{-1}(\Phi(a_i) + 2\varepsilon) \le a_i + \frac{\delta}{\sum_{j=1}^m \alpha_j}.$$

The second condition in (3.3.12) implies $P_t h \ge \Phi(\sum \alpha_j a_j)$ or $\Phi^{-1} \circ P_t h \ge \sum \alpha_j a_j$. Thus, when $|x|$ is sufficiently large we have

$$C(t, x) = \Phi^{-1} \circ P_t h - \sum \Phi^{-1} \circ P_t f_i(x_i) \ge -\delta$$

which yields (3.3.9) since $\delta > 0$ is arbitrary.

Now we assume that the conditions in (3.3.8) are satisfied for C. The condition $\nabla C = 0$ implies $\nabla F_i = \nabla H$. Thus

$$I_2 = |\nabla H|^2 \left[\sum_{i=1}^m \alpha_i F_i - H\right] = -|\nabla H|^2 C \ge 0 \tag{3.3.17}$$

if $C \le 0$. Now we explore the condition $\text{Hess}(C) \ge 0$. This means that for any matrix V of the same size, we have $\text{Tr}(V\,\text{Hess}(C)\,V^*) \ge 0$. Choose V to have the following block form $V = (v_{ij} I_d)_{1\le i,j\le m}$, where $v_{ij}, 1 \le i, j \le m$ are real numbers to be determined later. Then

$$\begin{aligned}
\mathrm{Tr}(V\,\mathrm{Hess}(C)\,V^*) &= \sum_{i,k,l=1}^m v_{ik} v_{il} \nabla_{x_k} \nabla^*_{x_l} C \\
&= \sum_{i,k,l=1}^m v_{ik} v_{il} \left[\alpha_k \alpha_l \mathrm{Tr}(\mathrm{Hess}(H)) - \delta_{kl} \alpha_k \mathrm{Tr}(\mathrm{Hess}(F_k))\right] \\
&= \sum_{i=1}^m \left(\sum_{k=1}^m v_{ik}\alpha_k\right)^2 \Delta H - \sum_{k=1}^m \left(\sum_{i=1}^m v_{ik}\right)^2 \alpha_k \Delta F_k \,.
\end{aligned}$$

Denote the vectors $v_k = (v_{1k}, \cdots, v_{mk})^T$, $k = 1, \cdots, m$. Then, the above coefficient of ΔH is $|\sum \alpha_k v_k|^2$ and the above coefficient of ΔF_k is $|v_k|^2$, which can be all equal to 1 by Lemma 3.1. This means that there are V such that $I_1 = \Delta H - \sum_{k=1}^m \alpha_k \Delta F_k = \mathrm{Tr}(V\,\mathrm{Hess}(C)\,V^*)$ is positive. Combining this with (3.3.16) and (3.3) we have that $\partial_t C \ge 0$. By Lemma 3.2 this proves that $C \ge 0$ for all $t \ge 0$ and the theorem is then proved. □

Theorem 3.6. *For any Borel sets $A_1, \cdots, A_m$ in $\mathbb{R}^d$ and for any positive $\alpha_1, \cdots, \alpha_m$ satisfying (3.3.13) we have*

$$\Phi^{-1} \circ \mu_d \left(\sum_{j=1}^m \alpha_j A_j\right) \ge \sum_{j=1}^m \alpha_j \Phi^{-1} \circ \mu_d (A_j) \,. \tag{3.3.18}$$

Proof. By a limiting argument we can assume that the Borel sets A_i's are compact. For any $1 \le i \le m$ and for any $\varepsilon > 0$, $-\infty < a < b < \infty$ there is a smooth function f_i such that $f_i = \Phi(b)$ on A_i and $f_i = \Phi(a)$ on $(A_i^\varepsilon)^c$ and $\Phi(a) \le f_i \le \Phi(b)$. Denote

$$a_0 = \max_{u_i = a \text{ or } \frac{b}{u} \ne (b,\cdots,b)} \sum \alpha_i u_i \quad \text{and} \quad b_0 = \sum \alpha_i b \,.$$

Choose a smooth function h such that $h = \Phi(b_0)$ on $\sum \alpha_i A_i$ and $h = \Phi(a)$ on $((\sum \alpha_i A_i^\varepsilon)^\varepsilon)^c$ and $\Phi(a_0) \le h \le \Phi(b_0)$. Since A_i are compact, it is possible to find the above functions. From the above definitions of f_i and h, we have

$$\Phi^{-1} \circ h \left(\sum_{j=1}^m \alpha_j x_j\right) \ge \sum_{j=1}^m \alpha_j \Phi^{-1} \circ f_j (x_j) \quad \forall\, x_i \in \mathbb{R}^d \,.$$

Thus, from Proposition 3.4 with $x_i = 0$ and $t = 1$, it follows

$$\Phi^{-1} \left(\int h \mu_d(dx)\right) \ge \sum \alpha_j \Phi^{-1} \left(\int f_i \mu_d(dx)\right) \,.$$

Fix $\varepsilon > 0$ and $b \in \mathbb{R}$. By the definitions of f_i and h, we see

$$\liminf_{a\to-\infty} f_i \ge \Phi(b) I_{A_i} \quad \text{and} \quad \limsup_{a\to-\infty} h \le \Phi(b_0) I_{\{(\sum \alpha_i A_i^\varepsilon)^\varepsilon\}} \,.$$

Hence, by Fatou lemma, we have

$$\Phi^{-1}\left(\Phi(b_0) I_{\{(\sum \alpha_i A_i^\varepsilon)^\varepsilon\}}\right) \ge \sum_{j=1}^m \alpha_j \Phi^{-1}\left(\Phi(b) I_{A_i}\right) \quad \forall\, x_i \in \mathbb{R}^d \,.$$

Letting b go to ∞ (hence, b_0 also goes to ∞) yields the theorem. □

If some A_i are convex sets, then the conditions on α_i's in the above theorem can be relaxed. In fact, we have the following more general result (see [Barthe and Huet (2009)]).

Theorem 3.7. *Let $A_1, \cdots, A_k$ be convex sets and let $A_{k+1}, \cdots, A_m$ be Borel sets. Let the positive numbers $\alpha_1, \cdots, \alpha_m$ satisfy*

$$\alpha_j - \sum_{i\ne j} \alpha_i \le 1 \quad \forall\, j = k+1, 2, \cdots, m \quad \text{and} \quad \sum_{j=1}^m \alpha_j \ge 1 \,. \tag{3.3.19}$$

Then

$$\Phi^{-1} \circ \mu_d \left(\sum_{j=1}^m \alpha_j A_j\right) \ge \sum_{j=1}^m \alpha_j \Phi^{-1} \circ \mu_d\left(A_j\right) \,. \tag{3.3.20}$$

The proof of this theorem is similar to the proof of Theorem 3.6. A straightforward consequence of this theorem is

Theorem 3.8. *For any convex sets $A_1, \cdots, A_m$ in $\mathbb{R}^d$ and for any positive $\alpha_1, \cdots, \alpha_m$ with $\sum_{j=1}^m \alpha_j \ge 1$ we have*

$$\Phi^{-1} \circ \mu_d \left(\sum_{j=1}^m \alpha_j A_j\right) \ge \sum_{j=1}^m \alpha_j \Phi^{-1} \circ \mu_d\left(A_j\right) \,. \tag{3.3.21}$$

A direct consequence of the above theorem is the following:

Theorem 3.9. *For any convex sets $A_1, \cdots, A_m$ in $\mathbb{R}^d$ and for any positive $\alpha_1, \cdots, \alpha_m$ with $\sum_{j=1}^m \alpha_j = 1$ we have*

$$\mu_d \left(\sum_{j=1}^m \alpha_j A_j\right) \ge \prod_{j=1}^m \left(\mu_d(A_j)\right)^{\alpha_j} \,. \tag{3.3.22}$$

Proof. It is straightforward to verify that the function Φ is increasing and log concave (on the positive line), namely,

$$\Phi(\alpha_1 x_1 + \cdots + \alpha_m x_m) \ge \prod_{j=1}^m \Phi(x_j)^{\alpha_j}, \quad \forall x_1, \cdots, x_m \ge 0 .$$

Thus, from the above inequality (3.3.21), it follows

$$\begin{aligned}
\mu_d\left(\sum_{j=1}^m \alpha_j A_j\right) &= \Phi \circ \Phi^{-1} \circ \mu_d\left(\sum_{j=1}^m \alpha_j A_j\right) \\
&\ge \Phi\left(\sum_{j=1}^m \alpha_j \Phi^{-1} \circ \mu_d(A_j)\right) \\
&\ge \prod_{j=1}^m \left(\Phi \circ \Phi^{-1} \circ \mu_d(A_j)\right)^{\alpha_j} \\
&= \prod_{j=1}^m \left(\mu_d(A_j)\right)^{\alpha_j} .
\end{aligned}$$

This proves the theorem. □

The following inequality is called the Anderson inequality.

Theorem 3.10. *Let μ_d be the standard Gaussian measure on $\mathbb{R}^d$. Then, for any convex Borel set A of $\mathbb{R}^d$ which is symmetric with respect to origin (this means that if $x \in A$, then $-x \in A$) and for any $a \in \mathbb{R}^d$, we have*

$$\mu_d(A) \ge \mu_d(A+a) . \tag{3.3.23}$$

Proof. It is easy to check by the convexity of A that

$$A \supseteq \frac{1}{2}(A+a) + \frac{1}{2}(A-a) .$$

Thus, from Theorem 3.9 that we are going to show we have

$$\mu_d(A) \ge \mu_d(A+a)^{1/2} \mu_d(A-a)^{1/2} .$$

By the symmetry of A, we have $-(A-a) = A+a$. Since the Gaussian measure is symmetric, we have $\mu_d(A+a) = \mu_d(A-a)$. Thus, $\mu_d(A) \ge \mu_d(A+a)$. The theorem is then proved. □

3.4 Hermite Polynomials

Definition 3.3. The Hermite polynomial $H_n(x)$ of degree n is defined as

$$H_n(x) = (-1)^n e^{\frac{x^2}{2}} \frac{d^n}{dx^n} e^{-\frac{x^2}{2}}, \quad x \in \mathbb{R}, \quad n = 0, 1, 2, \cdots \tag{3.4.1}$$

The generating function is defined as $\sum_{n=0}^{\infty} \frac{t^n}{n!} H_n(x)$.

Proposition 3.5. *The generating function of the Hermite polynomials is*

$$G(t, x) := e^{tx - \frac{t^2}{2}} = \sum_{n=0}^{\infty} \frac{t^n}{n!} H_n(x), \quad \forall\, t, x \in \mathbb{R}. \tag{3.4.2}$$

Proof. By the definition of the Hermite polynomials and by the Taylor expansion formula, we have

$$\begin{aligned}
\sum_{n=0}^{\infty} \frac{t^n}{n!} H_n(x) &= \sum_{n=1}^{\infty} \frac{t^n}{n!} (-1)^n e^{\frac{x^2}{2}} \frac{d^n}{dx^n} e^{-\frac{x^2}{2}} \\
&= e^{\frac{x^2}{2}} \sum_{n=1}^{\infty} \frac{t^n}{n!} \frac{d^n}{dt^n}\Big|_{t=0} e^{-\frac{(x-t)^2}{2}} \\
&= e^{\frac{x^2}{2}} e^{-\frac{(x-t)^2}{2}} = e^{tx - \frac{t^2}{2}}.
\end{aligned}$$

This proves the proposition. □

Remark 3.2. Quite often, it is more convenient to use the following complex generating function:

$$e^{\imath tx + \frac{t^2}{2}} = \sum_{n=0}^{\infty} \frac{\imath^n t^n}{n!} H_n(x), \quad \forall\, t, x \in \mathbb{R}, \tag{3.4.3}$$

where we recall that $\imath = \sqrt{-1}$ is the imaginary number.

We can write

$$\begin{aligned}
e^{tx - \frac{t^2}{2}} = e^{tx} e^{-\frac{t^2}{2}} &= \sum_{m=0}^{\infty} \frac{t^m x^m}{m!} \sum_{m=0}^{\infty} \frac{(-1)^k t^{2k}}{2^k k!} \\
&= \sum_{n=0}^{\infty} \sum_{2k+m=n} \frac{(-1)^k t^n x^m}{2^k m! k!}.
\end{aligned}$$

Comparing the coefficient of t^n with that of(3.4.2), we have

$$H_n(x) = \sum_{k \le n/2} \frac{(-1)^k n!}{2^k k! (n-2k)!} x^{n-2k}, \quad n = 0, 1, 2, \cdots \tag{3.4.4}$$

Thus, we see that $H_n(x)$ is really a polynomial of degree n and the coefficient of x^n is 1. For example, we have the first a few Hermite polynomials:

$$H_0(x) = 1\,, \quad H_1(x) = x\,, \quad H_2(x) = x^2 - 1\,, \quad H_3(x) = x^3 - 3x\,, \quad \cdots \tag{3.4.5}$$

From the generating function, we also have

$$e^{\frac{t^2}{2}} \sum_{m=0}^{\infty} \frac{t^m}{m!} H_m(x) = e^{\frac{t^2}{2}} e^{tx - \frac{t^2}{2}} = e^{tx}\,.$$

This can be written as

$$\sum_{n=0}^{\infty} \frac{t^n x^n}{n!} = \sum_{m=0}^{\infty} \frac{t^m}{m!} H_m(x) \sum_{k=0}^{\infty} \frac{t^{2k}}{2^k k!}\,.$$

From here we obtain that

$$x^n = \sum_{k \le n/2} \frac{n!}{2^k k!(n-2k)!} H_{n-2k}(x)\,, \quad n = 0, 1, 2, \cdots \tag{3.4.6}$$

Recall that $\mu(dx) = \mu_1(dx) = \frac{1}{\sqrt{2\pi}} e^{-\frac{|x|^2}{2}} dx$ is the standard Gaussian measure on $\mathbb{R}$. We also denote

$$L^p = L^p(\mathbb{R}, \mu) = \Big\{ f : \mathbb{R} \to \mathbb{R} \quad \text{is measurable and such that}$$

$$\|f\|_p^p = \int_{\mathbb{R}} |f(x)|^p \mu(dx) < \infty \Big\}\,.$$

It is clear that for any $p \ge 1$, L^p is a Banach space. When $p = 2$, L^2 is a Hilbert space with the scalar product:

$$\langle f, g \rangle = \mu(fg) := \int_{\mathbb{R}} f(x) g(x) \mu(dx)\,.$$

We want to compute $\langle H_n\,, H_m \rangle$. Using the explicit form of the generating function, we see for any $s, t \in \mathbb{R}$, we have

$$\int_{\mathbb{R}} G(t,x) G(s,x) \mu(dx) = \frac{1}{\sqrt{2\pi}} \int_{\mathbb{R}} e^{tx - \frac{t^2}{2}} e^{sx - \frac{s^2}{2}} e^{-\frac{x^2}{2}} dx = e^{st}\,.$$

This implies

$$\sum_{n,m=0}^{\infty} \frac{t^n s^m}{n! m!} \int_{\mathbb{R}} H_n(x) H_m(x) \mu(dx) = \sum_{k=0}^{\infty} \frac{s^k t^k}{k!}\,. \tag{3.4.7}$$

Comparing the coefficients of $s^n t^m$, $m, n = 0, 1, 2, \cdots$, we obtain

Proposition 3.6. *The Hermite polynomials $\{H_n(x)\,, n = 0, 1, 2, \cdots\}$ form an orthogonal basis of $L^2(\mathbb{R}, \mu)$. Moreover, we have*

$$\int_{\mathbb{R}} H_n(x) H_m(x) \mu(dx) = \begin{cases} n! & \text{if } n = m \\ 0 & \text{if } n \ne m\,. \end{cases} \tag{3.4.8}$$

Proof. The equality (3.4.8) follows from the comparison of the coefficients of $s^m t^n$ in (3.4.7). To show the set of all Hermite polynomials constitutes a basis, it suffices, by (3.4.8), to show that the set of Hermite polynomials is complete (or total) in L^2. Namely, we need to show that if $f \in L^2$ and $\langle f, H_n \rangle = 0$ for all $n = 0, 1, 2, \cdots$, then $f = 0$ a.s.

In fact, if $f \in L^2$ and $\langle f, H_n \rangle = 0$ for all $n = 0, 1, 2, \cdots$, then (since $e^{tx} \in L^2$ for all $t \in \mathbb{R}$)

$$\int_{\mathbb{R}} e^{\imath tx - \frac{x^2}{2}} f(x)dx = \sqrt{2\pi} \int_{\mathbb{R}} e^{\imath tx} f(x)\mu(dx) = \sqrt{2\pi} e^{-\frac{t^2}{2}} \int_{\mathbb{R}} e^{\imath tx + \frac{t^2}{2}} f(x)\mu(dx)$$
$$= \sqrt{2\pi} e^{\frac{t^2}{2}} \sum_{k=0}^{\infty} \frac{(-1)^k t^k}{k!} \int_{\mathbb{R}} f(x) H_k(x)\mu(dx) = 0\,.$$

Thus, the Fourier transform of $f(x)e^{-\frac{x^2}{2}}$ is 0, which implies that $f(x) = 0$ a.s. □

Proposition 3.7. *The set of all normalized Hermite polynomials*

$$\left\{ \frac{1}{\sqrt{n!}} H_n(x)\,, n = 0, 1, 2, \cdots \right\}$$

forms an orthonormal basis of $L^2(\mathbb{R}, \mu)$. Thus, for any square integrable function $f : \mathbb{R} \to \mathbb{R}$ (such that $f \in L^2(\mathbb{R}\,, \mu)$), we have

$$f(x) = \sum_{i=0}^{\infty} \frac{1}{n!} a_n(f) H_n(x)\,, \tag{3.4.9}$$

where the above convergence is in the sense of L^2 and

$$a_n(f) = \int_{\mathbb{R}} f(x) H_n(x)\mu(dx)\,. \tag{3.4.10}$$

Moreover, we have

$$\|f\|_2^2 = \sum_{i=0}^{\infty} \frac{[a_n(f)]^2}{n!}\,. \tag{3.4.11}$$

If f is differentiable up to order n, then integrating by parts n times yields

$$\int_{\mathbb{R}} f(x) H_n(x)\mu(dx) = (-1)^n \int_{\mathbb{R}} f(x) e^{\frac{x^2}{2}} \frac{d^n}{dx^n} e^{-\frac{x^2}{2}} \mu(dx)$$
$$= (-1)^n \frac{1}{\sqrt{2\pi}} \int_{\mathbb{R}} f(x) \frac{d^n}{dx^n} e^{-\frac{x^2}{2}} dx$$
$$= \frac{1}{\sqrt{2\pi}} \int_{\mathbb{R}} f^{(n)}(x) e^{-\frac{x^2}{2}} dx$$
$$= \int_{\mathbb{R}} f^{(n)}(x)\mu(dx)\,.$$

We can state above computation as

Proposition 3.8. *If f is differentiable up to order n, then*

$$\int_{\mathbb{R}} f(x)H_n(x)\mu(dx) = \int_{\mathbb{R}} f^{(n)}(x)\mu(dx)\,. \tag{3.4.12}$$

Differentiating the expression (3.4.2) with respect to x and comparing the coefficient of t^n, we obtain

$$\frac{d}{dx}H_n(x) = nH_{n-1}(x)\,. \tag{3.4.13}$$

Recursively, we have

$$\begin{aligned}\frac{d^k}{dx^k}H_n(x) &= n(n-1)\cdots(n-k+1)H_{n-k}(x)\\ &= \frac{n!}{(n-k)!}H_{n-k}(x)\,.\end{aligned} \tag{3.4.14}$$

Differentiating the expression (3.4.2) with respect to t, we have

$$\frac{d}{dt}e^{tx-\frac{t^2}{2}} = \sum_{n=1}^{\infty}\frac{t^{n-1}}{(n-1)!}H_n(x)\,.$$

On the other hand, it is easy to see

$$\frac{d}{dt}e^{tx-\frac{t^2}{2}} = (x-t)e^{tx-\frac{t^2}{2}} = (x-t)\sum_{n=0}^{\infty}\frac{t^n}{n!}H_n(x)\,.$$

Therefore, we have

$$(x-t)\sum_{n=0}^{\infty}\frac{t^n}{n!}H_n(x) = \sum_{n=1}^{\infty}\frac{t^{n-1}}{(n-1)!}H_n(x)\,.$$

Comparing the coefficient of t^n we have

$$H_{n+1}(x) = xH_n(x) - nH_{n-1}(x)\,, \quad n = 1\,,2\,,\cdots \tag{3.4.15}$$

This implies that (we denote $H_{-1}(x) = 0$)

Proposition 3.9. *For all $n = 1, 2\,, 3\,, \cdots$,*

$$H_n(x) = xH_{n-1}(x) - (n-1)H_{n-2}(x)\,. \tag{3.4.16}$$

Using (3.4.13), we can write

$$H_n(x) = x\frac{1}{n}\frac{d}{dx}H_n(x) - \frac{1}{n}\frac{d^2}{dx^2}H_n(x)\,, \quad n = 1\,,2\,,\cdots \tag{3.4.17}$$

Or H_n satisfies the following equation.

$$\frac{d^2}{dx^2}H_n(x) - x\frac{d}{dx}H_n(x) + nH_n(x) = 0\,, \quad n = 1\,, 2\,, \cdots$$

Denote $L = \frac{d^2}{dx^2} - x\frac{d}{dx}$. Then, the above identity means $LH_n = nH_n$. Since $\{H_n(x)\,, n = 0, 1, 2, \cdots\}$ is orthogonal basis of L^2, we have

Theorem 3.11. *The operator* $L = \frac{d^2}{dx^2} - x\frac{d}{dx}$ *has all the eigenvalues* $\lambda_n = 0\,, 1\,, 2\,, \cdots$ *with the corresponding eigenvectors* $H_n(x)$. *Namely,*

$$\frac{d^2}{dx^2}H_n(x) - x\frac{d}{dx}H_n(x) + nH_n(x) = 0\,, \quad n = 0\,, 1\,, 2\,, \cdots \tag{3.4.18}$$

Proposition 3.10. *We have for all* $x, y, a \in \mathbb{R}$,

$$H_n(x+y) = \sum_{k=0}^{n} \binom{n}{k} y^{n-k} H_k(x) \tag{3.4.19}$$

$$= \sum_{2l+k\le n} \frac{n!}{2^l l! k! (n-k-2l)!} H_k(x) H_{n-k-2l}(y)\,; \tag{3.4.20}$$

$$H_n(ax) = \sum_{k\le n/2} \frac{n! a^{n-2k}(a^2-1)^k}{2^{n-2k} k! (n-2k)!} H_{n-2k}(x)\,. \tag{3.4.21}$$

Proof. Using

$$\sum_{n=0}^{\infty} \frac{t^n}{n!} H_n(x+y) = e^{t(x+y)-\frac{t^2}{2}} = e^{tx-\frac{t^2}{2}} e^{ty} = \sum_{k=0}^{\infty} \frac{t^k}{k!} H_k(x) \sum_{l=0}^{\infty} \frac{t^l y^l}{l!}$$

and comparing the coefficient of t^n we obtain (3.4.19). Using

$$\sum_{n=0}^{\infty} \frac{t^n}{n!} H_n(x+y) = e^{t(x+y)-\frac{t^2}{2}} = e^{tx-\frac{t^2}{2}} e^{ty-\frac{t^2}{2}} e^{\frac{t^2}{2}}$$

$$= \sum_{k=0}^{\infty} \frac{t^k}{k!} H_k(x) \sum_{k=0}^{\infty} \frac{t^m}{m!} H_m(y) \sum_{l=0}^{\infty} \frac{t^{2l}}{2^l l!}$$

and comparing the coefficient of t^n we obtain (3.4.20). Using

$$\sum_{n=0}^{\infty} \frac{t^n}{n!} H_n(ax) = e^{tax-\frac{t^2}{2}} = e^{tax-\frac{a^2t^2}{2}} e^{\frac{(a^2-1)t^2}{2}}$$

$$= \sum_{k=0}^{\infty} \frac{a^k t^k}{k!} H_k(x) \sum_{l=0}^{\infty} \frac{(a^2-1)^l t^{2l}}{2^l l!}$$

and comparing the coefficient of t^n we obtain (3.4.21). $\square$

Proposition 3.11. *If* $0 \le \alpha \le 1$, *then*

$$H_n(\alpha x + \sqrt{1-\alpha^2}y) = \sum_{i=0}^{n} \binom{n}{k} \alpha^k (1-\alpha^2)^{(n-k)/2} H_k(x) H_{n-k}(y) . \quad (3.4.22)$$

Proof. Denote $\beta = \sqrt{1-\alpha^2}$. Then

$$\sum_{n=0}^{\infty} \frac{t^n}{n!} H_n(\alpha x + \sqrt{1-\alpha^2}y) = e^{t\alpha x + \beta y - \frac{t^2}{2}} = e^{t\alpha x - \frac{t^2}{2}} e^{t\beta y - \frac{t^2}{2}}$$

$$= \sum_{k=0}^{\infty} \frac{t^k \alpha^k}{k!} H_k(x) \sum_{l=0}^{\infty} \frac{t^l \beta^l}{l!} H_l(y) .$$

Comparing the coefficient of t^n yields (3.4.22). □

Proposition 3.12. *The Hermite polynomials can also be represented as*

$$H_n(x) = \frac{(-\iota)^n e^{\frac{x^2}{2}}}{\sqrt{2\pi}} \int_{\mathbb{R}} \exp\left\{-\frac{t^2}{2} + \iota t x\right\} t^n dt . \quad (3.4.23)$$

Proof. Denote the right hand side of (3.4.23) by $\tilde{H}_n(x)$. Then

$$\sum_{n=0}^{\infty} \tilde{H}_n(x) \frac{s^n}{n!} = \frac{e^{\frac{x^2}{2}}}{\sqrt{2\pi}} \int_{\mathbb{R}} \exp\left\{-\frac{t^2}{2} + \iota t x - \iota s t\right\} dt$$

$$= e^{\frac{x^2}{2}} \exp\left\{-\frac{(x-s)^2}{2}\right\} = G(t,x) .$$

This shows that $H_n(x) = \tilde{H}_n(x)$. □

Proposition 3.13. *We have for all* $x, y \in \mathbb{R}$, $s \in (-1,1)$

$$\sum_{n=0}^{\infty} H_n(x) H_n(y) \frac{s^n}{n!} = \frac{1}{\sqrt{1-s^2}} \exp\left\{\frac{2sxy - s^2(x^2+y^2)}{2(1-s^2)}\right\} . \quad (3.4.24)$$

Proof. From (3.4.23), it follows

$$\sum_{n=0}^{\infty} H_n(x) H_n(y) \frac{s^n}{n!}$$

$$= e^{\frac{x^2}{2} + \frac{y^2}{2}} \int_{\mathbb{R}^2} \exp\left\{-\frac{t_1^2}{2} + \iota t_1 x - \frac{t_2^2}{2} + \iota t_2 y - s t_1 t_2\right\} dt_1 dt_2$$

$$= e^{\frac{x^2}{2} + \frac{y^2}{2}} \int_{\mathbb{R}^2} \exp\left\{-\frac{1}{2} t^T A^{-1} t + \iota t_1 x + \iota t_2 y\right\} dt ,$$

where $t = (t_1, t_2)$, $dt = dt_1 dt_2$, and

$$A = \begin{pmatrix} 1 & -s \\ -s & 1 \end{pmatrix}^{-1} = \frac{1}{1-s^2} \begin{pmatrix} 1 & s \\ s & 1 \end{pmatrix} .$$

Thus, from (3.1.3) we have

$$\sum_{n=0}^{\infty} H_n(x)H_n(y)\frac{s^n}{n!} = \frac{1}{\sqrt{1-s^2}} e^{\frac{x^2}{2}+\frac{y^2}{2}} \exp\left\{-(x,y)A\begin{pmatrix} x \\ y \end{pmatrix}\right\},$$

which is simplified to (3.4.24). □

Remark 3.3. It is known that if $\{e_n, n \geq 1\}$ is an orthonormal basis of Hilbert space $\mathbb{H}$, then formally $\sum_{n=1}^{\infty} e_n \otimes e_n$ (see Section 5.1 for notation) is the expansion of the identity operator. This means $\sum_{n=0}^{\infty} \frac{1}{n!} H_n(x)H_n(y)$ (which is the left hand side of (3.4.24) when $s = 1$) should converge to the Dirac delta function $\delta(x-y)$ (in the sense of distribution). The identity (3.4.24) shows that although the identity does not hold true when $s = 1$ in the classical sense, it does hold true when $|s| < 1$. This can be used as an approximation of identity (as $s \to 1$).

Theorem 3.12. *All the zeros of a Hermite polynomial H_n of degree $n \geq 1$ are real and distinct.*

Proof. Fix an positive integer n. First, we claim that $H_n(x)$ must have a real zero. Otherwise, $H_n(x)$ has a constant sign (this means that it is always non-positive or always non-negative) and nonzero, which is impossible since $\int_{\mathbb{R}} H_n(x)\mu(dx) = 0$. Since all the coefficients of H_n are real, if $x_0 = a + \iota b$ is a zero of H_n, then its conjugate $\bar{x}_0 = a - \iota b$ is also a zero of H_n. Then, H_n has a factor of the form $(x - x_0)(x - \bar{x}_0) = (x - x_0)\overline{(x - x_0)} \geq 0$. Let $x_1, \cdots, x_l$ be the real zeros of $H_n(x)$. Then, $H_n(x)(x - x_1)\cdots(x - x_l)$ must have a constant sign. If $l < n$, then $(x - x_1)\cdots(x - x_l) = \sum_{j=0}^{n-1} a_j H_j(x)$ for some constants a_j. This implies

$$\int_{\mathbb{R}} H_n(x)(x - x_1)\cdots(x - x_l)\mu(dx) = \sum_{j=0}^{n-1} a_j \int_{\mathbb{R}} H_n(x)H_j(x)\mu(dx) = 0\,.$$

This is impossible since $H_n(x)(x - x_1)\cdots(x - x_l)$ has constant sign. Thus, we have $l = n$.

If x_0 is a multiple zero of $H_n(x)$, then $\frac{H_n(x)}{(x-x_0)^2}$ is a polynomial of degree $n - 2$. Then

$$\int_{\mathbb{R}} \left(\frac{H_n(x)}{(x - x_0)}\right)^2 \mu(dx) = \int_{\mathbb{R}} H_n(x)\frac{H_n(x)}{(x - x_0)^2}\mu(dx) = 0\,.$$

This is impossible. □

As an application we can show the following Turán's inequality.

Theorem 3.13. *For any $n \geq 1$, the following Turán determinant is positive:*

$$H_n^2(x) - H_{n+1}(x)H_{n-1}(x) > 0\,. \tag{3.4.25}$$

Proof. From the previous theorem, it follows that $H_{n+1}(x) = (x - x_1)\cdots(x-x_n)(x-x_{n+1})$. Thus,

$$\frac{(n+1)H_n(x)}{H_{n+1}(x)} = \frac{H'_{n+1}(x)}{H_{n+1}(x)} = \sum_{i=1}^{n+1}\frac{1}{x-x_i}\,. \tag{3.4.26}$$

However, it is easy to verify

$$\frac{(n+1)nH_{n-1}(x)H_{n+1}(x) - (n+1)^2H_n^2(x)}{H_{n+1}^2(x)} = \left(\frac{(n+1)H_n(x)}{H_{n+1}(x)}\right)'$$

$$= \left(\frac{H'_{n+1}(x)}{H_{n+1}(x)}\right)' = -\sum_{i=1}^{n+1}\frac{1}{(x-x_i)^2}\,. \tag{3.4.27}$$

The Cauchy-Schwartz inequality yields

$$\left(\frac{(n+1)H_n(x)}{H_{n+1}(x)}\right)^2 = \left(\sum_{i=1}^{n+1}\frac{1}{x-x_i}\right)^2 < (n+1)\sum_{i=1}^{n+1}\frac{1}{(x-x_i)^2}$$

$$\leq -(n+1)\frac{(n+1)nH_{n-1}(x)H_{n+1}(x) - (n+1)^2H_n^2(x)}{H_{n+1}^2(x)}\,. \tag{3.4.28}$$

This simplifies to (3.4.25). □

Remark 3.4. In fact, the proofs of the above two theorems are also valid for any orthogonal polynomials satisfying the property $H'_n(x) = nH_{n-1}(x)$ (see [Szegö (1959)] and [Szegö (1948)]).

Proposition 3.14. *Let $X_1, \cdots, X_n$ be a mean zero Gaussian random variables. Let $\mathcal{F}_k = \sigma(X_1, X_2, \cdots, X_k)$. Then, $Y = \mathbb{E}(X_n|\mathcal{F}_k)$ is a Gaussian and $Z = X_n - Y$ is independent of Y.*

Proposition 3.15. *Let Y and Z be two centered Gaussian random variables such that Y is measurable with respect to a σ-algebra $\mathcal{G} \subset \mathcal{F}$ and Z is independent of $\mathcal{G}$. Assume that $\mathbb{E}[Y^2] = \mathbb{E}[Z^2] = 1$. Then, for any $q \geq 1$, and any real numbers a, b such that $a^2 + b^2 = 1$, we have*

$$\mathbb{E}[H_q(aY + bZ)|\mathcal{G}] = a^q H_q(Y). \tag{3.4.29}$$

Proof. Apply identity (3.4.22) to $H_q(aY + bZ)$. Then identity (3.4.29) follows easily from the fact that Y is $\mathcal{G}$-measurable, Z is independent from $\mathcal{G}$ and Hermite polynomials have 0 mean under a centered Gaussian measure except for $H_0 \equiv 1$. □

3.5 Spectral Gap and Logarithmic Sobolev Inequalities

Sobolev inequality is a basic tool in classical analysis with respect to Lebesgue measure. There is no exact analogue for Gaussian measure. Some important contributions along this direction are the Nelson's hypercontractive inequality ([Nelson (1966, 1973)]), Gross's logarithmic inequality ([Gross (1975)], see also [Bonami (1968, 1970)]), and spectral gap inequality (Poincaré inequality). In this section, we shall present a more general inequality obtained in [Driver and Hu (1996)] (in more general case of diffusions) and then we explain how it leads to the logarithmic and spectral gap inequality.

The inequalities that we are going to prove will be independent of dimension. Thus, they can be extended in a straightforward way to infinite dimension (e.g. abstract Wiener space), which is left to readers.

First, we prove a simple lemma.

Lemma 3.3. *Suppose that $(V, (\cdot,\cdot))$ is an inner product space of dimension larger than or equal to one and $a, b, c \in \mathbb{R}$. Then,*

$$a(x,x) + b(x,y) + c(y,y) \geq 0 \quad \forall x, y \in V \tag{3.5.1}$$

iff

$$a \geq 0, c \geq 0, \qquad \text{and} \qquad b^2 \leq 4ac. \tag{3.5.2}$$

Proof. Assume that (3.5.1) holds. Choose $x \in V$ such that $|x| \equiv \sqrt{(x,x)} = 1$ and set $y = kx$, where $k \in \mathbb{R}$. Then, (3.5.1) implies that $p(k) \equiv a + bk + ck^2 \geq 0$ for all $k \in \mathbb{R}$. This clearly implies that $a, c \geq 0$ and $b^2 - 4ac \leq 0$, since otherwise p would have two distinct real roots and hence $\min_k p(k) < 0$. Hence (3.5.1) implies (3.5.2).

Now assume that (3.5.2) holds, then as above the polynomial $q(k) = a - bk + ck^2$ is non-negative. Hence if $|x| \neq 0$, then

$$\begin{aligned} a(x,x) + b(x,y) + c(y,y) &\geq a|x|^2 - |b||(x,y)| + c|y|^2 \\ &\geq a|x|^2 - |b||x||y| + c|y|^2 \\ &= |x|^2 q(|y|/|x|) \geq 0. \end{aligned}$$

Hence (3.5.2) implies (3.5.1). □

Let us recall that for a function $f : \mathbb{R}^d \to \mathbb{R}$, we denote

$$\mu_d(f) = (2\pi)^{-d/2} \int_{\mathbb{R}^d} \exp\left\{-|x|^2/2\right\} f(x)dx = \mathbb{E}(f(X)),$$

where X is a d-dimensional standard Gaussian random variable.

Theorem 3.14. *Let X be a d-dimensional standard Gaussian variable (whose probability law is $\mu_d(dx) = (2\pi)^{-d/2}\exp\left\{-\frac{|x|^2}{2}\right\}dx$) and let $\varphi : \mathbb{R}_+ \to \mathbb{R}$ be a twice continuously differentiable function satisfying*

$$\varphi^{(4)} \geq 0,\ \varphi'' \geq 0,\ \text{and } 2[\varphi^{(3)}]^2 \leq \varphi'' \cdot \varphi^{(4)}\,. \tag{3.5.3}$$

Then, for any positive and continuously differentiable function $f : \mathbb{R}^d \to \mathbb{R}$,

$$\mathbb{E}\left[\varphi(f(X))\right] \leq \varphi\left(\mathbb{E}\left[f(X)\right]\right) + \frac{1}{2}\mathbb{E}\left[\varphi''(f(X))|\nabla f(X)|^2\right]. \tag{3.5.4}$$

Remark 3.5. One can replace the conditions on ϕ in the theorem by the following: φ is defined on the range of f such that (3.5.3) is satisfied.

Proof. We follow the idea in [Driver and Hu (1996)]. Denote the heat kernel $p_t(x) = (2\pi t)^{-d/2}e^{-\frac{|x|^2}{2t}}, x \in \mathbb{R}^d, t > 0$ and denote $P_tf(x) = \int_{\mathbb{R}^d} p_t(x-y)f(y)dy$. Then, we see clearly

$$\begin{cases} P_1f(0) = \mathbb{E}(f(X)) \quad \text{and} \quad P_0f(x) = f(x) \\ \dfrac{\partial}{\partial t}p_t(x) = \dfrac{1}{2}\Delta p_t(x)\,. \end{cases}$$

Let φ be a function of four time continuously differentiable. Consider the function

$$a(t) = P_{1-t}\varphi(P_tf)\,, \quad 0 \leq t \leq 1\,.$$

Denote

$$f_t = P_tf\,, \quad |\nabla f_t|^2 = \sum_{i=1}^{d}\left(\partial_{x_i}f_t\right)^2\,.$$

Taking the derivative of $a(t)$ with respect to t and using the properties of the heat kernel, we have

$$\begin{aligned} a'(t) &= (\frac{d}{dt}P_{1-t})\varphi(f_t) + P_{1-t}\left[\varphi'(f_t)\frac{d}{dt}f_t\right] \\ &= -\frac{1}{2}\Delta P_{1-t}\varphi(f_t) + \frac{1}{2}P_{1-t}\left[\varphi'(f_t)\Delta f_t\right] \\ &= \frac{1}{2}P_{1-t}\left[-\Delta\left(\varphi(f_t)\right) + \varphi'(f_t)\Delta f_t\right] = -\frac{1}{2}P_{1-t}\left[\varphi''(f_t)|\nabla f_t|^2\right]. \end{aligned}$$

Differentiating again with respect to t we have

$$-2a''(t) = \left(\frac{d}{dt}P_{1-t}\right)\left[\varphi''(f_t)|\nabla f_t|^2\right] + P_{1-t}\left[\frac{d}{dt}\left\{\varphi''(f_t)|\nabla f_t|^2\right\}\right]$$

$$= P_{1-t}\left[-\frac{1}{2}\Delta\left(\varphi''(f_t)|\nabla f_t|^2\right) + \frac{1}{2}\varphi^{(3)}(f_t)\Delta f_t|\nabla f_t|^2 + \varphi''(f_t)\langle\nabla f_t\,,\nabla\Delta f_t\rangle\right]$$

$$= -\frac{1}{2}P_{1-t}\Bigg[\varphi^{(3)}(f_t)\Delta f_t|\nabla f_t|^2 + \varphi^{(4)}(f_t)|\nabla f_t|^4 + 2\varphi''(f_t)\mathrm{Tr}\left(\nabla^2 f_t\nabla^2 f_t\right)$$
$$+4\varphi^{(3)}(f_t)\mathrm{Tr}\left(\{\nabla f_t\otimes\nabla f_t\}\nabla^2 f_t\right) + 2\varphi''(f_t)\langle\nabla f_t\,,\nabla\Delta f_t\rangle$$
$$-\varphi^{(3)}(f_t)\Delta f_t|\nabla f_t|^2 - 2\varphi''(f_t)\langle\nabla f_t\,,\nabla\Delta f_t\rangle\Bigg]\,.$$

This simplifies to

$$a''(t) = \frac{1}{4}P_{1-t}\Bigg[\varphi^{(4)}(f_t)|\nabla f_t|^4 + 2\varphi''(f_t)\mathrm{Tr}\left(\nabla^2 f_t\nabla^2 f_t\right)$$
$$+4\varphi^{(3)}(f_t)\mathrm{Tr}\left(\{\nabla f_t\otimes\nabla f_t\}\nabla^2 f_t\right)\Bigg]\,.$$

By (3.3), the term inside above $[\ ]$ is positive if $\varphi''\geq 0$, $\varphi^{(4)}\geq 0$ and $2\left[\varphi^{(3)}\right]^2\leq\varphi''\varphi^{(4)}$. Under these conditions on ϕ, we see $a''(t)\geq 0$ and hence we have $a(1)-a(0)\geq a'(0)$. We first consider f_t and so on as a function of t and x and then evaluate $a(1)$, $a(0)$ and $a'(0)$ at $x=0$. We have

$$a(1) = \varphi\left(\mathbb{E}\left[f(X)\right]\right)\,,\quad a(0) = \mathbb{E}\left[\varphi(f(X))\right]\,,$$
$$a'(0) = -\frac{1}{2}\mathbb{E}\left[\varphi''(f(X))|\nabla f(X)|^2\right]\,.$$

Substituting the above identities into $a(1)-a(0)\geq a'(0)$ yields (3.5.4). □

Corollary 3.1.

(i) If we take $\varphi(x)=x^2$, then the condition (3.5.3) *is verified. We have the following spectral gap inequality (Poincaré inequality):*

$$\mathbb{E}\left[f(X)\right]^2 \leq \left[\mathbb{E}|f(X)|\right]^2 + \mathbb{E}\left[|\nabla f(X)|^2\right]\,. \tag{3.5.5}$$

(ii) If we take $\varphi(x)=x\log x$, then the condition (3.5.3) *is also verified. In this case the inequality states that for all $f>0$, we have the following logarithmic Sobolev inequality:*

$$\mathbb{E}\left[f(X)\log f(X)\right] \leq \mathbb{E}\left[f(X)\right]\log\left(\mathbb{E}\left[f(X)\right]\right) + \frac{1}{2}\mathbb{E}\left[\frac{1}{f(X)}|\nabla f(X)|^2\right]\,. \tag{3.5.6}$$

If we replace f by f^2, then

$$\mathbb{E}\left[|f(X)|^2 \log|f(X)|\right] \leq \frac{1}{2}\mathbb{E}\left([f(X)]^2\right)\log\left(\mathbb{E}\,[f(X)]^2\right)+\mathbb{E}\left[|\nabla f(X)|^2\right]. \tag{3.5.7}$$

It is interesting to have a close look at condition (3.5.3). It is worthy noting that when the last inequality in (3.5.3) becomes an equality, the solution to the equation is $\phi(x) = x^2$ (spectral gap inequality) and $\phi(x) = x \log x$ (logarithmic Sobolev inequality). We state this observation as the following proposition.

Proposition 3.16. *The solutions to the differential equation*

$$2[\phi^{(3)}]^2 = \phi'' \cdot \phi^{(4)} \tag{3.5.8}$$

are

$$\phi(x) = Bx^2 + Cx + D, \tag{3.5.9}$$

or

$$\phi(x) = A^{-2}(Ax + B)\log|Ax + B| + Cx + D \tag{3.5.10}$$

for some constants A, B, C, and D, such that $A \neq 0$.

Proof. Suppose that ϕ is a solution to (3.5.8) and set $k(x) \equiv \phi''(x)$. Then, k solves the ODE:

$$2[k']^2 = k \cdot k''. \tag{3.5.11}$$

One way to solve this equation is to have $k' \equiv 0$, in which case $k(x)$ has a constant value, i.e. $\phi''(x)$ is a constant. Hence $\phi(x) = Bx^2 + Cx + D$ for some constants B, C, and D.

Now assume that $k' \neq 0$ and we are on an interval in $\mathbb{R}$ where $k \neq 0$. In this case Eq. (3.5.11) may be written as:

$$2k'/k = k''/k'. \tag{3.5.12}$$

Integrating this last equation gives:

$$\log|k'| = \log|k|^2 + \text{const.}$$

and hence $k' = \tilde{A}k^2$, for some constant $\tilde{A}$. From this differential equation we learn that $k(x) = (Ax + B)^{-1}$ for some constants A and B. We may and do assume that $A \neq 0$, otherwise we will be back in the case where $k' \equiv 0$. Integrating the equation

$$\phi''(x) = k(x) = (Ax + B)^{-1}$$

gives (3.5.10). □

3.6 Variance Identity and Inequality

In this section we present the variance inequality. Denote $p_t(x,y) = (2\pi t)^{-d/2}\exp(-|x-y|^2/(2t))$ and

$$\nabla^n p_t(x,y) = \frac{\partial^n}{\partial x_{i_1}\cdots\partial x_{i_n}} P_t(x,y),$$

$$\nabla^n P_t f = \nabla^n P_t f(x) = \int_{\mathbb{R}^d} \nabla^n p_t(x,y) f(y)dy\,,$$

$$|f^{(n)}|^2 = |\nabla^n f|^2 = \sum_{i_1,\cdots,i_n=1}^{d} \left(\partial_{i_1,\cdots,i_n} f\right)^2\,.$$

Lemma 3.4. *Let $n \ge 1$ and $f \in C_b^\infty(\mathbb{R}^d)$ (the set of all smooth functions with bounded derivatives). Then,*

$$\int_0^1 \frac{t^{n-1}}{(n-1)!} P_t\left(|\nabla^n P_{1-t} f|^2\right) dt$$
$$= \frac{1}{n!} P_t\left(|f^{(n)}|^2\right) - \int_0^1 \frac{t^n}{n!} P_t\left(|\nabla^{n+1} P_{1-t} f|^2\right) dt\,. \tag{3.6.1}$$

Proof. First, we compute

$$\begin{aligned}\frac{d}{dt} P_t\left(|\nabla^n P_{1-t} f|^2\right) &= \left(\frac{d}{dt} P_t\right)\left(|\nabla^n P_{1-t} f|^2\right) + 2P_t\left([\nabla^n P_{1-t} f][\nabla^n \frac{d}{dt} P_{1-t} f]\right)\\ &= \frac{1}{2} P_t\left(\Delta\left\{|\nabla^n P_{1-t} f|^2\right\}\right) - P_t\left([\nabla^n P_{1-t} f][\nabla^n \Delta P_{1-t} f]\right)\\ &= P_t\left(|\nabla^{n+1} P_{1-t} f|^2\right)\,. \end{aligned} \tag{3.6.2}$$

Applying the integration by parts formula to the left hand side of (3.6.1), we obtain

$$\int_0^1 \frac{t^{n-1}}{(n-1)!} P_t\left(|\nabla^n P_{1-t} f|^2\right) dt$$
$$= \frac{t^n}{n!} P_t\left(|\nabla^n P_{1-t} f|^2\right)\Big|_0^1 - \int_0^1 \frac{t^n}{n!}\frac{d}{dt} P_t\left(|\nabla^n P_{1-t} f|^2\right) dt\,. \tag{3.6.3}$$

Now we apply (3.6.2) to the above derivative term to obtain the lemma. □

Theorem 3.15. *Let X be a d-dimensional centered normalized Gaussian random vector and $f \in C_b^\infty$. Then*

$$\mathbb{E}|f(X) - \mathbb{E}f(X)|^2 = \sum_{k=1}^n \frac{(-1)^{k+1}}{k!}\mathbb{E}\left(|f^{(k)}(X)|^2\right)$$
$$+(-1)^{k+2}\int_0^1 \frac{t^n}{n!} P_t\left(|\nabla^{n+1} P_{1-t} f|^2\right)(0)dt\,. \tag{3.6.4}$$

Proof. The case $n = 1$ is a consequence of inequality (3.5.5). To pass from n to $n+1$, we need only a simple application of Lemma 3.4 (formula (3.6.1)). □

Now the following variance inequality ([Houdré and Kagan (1995)]) is an easy consequence of Theorem 3.15 .

Theorem 3.16. *Let X be a d-dimensional centered normalized Gaussian random vector and let f be a sufficiently differentiable function from $\mathbb{R}^d$ to $\mathbb{R}$ with bounded derivatives. Then,*

$$\sum_{k=1}^{2n} \frac{(-1)^{k+1}}{k!} \mathbb{E}\left(|f^{(k)}(X)|^2\right) \le \left(|f(X) - \mathbb{E}f(X)|^2\right)$$

$$\le \sum_{k=1}^{2n-1} \frac{(-1)^{k+1}}{k!} \mathbb{E}\left(|f^{(k)}(X)|^2\right). \tag{3.6.5}$$

3.7 Correlation Inequality

Theorem 3.17. *If f, $g : \mathbb{R}^d \to \mathbb{R}$ are convex and if one of these two functions is symmetric (symmetric means $f(-x) = f(x)$ $\forall x \in \mathbb{R}^d$), then*

$$\mu_d(fg) \ge \mu_d(f)\mu_d(g). \tag{3.7.1}$$

Proof. Without loss of generality, we can assume that f and g are twice continuously differentiable. For any given fixed $T > 0$, we shall use the following notations.

$$\begin{cases} p_t(x) = (2\pi t)^{-d/2} \exp\left\{-|x|^2/(2t)\right\}, \\ P_t f(x) = \int_{\mathbb{R}^d} p_t(x-y) f(y) dy, \\ F_t = P_t f = P_t f(x), G_t = P_t g = P_t g(x) \\ h_t = P_{T-t}(F_t G_t) \end{cases} \tag{3.7.2}$$

Then,

$$\begin{aligned} h_t' &:= \frac{d}{dt} h_t = \left(\frac{d}{dt} P_{T-t}\right)(F_t G_t) + P_{T-t}\left(G_t \frac{d}{dt} F_t + \left(F_t \frac{d}{dt} G_t\right)\right) \\ &= -\frac{1}{2} P_{T-t}\left(\Delta\left[F_t G_t\right]\right) + \frac{1}{2} P_{T-t}\left(G_t \Delta F_t + F_t \Delta G_t\right) \\ &= -P_{T-t}\left(\langle \nabla F_t, \nabla G_t\rangle\right). \end{aligned}$$

Continuing to compute the second derivative of h_t we have

$$\begin{aligned} h_t'' &= \frac{1}{2} P_{T-t} \left(\Delta \left[\langle \nabla F_t \,, \nabla G_t \rangle \right] - \langle \nabla \Delta F_t \,, \nabla G_t \rangle - \langle \nabla F_t \,, \nabla \Delta G_t \rangle \right) \\ &= P_{T-t} \left(\nabla^2 F_t \,, \nabla^2 G_t \rangle \right) . \end{aligned}$$

We are only interested in the values of h_t, h_t' and h_t'' at $x = 0$ still denoted by h_t, h_t' and h_t''. We also let $T = 1$. The condition that at least one of f and g is symmetric implies $\mu_d(\nabla f)\mu_d(\nabla g) = 0$. From the above computation we have

$$\begin{cases} h_0 = \mu_d(fg) \,, \quad h_1 = \mu_d(f)\mu_d(g) & (3.7.3) \\ h_1' = \mu_d(\nabla f)\mu_d(\nabla g) = 0 \,, \quad h_t'' = P_{1-t} \left(\nabla^2 F_t \,, \nabla^2 G_t \rangle \right) \geq 0 \,. & (3.7.4) \end{cases}$$

By Taylor expansion and then by (3.7.4), we have for some $\theta \in [0, 1]$,

$$h_0 = h_1 - h_1' + \frac{1}{2} h_\theta'' \geq h_1$$

which is (3.7.1) by (3.7.3). □

Remark 3.6. Inequality (3.7.1) was first obtained in [Hu (1997)]. There are some improvements of the inequality [Hargé (1999)], [Hargé (2005)]. See also [Hu (2000)]. We state a result in [Hargé (1999)].

Theorem 3.18. *Denote* $\mu_d(dx) = (2\pi)^{-d/2} e^{-\frac{|x|^2}{2}} dx$. *Let* $g : \mathbb{R}^d \to \mathbb{R}$ *be log-concave and even and let* f *be a differentiable such that* $f' \leq 0$ *on* $\mathbb{R}_+$. *Let* C *be a symmetric positive definite matrix. Then,*

$$\int_{\mathbb{R}^d} f\left(\langle C^{-1} x \,, x \rangle \right) g(x) \mu_d(dx) \geq \int_{\mathbb{R}^d} f\left(\langle C^{-1} x \,, x \rangle \right) \mu_d(dx) \int_{\mathbb{R}^d} g(x) \mu_d(dx) \,. \tag{3.7.5}$$

Proof. It should be possible to use the same approach as in the proof of Theorem 3.17. However, we follow the method of [Hargé (1999)] to use the Ornstein-Uhlenbeck semigroup instead. Denote (the Mehler formula for the Ornstein-Ulenbeck semigroup)

$$P_t^C f(x) = \int_{\mathbb{R}^d} f\left(\exp\left[-\frac{t}{2} C \right] x - \left(I - \exp\left[-\, tC \right] \right)^{1/2} y \right) \mu_C(dy) \,,$$

where

$$\mu_C(dx) = \sqrt{\det(C)} (2\pi)^{-d/2} \exp\left\{ -\frac{1}{2} \langle Cx, x \rangle \right\} .$$

Introduce also

$$L_C f = \frac{1}{2} \left(\Delta f - \langle Cx, \nabla f \rangle \right) .$$

Then, we have

$$\frac{d}{dt}P_t^C f = L_C P_t^C f\,. \tag{3.7.6}$$

We see easily that

$$P_0 f = f\,,\quad \lim_{t\to\infty} P_t f(x) \to \mu_C(f) = \int_{\mathbb{R}^d} f(x)\mu_C(dx)\,.$$

Now we turn to prove the theorem. Denote $\bar{g}(y) = g(C^{1/2}y)$. Then, $\bar{g}$ is log-concave and even. Make the substitution $y = C^{-1/2}x$. The inequality (3.7.5) becomes

$$\int_{\mathbb{R}^d} f(|y|^2) g\left(C^{1/2}y)\right) \mu_C(dy) \ge \int_{\mathbb{R}^d} f(|y|^2)\mu_C(dy) \int_{\mathbb{R}^d} g\left(C^{1/2}y\right)\mu_C(dy)\,. \tag{3.7.7}$$

Set

$$h_t = \mu_C\left[f(|y|^2)P_t^C(\bar{g})(y)\right]\,.$$

Then, it is easy to see that

$$\lim_{t\to\infty} h_t = \mu_C\left[f(|y|^2)\right]\mu_C(\bar{g})\,,\quad \lim_{t\to 0} h_t = \mu_C\left[f(|y|^2)\bar{g}(y)\right]\,.$$

Thus, to prove the theorem or to prove (3.7.7), it suffices to show $h_t' \le 0$. Now

$$\begin{aligned} h_t' &= \mu_C\left[f(|y|^2)L_C P_t^C \bar{g}(y)\right] \\ &= -\frac{1}{2}\mu_C\left[\langle \nabla f(|y|^2)\,, \nabla P_t^C \bar{g}\rangle\right] \\ &= -\mu_C\left[f'(|y|^2)\langle y\,, \nabla P_t^C \bar{g}\rangle\right]\,. \end{aligned}$$

By Theorem 3.3, the function $y \to P_t^C \bar{g}(y)$ is a log-concave, even, and positive. Then, there is a convex and even function G such that $P_t^C\bar{g}(y) = e^{-G}$. Moreover, since G is even, $\nabla G(y)$ is odd. In particular, $\nabla G(0) = 0$. Thus,

$$\nabla G(y) = \nabla G(y) - \nabla G(0) = \int_0^1 \nabla^2 G(\theta y) y d\theta\,.$$

Thus, we have by the convexity of G that

$$\langle y\,, \nabla G(y)\rangle = \int_0^1 \langle y\,, \nabla^2 G(\theta y) y\rangle d\theta \ge 0\,.$$

Hence,

$$h_t' = -\mu_C\left[f'(|y|^2)\langle y\,, \nabla e^{-G}\rangle\right] = \mu_C\left[f'(|y|^2)e^{-G}\langle y\,, \nabla G(y)\rangle\right] \le 0$$

since $f' \le 0$. This proves the theorem. □

Remark 3.7. If μ is the standard Gaussian measure on $\mathbb{R}^d$ and if A and B are two convex and symmetric sets (A is called symmetric if $x \in A$ implies $-x \in A$), then it has been conjectured

$$\mu_d(AB) \geq \mu_d(A)\mu_d(B) . \tag{3.7.8}$$

The case $d = 1$ is obvious. Inequality (3.7.8) is proved in the case $d = 2$ by Pitt [Pitt (1977)]. Recently, the general case is proved by [Royen (2014)] and a simplified presentation can be found in [Latała and Matlak (2015)].

3.8 Hypercontractivity

If $f : \mathbb{R}^d \to \mathbb{R}$ is a square integrable function with respect to the standard Gaussian measure $\mathbb{R}^d$, then we can define for any $\alpha \in (-1, 1)$, the following second quantization operator

$$\Gamma(\alpha)f(x) = \int_{\mathbb{R}^d} f(\alpha x + \sqrt{1-\alpha^2}y)\mu(dy) , \tag{3.8.1}$$

where $\mu(dy) = \mu_d(dy) = (2\pi)^{-d/2}e^{-\frac{|y|^2}{2}}dy$ is the standard Gaussian measure on $\mathbb{R}^d$. This is called the Mehler formula for the second quantization operator $\Gamma(\alpha)$ (Ornstein-Uhlenbeck semigroup).

When $d = 1$, from (3.4.22), we see that

$$\Gamma(\alpha)H_n(x) = \sum_{i=0}^{n} \binom{n}{k} \alpha^k (1-\alpha^2)^{(n-k)/2} H_k(x) \int_{\mathbb{R}} H_{n-k}(y)\mu(dy) = \alpha^n H_n(x) ,$$

since $\int_{\mathbb{R}} H_k(y)\mu(dy) = 0$ for all $k \geq 1$. If $f \in L^2(\mathbb{R}, \mu)$, then

$$f(x) = \sum_{n=0}^{\infty} a_n H_n(x) \quad \text{for some constants } f_n \text{ such that } \textstyle\sum_{n=0}^{\infty} n!|a_n|^2 < \infty .$$

For this expansion of f, it is easy to see

$$\Gamma(\alpha)f(x) = \sum_{n=0}^{\infty} a_n \alpha^n H_n(x) .$$

In general, if $f \in L^2(\mathbb{R}^d, \mu)$, then

$$f = \sum_{n=0}^{\infty} f_n(x) ,$$

where $f_n(x)$ is finite linear combinations of $H_{n_1}(x_1)\cdots H_{n_d}(x_d)$ such that $n_1 + \cdots + n_d = n$. For such expansion of f we have

$$\Gamma(\alpha)f(x) = \sum_{n=0}^{\infty} \alpha^n f_n(x) .$$

We can also define $\Gamma(\alpha)$ for vector valued function and the following proposition is obvious from the Mehler formula (3.8.1).

Proposition 3.17. *Let $f = (f_1, \cdots, d_m) \in L^2(\mathbb{R}^d, \mu)$. Then,*

$$|\Gamma(\alpha)f(x))| \le \Gamma(\alpha)(|f|)(x) \quad \text{for all } x \in \mathbb{R}^d . \tag{3.8.2}$$

Next, we shall deal with the hypercontractivity inequality. This inequality can follow from the logarithmic inequality. There are numerous proofs of this inequality. An elegant probabilistic proof is due to [Neveu (1976)] (see also [Dellacherie, Maisonneuve and Meyer (1987)], p. 239-241). Here we simplify the argument in [Hu (2000)] to give a direct proof using idea analogous to the proof of the logarithmic inequality (3.5.4). First, we have the following theorem (see [[Hu (2000)]]).

Theorem 3.19. *Let μ be a standard Gaussian measure on $\mathbb{R}^d$. Denote $\mu(f) = \int_{\mathbb{R}^d} f(x)\mu(dx)$. Let α be a number in $(-1, 1)$. Let φ and ψ be two positive concave functions on $\mathbb{R}_+$ such that*

$$\begin{aligned} &\varphi(x) \ge 0 , \quad \psi(x) \ge 0 , \quad \varphi''(x) \le 0 , \quad \psi''(x) \le 0 , \\ &\alpha^2 \left(\varphi'(x)\psi'(y)\right)^2 \le \varphi(x)\psi(y)\varphi''(x)\psi''(y) \end{aligned} \tag{3.8.3}$$

for al $x, y \in \mathbb{R}_+$. Then, for all positive measurable bounded functions f and g, the following inequality holds

$$\mu \{\psi(g)\, \Gamma(\alpha)(\varphi(f))\} \le \varphi(\mu(f))\psi(\mu(g)) . \tag{3.8.4}$$

Proof. Let us define the semigroup

$$P_t f(x) = \mathbb{E} f(B_t + x) = \int_{\mathbb{R}^d} p_t(x-y) f(y) dy ,$$

where $p_t(x) = (2\pi t)^{-d/2} \exp\left\{-|x|^2/2t\right\}$. We also define

$$\begin{aligned} P_{\alpha,t}(f \times g)(x) &= \mathbb{E}\left[f(\alpha B_t + \beta B_t' + x) g(B_t + x)\right] \\ &= \int_{\mathbb{R}^{2d}} p_t(y) p_t(z) f(\alpha y + \beta z + x) g(y + x) dy dz , \end{aligned}$$

where B_t and B_t' are two independent Brownian motions and $\beta = \sqrt{1-\alpha^2}$. To simplify notation we also use $P_t f = P_t f(x)$ and $P_{\alpha,t}(f \times g) = P_{\alpha,t}(f \times$

$g)(x)$. Then, it is easy to verify that

$$\begin{aligned}
\frac{d}{dt}P_{\alpha,t}(f\times g) &= \int_{\mathbb{R}^{2d}}\left[p_t(z)\frac{d}{dt}p_t(y)+p_t(y)\frac{d}{dt}p_t(z)\right]f(\alpha y+\beta z)g(y)dydz\\
&= \int_{\mathbb{R}^{2d}}\left[p_t(z)\frac{1}{2}\Delta_y p_t(y)+p_t(y)\frac{1}{2}\Delta_z p_t(z)\right]f(\alpha y+\beta z)g(y)dydz\\
&= \int_{\mathbb{R}^{2d}}p_t(y)p_t(z)\frac{1}{2}(\Delta_y+\Delta_z)\left[f(\alpha y+\beta z)g(y)\right]dydz\\
&= \frac{1}{2}\int_{\mathbb{R}^{2d}}p_t(y)p_t(z)\Big[(\beta^2+\alpha^2)f''(\alpha y+\beta z)g(y)\\
&\qquad +2\alpha f'(\alpha y+\beta z)g'(y)+f(\alpha y+\beta z)g''(y)\Big]dydz\\
&= \frac{1}{2}P_{\alpha,t}\left[f''\times g+2\alpha f'\times g'+f\times g''\right]. \qquad (3.8.5)
\end{aligned}$$

Consider the function

$$h_t = P_{\alpha,t}(\varphi(P_{1-t}f)\times\psi(P_{1-t}g)),\quad 0\le t\le 1. \qquad (3.8.6)$$

[$f\ge 0$ implies that $P_{1-t}f\ge 0$. Thus, the above h_t is well-defined.] We are going to compute the derivative of h_t. To simplify notations, we denote $F_t = P_{1-t}f$, $G_t = P_{1-t}g$, $\varphi_t = \varphi(P_{1-t}f)$ and $\varphi_t' = \varphi'(P_{1-t}f)$ and so on. We also use $\langle\nabla f,\times\nabla g\rangle = \sum_{i=1}^d \partial_i f\times\partial_i g$. Then

$$\begin{aligned}
\frac{d}{dt}h_t &= \left(\frac{d}{dt}P_{\alpha,t}\right)(\varphi_t\times\psi_t)) + P_{\alpha,t}\left(\frac{d\varphi_t}{dt}\times\psi_t\right)+P_{\alpha,t}\left(\varphi_t\times\frac{d\psi_t}{dt}\right)\\
&= \frac{1}{2}P_{\alpha,t}\left\{\Delta\varphi_t\times\psi_t+2\alpha\langle\nabla\varphi_t,\times\nabla\psi_t\rangle+\varphi_t\times\Delta\psi_t+2\frac{d\varphi_t}{dt}\times\psi_t+2\varphi_t\times\frac{d\psi_t}{dt}\right\}\\
&= \frac{1}{2}P_{\alpha,t}\Big\{\varphi_t'\Delta F_t\times\psi_t+\varphi_t''|\nabla F_t|^2\times\psi_t+2\alpha(\varphi_t'\times\psi_t')\langle\nabla F_t,\times\nabla G_t\rangle\\
&\qquad +\varphi_t\times\psi_t'\Delta G_t+\varphi_t\times\psi_t''|\nabla G_t|^2-\varphi_t'\Delta F_t\times\psi_t-\varphi_t\times\varphi_t'\Delta G_t\Big\}\\
&= \frac{1}{2}P_{\alpha,t}\left\{\varphi_t''|\nabla F_t|^2\times\psi_t+2\alpha(\varphi_t'\times\psi_t')\langle\nabla F_t,\times\nabla G_t\rangle+\varphi_t\times\psi_t''|\nabla G_t|^2\right\}.
\end{aligned}$$

It is clear that under the conditions (3.8.3), the above sum inside the above { } is negative by Lemma 3.3. Thus, h_t is a decreasing function. Namely, $h_1\le h_0$. But

$$h_1 = \mu(\Gamma_\alpha\varphi(f)\psi(g))\quad h_0 = \varphi(\mu(f))\psi(\mu(g)).$$

The theorem is then proved. □

Theorem 3.20.

(i) If α is a real number such that $|\alpha| \le \sqrt{\frac{p-1}{q-1}}$ with $1 < p \le q < \infty$, then

$$\|\Gamma(\alpha)f\|_q \le \|f\|_p \quad \forall\, f \in L^p(\mathbb{R}^d)\,. \tag{3.8.7}$$

(ii) If there is a finite positive constant C such that

$$\|\Gamma(\alpha)f\|_q \le C\|f\|_p \quad \forall\, f \in L^p(\mathbb{R}^d)\,,$$

then $|\alpha| \le \sqrt{\frac{p-1}{q-1}}$.

Proof. Let us consider (i). Applying Theorem 3.19 to $\varphi(x) = x^{1/p}$ and $\psi(x) = x^{1/r}$, where r is the conjugate number of q, we see that the conditions (3.8.3) are satisfied if and only if $\alpha^2 \le (p-1)(r-1) = \frac{p-1}{q-1}$. Now we apply the inequality (3.8.4) (with f replaced by $\varphi^{-1}(|f|)$ and g replaced by $\psi^{-1}(|g|)$) to obtain

$$\begin{aligned}|\mu((\Gamma(\alpha)f)g)| &\le \mu((\Gamma(\alpha)|f|)\,|g|)\\ &\le \phi(\mu(\phi^{-1}(|f|)))\,\psi(\mu(\psi^{-1}(|g|)))\\ &\le \|f\|_p\|g\|_r\,.\end{aligned}$$

Since this inequality holds true for all $g \in L^r(\mathbb{R}^d)$ we see that $\|\Gamma(\alpha)f\|_q \le \|f\|_p$, proving (i).

To show (ii) we take $f_a(x) = e^{\langle a,x\rangle - \frac{|a|^2}{2}}$. Then $\Gamma(\alpha)f_a(x) = e^{\alpha\langle a,x\rangle - \frac{\alpha^2|a|^2}{2}}$. It is easy to check that $\|\Gamma(\alpha)f_a(x)\|_q = e^{\alpha^2 a^2(q-1)/2}$ and $\|f\|_p = e^{a^2(p-1)/2}$. $\|\Gamma(\alpha)f_a(x)\|_q \le C\|f\|_p$ immediately implies $\alpha^2(q-1) \le p-1$. Then, the theorem is proved. □

3.9 Hermite Polynomials in Physics and Hermite Functions

The Hermite polynomials used in physics are slightly different from the ones introduced in the previous section (which we call probabilistic Hermite polynomials when we need to stress the difference). Now we give a brief presentation of these polynomials.

The physical Hermite polynomials are defined as

$$\mathcal{H}_n(x) = (-1)^n e^{x^2}\frac{d^n}{dx^n}e^{-x^2} = \sum_{k\le n/2}\frac{(-1)^k n!}{k!(n-2k)!}(2x)^{n-2k}\,,\quad n = 0,1,2,\cdots \tag{3.9.1}$$

The first few of these polynomials are

$$\mathcal{H}_0(x) = 1 \qquad \mathcal{H}_1(x) = 2x\,,$$
$$\mathcal{H}_2(x) = 4x^2 - 2\,, \quad \mathcal{H}_3(x) = 8x^3 - 12x\,, \quad \cdots$$

We list some properties of the physical Hermite polynomials whose proofs are similar to those of probabilistic Hermite polynomials and we shall omit them.

Theorem 3.21.

(i) *The generating function of these Hermite polynomials are*

$$e^{2xt-t^2} = \sum_{n=0}^{\infty} \mathcal{H}_n(x) \frac{t^n}{n!}\,. \tag{3.9.2}$$

(ii) *$\mathcal{H}_n(x)$ can be represented by*

$$\mathcal{H}_n(x) = \frac{2^n(-\iota)^n e^{x^2}}{\sqrt{\pi}} \int_{\mathbb{R}} e^{-t^2+2\iota tx} t^n dt\,, n = 0, 1, 2, \cdots \tag{3.9.3}$$

(iii)

$$\frac{d}{dx}\mathcal{H}_n(x) = 2n\mathcal{H}_{n-1}(x)\,. \tag{3.9.4}$$

(iv) *The recursive relation is*

$$\mathcal{H}_{n+1}(x) - 2x\mathcal{H}_n(x) + 2n\mathcal{H}_{n-1}(x) = 0\,. \tag{3.9.5}$$

(v) *The equation satisfied by the physical Hermite polynomials is*

$$\mathcal{H}_n''(x) - 2x\mathcal{H}_n'(x) + 2n\mathcal{H}_n(x) = 0 \tag{3.9.6}$$

which means that the operator $\frac{d^2}{dx^2} - 2x\frac{d}{dx}$ has eigenvalues $2n$ and eigenvectors $\{\mathcal{H}_n(x)\,, n = 0, 1, 2, \cdot\}$, where $n = 0, 1, 2, \cdots$

(vi)

$$\frac{1}{\sqrt{\pi}} \int_{\mathbb{R}} \mathcal{H}_n(x)\mathcal{H}_m(x) e^{-x^2} dx = 2^n n! \delta_{nm}\,, \qquad n, m = 0, 1, 2, \cdots \tag{3.9.7}$$

(vii) $\left\{\frac{1}{2^{n/2}\sqrt{n!}}\mathcal{H}_n(x)\,,\ n = 0, 1, 2, \cdots\right\}$ *is a complete orthonormal basis of* $L^2(\mathbb{R}, \frac{1}{\sqrt{\pi}} e^{-x^2} dx)$.

(viii) *For any $|t| < 1$, we have*

$$\sum_{n=0}^{\infty} \frac{\mathcal{H}_n(x)\mathcal{H}_n(y)}{2^n n!} t^n = (1-t^2)^{-1/2} \exp\left\{\frac{2xyt - (x^2+y^2)t^2}{1-t^2}\right\}\,, \tag{3.9.8}$$

for all $x, y \in \mathbb{R}$. In particular, we have

$$\sum_{n=0}^{\infty} \frac{[\mathcal{H}_n(x)]^2}{2^n n!} t^n = (1-t^2)^{-1/2} \exp\left\{\frac{2x^2 t}{1+t}\right\}\,, \qquad \forall\, x \in \mathbb{R}\,. \tag{3.9.9}$$

The relation between the physical Hermite polynomials and the probabilistic Hermite polynomials is

$$\mathcal{H}_n(x) = 2^{n/2} H_n(\sqrt{2}x)\,, \quad H_n(x) = 2^{-n/2}\mathcal{H}_n(x/\sqrt{2})\,. \tag{3.9.10}$$

We introduce the following Hermite functions

$$\mathfrak{H}_n(x) = e^{-x^2/2}\mathcal{H}_n(x)\,, \quad n = 0, 1, 2, \cdots \tag{3.9.11}$$

Since $\{\mathcal{H}_n(x)\}$ is an orthogonal basis of $L^2(\mathbb{R}, e^{-x^2}dx/\sqrt{\pi})$,

$$\{\mathfrak{H}_n(x)\,, n = 0, 1, \cdots\}$$

is an orthogonal basis of $L^2(\mathbb{R}, dx)$. Many properties of $\mathfrak{H}_n(x)$ can be obtained from the properties of $\mathcal{H}_n$. For example, we have

$$\sum_{n=0}^{\infty} \mathfrak{H}_n(x)\frac{t^n}{n!} = e^{2tx - t^2 - \frac{x^2}{2}}\,. \tag{3.9.12}$$

It is easy to verify that

$$\begin{aligned}
\frac{d}{dx}\mathfrak{H}_n(x) &= -x\mathfrak{H}_n(x) + 2n\mathfrak{H}_{n-1}(x)\,; \\
\frac{d^2}{dx^2}\mathfrak{H}_n(x) &= (x^2 - 1)\mathfrak{H}_n(x) - 4nx\mathfrak{H}_{n-1}(x) + 4n(n-1)\mathfrak{H}_{n-2}(x) \\
&= (x^2 - 1)\mathfrak{H}_n(x) + 2n(-2x\mathfrak{H}_{n-1} + 2(n-1)\mathfrak{H}_{n-2}(x)) \\
&= (x^2 - 1 - 2n)\mathfrak{H}_n(x)\,,
\end{aligned}$$

where the last identity follows from Theorem 3.21, part (iv). This yields

$$\left[-\frac{d^2}{dx^2} + x^2\right]\mathfrak{H}_n(x) = (2n+1)\mathfrak{H}_n(x)\,. \tag{3.9.13}$$

Thus, we have

Proposition 3.18. *The operator $-\frac{d^2}{dx^2} + x^2$ has the eigenvalues $2n+1$ and the corresponding eigenfunctions $\mathfrak{H}_n(x)$, $n = 0, 1, 2, \cdots$, which consists of a complete orthogonal system of $L^2(dx)$.*

Remark 3.8. The above identity means that $\mathfrak{H}_n$ is an eigenfunction of the harmonic oscillator (operator) $-\frac{d^2}{dx^2} + x^2$ corresponding to the eigenvalue $(2n+1)$.

The equation (3.9.13) can be written as

$$\mathfrak{H}_n''(x) + (2n+1)\mathfrak{H}_n(x) = x^2\mathfrak{H}_n(x)\,. \tag{3.9.14}$$

The solution to the above equation (3.9.14) can be represented by

$$\mathfrak{H}_n(x) = \mathfrak{H}_n(0)\cos(\sqrt{2n+1}x) + \mathfrak{H}_n'(0)\frac{\sin(\sqrt{2n+1}x)}{\sqrt{2n+1}} + \frac{1}{\sqrt{2n+1}}\int_0^x y^2\mathfrak{H}_n(y)\sin(\sqrt{2n+1}(x-y))dy\,.$$

Now from (3.9.11) and then from (3.9.1) it follows

$$\mathfrak{H}_n(0) = \mathcal{H}_n(0) = \begin{cases} \frac{(-1)^m(2m)!}{m!} & \text{if } n = 2m \\ 0 & \text{if } n = 2m+1 \end{cases}$$

and

$$\mathfrak{H}_n'(0) = \mathcal{H}_n'(0) = \begin{cases} \frac{2(-1)^m(2m+1)!}{m!} & \text{if } n = 2m+1 \\ 0 & \text{if } n = 2m\,. \end{cases}$$

Denote

$$\alpha_n = \begin{cases} \frac{(-1)^m(2m)!}{m!} & \text{if } n = 2m \\ \frac{2(-1)^m(2m+1)!}{m!(4m+2)} & \text{if } n = 2m+1 \end{cases} \tag{3.9.15}$$

Then

$$\mathfrak{H}_n(x) = \alpha_n\left[\cos(\sqrt{2n+1}x - \frac{n\pi}{2}) + r_n(x)\right],$$

where

$$r_n(x) = \frac{1}{\alpha_n\sqrt{2n+1}}\int_0^x y^2\mathfrak{H}_n(y)\sin(\sqrt{2n+1}(x-y))dy\,.$$

By (3.9.7), we have

$$\begin{aligned} |r_n(x)| &\le \frac{1}{|\alpha_n|\sqrt{2n+1}}(\int_0^x y^4dy)^{1/2}\left(\int_{-\infty}^{\infty}\mathfrak{H}_n^2(y)dy\right)^{1/2} \\ &= \frac{1}{|\alpha_n|\sqrt{5(2n+1)}}x^{5/2}\left(\int_{-\infty}^{\infty}H_n^2(y)e^{-y^2}dy\right)^{1/2} \\ &\le \frac{2^{n/2}\sqrt[4]{\pi}\sqrt{n!}}{|\alpha_n|\sqrt{10(2n+1)}}|x|^{5/2}\,. \end{aligned}$$

From the Stirling formula, we have

$$\alpha_n \approx n^{(n+1)/2}e^{-n/2}n^{n/2}\,.$$

Therefore,

$$|r_n(x)| \preceq n^{-1/4}|x|^{5/2},$$

where and throughout the book the notation $a_n \approx b_n$ means that $\lim_n \frac{a_n}{b_n} = 1$ and the notation $a_n \preceq b_n$ means there is a constant C such that $a_n \le Cb_n$. Thus, we have

Proposition 3.19. *When $n \to \infty$, we have*

$$\mathfrak{H}_n(x) \approx 2^{(n+1)/2} e^{-n/2} n^{n/2} \cos(\sqrt{2n+1}x - \frac{n\pi}{2}) \tag{3.9.16}$$

or

$$\mathcal{H}_n(x) \approx 2^{(n+1)/2} e^{-n/2} n^{n/2} e^{-x^2/2} \cos(\sqrt{2n+1}x - \frac{n\pi}{2}) . \tag{3.9.17}$$

3.10 Segal-Bargmann Space and Complex Hermite Polynomials

Let U be a non empty open subset of $\mathbb{C}^d = \mathbb{R}^{2d}$ (the d-dimensional complex space). A function $F : U \to \mathbb{C}$ is called holomorphic (complex analytic) if it is continuous and holomorphic in each of its variables when other variables are fixed.

Definition 3.4. Let α be a positive function on U. We define

$$\mathcal{HL}^2(U,\alpha) := \Big\{ F : U \to \mathbb{C} \;\; \text{holomorphic such that} \quad \|F\|^2_{L^2(U,\alpha)} := \int_U |F(z)|^2 \alpha(z) dz \Big\}, \tag{3.10.1}$$

where dz is the Lebesgue measure on $\mathbb{C}^d$.

Theorem 3.22. *Let $\alpha : U \to \mathbb{C}$ be continuous and strictly positive on U. Then,*

(i) *For any compact set $U_0 \subseteq U$, there is a constant C_{U_0} such that*

$$|F(z)| \le C_{U_0} \|F\|_{L^2(U,\alpha)} \quad \forall F \in \mathcal{HL}^2(U,\alpha) , \quad \forall z \in U_0 . \tag{3.10.2}$$

(ii) *$\mathcal{HL}^2(U,\alpha)$ is a closed subset $L^2(U,\alpha)$.*

Proof. Let us first prove (i) when $d = 1$. Since $z \in U$ there is a positive number $s > 0$ such that $B(z,s) = \{v \in \mathbb{C} , \; |v - z| < s\}$ is contained in U. Since F is holomorphic, we have

$$F(v) = F(z) + \sum_{k=1}^{\infty} a_k (v-z)^k , \quad \forall\, v \in B(z,s) .$$

Integrating the above identity over $B(z,s)$ and noticing $\int_{B(z,s)}(v-z)^k dv = 0$ for all $k \geq 1$, we see that

$$F(z) = \frac{1}{\pi s^2}\int_{B(z,s)} F(v)dv\,. \tag{3.10.3}$$

Since α is continuous and strictly positive in $\overline{B(z,s)}$ it has minimum $\alpha_0 \in (0,\infty)$ and maximum $\alpha_1 \in (0,\infty)$ on $\overline{B(z,s)}$. Thus, the Cauchy-Schwarz inequality yields

$$\begin{aligned}
|F(z)| &\leq \frac{1}{\alpha_0 \pi s^2}\int_{B(z,s)} |F(v)|\alpha(v)dv \\
&\leq \frac{1}{\alpha_0 \pi s^2}\left(\int_{B(z,s)} |F(v)|^2\alpha(v)dv\right)^{1/2}\left(\int_{B(z,s)} \alpha(v)dv\right)^{1/2} \\
&\leq \frac{\sqrt{\alpha_1}}{\alpha_0\sqrt{\pi}s}\left(\int_{B(z,s)} |F(v)|^2\alpha(v)dv\right)^{1/2} \\
&\leq \frac{\sqrt{\alpha_1}}{\alpha_0\sqrt{\pi}s}\|F\|_{L^2(U,\alpha)}\,.
\end{aligned} \tag{3.10.4}$$

Notice that the constant $C_z := \frac{\sqrt{\alpha_1}}{\alpha_0\sqrt{\pi}s}$ depends only on the radius s of the ball $B(z,s)$ which can be contained in U and the maximum and minimum of α on the ball. If U_0 is a compact subset of U, then we can find a common s_0 such that for any $z \in U_0$, the ball $B(z,s_0)$ is contained in U and the closure U_1 of $\cup_{z\in U_0} B(z,s_0)$ is also contained in U. Thus, we can find a constant C_{U_0} (one can take $C_{U_0} = \frac{\sqrt{\max_{z\in U_1}\alpha(z)}}{\min_{z\in U_1}\alpha(z)\sqrt{\pi}s_0}$) such that the inequality (3.10.2) holds. For general dimension, analogous to (3.10.3), we can easily establish

$$F(z) = \frac{1}{\pi^d s^{2d}}\int_{B_1(z_1,s)\times\cdots\times B_d(z_d,s)} F(v)dv\,, \tag{3.10.5}$$

where $z = (z_1, \cdots, z_d)$, which can be used to prove (3.10.2).

If $F_n \in \mathcal{H}L^2(U,\alpha)$ is a Cauchy sequence in $L^2(U,\alpha)$, then it converges to a function $F \in L^2(U,\alpha)$ in $L^2(U,\alpha)$ norm. By (3.10.2), for any compact set $U_0 \subseteq U$, F_n is a Cauchy sequence with respect to the maximum norm on U_0. Therefore, F_n converges to a function uniformly on U_0 and this function must be F. By the Morera's theorem, a locally uniform limit of holomophic function is always holomorphic, we see that F is also holomorphic. The theorem is proved. □

Theorem 3.23 (Existence of Reproducing Kernel). *Let U be a nonempty open set in $\mathbb{C}^d$ and let α be a continuous and strictly positive function on U. Then, there exists a function $K(z,w)$, $z,w \in U$ such that*

(i) $K(z,w)$ is holomorphic in z and antiholomorphic in w, and satisfies

$$\overline{K(z,w)} = K(w,z)\,.$$

(ii) For any $z, u \in U$,

$$\int_U K(z,w)K(w,u)\alpha(w)dw = K(z,u)\,. \tag{3.10.6}$$

(iii) Define

$$PF(z) = \int_U K(z,w)F(w)\alpha(w)dw\,, \quad \forall\ F \in L^2(U,\alpha)\,. \tag{3.10.7}$$

Then, $P : L^2(U,\alpha) \to \mathcal{H}L^2(U,\alpha)$ is a projection. In particular, when $F \in \mathcal{H}L^2(U,\alpha)$, then

$$F(z) = PF(z) = \int_U K(z,w)F(w)\alpha(w)dw\,, \quad \forall z \in U\,. \tag{3.10.8}$$

(iv) For all $z \in U$,

$$|F(z)| \le \sqrt{K(z,z)}\|F\|_{L^2(U,\alpha)}$$

and $\sqrt{K(z,z)}$ is the best constant in the above inequality.

(v) Given any $z \in U$, if $\phi_z(\cdot) \in \mathcal{H}L^2(U,\alpha)$ satisfies

$$F(z) = \int_U \overline{\phi_z(w)}F(w)\alpha(w)dw\,, \quad \forall F \in \mathcal{H}L^2(U,\alpha)\,,$$

then $\overline{\phi_z(w)} = k(z,w)$.

Proof. Since the operator "value at z" is bounded with respect to $L^2(U,\alpha)$ norm (see (3.10.2)), by Riesz representation theorem applied to the Hilbert space $\mathcal{H}L^2(U,\alpha)$, for any $z \in U$, there is a $\Phi_z \in \mathcal{H}L^2(U,\alpha)$ such that

$$F(z) = \langle \Phi_z, F\rangle_{L^2(U,\alpha)} = \int_U \overline{\Phi_z(w)}F(w)\alpha(w)dw\,, \quad \forall\ F \in \mathcal{H}L^2(U,\alpha)\,. \tag{3.10.9}$$

We take $k(z,w) = \overline{\Phi_z(w)}$, which is anti-holomorphic in w. Since $\Phi_z(\cdot) \in \mathcal{H}L^2(U,\alpha)$, we see

$$\overline{k(z,w)} = \Phi_z(w) = \langle \Phi_w, \Phi_z\rangle = \overline{\langle \Phi_z, \Phi_w\rangle} = \overline{\Phi_w(z)} = k(w,z)\,.$$

This proves part (i). The equation (3.10.9) applied to $F(w) = k(w,u)$ becomes $k(z,u) = \int_U k(z,w)k(w,u)dw$ which is part (ii).

We use the notation (3.10.7). If $F \in \mathcal{H}L^2(U,\alpha)^\perp$, then

$$PF(z) = \langle \Phi_z, F\rangle_{L^2(U,\alpha)} = 0$$

since $\Phi_z \in \mathcal{H}L^2(U,\alpha)$. Thus, P is a projection to $\mathcal{H}L^2(U,\alpha)$ and part (iii) is proved.

From Cauchy-Schwarz inequality, we have

$$|F(z)| \le |\langle \Phi_z, F\rangle_{\mathcal{H}L^2(U,\alpha)}| \le \|\Phi_z\|_{L^2(U,\alpha)}\|F\|_{L^2(U,\alpha)}\,.$$

But $\|\Phi_z\|^2_{L^2(U,\alpha)} = \int_U k(z,w)k(w,z)\alpha(dw) = k(z,z)$. This proves part (iv).

If ϕ_z satisfies the properties of part (v), then we see $\langle \Phi_z - \phi_z, F\rangle = 0$ for all $F \in \mathcal{H}L^2(U,\alpha)$. But $\Phi_z - \phi_z \in \mathcal{H}L^2(U,\alpha)$. This implies $\Phi_z = \phi_z$. The theorem is proved. □

Definition 3.5. If one takes $\alpha_t(z) = (\pi t)^{-d}\exp(-|z|^2/t)$, where $|z|^2 = |z_1|^2 + \cdots + |z_d|^2$ and $t > 0$, then $\mathcal{H}L^2(\mathbb{C}^d, \alpha_t)$ is called a Segal-Bargmann space. If $t = 1$, then we denote $\mathcal{H}L^2(\mathbb{C}^d,\alpha_1)$ by $\mathcal{H}L^2(\mathbb{C}^d)$ and $L^2(\mathbb{C}^d,\alpha_1)$ by $L^2(\mathbb{C}^d)$.

We shall further study $L^2(\mathbb{C}^d)$ and $\mathcal{H}L^2(\mathbb{C}^d)$. First, consider the case $d=1$. Denote $\alpha(z) = (\pi)^{-1}\exp(-|z|^2)\,,\ z \in \mathbb{C}$.

The complex Hermite polynomials $\{H_{m,n}(z,\bar z)\,, m,n = 0,1,2,\cdots\}$ are defined as

$$H_{m,n}(z,\bar z) = \sum_{k=0}^{m\wedge n} (-1)^k \binom{m}{k}\binom{n}{k} z^{m-k}\bar z^{n-k}\,. \tag{3.10.10}$$

It is easy to check that their exponential generating function is

$$\sum_{m,n=0}^{\infty} H_{m,n}(z,\bar z)\frac{u^m v^n}{m!n!} = e^{uz+v\bar z - uv}\,. \tag{3.10.11}$$

From this generating formula and using $\frac{1}{\sqrt{\pi}}\int_{\mathbb{R}} e^{tx-x^2}dx = e^{t^2/4}$, we see that

$$\begin{aligned}
&\sum_{m,n,m',n'=0}^{\infty} \int_{\mathbb{C}} \overline{H_{m,n}(z,\bar z)} H_{m',n'}(z,\bar z)\alpha(z)dz \frac{u^m v^n s^{m'} t^{n'}}{m!n!m'!n'!}\\
&= \sum_{m,n,m',n'=0}^{\infty} \int_{\mathbb{C}} H_{n,m}(z,\bar z) H_{m',n'}(z,\bar z)\alpha(z)dz \frac{u^m v^n s^{m'} t^{n'}}{m!n!m'!n'!}\\
&= \frac{1}{\pi}\int_{\mathbb{C}} e^{vz+u\bar z+sz+t\bar z-|z|^2-uv-st}dz\\
&= \frac{1}{\pi}\int_{\mathbb{R}^2} e^{(u+s+v+t)x+\iota y(v+s-u-t)-|x|^2-|y|^2-uv-st}dxdy\\
&= e^{vt+su} = \sum_{k_1,k_2}^{\infty} \frac{(vt)^{k_1}(us)^{k_2}}{k_1!k_2!}\,.
\end{aligned}$$

Comparing the coefficients of $u^m v^n s^{m'} t^{n'}$, we see that

$$\frac{1}{\pi}\int_{\mathbb{C}} \overline{H_{m,n}(z,\bar{z})} H_{m',n'}(z,\bar{z}) e^{-|z|^2} dz = m!n!\delta_{m,m'}\delta_{n,n'} . \tag{3.10.12}$$

Since any function of x and y can be represented as a function of $z = x + \iota y$ and $\bar{z} = x - \iota y$ and the Hermite polynomials are complete in L^2, we see that the set of all linear combinations of complex Hermite polynomials are dense $L^2(\mathbb{C})$.

When $n = 0$, the complex Hermite polynomials are

$$H_{m,0}(z,\bar{z}) = z^m . \tag{3.10.13}$$

Since a holomorphic function on $\mathbb{C}$ can be approximated by polynomials we see the set of Hermite polynomials $\{H_{m,0}(z,\bar{z}) = z^m , m = 0, 1, 2, \cdots\}$ spans a dense set of $\mathcal{H}L^2(\mathbb{C})$. Summarizing we have

Proposition 3.20. *The set of complex Hermite polynomials*

$$\{H_{m,n}(z,\bar{z}) , m, n = 0, 1, \cdots\}$$

is a complete orthogonal system of $L^2(\mathbb{C})$. The set of complex Hermite polynomials

$$\{H_{m,0}(z) = z^m, m = 0, 1, \cdots\}$$

is a complete orthogonal system of $\mathcal{H}L^2(\mathbb{C})$.

For complex Hermite polynomials of d dimensions, we define

$$H_{m_1,\cdots,m_d,n_1,\cdots,n_d}(z_1,\cdots,z_d,\bar{z}_1,\cdots,\bar{z}_d) = H_{m_1,n_1}(z_1,\bar{z}_1)\cdots H_{m_d,n_d}(z_d,\bar{z}_d) .$$

The extension of the above result to any finite dimension is straightforward.

Proposition 3.21. *The set of complex Hermite polynomials*

$$\{H_{m_1,\cdots,m_d,n_1,\cdots,n_d}(z_1,\cdots,z_d,\bar{z}_1,\cdots,\bar{z}_d) , m_1,\cdots,m_d,n_1,\cdots,n_d = 0,1,\cdots\}$$

is a complete orthogonal system of $L^2(\mathbb{C}^d)$. The sct of complex Hermite polynomials

$$\{H_{m_1,\cdots,m_d,0,\cdots,0}(z_1,\cdots,z_d,\bar{z}_1,\cdots,\bar{z}_d) = z_1^{m_1}\cdots z_d^{m_d} , m_1,\cdots,m_d = 0,1,\cdots\}$$

is a complete orthogonal system of $\mathcal{H}L^2(\mathbb{C}^d)$.

Proposition 3.22. *The reproducing kernel of $\mathcal{H}L^2(\mathbb{C}^d)$ is $k(z,w) = e^{\bar{w}z} = e^{\bar{w}_1 z_1 + \cdots + \bar{w}_d z_d}$.*

Proof. By Theorem 3.23, part (v) it suffices to show (3.10.8). By Proposition 3.21, it suffices to show (3.10.8) for $F(z) = z_1^{m_1}\cdots z_d^{m_d}$. For simplicity, consider only the case $d = 1$. We have

$$\frac{1}{\pi}\int_{\mathbb{C}} w^m e^{\bar{w}z} e^{-|w|^2} dw = \sum_{k=0}^{\infty} \frac{z^k}{\pi k!}\int_{\mathbb{C}} w^m (\bar{w})^k e^{-|w|^2} dw = z^m$$

which is (3.10.8) when $F(z) = z^m$, proving the proposition. □

3.11 Segal-Bargmann Transform

Definition 3.6. The transformation $A : L^2(\mathbb{R}^d, dx) \to \mathcal{H}L^2(\mathbb{C}^d)$ defined by

$$Af(z) = (\pi)^{-d/4} \int_{\mathbb{R}^d} e^{(-z^2+2\sqrt{2}xz-x^2)/2} f(x)dx \tag{3.11.1}$$

is called the Segal-Bargmann transformation, where $z^2 = z_1^2 + \cdots + z_d^2$ and $xz = x_1 z_1 + \cdots + x_d z_d$.

The Segal-Bargmann transformation was first introduced by Bargmann [Bargmann (1961)] and [Segal (1960)] in the study of quantum theory.

In quantum mechanics, the position operator X_k, $k = 1, \ldots, d$, is the multiplication operator

$$X_k f(x) = x_k f(x)\,,$$

and the momentum operator P_k is given by

$$P_k f(x) = -\iota \frac{\partial}{\partial x_k} f(x)\,.$$

[We take the Planck constant $\hbar = 1$.] It is obvious that X_k and P_k are unbounded linear operators on $L^2(\mathbb{R}^d, dx)$. We do not specify the domains of these operators.

It is easy to verify that the above operators satisfy the following commutation relations (we denote $[A, B] = AB - BA$).

$$\begin{aligned} [X_k, P_j] &= \iota \delta_{jk} I \\ [X_k, X_j] &= 0 \\ [P_k, P_j] &= 0\,. \end{aligned}$$

These relations are called the canonical commutation relations in quamtum physics. From these position and momentum operators we introduce the following annihilation and creation operators

$$\begin{aligned} a_k &= \frac{X_k + \iota P_k}{\sqrt{2}} \\ a_k^* &= \frac{X_k - \iota P_k}{\sqrt{2}}\,. \end{aligned}$$

The commutation relation of a_k and a_j^* is

$$\begin{aligned} [a_k, a_j^*] &= \frac{1}{2}[X_k + \iota P_k, X_j - \iota P_j] \\ &= \frac{1}{2}\left\{[X_k, X_j] - \iota[X_k, P_j] + \iota[P_k, X_j] + [P_k, P_j]\right\} \\ &= \delta_{kj} I\,. \end{aligned}$$

We can do the similar computation for $[a_k, a_k]$ and $[a_k^*, a_k^*]$ to obtain

$$\begin{aligned}[a_k, a_j^*] &= \delta_{kj} I\,;\\ [a_k, a_j] &= 0\,;\\ [a_k^*, a_j^*] &= 0\,.\end{aligned}$$

The advantage of the operators a_k and a_k^* is that they satisfy the same canonical commutation relations and they are adjoint with each other on $L^2(\mathbb{R}^d, dx)$ (with respect to the Lebesgue measure). And X_k, P_k are not adjoint with each other $L^2(\mathbb{R}^d, dx)$. The following theorem states that if we extend X_k and P_k to the space of complex functions, then they are adjoint with each other.

When dimension $d = 1$, we use $a^* = \frac{X - \iota P}{\sqrt{2}}$ and $a = \frac{X + \iota P}{\sqrt{2}}$ to denote the creation and annihilation operators. The operator a and a^* have some nice properties.

Proposition 3.23. *Let $\mathcal{H}_n$ be the Hermite functions defined by* (3.9.11). *Then,*

$$a^* \mathcal{H}_n(x) = \frac{1}{\sqrt{2}} \mathcal{H}_{n+1}(x)\,; \tag{3.11.2}$$

$$a \mathcal{H}_n(x) = \sqrt{2} n \mathcal{H}_{n-1}(x)\,. \tag{3.11.3}$$

Proof. We use the generating function (3.9.12).

$$\begin{aligned}\sum_{n=0}^{\infty} a\mathcal{H}_n(x)\frac{t^n}{n!} &= \frac{(x + \frac{d}{dx})}{\sqrt{2}} e^{2tx - t^2 - \frac{x^2}{2}} = \frac{(x + \frac{d}{dx})}{\sqrt{2}} e^{2tx - t^2 - \frac{x^2}{2}}\\ &= \sqrt{2} t e^{2tx - t^2 - \frac{x^2}{2}} = \sqrt{2} \sum_{n=1}^{\infty} \mathcal{H}_{n-1}(x) \frac{t^n}{(n-1)!}\end{aligned}$$

This proves (3.11.3).

Now we apply a^* to the generating function and we can obtain the following identity in a similar way:

$$\begin{aligned}\sum_{n=0}^{\infty} a^*\mathcal{H}_n(x)\frac{t^n}{n!} &= \frac{(x - \frac{d}{dx})}{\sqrt{2}} e^{2tx - t^2 - \frac{x^2}{2}}\\ &= \sum_{n=1}^{\infty} \left[\sqrt{2} x \mathcal{H}_{n-1}(x) - \sqrt{2} n \mathcal{H}_{n-1}(x)\right] \frac{t^n}{n!}\,.\end{aligned}$$

Thus, we have by (3.9.5)

$$a^* \mathcal{H}_n(x) = \sqrt{2} x \mathcal{H}_{n-1}(x) - \sqrt{2} n \mathcal{H}_{n-1}(x) = \frac{1}{\sqrt{2}} \mathcal{H}_{n+1}(x)\,.$$

This proves the proposition. □

Remark 3.9. The operator a^* maps a Hermite function of degree n (corresponding to tensor product of order n) to a Hermite function of degree $n+1$ (a higher order). Since in physics a tensor product of order n represents the space n particles live, a^* is called a creation operator. Similar explanation for annihilation operator a.

Denote $\tilde{\mathcal{H}}_n(x) = \pi^{-1/4}2^{-n/2}\mathcal{H}_n(x)$.

Theorem 3.24. *The Segal-Bargmann transformation A has the following properties*

(i) For all $f \in L^2(\mathbb{R}^d, dx)$, $Af(z)$ is well-defined and as a function $z \in \mathbb{C}^d$ is holomorphic.
(ii) A is unitary from $L^2(\mathbb{R}^d, dx)$ to $\mathcal{H}L^2(\mathbb{C}^d)$.
(iii) For every $k = 1, \cdots, d$,

$$Aa_kA^{-1} = \frac{\partial}{\partial z_k}; \tag{3.11.4}$$

$$Aa_k^*A^{-1} = z_k. \tag{3.11.5}$$

Proof. We follow [Bargmann (1961)], [Hall (2000)]. For simplicity, we consider the case $d = 1$. The general dimension case is similar. The integral defining the operator A is convergent since the integral kernel $e^{(-z^2+2\sqrt{2}xz-x^2)/2}$ is in $L^2(\mathbb{R}, dx)$ for any fixed $z \in \mathbb{C}$. The holomophicity can be proved by using the Morera's theorem.

To prove the parts (ii) and (iii), we shall first prove

$$Aa = A\frac{X + \iota P}{\sqrt{2}} = \frac{\partial}{\partial z}A; \tag{3.11.6}$$

$$Aa^* = A\frac{X - \iota P}{\sqrt{2}} = zA, \tag{3.11.7}$$

for nice functions, where we recall that X is the position operator (multiply by x) and $P = -\iota\frac{d}{dx}$ is the momentum operator. If f is nice (smooth, decay rapidly at infinity so that we can use integration by parts, differentiation under the integral sign), then

$$\begin{aligned}\frac{d}{dz}Af(z) &= (\pi)^{-1/4}\int_{\mathbb{R}}\frac{d}{dz}e^{(-z^2+2\sqrt{2}xz-x^2)/2}f(x)dx \\ &= (\pi)^{-1/4}\int_{\mathbb{R}}\left(-z+\sqrt{2}x\right)e^{(-z^2+2\sqrt{2}xz-x^2)/2}f(x)dx \\ &= -zAf(z) + \sqrt{2}A[xf](z).\end{aligned}$$

Or we can write

$$\frac{d}{dz}A = -zA + \sqrt{2}AX\,, \tag{3.11.8}$$

On the other hand, using the integration by parts we have

$$\begin{aligned} A\frac{d}{dx}f(z) &= (\pi)^{-1/4}\int_{\mathbb{R}} e^{(-z^2+2\sqrt{2}xz-x^2)/2}\frac{df}{dx}dx \\ &= -(\pi)^{-1/4}\int_{\mathbb{R}}(\sqrt{2}z-x)e^{(-z^2+2\sqrt{2}xz-x^2)/2}f(x)dx\,. \end{aligned}$$

This can be written as

$$\imath AP = A\frac{d}{dx} = -\sqrt{2}zA + AX\,, \tag{3.11.9}$$

Multiplying (3.11.8) by $\sqrt{2}$ and subtracting the resulted equation by (3.11.9) yield (3.11.6). Substituting $\frac{d}{dz}A$ given by (3.11.6) to (3.11.8) yields (3.11.7).

Let $f_0(x) = \pi^{-1/4}e^{-x^2/2} = \pi^{-1/4}\mathcal{H}_0(x)$. It is straightforward to verify that

$$Af_0(x) = 1\,.$$

Now by the (3.11.7), we have

$$A((a^*)^n f_0) = z^n Af_0 = z^n\,.$$

But according to (3.11.2), we see that $\tilde{\mathcal{H}}_n(x) := (a^*)^n f_0(x) = 2^{-n/2}\pi^{-1/4}\mathcal{H}_n(x)$. From (3.9.7), it follows that $\int_{\mathbb{R}}\tilde{\mathcal{H}}_n(x)\tilde{\mathcal{H}}_m(x)dx = n!\delta_{nm}$. This together with (3.10.12) yields that A is a unitary operator from $L^2(\mathbb{R}, dx)$ to $\mathcal{H}L^2(\mathbb{R})$. □

Remark 3.10. One important result of Bargmann and Fock is the characterization of the image of the Schwartz space of rapidly decreasing smooth functions on $\mathbb{R}^d$ under the Segam-Bargmann transformation A defined by (3.11.1). It is known that $F = Af$ for some Schwartz rapidly decreasing smooth function f if and only if

$$|F(x+\imath y)|^2 \le A_n\frac{e^{y^2}}{[1+(x^2+y^2)]^{2n}}\,, \quad \text{for some sequence } A_n \quad n = 1, 2, \cdots$$

We refer to [Bargmann (1961); Olafsson (2014)] for more discussion.

Chapter 4

Gaussian Measures on Banach Space

4.1 Random Variables in Banach Space

In (harmonic) analysis Lebesgue measure is the most important measure on $\mathbb{R}^d$. One basic feature of the Lebesgue measure is its translation invariance. Namely, if A is a Borel set of $\mathbb{R}^d$, then the Lebesgue measures of A and $A+x$ are the same for any $x \in \mathbb{R}^d$. However, the Lebesgue measure cannot be extended to infinite dimensional case.

Proposition 4.1. *If $\mathbb{B}$ is an infinite dimensional separable Banach space, then the only locally finite and translation-invariant Borel measure μ on $\mathbb{B}$ is the trivial measure, namely, $\mu(A) = 0$ for every measurable set A. Equivalently, every translation-invariant measure that is not identically zero assigns infinite measure to all open subsets of $\mathbb{B}$.*

Proof. If there is an open set O such that it has a finite measure, then there is a $\delta > 0$ such that an open ball $B(x, \delta)$ of center x and radius ε has finite measure. Because the space is infinite dimensional, one can construct an infinite sequence of disjoint open balls $B(x_i, \delta/4)$, $i = 1, 2, \cdots$, of radius $\delta/4$ which are contained in the open ball $B(x, \delta)$. By translation invariant property each of these balls $B(x_i, \delta/4)$ has the same measure, and since the infinite sum of their measures is finite, each of the balls $B(x_i, \delta/4)$ must have measure 0. On the other hand, since the space is separable, it can be covered with a countable collection of balls of the form $B(y_i, \delta/4)$. Thus, the whole space must have measure 0. □

For more discussion we refer to [Hunt, and Sauer and Yorke (1992)].

Let $\mathbb{B}$ be a (separable) Banach space with norm $\|\cdot\| = \|\cdot\|_{\mathbb{B}}$. A continuous linear functional ℓ on $\mathbb{B}$ is denoted $\ell(x) = \langle x, \ell \rangle$, $x \in \mathbb{B}$. The dual $\mathbb{B}'$ of $\mathbb{B}$ is the Banach space of all continuous linear functional ℓ with the norm

$$\|\ell\|_{\mathbb{B}'} = \sup_{x\in\mathbb{B}\,,\|x\|\le 1} \langle x, \ell\rangle\,.$$

The Borel σ-algebra $\mathcal{B}(\mathbb{B})$ of $\mathbb{B}$ is the σ-algebra generated by (the smallest σ-algebra contains) the set of all open sets of $\mathbb{B}$.

Theorem 4.1. *Let $\mathcal{C}$ denote the algebra of all cylindrical sets*

$$\left\{x \in \mathbb{B}\,,\ (\langle x, \ell_1\rangle\,,\cdots,\langle x,\ell_d\rangle) \in A\,,\quad \ell_i \in \mathbb{B}'\,, i = 1,\cdots,d,\quad A \in \mathcal{B}(\mathbb{R}^d)\right\}\,.$$

Let $\mathcal{B}_C(\mathbb{B})$ be the σ-algebra generated by the cylindrical sets of $\mathcal{C}$. Then, $\mathcal{B}_C(\mathbb{B}) = \mathcal{B}(\mathbb{B})$.

Proof. It is obvious that $\mathcal{B}_C(\mathbb{B}) \subseteq \mathcal{B}(\mathbb{B})$. We want to show $\mathcal{B}(\mathbb{B}) \subseteq \mathcal{B}_C(\mathbb{B})$. For any $r > 0$, denote $B_r = \{x \in \mathbb{B}\,;\ \|x\| \le r\}$. We will show that $B_r \in \mathcal{B}_C(\mathbb{B})$. To this end, we let $\{x_n\}$ be a countable dense subset of $\mathbb{B}$. From the Hahn-Banach extension theorem (see [Yosida (1980)]) we can find $z_n \in \mathbb{B}'$ such that

$$\|z_n\|_{\mathbb{B}'} = 1\,,\quad \langle x_n, z_n\rangle = \|x_n\|\,,\quad n = 1, 2, \cdots\,. \tag{4.1.1}$$

We claim

$$B_r = \cap_{n=1}^{\infty}\{x\,;\ \langle x, z_n\rangle \le r\}\,. \tag{4.1.2}$$

In fact, it is obvious that $B_r \subseteq \cap_{n=1}^{\infty}\{x\,;\ \langle x, z_n\rangle \le r\}$. Now, let $x \notin B_r$. Since $\{x_n\}$ is a dense set of $\mathbb{B}$, there is a $x_{n_0} \in \mathbb{B}$ such that

$$\|x - x_{n_0}\| < \frac{1}{2}(\|x\| - r)\,.$$

Thus

$$\|x_{n_0}\| \ge \|x\| - \|x - x_{n_0}\| > \frac{1}{2}(\|x\| + r)$$

which implies

$$|\langle x\,, z_{n_0}\rangle - \|x_{n_0}\|| = |\langle x\,, z_{n_0}\rangle - \langle x_{n_0}\,, z_{n_0}\rangle|| \le \|z_{n_0}\|_{B^*}\|x_{n_0} - x\| < \frac{1}{2}(\|x\| - r)\,.$$

Thus, we have

$$\langle x\,, z_{n_0}\rangle \ge \|x_{n_0}\| - |\langle x\,, z_{n_0}\rangle - \|x_{n_0}\|| > \frac{1}{2}(\|x\| + r) - \frac{1}{2}(\|x\| - r) = r\,.$$

This means $x \notin \cap_{n=1}^{\infty}\{x\,;\ \langle x, z_n\rangle \le r\}$. This proves (4.1.2) and hence $B_r \in \mathcal{B}_c(\mathbb{B})$). Since $\mathcal{B}_C(\mathbb{B})$ is translation invariant, we see that for any $x_0 \in \mathbb{B}$ and any $r > 0$, $B_{x_0,r} := \{x \in \mathbb{B}\,;\ \|x - x_0\| \le r\}$ is in $\mathcal{B}_C(\mathbb{B})$. Since $\mathcal{B}(\mathbb{B})$ is generated by $B_{x_0,r}\,,\ x_0 \in \mathbb{B}\,, r > 0$, we prove $\mathcal{B}(\mathbb{B}) \subseteq \mathcal{B}_C(\mathbb{B})$. □

Definition 4.1. Let $(\Omega, \mathcal{F}, \mathbb{P})$ be a probability space and let $\mathbb{B}$ be a Banach space. A $\mathbb{B}$-valued random variable is a strongly measurable mapping $X : \Omega \to \mathbb{B}$. This means that for any Borel set $A \in \mathcal{B}(\mathbb{B})$, we have $\{\omega \in \Omega\,,\ X(\omega) \in A\} \in \mathcal{F}$. The (probability) law of X is a probability measure on the measurable space $(\mathbb{B}, \mathcal{B}(\mathbb{B}))$ defined by

$$\mu_X(A) = \mathbb{P}\left(X \in A\right)\,, \ \forall A \in \mathcal{B}(\mathbb{B})\,.$$

The law of X is also denoted by $P \circ X^{-1}$.

Example 4.1. If $X(\omega) = x_0$, where x_0 is an element of $\mathbb{B}$, then $\mu_X(A) = I_A(x_0)\,,\ \forall A \in \mathcal{B}(\mathbb{B})$. We denote this measure by δ_{x_0}. When $x_0 = 0$, then we denote $\delta = \delta_0$.

Definition 4.2. If μ is any measure on $\mathbb{B}$, then for any $\ell \in \mathbb{B}'$, the (continuous) linear functional $\ell(x) = \langle x, \ell\rangle$ can be considered as a random variable. We denote this random variable by $\tilde{\ell}$. The characteristic functional of the measure μ, as a functional from $\mathbb{B}'$ to $\mathbb{C}$, can be defined as follows.

$$\hat{\mu}(\ell) = \mathbb{E}\left[e^{\iota\tilde{\ell}}\right] = \int_{\mathbb{B}} e^{\iota\langle x,\ell\rangle}\mu(dx)\,.$$

Let X be a $\mathbb{B}$ valued random variable. Then, the Fourier transform or characteristic functional of X (or μ_X) is defined as a mapping from $\mathbb{B}'$ to $\mathbb{C}$ as $\hat{\mu}_X(\ell)$, $\ell \in \mathbb{B}'$.

Lemma 4.1. *If there is $r > 0$ such that $\hat{\mu}(\ell) = 1$ for all $\|\ell\|_{\mathbb{B}'} < r$, then μ is concentrated on 0. Namely, $\mu = \delta$.*

Proof. Denote $\phi(t) = \hat{\mu}(t\ell)$, $t \in \mathbb{R}$. Then, $\phi(t)$ is a characteristic function in $\mathbb{R}$ and $\phi(t) = 0$ for all $t < r/\|\ell\|_{\mathbb{B}'}$. For any $s, t \in \mathbb{R}$, we have

$$\begin{aligned}|\phi(t) - \phi(s)| &= |\mathbb{E}e^{it\tilde{\ell}} - \mathbb{E}e^{is\tilde{\ell}}| = |\mathbb{E}e^{is\tilde{\ell}}\left[e^{i(t-s)\tilde{\ell}} - 1\right]|\\ &\le \mathbb{E}\left|e^{i(t-s)\tilde{\ell}} - 1\right| \le \sqrt{\mathbb{E}\left|e^{i(t-s)\tilde{\ell}} - 1\right|^2}\\ &= \sqrt{2 - \phi(t-s) - \bar{\phi}(t-s)} \le \sqrt{2|1 - \phi(t-s)|}\,.\end{aligned}$$

This inequality can be used to show that $\phi(t) = 1$ for all t. Thus, $\hat{\mu}(\ell) = \hat{\delta}(\ell)$. This means that the characteristic functions of two probability measures μ and δ are the same, which implies $\mu = \delta$. □

Definition 4.3. A Borel random variable X with value in a Banach space $\mathbb{B}$ is called *Radon* or *tight* if for any $\varepsilon > 0$, there is a compact subset K_ε of $\mathbb{B}$ such that

$$\mathbb{P}(X \in K_\varepsilon) > 1 - \varepsilon\,. \tag{4.1.3}$$

Definition 4.4. Let $\mathcal{P}$ be a set of probability measures on $\mathbb{B}$ (or a set of $\mathbb{B}$-valued random variables). $\mathcal{P}$ is called (uniformly) tight if for any $\varepsilon > 0$, there is a compact set K_ε such that

$$\mu(K_\varepsilon) > 1 - \varepsilon \quad \forall\, \mu \in \mathcal{P}\,. \tag{4.1.4}$$

Theorem 4.2. *Let $\{X_n\,, n = 1, 2, \cdots\}$ be a sequence of independent $\mathbb{B}$-valued random variables and set $S_n = \sum_{i=1}^n X_i$. If the set of measures $\{\mu_{S_n}\}$ is uniformly tight, then there exists a sequence c_n such that $S_n - c_n$ converges a.s.*

Proof. Let $\{Y_n\}$ be an independent copy of $\{X_n\}$. Then, $X_1 - Y_1, X_2 - Y_2, \cdots$ are independent. Define

$$T_n = \sum_{i=1}^n Y_i\,, \quad U_n = \sum_{i=1}^n (X_i - Y_i) = S_n - T_n\,.$$

From our assumption, for any $\varepsilon > 0$, there is a compact set K_ε of $\mathbb{B}$ such that

$$\mu_{S_n}(K_\varepsilon) > 1 - \varepsilon\,, \quad \forall\, n\,.$$

Set $\tilde{K}_\varepsilon = K_\varepsilon - K_\varepsilon$. Then, $\tilde{K}_\varepsilon$ is also compact set of $\mathbb{B}$. We have

$$\begin{aligned}
\mu_{U_n}(\tilde{K}_\varepsilon) &= \mathbb{P}(S_n - T_n \in K_1) \\
&\geq \mathbb{P}(S_n \in K_\varepsilon, T_n \in K_\varepsilon) \geq 1 - \mathbb{P}(S_n \notin K_\varepsilon) - \mathbb{P}(T_n \notin K_\varepsilon) \\
&= 1 - 2\mu_{S_n}(K_\varepsilon^c) > 1 - 2\varepsilon\,.
\end{aligned}$$

This means that $\{\mu_{U_n}\}$ is also uniformly tight.

The characteristic function of μ_{U_n} is

$$\begin{aligned}
\hat{\mu}_{U_n}(\ell) &= \mathbb{E}\exp\left\{\iota \sum_{i=1}^n \langle X_i - Y_i, \ell\rangle\right\} \\
&= \prod_{i=1}^n |\mathbb{E}\exp\{\iota\langle X_i\,, \ell\rangle\}|^2\,.
\end{aligned}$$

Since $0 \leq |\mathbb{E}\exp\{\iota\langle X_i, \ell\rangle\}|^2 \leq 1$, $\lim_{n\to\infty} \hat{\mu}_{U_n}(\ell)$ exists for every $\ell \in \mathbb{B}'$. Since μ_{U_n} is tight, it is conditionally compact. Thus, it suffices to show that any two convergent subsequence $\mu_{U_{n'}}$ and $\mu_{U_{n''}}$ of μ_{U_n} have the same limit. Denote the limit of $\mu_{U_{n'}}$ by μ' and that of $\mu_{U_{n''}}$ by μ''. The characteristic functional of $\mu_{U_{n'}}$ converges to $\hat{\mu}'$ and that of $\mu_{U_{n''}}$ converges to $\hat{\mu}''$. But the limit of μ_{U_n} exists. So $\hat{\mu}' = \hat{\mu}''$. This implies $\mu' = \mu''$. The theorem is proved. □

For more discussion of Banach valued random variables we refer to [Diestel, Jarchow and Tonge (1995)], [Ledoux and Talagrand (2011)], and [Vakhania, Tarieladze and Chobanyan (1987)]. Before we continue, we state the following well-known result in probability theory.

Theorem 4.3 (Khintchine inequality). *Let $(\Omega, \mathcal{F}, \mathbb{P})$ be a probability space and let $\gamma_k(\omega)$, $k = 1, 2, \cdots$ be a sequence of iid random variables such that $\mathbb{P}(\gamma_k = 1) = \mathbb{P}(\gamma_k = -1) = 1/2$. Then, there are constants c_p and C_p such that*

$$c_p \left(\sum_{k=1}^{\infty} |a_k|^2 \right)^{p/2} \le \mathbb{E} \left| \sum_{k=1}^{\infty} a_k \gamma_k(\omega) \right|^p \le C_p \left(\sum_{k=1}^{\infty} |a_k|^2 \right)^{p/2} \tag{4.1.5}$$

for any sequence of real numbers $a_1, a_2, \cdots$ such that $\sum_{k=0}^{\infty} |a_k|^2 < \infty$.

Proof. For any fixed positive integer N we have for any $t \in \mathbb{R}$

$$\mathbb{E} e^{t \sum_{k=1}^N a_k \gamma_k(\omega)} = \prod_{k=1}^N \frac{1}{2} \left(e^{t a_k} + e^{-t a_k} \right) \le \prod_{k=1}^N e^{\frac{t^2 a_k^2}{2}} = e^{\frac{t^2}{2} \sum_{k=1}^N a_k^2}$$

since $\frac{1}{2} \left(e^x + e^{-x} \right) \le e^{\frac{x^2}{2}}$ for all $x \in \mathbb{R}$.

Therefore for any $\lambda > 0$, we have

$$\mathbb{P} \left(\sum_{k=0}^{\infty} a_k \gamma_k(\omega) \ge \lambda \right) \le e^{-t\lambda} \mathbb{E} e^{t \sum_{k=1}^N a_k \gamma_k(\omega)} \le e^{-t\lambda + \frac{t^2}{2} \sum_{k=1}^N a_k^2} \quad \forall\, t > 0 \,.$$

Taking $t = \dfrac{\lambda}{2 \sum_{k=1}^N a_k^2}$, we have

$$\mathbb{P} \left(\sum_{k=0}^{\infty} a_k \gamma_k(\omega) \ge \lambda \right) \le \exp \left\{ - \frac{\lambda^2}{2 \sum_{k=1}^N a_k^2} \right\} .$$

Thus

$$\mathbb{P} \left(\left| \sum_{k=0}^{\infty} a_k \gamma_k(\omega) \right| \ge \lambda \right) \le 2 \exp \left\{ - \frac{\lambda^2}{2 \sum_{k=1}^N a_k^2} \right\} .$$

Using $\mathbb{E}|X|^p = p \int_0^{\infty} \lambda^{p-1} \mathbb{P}(|X| \ge \lambda) d\lambda$, we have

$$\begin{aligned} \mathbb{E} \left| \sum_{k=1}^{\infty} a_k \gamma_k(\omega) \right|^p &\le 2p \int_0^{\infty} \lambda^{p-1} \exp \left\{ - \frac{\lambda^2}{2 \sum_{k=1}^N a_k^2} \right\} d\lambda \\ &= 2^{2+\frac{p}{2}} p \Gamma \left(\frac{p}{2} \right) \left(\sum_{k=1}^N a_k^2 \right)^{p/2} . \end{aligned}$$

Letting $N \to \infty$ proves the upper bound in the theorem.

Now we show the lower bound. Let p' be the conjugate number of p. By the Hölder inequality, we have

$$\sum_{k=1}^N a_k^2 = \mathbb{E}\left|\sum_{k=1}^N a_k\gamma_k(\omega)\right|^2 \le \left(\mathbb{E}\left|\sum_{k=1}^N a_k\gamma_k(\omega)\right|^{p'}\right)^{1/p'} \left(\mathbb{E}\left|\sum_{k=1}^N a_k\gamma_k(\omega)\right|^{p}\right)^{1/p}$$

$$\le C_{p'}\left(\sum_{k=1}^N a_k^2\right)^{1/2} \left(\mathbb{E}\left|\sum_{k=1}^N a_k\gamma_k(\omega)\right|^{p}\right)^{1/p}.$$

Dividing $\left(\sum_{k=1}^N a_k^2\right)^{1/2}$ and letting $N \to \infty$ yields the lower bound of the theorem. □

Remark 4.1. The above proof also gives two constants $C_p = 2^{2+\frac{p}{2}} p\Gamma\left(\frac{p}{2}\right)$ and $c_p = \dfrac{1}{2^{2+\frac{p'}{2}} p'\Gamma\left(\frac{p'}{2}\right)}$, where p' is the conjugate number of p. For the best c_p and C_p see [Haagerup (1981)] and [Nazarov, Podkorytov, Ball and Haagerup (2000)].

Remark 4.2. If we define $\xi_n = \sum_{k=1}^n a_k\gamma_k(\omega)\,, n = 1, 2, \cdots$, then $\{\xi_n, n = 1, 2, \cdots\}$ is a martingale with respect to $\mathcal{G}_n = \sigma(\gamma_1, \cdots, \gamma_n)$. The above inequality is then a simple consequence of the martingale inequality.

Now we give an infinite dimensional version of the above inequality.

Lemma 4.2 (Lévy's inequality). *Let $X_1, \cdots, X_N$ be independent Banach space $\mathbb{B}$-valued symmetric random variables and let $S_n := \sum_{k=1}^n X_k$, $n = 1, 2, \cdots, N$. Then, for all $r > 0$, we have*

$$\mathbb{P}\left\{\max_{1\le n\le N} \|S_n\| > r\right\} \le 2\mathbb{P}\left\{\|S_N\| > r\right\}. \tag{4.1.6}$$

Proof. We denote

$$A_n := \{\|S_1\| \le r\,, \cdots, \|S_{n-1}\| \le r\,, \|S_n\| > r\}$$

and $A := \{\max_{1\le n\le N} \|S_n\| > r\}$. Then, $A_1, \cdots, A_N$ are disjoint and $A = \cup_{n=1}^N A_n$. The identity $S_n = \frac{1}{2}\left(S_N + (2S_n - S_N)\right)$ implies

$$\{\|S_n\| > r\} \subseteq \{\|S_N\| > r\} \cup \{\|2S_n - S_N\| > r\}.$$

It is easy to check that $(X_1, \cdots, X_N) \overset{d}{=} (X_1, \cdots, X_n, -X_{n+1}, \cdots, -X_n)$. This implies that

$$(X_1, \cdots, X_n, S_N) \overset{d}{=} (X_1, \cdots, X_n, 2S_n - S_N).$$

Thus

$$\begin{aligned}\mathbb{P}(A_n) &\le \mathbb{P}(A_n \cap \{\|S_N\| > r\}) + \mathbb{P}(A_n \cap \{\|2S_n - S_N\| > r\}) \\ &= 2\mathbb{P}(A_n \cap \{\|S_N\| > r\})\,.\end{aligned}$$

Summing over n we have

$$\mathbb{P}(A) = \sum_{n=1}^{N} \mathbb{P}(A_n) \le 2\mathbb{P}(A_n \cap \{\|S_N\| > r\}) = 2\mathbb{P}(\|S_N\| > r)\,,$$

proving the lemma. □

Lemma 4.3. *Let $\mathbb{B}$ be a Banach space. For all $x_1, x_2, \cdots, x_N$ in $\mathbb{B}$ and $r > 0$ we have*

$$\mathbb{P}\left(\left\|\sum_{n=1}^{N} x_n \gamma_n(\omega)\right\| > 2r\right) \le 4\mathbb{P}\left(\left\|\sum_{n=1}^{N} x_n \gamma_n(\omega)\right\| > r\right)^2. \tag{4.1.7}$$

Proof. Denote $S_n := \sum_{k=1}^{n} x_k \gamma_k(\omega)\,, n = 1, 2, \cdots, N$ and

$$A_n := \{\|S_1\| \le r\,, \cdots, \|S_{n-1}\| \le r\,, \|S_n\| > r\}\,.$$

Then

$$A_n \cap \{\|S_N\| > 2r\} \subseteq \{\|S_N - S_{n-1}\| > r\} \quad \forall\ n = 1, 2, \cdots, N\,. \tag{4.1.8}$$

It is easy to verify that $(\gamma_1(\omega), \cdots, \gamma_n(\omega), \gamma_{n+1}(\omega)\,, \cdots, \gamma_N(\omega))$ and $(\gamma_1(\omega), \cdots, \gamma_n(\omega), \gamma_n(\omega)\gamma_{n+1}(\omega)\,, \cdots, \gamma_n(\omega)\gamma_N(\omega))$ are identically distributed. Therefore, we have

$$\begin{aligned}\mathbb{P}\left(A_n \cap \{\|S_N - S_{n-1}\| > r\}\right) &= \mathbb{P}\left(A_n \cap \left\{\left\|\gamma_n(\omega)\sum_{k=n}^{N} x_k\gamma_k(\omega)\right\| > r\right\}\right) \\ &= \mathbb{P}\left(A_n \cap \left\{\left\|x_n + \sum_{k=n+1}^{N} x_k\gamma_n(\omega)\gamma_k(\omega)\right\| > r\right\}\right) \\ &= \mathbb{P}\left(A_n \cap \left\{\left\|x_n + \sum_{k=n+1}^{N} x_k\gamma_k(\omega)\right\| > r\right\}\right) \\ &= \mathbb{P}\left(A_n \cap \{\|x_n + S_N - S_n\| > r\}\right)\,.\end{aligned}$$

Similarly, we have

$$\mathbb{P}\{\|S_N - S_{n-1}\| > r\} = \mathbb{P}\{\|x_n + S_N - S_n\| > r\}\,.$$

Using the fact that A_n and $S_N - S_n$ are independent and also using (4.1.8) we have

$$\begin{aligned}&\mathbb{P}\left(A_n \cap \{\|S_N\| > 2r\}\right) \le \mathbb{P}\left(A_n \cap \{\|S_N - S_{n-1}\| > r\}\right) \\ &\le \mathbb{P}(A_n)\,\mathbb{P}\{\|x_n + S_N - S_n\| > r\} = \mathbb{P}(A_n)\,\mathbb{P}\{\|S_N - S_{n-1}\| > r\}\,.\end{aligned}$$

Now $S_N - S_{n-1} = X_N + X_{N-1} + \cdots + X_n$. The Lévy's inequality with the change of the order of summation reads $\mathbb{P}\{\|S_N - S_{n-1}\| > r\} \le 2\mathbb{P}(S_N > r)$. Thus

$$\mathbb{P}(A_n \cap \{\|S_N\| > 2r\}) \le 2\mathbb{P}(A_n)\mathbb{P}(S_N > r).$$

From the definition of A_n, we see that

$$\{\|S_N\| > 2r\} \subseteq \cup_{n=1}^N A_n = \left\{\max_{1\le n\le N} \|S_n\| > r\right\}.$$

Hence, we have

$$\begin{aligned}\mathbb{P}(\|S_N\| > 2r) &= \sum_{n=1}^N \mathbb{P}(A_n \cap \{\|S_N\| > 2r\}) \le 2\sum_{n=1}^N \mathbb{P}(A_n)\mathbb{P}(S_N > r)\\ &= 2\mathbb{P}\left\{\max_{1\le n\le N} \|S_n\| > r\right\}\mathbb{P}(S_N > r) \le 4\left[\mathbb{P}(S_N > r)\right]^2.\end{aligned}$$

This proves the lemma. □

Theorem 4.4 (Kahane-Khintchine inequality). *For all $1 \le p, q < \infty$ there exists a constant $K_{p,q}$, depending only on p and q, such that for all finite sequence $x_1, x_2, \cdots, x_N \in \mathbb{B}$, we have*

$$\left(\mathbb{E}\left\|\sum_{n=1}^N \gamma_n(\omega)x_n\right\|^p\right)^{1/p} \le K_{p,q}\left(\mathbb{E}\left\|\sum_{n=1}^N \gamma_n(\omega)x_n\right\|^q\right)^{1/q}. \qquad (4.1.9)$$

Proof. Denote $S_N := \sum_{n=1}^N \gamma_n(\omega)x_n$ and we can assume that $\mathbb{E}\|S_N\| = 1$ since the inequality to be proved is homogeneous. Let $j \ge 1$ be any integer such that $2^j > p$. Repeatedly applying Lemma 4.3, we have for any $r > 0$,

$$\mathbb{P}\{\|S_N\| > 2^j r\} \le 4^{2^j-1}\left[\mathbb{P}\{\|S_N\| > r\}\right]^{2^j} \le 4^{2^j-1}r^{-2^j},$$

where the last inequality follows from Chebyshev's inequality since

$$\mathbb{P}\{\|S_N\| > r\} \le \frac{\mathbb{E}\|S_N\|}{r} \le r^{-1}.$$

Or we can write

$$\mathbb{P}\{\|S_N\| > t\} \le 2^{(j+2)2^j-2}t^{-2^j}, \quad \forall\ t > 0.$$

Now we can use the similar argument as in the proof of scalar case.

$$\begin{aligned}\mathbb{E}\|S_n\|^p &= p\int_0^\infty t^{p-1}\mathbb{P}\{\|S_N\| > t\}dt \le 1 + p\int_1^\infty t^{p-1}\mathbb{P}\{\|S_N\| > t\}dt\\ &\le 1 + p2^{(j+2)2^j-2}\int_1^\infty t^{p-1-2^j}dt = 1 + \frac{p2^{(j+2)2^j-2}}{2^j - p}.\end{aligned}$$

We can choose any j such that $2^j > p$. □

Remark 4.3. Let $x = 2^j$. Then, $\frac{2^{(j+2)2^j - 2}}{2^j - p} = \frac{x4^x}{4(x-p)} =: h(x)$. We may minimize $h(x)$ on the interval (p, ∞) to obtain a bound for $K_{p,q}$. But certainly this may not be the best constant.

If we choose $\mathbb{B}$ to be the Hilbert space

$$\ell^2 = \left\{ a = (a_1, a_2, \cdots,), \quad \|a\|_2^2 := \sum_{k=1}^{\infty} a_k^2 < \infty \right\}.$$

then we have the following

Corollary 4.1. *Let $1 < p < \infty$. If $\{a_{kl}\}_{\{1\le k,l<\infty\}}$ is a sequence of real numbers, then there are positive constants c_p and C_p such that*

$$c_p \left(\sum_{k,l=1}^{\infty} a_{kl}^2 \right)^{p/2} \le \mathbb{E} \left[\sum_{k=1}^{\infty} \left(\sum_{k=l}^{\infty} a_{kl}\gamma_l(\omega) \right)^2 \right]^{p/2} \le C_p \left(\sum_{k,l=1}^{\infty} a_{kl}^2 \right)^{p/2}. \tag{4.1.10}$$

Proof. Let $\bar{a}_l = (a_{1l}, a_{2l}, \cdots)$, $l = 1, 2, \cdots$. Applying Theorem 4.4 to $\sum_{l=1}^{\infty} \bar{a}_l \gamma_l(\omega)$ yields the corollary. □

The following corollary will be useful later.

Corollary 4.2. *Let $1 < p < \infty$. If $\{a_{kl}\}_{\{1\le k,l\le N\}}$ is a non-negative definite symmetric matrix for any $N \ge 1$, namely,*

$$\sum_{k,l=1}^{N} a_{kl}\xi_k\xi_l \ge 0 \quad \forall \text{ real numbers } \quad \xi_k, \ k = 1, 2, \cdots, N,$$

then

$$c_p \left(\sum_{k=1}^{\infty} a_{kk}^2 \right)^{p/2} \le \mathbb{E} \left[\sum_{k,l=l}^{\infty} a_{kl}\gamma_k(\omega)\gamma_l(\omega) \right]^{p/2} \le C_p \left(\sum_{k=1}^{\infty} a_{kk}^2 \right)^{p/2}. \tag{4.1.11}$$

Proof. Fix $N \ge 1$. Since $A = \{a_{kl}\}_{\{1\le k,l\le N\}}$ is non-negative symmetric definite there is a symmetric and non-negative definite matrix $B = \{b_{kl}\}_{\{1\le k,l\le N\}}$ such that $A = B^2$. Namely, $a_{kl} = \sum_{j=1}^{N} b_{kj}b_{jl} = \sum_{j=1}^{N} b_{kj}b_{lj}$. Applying the inequality (4.1.10) to the sequence b_{kl}, we have

$$c_p \left(\sum_{k,l=1}^{N} b_{kl}^2 \right)^{p/2} \le \mathbb{E} \left[\sum_{k=1}^{N} \left(\sum_{k=l}^{\infty} b_{kl}\gamma_l(\omega) \right)^2 \right]^{p/2} \le C_p \left(\sum_{k,l=1}^{N} b_{kl}^2 \right)^{p/2}.$$

However,

$$\sum_{k,l=1}^{N} b_{kl}^2 = \sum_{k=1}^{\infty} a_{kk}^2$$

$$\left(\sum_{k=l}^{N} b_{kl}\gamma_l(\omega)\right)^2 = \sum_{l,j=1}^{\infty} a_{lj}\gamma_l(\omega)\gamma_j(\omega)\,.$$

Thus

$$c_p\left(\sum_{k=1}^{N} a_{kk}^2\right)^{p/2} \le \mathbb{E}\left[\sum_{k,l=l}^{N} a_{kl}\gamma_k(\omega)\gamma_l(\omega)\right]^{p/2} \le C_p\left(\sum_{k=1}^{N} a_{kk}^2\right)^{p/2}\,.$$

Letting $N \to \infty$ proves the corollary. □

Theorem 4.5 (Itô-Nisio). *Let X_n, $n = 1, 2, \cdots$ be a sequence of independent $\mathbb{B}$-valued random variables and let*

$$S_n = \sum_{i=1}^{n} X_n\,, \quad \mu_n = \text{The probability law of } S_n\,. \tag{4.1.12}$$

The following statements are equivalent.

(a) S_n converges almost surely.
(b) S_n converges in probability.
(c) S_n converges in law.

Proof. The proof of "(a) ⇒ (b) ⇒ (c)" can be found in any basic probability course. We need to show "(b)⇒ (a)" and "(c)⇒ (b)".

First we prove "(b)⇒ (a)". Suppose S_n converges in probability. Denote $S_{m,n} = S_n - S_m = \sum_{k=m+1}^{n} X_k$. We have for any $\varepsilon > 0$, there exists a N_ε such that when $n > m \ge N_\varepsilon$,

$$\mathbb{P}(\|S_{m,n}\| > \varepsilon) < \varepsilon\,.$$

For any $m < k \le n$, it is obvious $\{\|S_{m,k}\| > 2\varepsilon\,;\ \|S_{k,n}\| \le \varepsilon\} \subseteq \{\|S_{m,n}\| > \varepsilon\}$. Thus, we obtain

$$\bigcup_{k=m+1}^{n}\left\{\max_{m<j\le k-1}\|S_{m,j}\| \le 2\varepsilon\,;\ \|S_{m,k}\| > 2\varepsilon\,;\ \|S_{k,n}\| \le \varepsilon\right\} \subseteq \{\|S_{m,n}\| > \varepsilon\}\,.$$

The sets in the above union are disjoint. Also using the independence, we have

$$\sum_{k=m+1}^{n} \mathbb{P}\left(\max_{m<j\le k-1}\|S_{m,j}\| \le 2\varepsilon\,;\ \|S_{m,k}\| > 2\varepsilon\right)\mathbb{P}\left(\|S_{k,n}\| \le \varepsilon\right)$$
$$\le \mathbb{P}\left(\|S_{m,n}\| > \varepsilon\right)\,.$$

Hence,

$$\sum_{k=m+1}^{n} \mathbb{P}\left(\max_{m<j\le k-1} \|S_{m,j}\| \le 2\varepsilon\,;\ \|S_{m,k}\| > 2\varepsilon \right) \le \frac{\mathbb{P}\left(\|S_{m,n}\| > \varepsilon\right)}{\min_{m<k\le n} \mathbb{P}\left(\|S_{k,n}\| \le \varepsilon\right)}\,.$$

The sum on the above left hand is $\mathbb{P}\left(\max_{m<j\le n}\|S_{m,j}\| > 2\varepsilon\right)$. Therefore, we have the following inequality

$$\mathbb{P}\left(\max_{m\le k\le n} \|X_{m+1} + \cdots + X_k\| > 2\varepsilon \right) \le \frac{\mathbb{P}\left(\|X_{m+1} + \cdots + X_n\| > \varepsilon\right)}{1 - \max\limits_{m\le k\le n} \mathbb{P}\left(\|X_{k+1} + \cdots + X_n\| \ge \varepsilon\right)}$$

for any $m < n$ and for any $\varepsilon > 0$. Now the Borel-Cantelli Lemma can be used to show the almost sure convergence of S_n.

Now we show "(c) ⇒ (b)". Assume S_n converges in law to S. Denote $\mu_{m,n}$ the law of $S_n - S_m = \sum_{k=m+1}^{n} X_k$. Since μ_n is convergent in law to a probability distribution μ in $\mathbb{B}$, $\{\mu_n\}$ is tight. This means that for any $\varepsilon > 0$, there is a compact subset K of $\mathbb{B}$ such that $\mu_n(K) > 1 - \varepsilon$. Let $K_1 = \{x - y\,;\ x\,, y \in K\}$. Thus, K_1 is also a compact subset of $\mathbb{B}$. Moreover, we have

$$\begin{aligned} \mu_{m,n}(K_1) &= \mathbb{P}\left(S_n - S_m \in K_1\right) \ge \mathbb{P}\left(S_n \in K\,, S_m \in K\right) \\ &\ge 1 - \mathbb{P}\left(S_n \in K^c\right) - \mathbb{P}\left(S_m \in K^c\right) = 1 - \mu_n(K^c) - \mu_m(K^c) \\ &> 1 - 2\varepsilon\,. \end{aligned} \tag{4.1.13}$$

Thus, $\{\mu_{m,n}\,;\ 1 \le m \le n\}$ is tight. To show (b), we need to show

$$\forall\, \varepsilon > 0 \quad \exists N \quad \forall n > m \ge N \quad \text{such that} \quad \mu_{m,n}(U_\varepsilon) > 1 - \varepsilon\,, \tag{4.1.14}$$

where U_ε is the ε-neighborhood of 0 in $\mathbb{B}$. If (4.1.14) does not hold, then

$$\exists\, \varepsilon > 0 \quad \forall N \quad \exists n(N) > m(N) \ge N \quad \text{such that} \quad \mu_{m(N),n(N)}(U_\varepsilon) \le 1 - \varepsilon\,. \tag{4.1.15}$$

Since $\left\{\mu_{m(N),n(N)}\,, N = 1, 2, \cdots\right\}$ is tight there exists probability measure ν on $(B\,, \mathcal{B}(\mathbb{B}))$ such that

$$\nu(U_\varepsilon) \le \liminf_{N\to\infty} \mu_{m(N),n(N)}(U_\varepsilon) \le 1 - \varepsilon\,. \tag{4.1.16}$$

On the other hand, we have

$$\begin{aligned} \hat{\mu}_{n(N)}(\xi) &= \mathbb{E}e^{i\langle S_{n(N)}\,,\xi\rangle} = \mathbb{E}e^{i\langle S_{m(N)}\,,\xi\rangle}\mathbb{E}e^{i\langle S_{n(N)} - S_{m(N)}\,,\xi\rangle} \\ &= \hat{\mu}_{m(N)}(\xi)\hat{\mu}_{m(N),n(N)}(\xi)\,. \end{aligned} \tag{4.1.17}$$

Letting $N \to \infty$ we see

$$\hat{\mu}(\xi) = \hat{\mu}(\xi)\hat{\nu}(\xi). \tag{4.1.18}$$

Since $\hat{\mu}(0) = 1$, there is $r_0 > 0$ such that $\hat{\mu}(\xi) \neq 0$ for all $|\xi| \le r_0$. From (4.1.18) this implies that $\hat{\nu}(\xi) = 0$ all $|\xi| \le r_0$. By Lemma 4.1 we have $\nu = \delta$ concentrates on 0, which contradicts (4.1.16). This means (4.1.14) holds. That is μ_n converges to μ in probability. □

Theorem 4.6 (Itô-Nisio). *Let X_n, $n = 1, 2, \cdots$ and S_n be defined as in Theorem 4.5 and let each random variable X_n be symmetrically distributed. Then the statements (a), (b), (c) in Theorem 4.5 and the following statements are equivalent.*

(d) $\{\mu_n\, , n = 1, 2, \cdots\}$ is tight.
(e) There exists a $\mathbb{B}$-valued random variable S such that $\langle S_n, \xi\rangle$ converges to $\langle S, \xi\rangle$ in probability for every $\xi \in \mathbb{B}^*$.
(f) There exists a probability measure μ on $\mathbb{B}$ such that $\hat{\mu}_n(\xi)$ converges to $\hat{\mu}(\xi)$ for every $\xi \in \mathbb{B}^*$.

Proof. It is obvious that (c) ⇒ (d) and (a) ⇒ (e) ⇒ (f). It is then suffices to show (f) ⇒ (e) ⇒ (d) ⇒ (a).

Suppose (d). By Theorem 4.2 there exists a $\{c_n\}$ such that $S_n - c_n$ converges a.s. Since X_n is symmetrically distributed and $X_1, X_2, \cdots$ are independent, we see that $(-X_1, -X_2, \cdots)$ has the same law as $(X_1, X_2, \cdots)$. Thus, $-S_n - c_n = \sum_{k=1}^n (-X_k) - c_n$ also converges a.s. This implies $S_n = ([S_n - c_n] - [-S_n - c_n])/2$ converges a.s., proving (a).

Let (e) hold. Fix an arbitrary $\xi_1\, , \cdots, \xi_p \in \mathbb{B}^*$. The condition (e) implies that the random vectors $\sigma_n = (\langle S_n\, , \xi_1\rangle\, , \cdots, \langle S_n\, , \xi_p\rangle)$ converge in probability to $\sigma = (\langle S\, , \xi_1\rangle\, , \cdots, \langle S\, , \xi_p\rangle)$. Since σ_n and $\sigma_m - \sigma_n$ $(m > n \ge 1)$ are independent, for any Borel sets $\Gamma_1\, , \Gamma_2 \in \mathcal{B}(\mathbb{R}^p)$, we have

$$\mathbb{P}(\sigma_n \in \Gamma_1\, ,\ \sigma_m - \sigma_n \in \Gamma_2) = \mathbb{P}(\sigma_n \in \Gamma_1)\mathbb{P}(\sigma_m - \sigma_n \in \Gamma_2).$$

Letting $m \to \infty$ and using the condition (e), we see

$$\mathbb{P}(\sigma_n \in \Gamma_1\, ,\ \sigma - \sigma_n \in \Gamma_2) = \mathbb{P}(\sigma_n \in \Gamma_1)\mathbb{P}(\sigma - \sigma_n \in \Gamma_2).$$

From Theorem 4.1 we see that S_n and $S - S_n$ are independent. Thus,

$$\mathbb{P}(S \in K) = \int_B \mathbb{P}(S_n + x \in K)\mathbb{P}(S - S_n \in dx).$$

This means that we can find an x_0 such that

$$\mathbb{P}(S \in K) \le \mathbb{P}(S_n + x_0 \in K).$$

From the symmetry of X_n, we see

$$\mathbb{P}(S \in K) \leq \mathbb{P}(S_n + x_0 \in K) = \mathbb{P}(-S_n + x_0 \in K)\,.$$

Denote $K_1 = (K - K)/2 := \{(x-y)/2\,, x\,, y \in K\}$. Then, we have

$$\begin{aligned}\mu_n(K_1) &= \mathbb{P}(S_n \in K_1) \geq \mathbb{P}(S_n + x_0 \in K, -S_n + x_0 \in K)\\ &\geq 1 - \mathbb{P}(S_n + x_0 \in K^c) - \mathbb{P}(-S_n + x_0 \in K^c)\\ &\geq 1 - 2\mathbb{P}(S \in K^c)\,.\end{aligned} \tag{4.1.19}$$

If K is compact, then K_1 is also compact. From the tightness of S we see for any $\varepsilon > 0$, there is a compact subset K_ε of $\mathbb{B}$ such that $\mathbb{P}(S \in K_\varepsilon^c) < \varepsilon$. From (4.1.19), it follows

$$\mu_n\left(K_{1,\varepsilon}^c\right) = 1 - \mu_n\left(K_{1,\varepsilon}\right) \leq 2\mathbb{P}(S \in K_\varepsilon^c) < 2\varepsilon\,,$$

where $K_{1,\varepsilon} = (K_\varepsilon - K_\varepsilon)/2)$. This implies the tightness of $\{\mu_n\,, n = 1, 2, \cdots\}$.

Finally, we prove "(f) $\Rightarrow$ (e)". Given $\xi \in \mathbb{B}^*$, the condition (f) implies that for any $t \in \mathbb{R}$, $\hat{\mu}_n(t\xi) = \mathbb{E}e^{it\langle S_n, \xi\rangle}$ converges to $\hat{\mu}(t\xi)$ which is a characteristic function of t. Thus, there is a random variable Y_ξ such that $\langle S_n, \xi\rangle$ converges to Y_ξ in law. Since $\langle S_n, \xi\rangle$ are sum of independent variables, this also implies that $\langle S_n, \xi\rangle$ converges to Y_ξ a.s. Let $\{z_1, z_2\,, \cdots\}$ be a sequence satisfying the property of (4.1.1). Since the union of countably many null sets is also a null set, we see that there is common null set such that outside of this set

$$\langle S_n, z_k\rangle \quad \text{converges to} \quad Y_{z_k} \quad a.s. \tag{4.1.20}$$

We need to show the existence of $S : \Omega \to \mathbb{B}$ such that $\langle S_n(\omega), z\rangle \to \langle S(\omega), z\rangle$.

To construct such S, we denote by $\mathbb{R}^\infty$ the countable product $\mathbb{R} \times \mathbb{R} \times \cdots$ and denote by $\mathcal{B}(\mathbb{R}^\infty)$ the σ-algebra generated by all cylindrical sets in $\mathbb{R}^\infty$. $\mathcal{B}(\mathbb{R}^\infty)$ is also the σ-algebra of all Borel subsets of $\mathbb{R}^\infty$ with respect to the product topology. It is easy to check the following fact by using the characteristic function.

$$\mu\left(\{\langle x, z_1\rangle\,, \langle x, z_1\rangle\,, \cdots\} \in \mathbb{B}\right) = \mathbb{P}\left(\{Y_{z_1}(\omega)\,, Y_{z_2}(\omega)\,, \cdots\} \in \mathbb{B}\right)\,. \tag{4.1.21}$$

Since μ is a probability measure, it is tight. This means that there exists an increasing sequence of compact subsets $K_n, n = 1, 2, \cdots$ such that $\mu(K_\infty) = 1$, where $K_\infty = \cup_{n=1}^\infty K_n$. Define the mapping $\theta : B \to \mathbb{R}^\infty$ as

$$\theta x = \left(\langle x\,, z_1\rangle\,, \langle x, z_2\rangle\,, \langle x, z_2\rangle\,, \cdots\right)\,.$$

Then, θ is continuous and one to one from $\mathbb{B}$ to $\theta\mathbb{B}$. In fact, if $\theta x = 0$, then

$$x \in \cap_{n=1}^{\infty} \{x\, ;\ \langle x\,, z_n\rangle \le r\} \quad \forall\, r > 0\,.$$

Thus, $x = 0$. It is easy to see that θK_n is compact and $\theta\big|_{K_n}$ has a continuous inverse map. Since $\theta K_\infty = \cap_{n=1}^{\infty}\theta K_n$ is a Borel subset of $\mathbb{R}^\infty$, the restriction of θ to K_∞ has an inverse which can be extended to a $(\mathcal{B}(\mathbb{R}^\infty), \mathcal{B}(\mathbb{B}))$-measurable map $\phi : \mathbb{R}^\infty \to \mathbb{B}$. It is clear that

$$x = \phi(\theta x) = \phi\left(\langle x\,, z_1\rangle\,, \langle x, z_2\rangle\,, \cdots\right) \quad \forall\, x \in K_\infty\,. \tag{4.1.22}$$

Since

$$\mu\left(\left(\langle x\,, z_1\rangle\,, \langle x, z_2\rangle\,, \cdots\right) \in \theta K_\infty\right) \ge \mu(K_\infty) = 1$$

we see from (4.1.21) that

$$\mathbb{P}\left(\left(Y_{z_1}(\omega)\,, Y_{z_2}(\omega)\,, \cdots\right) \in \theta K_\infty\right) = 1\,.$$

Now we define

$$S(\omega) = \phi\left(Y_{z_1}(\omega)\,, Y_{z_2}(\omega)\,, \cdots\right)\,. \tag{4.1.23}$$

This is well-defined since $\left(Y_{z_1}(\omega)\,, Y_{z_2}(\omega)\,, \cdots\right) \in \theta K_\infty$ and ϕ is well-defined on θK_∞.

For any $\xi \in \mathbb{B}^*$, we have

$$\begin{aligned}
\mu\left(\langle x, \xi\rangle = \langle \phi\left(\langle x\,, z_1\rangle\,, \langle x, z_2\rangle\,, \cdots\right)\,, \xi\rangle\right) \\
&\ge \mu\left(x = \phi\left(\langle x\,, z_1\rangle\,, \langle x, z_2\rangle\,, \cdots\right)\right) \\
&\ge \mu\left(x = \phi\left(\langle x\,, z_1\rangle\,, \langle x, z_2\rangle\,, \cdots\right)\,,\ x \in K_\infty\right) \\
&= \mu(K_\infty) = 1\,,
\end{aligned} \tag{4.1.24}$$

where the last second equality follows from (4.1.22). This identity together with (4.1.21) gives

$$\mathbb{P}(Y_z = \langle \phi(Y_{z_1}\,, Y_{z_2}\,, \cdots)\,, z\rangle) = 1\,. \tag{4.1.25}$$

By the definition of S, we have $\mathbb{P}(Y_z = \langle S(\omega)\,, z\rangle) = 1$. This proves "(f) $\Rightarrow$ (e)". The theorem is then proved. □

4.2 Abstract Wiener Space

Let $\mathbb{B}$ be a Banach space with the norm $\|x\|$ and let $\mathbb{B}'$ be the dual of $\mathbb{B}$, namely, the space of all continuous linear functionals on $\mathbb{B}$. The dual $\mathbb{B}'$ is also a Banach space with norm

$$\|\ell\| = \|\ell\|_{\mathbb{B}'} = \sup_{x\in\mathbb{B}\,;\|x\|\le 1} |\langle x, \ell\rangle|\,.$$

We use $\langle x , \ell \rangle$ to denote the pairing between $\mathbb{B}$ and $\mathbb{B}'$, where $x \in \mathbb{B}$ and $\ell \in \mathbb{B}'$. Let $\mathcal{K}(\mathbb{B}')$ be the set of all finite dimensional subspaces of $\mathbb{B}'$. A cylindrical set of $\mathbb{B}$ based on $K \in \mathcal{K}(\mathbb{B}')$ is

$$C_{E,\ell_1,\cdots,\ell_n} = \{x \in \mathbb{B} : (\langle x, \ell_1 \rangle , \cdots , \langle x, \ell_n \rangle \in E\} , \tag{4.2.1}$$

where $\ell_1 , \cdots , \ell_n \in K$ and E is a Borel subset of $\mathbb{R}^n$.

Denote by $\mathcal{C}(K)$ the σ-algebra of all cylindrical sets of $\mathbb{B}$ based on K. Let $\mathcal{R}(\mathbb{B}) = \cup_{K\in\mathcal{K}}\mathcal{C}(K)$.

Definition 4.5. A cylindrical measure on $\mathbb{B}$ is a non-negative function μ defined on $\mathcal{R}(\mathbb{B})$ such that for any $K \in \mathcal{K}(\mathbb{B}')$, when μ is restricted on $\mathcal{C}(K)$ it is σ-additive.

For a cylindrical measure μ its characteristic functional

$$\hat{\mu}(\ell) = \int_{\mathbb{B}} e^{\iota\langle x,\ell\rangle} \mu(dx) , \quad \ell \in \mathbb{B}'$$

is well-defined since when restricted to $\mathcal{C}(\ell)$, μ is a σ-additive measure.

Example 4.2. Let $\mathbb{H}$ be a separable Hilbert space with scalar product $\langle \cdot , \cdot \rangle$ and the norm $| \cdot |$. Let $\mathcal{C}(\mathbb{H})$ be the set of all subsets C of $\mathbb{H}$ of the form

$$C_{E,h_1,\cdots,h_n} = \{h \in \mathbb{H}; \ (\langle h, h_1 \rangle , \cdots , \langle h, h_n \rangle) \in E\} \tag{4.2.2}$$

for some positive integer n, linearly independent $h_1, \cdots , h_n \in \mathbb{H}$, and $E \in \mathcal{B}(\mathbb{R}^n)$.

A centered cylindrical Gaussian measure on $\mathbb{H}$ is a non-negative function μ defined for on every $C_{E,h_1,\cdots,h_n}$ of the form (4.2.2) and is given by

$$\mu(C_{E,h_1,\cdots,h_n}) = (2\pi)^{-n/2} \det(A)^{-1/2} \int_E e^{-\frac{\langle x, A^{-1} x \rangle}{2}} dx , \tag{4.2.3}$$

where the $n \times n$ matrix A is given by $A = (\langle h_i , h_j \rangle)_{1\le i ,j\le n}$.

For any fixed $h_1, \cdots , h_n$, the random vector $(\langle \cdot, h_1 \rangle , \cdots , \langle \cdot, h_1 \rangle)$ is centered Gaussian with covariance matrix $A = (\langle h_i , h_j \rangle)_{1\le i ,j\le n}$. This means

$$\mathbb{E}(\langle \cdot, h_i \rangle) = 0 , \quad \mathbb{E}\left(\langle \cdot, h_i \rangle \langle \cdot, h_j \rangle\right) = \langle h_i, h_j \rangle , \quad \forall 1 \le i , j \le n .$$

For this Gaussian cylindrical Gaussian measure

$$\hat{\mu}(h) = e^{-\frac{|h|^2}{2}} .$$

The above cylindrical measure is not a (σ-additive) measure. We can show this by contradiction. Assume it is a measure. Let $\{e_1, e_2, \cdots\}$ be an orthonormal basis of $\mathbb{H}$. Then, the collection of all balls

$$A_k = B(e_k, 1/4) = \{h \in \mathbb{H}; \ |h - e_k|_{\mathbb{H}} < 1/4\} , \quad k = 1, 2, \cdots$$

are disjoint and $\mu(A_k) > 0$. Thus, $\mu(\cup_k A_k) = \sum_k \mu(A_k) = \infty$. But $\cup_k A_k$ is contained in a ball of radius 2 and then we have that $\mu(B(0,2)) = \infty$. In a similar way, one can show that for any open set A, $\mu(A) = \infty$ if μ is σ-additive.

In application, usually, it is relatively easier to describe a cylindrical measure on a Hilbert space $\mathbb{H}$ (one needs only to specify the finite dimensional distribution of the probability measure) such as the Gaussian measure described above. The problem is if it is possible to extend this cylindrical measure to a true σ-additive measure. In many interesting case, such σ-additive measure exists only in a larger Banach space. To describe such Banach space, we need to introduce the following concept of measurable norm.

Definition 4.6. Let $\mathbb{H}$ be a Hilbert space and let μ be a cylindrical probability measure on $\mathbb{H}$. Let $\mathcal{P}$ denote the set of all finite dimensional projections in $\mathbb{H}$. For any $P \in \mathcal{P}$, $\{x \in \mathbb{H}, Px \in E\}$ is a cylindrical set of $\mathbb{H}$. A norm $\|\cdot\|$ on $\mathbb{H}$ is called a measurable norm if for any $\varepsilon > 0$ there is a $P_\varepsilon \in \mathcal{P}$ such that for any $P \in \mathcal{P}$, orthogonal to P_ε,

$$\mu\left(x \in \mathbb{H}, \|Px\| > \varepsilon\right) < \varepsilon\,.$$

Let $\mathbb{H}$ be a Hilbert space continuously and densely embedded in $\mathbb{B}$. We identify $\mathbb{H}$ as a subset of $\mathbb{B}$. The scalar product on $\mathbb{H}$ is denoted by $\langle h_1\,, h_2\rangle$ and the Hilbert norm is $|h| = \langle h\,, h\rangle^{1/2}$. The norm of $\mathbb{B}$ is denoted by $\|\cdot\|$. An element $\ell \in \mathbb{B}'$ is a continuous linear functional on $\mathbb{B}$. When restricted to $\mathbb{H}$, it is also a continuous linear functional on $\mathbb{H}$. Thus, ℓ can be identified as an element of $\mathbb{H}' = \mathbb{H}$ (We identify $\mathbb{H}$ with $\mathbb{H}'$ by $\mathbb{H} \ni h \leftrightarrow \langle\cdot, h\rangle \in \mathbb{H}'$). Thus, we have $\mathbb{B}' \subseteq \mathbb{H}' = \mathbb{H} \subseteq \mathbb{B}$. If $\ell \in \mathbb{B}'$ and $x \in \mathbb{H} \subseteq \mathbb{B}$, then $\langle x\,, \ell\rangle$ can be considered as the value of the linear functional ℓ at $x \in \mathbb{B}$ and it can be also considered as scalar product between $x \in \mathbb{H}$ and $\ell \in \mathbb{H}' = \mathbb{H}$. Thus, the use of $\langle\cdot\,,\cdot\rangle$ both for the pairing and for the scalar product is legitimate.

Let μ be cylindrical measure on the Hilbert space $\mathbb{H}$. Since $\mathbb{B}' \subseteq \mathbb{H}$, we see $\mathcal{K}(\mathbb{B}') \subseteq \mathcal{K}(\mathbb{H})$. Thus, for any cylindrical set C of $\mathbb{B}$ of the form (4.2.1), $C \cap \mathbb{H}$ is a cylindrical set of $\mathbb{H}$. We can extend μ to be a cylindrical measure of $\mathbb{B}$ by $\mu^*(C) = \mu(C \cap \mathbb{H})$. We shall identify μ^* with μ and use the triple $(\mathbb{B}, \mathbb{H}, \mu)$ to denote the three basic objects described as above.

Theorem 4.7. *Let $(\mathbb{B}, \mathbb{H}, \mu)$ be a basic triple described as above. Namely, $\mathbb{B}$ is a separable Banach space and $\mathbb{H}$ is a separable Hilbert space continuously and densely embedded in $\mathbb{B}$. μ is cylindrical measure on $\mathbb{H}$ extended to a cylindrical measure on $\mathbb{B}$ in a canonical way described above.*

(i) Assume the characteristic function $\hat{\mu}(\ell)\,, \ell \in \mathbb{H}$ is continuous on $\mathbb{H}$. If the norm $\|\cdot\|$ of $\mathbb{B}$ is measurable, then the extension of μ to $\mathcal{R}(\mathbb{B})$ is σ-additive.

(ii) If μ is a centered Gaussian cylindrical measure on $\mathbb{H}$ and the extension of μ to $\mathcal{R}(\mathbb{B})$ is σ-additive, then the norm $\|\cdot\|$ of $\mathbb{B}$ is measurable.

Proof. First, we prove (i). Since μ is a cylindrical measure on $\mathbb{H}$, from the Kolmogorov's extension theorem, there is a probability space $(\Omega, \mathcal{F}, P)$ and a linear mapping F from $\mathbb{H}$ to $L^1(\Omega, \mathcal{F}, P)$ such that for any integer $n \geq 1$, $\ell_1, \cdots, \ell_n \in \mathbb{H}$ and $E \in \mathcal{B}(\mathbb{R}^n)$ we have

$$\begin{aligned}&\mu\{x \in \mathbb{H};\ (\langle x, \ell_1\rangle, \cdots, \langle x, \ell_1\rangle) \in E\}\\&\quad = P\{\omega \in \Omega;\ (F(\ell_1)(\omega), \cdots, F(\ell_n)(\omega)) \in E\}\,.\end{aligned} \tag{4.2.4}$$

Let $P \in \mathcal{P}$ and let $(h_1, \cdots, h_n)$ be an ONB (orthonormal basis) of $P(\mathbb{H})$. Define the following $\mathbb{H}$-valued random variable

$$F(P) := \sum_{j=1}^{n} F(h_j)h_j\,. \tag{4.2.5}$$

It is easy to see by the linearity of F

$$\langle F(P), h\rangle = F(Ph) \quad \forall\, h \in \mathbb{H}\,. \tag{4.2.6}$$

Thus, $F(P)$ is independent of the choice of the ONB $(h_1, \cdots, h_n)$. For any $C \in \mathcal{C}(P\mathbb{H})$, there is some $E \in \mathcal{B}(\mathbb{R}^n)$ such that

$$C = \{x \in \mathbb{H}\ :\ (\langle x, h_1\rangle, \cdots, \langle x, h_n\rangle \in E\}\,.$$

Thus, we see easily that

$$\{\omega \in \Omega;\ F(P)(\omega) \in C\} = \{\omega \in \Omega;\ (F(h_1)(\omega), \cdots; F(h_n)(\omega)) \in E\}\,.$$

This combined with (4.2.4) yields

$$P\{\omega \in \Omega;\ F(P)(\omega) \in C\} = \mu(C) \quad \forall\, C \in \mathcal{C}(P\mathbb{H})\,. \tag{4.2.7}$$

From the definition of the measurable norm it is easy to see that there exists an increasing sequence $P_n \in \mathcal{P}$ such that $P_n \uparrow I$ and for any $\varepsilon > 0$,

$$\lim_{n, m\to\infty} \mu\{x \in \mathbb{H};\ \|P_n x - P_n x\| > \varepsilon\} = 0\,. \tag{4.2.8}$$

From (4.2.8) we can find an orthonormal basis (ONB) $\{e_1, \cdots, e_n, \cdots\}$ and a subsequence $k_1, k_2, \cdots$ such that

$$P\left(\|F(e_1)e_1 + \cdots + F(e_{k_1})e_{k_1}\| \geq 1/2\right) < 1/2\,,$$

$$P\left(\|F(e_{k_{i-1}+1})e_{k_{i-1}+1} + \cdots + F(e_{k_i})e_{k_i}\| \geq 1/2^i\right) < 1/2^i\,,$$

By the Borel-Cantelli lemma it follows that

$$F(e_1)e_1 + \cdots + F(e_{k_1})e_{k_1} + \cdots + F(e_{k_{i-1}+1})e_{k_{i-1}+1} + \cdots + F(e_{k_i})e_{k_i} + \cdots$$

is convergent almost surely to a $\mathbb{B}$-valued random variable Z. This random variable Z induces a probability measure on $\mathbb{B}$: $\mu'(A) = P \circ Z^{-1}(A)$.

It is easy to verify that for any $y_1, \cdots, y_d \in \mathbb{B}'$,

$$\begin{aligned}
&\mu\left(\{x \in \mathbb{B}, (\langle x, y_1\rangle, \cdots, \langle x, y_d\rangle) \in A\}\right) = P\left(\{(F(y_1) \cdots, F(y_d)) \in A\}\right)\\
&\quad = P(\omega \in \Omega\,,\ \langle Z, y_1\rangle, \cdots, \langle Z, y_d\rangle) \in A\}\\
&\quad = \mu'\left(\{x \in \mathbb{B}, (\langle x, y_1\rangle, \cdots, \langle x, y_d\rangle) \in A\}\right).
\end{aligned}$$

This shows that $\mu = \mu'$ on the cylindrical sets which implies that μ can be extended to a σ-additive measure μ' on $\mathbb{B}$.

Now we prove (ii). In this case we can choose $(\Omega, \mathcal{F}, P) = (B\,, \mathcal{B}(\mathbb{B}), \mu)$. Let $P_n \in \mathcal{P}$ be an increasing sequence such that $P_n \uparrow I$. Let ξ_k be defined as

$$\xi_0 = 0\,, \quad \xi_1 = F(P_1)\,, \quad \cdots, \quad \xi_k = F(P_k) - F(P_{k-1})\,,\ k = 2, 3, \cdots.$$

Since μ is cylindrical Gaussian, and $P_k - P_{k-1}$, $k = 1, 2\cdots$, are orthogonal, we see that $\xi_k, k = 1, 2, \cdots$ are independent symmetric $\mathbb{B}$-valued random variables. For any $h \in \mathbb{B}'$, $P_n h \to h$ in $\mathbb{H}$. we have

$$\lim_{n\to\infty} \mathbb{E}\left[e^{\iota \sum_{j=1}^n \langle \xi_i, h\rangle}\right] = \lim_{n\to\infty} \mathbb{E}\left[e^{\iota\langle F(P_n), h\rangle}\right] = \lim_{n\to\infty} \mathbb{E}\left[e^{\iota F(P_n h)}\right] = \hat{\mu}(h)\,.$$

From the Ito-Nisio theorems 4.5-4.6 it follows that $\sum_{j=1}^n \xi_j = F(P_n)$ converges almost surely to a $\mathbb{B}$-valued random variable ξ. This also implies (4.2.8). □

Definition 4.7. Let $\mathbb{H}$ be a separable Hilbert space and let $\mathbb{B}$ be a Banach space such that $\mathbb{H}$ is continuously embedded in $\mathbb{B}$. A probability measure $\mathbb{P}$ on the measurable space $(\mathbb{B}, \mathcal{B}(\mathbb{B}))$ is called an *abstract Wiener measure* (the triple $(\mathbb{B}, \mathbb{H}, \mathbb{P})$ is called an abstract Wiener space) if for any $\ell \in \mathbb{B}' \subseteq \mathbb{H}' = \mathbb{H} \subseteq \mathbb{B}$, we have

$$\mathbb{E}\left[e^{\iota\langle \cdot, \ell\rangle}\right] = \int_{\mathbb{B}} e^{\iota\langle x, \ell\rangle}\mathbb{P}(dx) = e^{-\frac{1}{2}|\ell|^2}\,, \quad \text{for any } \ell \in \mathbb{B}' \subseteq \mathbb{H}' = \mathbb{H} \subseteq \mathbb{B}\,. \tag{4.2.9}$$

From this identity for any $\ell \in \mathbb{B}'$, $\langle \cdot, \ell\rangle$ is a mean 0 Gaussian random variable with variance $|\ell|$. We denote this random variable by $\tilde{\ell}$. For any element $h \in \mathbb{H}$, we can find a sequence $\ell_n \in \mathbb{B}'$ such that $\lim_{n\to\infty} |\ell_n - h| = 0$. The sequence of Gaussian random variables $\left\{\tilde{\ell}_n\right\}$ is a Cauchy sequence in $L^2(\mathbb{B})$. Its limit, denoted by $\tilde{h}$, is also a Gaussian random variable with mean 0 and variance $|h|^2$.

We shall denote

$$L^p(\mathbb{B}) = L^p(\mathbb{B}, \mathbb{P}) = L^p(\mathbb{B}, \mathbb{H}, \mathbb{P}) = \left\{F : \mathbb{B} \to \mathbb{R}\,;\ \|F\|_p := (\mathbb{E}|F|^p)^{1/p} < \infty\right\}.$$

Lemma 4.4. *Let $X = (X_1, \cdots, X_d)^T$ be a random vector. Assume that there is a vector $\xi = (\xi_1, \cdots, \xi_d)^T$ in $\mathbb{R}^d$ and positive definite matrix $A = (a_{ij})_{1\le i,j\le d}$ such that for any vector $t = (t_1, \cdots, t_d)$, the random variable $t^T X = t_1 X_1 + \cdots + t_d X_d$ is a Gaussian random variable with mean $t^T\xi = \sum_{i=1}^d t_i\xi_i$ and variance $t^T At = \sum_{i,j=1}^d a_{ij}t_jt_j$. Then, X is a (jointly) Gaussian random vector with mean $\xi = \mathbb{E}(X)$ and covariance matrix $A = \mathbb{E}(XX^T)$.*

Proof. The condition can be interpreted as that the characteristic function of X is that of Gaussian with mean ξ and covariance A. □

If $h_1 \cdots, h_d$ are elements in $\mathbb{H}$, then their linear combination $h = \alpha_1 h_1 + \cdots + \alpha_d h_d$, where $\alpha_1, \cdots, \alpha_d \in \mathbb{R}$, is also an element in $\mathbb{H}$. Thus, we have

$$\mathbb{E}\left[e^{\iota\sum_{j=1}^d \alpha_j\tilde{h}_j}\right] = \mathbb{E}\left[e^{\iota\widetilde{\sum_{j=1}^d \lambda_j\ell_j}}\right] = e^{-\frac{1}{2}\left|\sum_{j=1}^d \lambda_j\ell_j\right|^2} = e^{-\frac{1}{2}\sum_{i,j=1}^d \alpha_i\alpha_j\langle h_i,h_j\rangle}.$$

This means that

$$\tilde{h} = (\tilde{h}_1, \cdots, \tilde{h}_d)^T \text{ is jointly Gaussian with mean } 0$$
$$\text{and covariance matrix} \quad \mathbb{E}(\tilde{h}\tilde{h}^T) = (\langle h_i, h_j\rangle)_{1\le i,j\le d}\,.$$

It is also possible to define a Gaussian measure on a Hilbert space with a covariance form other than thc scalar product $\langle\cdot,\cdot\rangle$. Namely, we may require $\mathbb{E}\left[e^{\iota\langle\cdot,\ell\rangle}\right] = e^{-\frac{1}{2}\langle\ell,Q\ell\rangle}$ for some linear operator Q.

From now on we shall work on an abstract Wiener space $(\mathbb{B}, \mathbb{H}, \mathbb{P})$.

Example 4.3. Let $\mathbb{B} = \mathbb{H} = \mathbb{R}^d$ and let μ be the standard Gaussian measure on $\mathbb{B}$. Then, $(\mathbb{B}, \mathcal{B}, \mu)$ is the d-dimensional standard Gaussian measure space.

Example 4.4. Consider a domain D of $\mathbb{R}^d$ (d dimensional Euclidean space). Together with some nice boundary conditions (if D is not the whole space $\mathbb{R}^d$) we can prove (see [Glimm and Jaffe (1987)]) that there is a kernel $K(x, y)$ such that the following equation (with mass $m = 1$) holds

$$(-\Delta + 1)K(x, y) = \delta(x - y)\,,$$

where Δ is the Dirichlet Laplacian on D. A Hilbert space of (generalized) functions can be determined by

$$\langle f\,, g\rangle = \int_D K(x,y)f(x)f(y)dxdy\,,$$

where f and g are two (generalized) real-valued functions. The Gaussian measure associated with this Hilbert space is useful in the Euclidean quantum field theory (see also [Hu and Kallianpur (1998)] and the references therein).

Theorem 4.8. *Given $h \in \mathbb{H}$, consider $T_h : \mathbb{B} \to \mathbb{B}$ given by $T_h x = x + h$. Define*

$$\tilde{\mu}(A) = \mu \circ T_h^{-1}(A) = \mu(\{x \in \mathbb{B}\,, x + h \in A\})\,.$$

If $h \in \mathbb{H}$, then $\tilde{\mu}$ and μ are absolutely continuous with each other and we have the following Radon-Nikodym derivative

$$\frac{d\hat{\mu}}{d\mu} = e^{\langle x,h\rangle - \frac{|h|^2}{2}}\,.$$

Proof. Let $h_1, \cdots, h_d$ be an orthonormal vectors in $\mathbb{H}$. Let $F(x) = f(\langle x, h_1\rangle\,, \cdots, \langle x, h_d\rangle)$ be a cylindrical function. Then

$$\begin{aligned}\int_{\mathbb{B}} F(x)\hat{\mu}(dx) &= \int_{\mathbb{B}} F(x+h)\mu(dx) = \int_{\mathbb{B}} f(y_1 + \hat{h}_1, \cdots, y_d + \hat{h}_d)\mu(dy)\\ &= \int_{\mathbb{R}^d} f(z_1, \cdots, z_d) \exp\left\{\sum_{i=1}^d \hat{h}_i z_i - \frac{\hat{h}_i^2}{2}\right\}\mu(dz)\\ &= \int_{\mathbb{B}} F(x) e^{\langle x,h\rangle - \frac{|h|^2}{2}}\mu(dx)\,.\end{aligned}$$

The theorem is proved since the functional of form $F(x) = f(\langle x, h_1\rangle\,, \cdots, \langle x, h_d\rangle)$ is dense in $L^p(\mathbb{B})$. □

Theorem 4.9. *Let $(\mathbb{B}, \mathbb{H}, \mathbb{P})$ be an abstract Wiener space. If $\varphi : \mathbb{B} \to \mathbb{R}$ is measurable seminorm, namely, φ satisfies*

$$\varphi(x+y) \le \varphi(x) + \varphi(y) \quad \forall\, x\,, y \in \mathbb{B}\,, \tag{4.2.10}$$

$$\varphi(\lambda x) = |\lambda|\varphi(x) \quad \forall\, \lambda \ge \mathbb{R}\,,\ x \in \mathbb{B}\,, \tag{4.2.11}$$

and for any $\varepsilon > 0$, there is a finite dimensional projection P_ε such that

$$\mathbb{P}\,(x\,;\ \phi(Px) > \varepsilon) < \varepsilon\,, \quad \forall\, P \perp P_\varepsilon\,, \tag{4.2.12}$$

then the restriction of φ on $\mathbb{H}$ is continuous. More precisely, there is a positive constant C such that

$$\varphi(x) \le C|x|_{\mathbb{H}}\,, \quad \forall\, x \in \mathbb{H}\,. \tag{4.2.13}$$

Proof. By the assumption that $\varphi(x)$ is measurable, there is $R > 0$ and finite dimensional projection P_0 such that

$$\mathbb{P}(\varphi(Px) \geq R) < 1/2 \quad \forall \text{ finite dimensional project } P \perp P_0 \,.$$

Since all norms in a finite dimensional space are equivalent and φ is a seminorm on $P_0\mathbb{H}$, there is a constant C_1 such that

$$\varphi(x) \leq C_1 |x|_{\mathbb{H}} \,, \quad \forall\, x \in P_0\mathbb{H} \,.$$

On the other hand, for any $z \in (P_0\mathbb{H})^{\perp}$ and $\varphi(z) \neq 0$, we define

$$P_z x = \left\langle x, \frac{z}{|z|_{\mathbb{H}}} \right\rangle \frac{z}{|z|_{\mathbb{H}}} \,.$$

Since $P_z \perp P_0$

$$\mathbb{P}\left(\left| \left\langle x, \frac{z}{|z|_{\mathbb{H}}} \right\rangle \right| \frac{\varphi(z)}{|z|_{\mathbb{H}}} \geq R \right) = \mathbb{P}(\varphi(P_z x) \geq R) < 1/2 \,.$$

This means

$$\mathbb{P}\left(\left| \left\langle x, \frac{z}{|z|_{\mathbb{H}}} \right\rangle \right| \geq \frac{R|z|_{\mathbb{H}}}{\varphi(z)} \right) < 1/2 \,.$$

But under $\mathbb{P}$, $\langle \cdot, \frac{z}{|z|_{\mathbb{H}}} \rangle$ is a standard Gaussian random variable, the above inequality can be written as

$$\frac{2}{\sqrt{2\pi}} \int_{\frac{R|z|_{\mathbb{H}}}{\varphi(z)}}^{\infty} e^{-\frac{x^2}{2}} dx < 1/2 \,.$$

If a is such that $\frac{2}{\sqrt{2\pi}} \int_a^{\infty} e^{-\frac{x^2}{2}} dx = 1/2$, then $0 < a < \infty$ and the above inequality implies $\frac{R|z|_{\mathbb{H}}}{\varphi(z)} > a$, and hence

$$\varphi(z) \leq \frac{R}{a} |z|_{\mathbb{H}} \quad \forall\, z \in (P_0\mathbb{H})^{\perp} \,.$$

Notice that the above inequality holds true when $\varphi(z) = 0$. Now for any $x \in \mathbb{H}$, we can write $x = y + z$, where $y \in P_0\mathbb{H}$ and $z \in (P_0\mathbb{H})^{\perp}$. Thus

$$\begin{aligned}
\varphi(x)^2 &\leq (\varphi(y) + \varphi(z))^2 \leq 2\left(\varphi(y)^2 + \varphi(z)^2\right) \\
&\leq 2C_1^2 |y|_{\mathbb{H}}^2 + \frac{2R^2}{a^2} |z|_{\mathbb{H}}^2 \leq \left[2C_1^2 + \frac{2R^2}{a^2} \right] \left[|y|_{\mathbb{H}}^2 + |z|_{\mathbb{H}}^2 \right] \\
&= \left[2C_1^2 + \frac{2R^2}{a^2} \right] |x|_{\mathbb{H}}^2 \,.
\end{aligned}$$

This proves the theorem. □

4.3 Canonical Wiener Space

Theorem 4.10. *Fix a time $T>0$. Let $\mathbb{H}$ be the following Cameron-Martin space*

$$\mathbb{H} := \{f : [0,T] \to \mathbb{R}\,;\ f(0) = 0\,;\ f \textit{ is absolutely continuous on } [0,T]\}$$

with the scalar product $\langle f, g\rangle = \int_0^T f'(t)g'(t)dt$. Let

$$\mathbb{B} = \{f : [0,T] \to \mathbb{R}\,;\ f(0) = 0\,;\ \ f \textit{ is continuous on } [0,T]\}$$

with the sup norm. Let $\mathbb{P}$ be the canonical Wiener measure on $\mathbb{B}$. Then, $(\mathbb{B}, \mathbb{H}, \mathbb{P})$ is an abstract Wiener space.

Proof. If $\ell : [0,T] \to \mathbb{R}$ is twice continuously differentiable, then for any $x \in \mathbb{H}$, $\langle x, \ell\rangle = \int_0^T X'(t)\ell'(t) = X(T)\ell'(T) - \int_0^T X(t)\ell''(t)dt$. From here it is easy to see that $\ell \in \mathbb{B}'$ and

$$\begin{aligned}
\mathbb{E}(\tilde{\ell})^2 &= \mathbb{E}\left[X(T)\ell'(T) - \int_0^T X(t)\ell''(t)dt\right]^2 \\
&= T\ell'(T)^2 - 2\ell'(T)\int_0^t t\ell''(t)dt + \int_0^T\int_0^T (s\wedge t)\ell''(s)\ell''(t)dsdt \\
&= \int_0^T \left|\dot{\ell}(t)\right|^2 dt = |\ell|_{\mathbb{H}}^2\,,
\end{aligned}$$

where x is considered as a standard Brownian motion. It is clear that the set of twice continuously differentiable functions are dense in $\mathbb{H}$. This means that the measure $\mathbb{P}$ is a cylindrical Gaussian measure associated with the Hilbert space $\mathbb{H}$. From the classical construction of the Wiener measure, we know that $\mathbb{P}$ is a σ-additive measure on $\mathbb{B}$. By Theorem 4.7, part (ii), we see that $\mathbb{P}$ is a measurable norm. Thus, $(\mathbb{B}, \mathbb{H}, \mathbb{P})$ is an abstract Wiener space.

In fact, we can prove that the sup norm $\|\cdot\|$ is a measurable norm directly without appealing to the classical construction of the Wiener measure. The remaining part of this section is to give a direct proof that $\|\cdot\|$ is a measurable norm.

On the above Cameron-Martin space, one can define the cylindrical Gaussian measure, denoted still by $\mathbb{P}$. For a cylindrical functional, one can compute the expectation, denoted by $\mathbb{E}$, with respect to this cylindrical measure $\mathbb{P}$. Let $\pi : 0 = t_0 < t_1 < \cdots < t_n = T$ be a partition of $[0,T]$. Let $h \in \mathbb{H}$ and we define $P^\pi h$ as the polygonal approximation of h.

$$h_\pi(t) = P^\pi h(t) = \sum_{k=0}^{n-1}\left[h(t_k) + (h(t_{k+1}) - h(t_k))\frac{t-t_k}{t_{k+1}-t_k}\right]\chi_{\{t_k\le t<t_{k+1}\}}\,.$$

Then

$$\langle x, h_\pi \rangle = \sum_{k=0}^{n-1} (x(t_{k+1}) - x_{t_k}) \frac{h(t_{k+1}) - h(t_k)}{t_{k+1} - t_k} .$$

Let

$$h_k(t) = \begin{cases} t & 0 \le t < t_k \\ t_k & t_k \le t \le T . \end{cases}$$

Then, we know that $\langle x, h_k \rangle = x(t_k)$, $k = 1, 2, \cdots, n$ are mean zero Gaussian random variables with the following covariance matrix $R = (t_i \wedge t_j)$. This also implies that $\{x(t_1), \cdots, x(t_n)\}$ is a martingale. Let

$$\pi' : \quad t_0 = t_{0,0} < t_{0,1} < \cdots < t_{0,l_0} < \\ t_1 = t_{1,0} < t_{1,1} < \cdots < t_{1,l_1} < \cdots < t_n = T$$

be another partition, finer than π. Then, we have

$$\begin{aligned} \|P^\pi x - P^{\pi'} x\| &= \sup_k \sup_j \sup_{t_{k,j} \le t \le t_{k,j+1}} \Bigg| x(t_k) + (x(t_{k+1}) - x(t_k)) \frac{t - t_k}{t_{k+1} - t_k} \\ &\quad - \left[x(t_{k,j}) + (x(t_{k,j+1}) - x(t_{k,j})) \frac{t - t_{k,j}}{t_{k,j+1} - t_{kj}} \right] \Bigg| \\ &= \sup_k \sup_j \sup_{t_{k,j} \le t \le t_{k,j+1}} \left| (x(t_{k+1}) - x(t_k)) \frac{t - t_k}{t_{k+1} - t_k} \right| \\ &\quad + \sup_k \sup_j \sup_{t_{k,j} \le t \le t_{k,j+1}} \left| (x(t_{k,j+1}) - x(t_{k,j})) \frac{t - t_{k,j}}{t_{k,j+1} - t_{kj}} \right| \\ &\quad + \sup_k \sup_j \sup_{t_{k,j} \le t \le t_{k,j+1}} |x(t_k) - x(t_{k,j})| \\ &= I_1 + I_2 + I_3 . \end{aligned}$$

Let us consider I_2 first. It is easy to see that

$$\sup_{t_{k,j} \le t \le t_{k,j+1}} \left| (x(t_{k,j+1}) - x(t_{k,j})) \frac{t - t_{k,j}}{t_{k,j+1} - t_{kj}} \right| \le |x(t_{k,j+1}) - x(t_{k,j})| .$$

Fix an arbitrary $p > 2$. We have

$$\begin{aligned} \mathbb{E}(I_2^p) &\le \mathbb{E} \sup_k \sup_j |x(t_{k,j+1}) - x(t_{k,j})|^p \\ &\le \mathbb{E} \sum_{k=0}^{n-1} \sum_{j=0}^{l_j} \mathbb{E} |x(t_{k,j+1}) - x(t_{k,j})|^p \\ &\le C_p \sum_{k=0}^{n-1} \sum_{j=0}^{l_j} |t_{k,j+1} - t_{k,j}|^{p/2} \\ &\le C_p |\pi'|^{p/2-1} \sum_{k=0}^{n-1} \sum_{j=0}^{l_j} |t_{k,j+1} - t_{k,j}| = C_p T |\pi'|^{p/2-1} . \end{aligned} \tag{4.3.1}$$

Similarly, we see that

$$\begin{aligned}\mathbb{E}\,(I_1^p) &\le \mathbb{E}\sup_k |x(t_{k+1}) - x(t_k)\,|^p \\ &\le C_p \sum_{k=0}^{n-1} |t_{k+1} - t_k|^{p/2} \\ &\le C_p|\pi|^{p/2-1} \sum_{k=0}^{n-1}\sum_{j=0}^{l_j} |t_{k+1} - t_k| = C_p T|\pi|^{p/2-1}\,. \end{aligned} \tag{4.3.2}$$

As a sequence indexed by $j = 0, 1, \cdots, l_j$, $x(t_k) - x(t_{k,j})$ is a martingale. Thus

$$\begin{aligned}\mathbb{E}\,(I_3^p) &\le \mathbb{E}\left(\sup_k \sup_j |x(t_k) - x(t_{k,j})|^p\right) \\ &\le \sum_{k=0}^{n-1} \mathbb{E}\left(\sup_j |x(t_k) - x(t_{k,j})|^p\right) \\ &\le C_p \sum_{k=0}^{n-1} \mathbb{E}\,(|x(t_k) - x(t_{k+1})|^p) \\ &\le C_p \sum_{k=0}^{n-1} |t_{k+1} - t_k|^{p/2} = C_p T|\pi|^{p/2-1}\,. \end{aligned} \tag{4.3.3}$$

Combining (4.3.1), (4.3.2) and (4.3.3), we see that for any $\varepsilon > 0$,

$$\begin{aligned}\mathbb{P}(\|P^\pi x - P^{\pi'} x\| \ge \varepsilon) &\le \frac{1}{\varepsilon^p}\mathbb{E}\left(\|P^\pi x - P^{\pi'} x\|^p\right) \\ &\le \frac{C_p}{\varepsilon^p}\left[\mathbb{E}\,(I_1^p) + \mathbb{E}\,(I_2^p) + \mathbb{E}\,(I_3^p)\right] \\ &\le \frac{C_p}{\varepsilon^p}\left[|\pi| + |\pi'|\right]\,. \end{aligned}$$

This shows that $\|x\| = \sup_{0\le t\le T} |x(t)|$ is a measurable norm on $\mathbb{H}$. □

4.4 Right Tail Estimate

Theorem 4.11. *Let S^{N-1} be the unit sphere in $\mathbb{R}^N$ equipped with the geodesic distance ρ and normalized Haar measure σ_{N-1}. Let A be a Borel set in S^{N-1} and let H be a cap (namely, a ball with respect to the geodesic distance). If $\sigma_{N-1}(A) = \sigma_{N-1}(H)$, then for any $r > 0$,*

$$\sigma_{N-1}(A_r) \ge \sigma_{N-1}(H_r)\,, \tag{4.4.1}$$

where $A_r = \left\{x \in S^{N-1};\ \rho(x, A) < r\right\}$ is the neighborhood of order r of A.

This theorem is an isoperimetric inequality on the sphere. We shall not give proof here. Interested readers may be referred to [Burago and Zalgaller (1988)].

Lemma 4.5. *Let $\sigma_{N-1}^{\sqrt{N}}$ be the uniform normalized Haar measure on the sphere $\sqrt{N}S^{N-1}$ of radius $\sqrt{N}$. Denote $\Pi_{N,d}$ the projection from $\sqrt{N}S^{N-1}$ to $\mathbb{R}^d$. Then, the sequence of measures on $\mathbb{R}^d$, $\left\{\Pi_{N,d}\left(\sigma_{N-1}^{\sqrt{N}}\right), N \geq d\right\}$, converges weakly as $N \to \infty$ to the standard Gaussian measure μ_d in $\mathbb{R}^d$.*

Proof. This lemma is usually credited to Poincaré. Here is a sketch of the proof. Let $Z_1, \cdots, Z_N$ be iid standard real valued Gaussian random variables. Then, the law of $\frac{\sqrt{N}}{\rho_N}(Z_1, \cdots, Z_N)$ on $\sqrt{N}S^{N-1}$ is $\sigma_{N-1}^{\sqrt{N}}$, where $\rho_N = \sqrt{Z_1^2 + \cdots + Z_n^2}$. The law of $\frac{\sqrt{N}}{\rho_N}(Z_1, \cdots, Z_d)$ is $\Pi_{N,d}\left(\sigma_{N-1}^{\sqrt{N}}\right)$. When $N \to \infty$, $\frac{\sqrt{N}}{\rho_N}$ converges to 1 by the strong law of large numbers. Thus, $\frac{\sqrt{N}}{\rho_N}(Z_1, \cdots, Z_d)$ converges in distribution to the law of $(Z_1, \cdots, Z_d)$, namely, the standard normal distribution, by the Slutsky theorem. □

We denote

$$\Phi(t) = \frac{1}{\sqrt{2\pi}} \int_{-\infty}^{t} e^{-\frac{u^2}{2}} du, \quad t \in [-\infty, \infty].$$

Let $\Phi^{-1}(t)$ be the inverse function of Φ and $\Psi(t) = 1 - \Phi(t)$.

Theorem 4.12. *Let A be a Borel set in $\mathbb{R}^d$ and let H be a half-space, namely, $H = \left\{x \in \mathbb{R}^d;\ \langle x, u\rangle < \lambda\right\}$ for some $u \in \mathbb{R}^d$ and $\lambda \in [-\infty, \infty]$. Let A_r be a neighborhood of order r of A, namely, $A_r = \left\{x \in \mathbb{R}^d, \rho(x, A) < r\right\}$ and let H_r be a neighborhood of order r of H. If $\mu_d(A) = \mu_d(H)$, then for any $r > 0$,*

$$\mu_d(A_r) \geq \mu_d(H_r) \quad \text{or equivalently} \quad \Phi^{-1}(\mu_d(A_r)) \geq \Phi^{-1}(\mu_d(A)) + r. \tag{4.4.2}$$

In particular, if $\mu_d(A) \leq 1/2$, then

$$1 - \mu_d(A_r) \leq \Psi(r) \leq \frac{1}{2} e^{-r^2/2}. \tag{4.4.3}$$

Proof. We can choose $\|u\| = 1$ with a proper choice of λ in the definition of H. Thus

$$\mu_d(H_r) = \mu_d\left(\langle x, u\rangle < \lambda + r\right) = \frac{1}{\sqrt{2\pi}} \int_{-\infty}^{\lambda + r} e^{-\frac{u^2}{2}} du.$$

This means that $\Phi^{-1}(\mu_d(H_r)) = \lambda + r$. Similarly, we have $\Phi^{-1}(\mu_d(H)) = \lambda$. Therefore,

$$\Phi^{-1}(\mu_d(H_r)) = \Phi^{-1}(\mu_d(H)) + r = \Phi^{-1}(\mu_d(A)) + r\,.$$

This proves the equivalence relation in (4.4.2).

Now we prove the theorem. Since $\Phi^{-1}(0) = -\infty$, we may assume $a = \Phi^{-1}(\mu_d(A)) > -\infty$. Let $b \in (-\infty, a)$. Then, by Poincaré lemma 4.5 we have for every sufficiently large $k \geq d$,

$$\sigma_{k-1}^{\sqrt{k}}\left(\Pi_{k,d}^{-1}(A)\right) > \sigma_{k-1}^{\sqrt{k}}\left(\Pi_{k,1}^{-1}((-\infty, b])\right)\,.$$

It is easy to verify that $\Pi_{k,d}^{-1}(A_r) \supseteq \left(\Pi_{k,d}^{-1}(A)\right)_r$, where the neighborhood of order r on the right hand side is understood with respect to the geodesic distance on $\sqrt{k}S^{k-1}$. Since $\Pi_{k,1}^{-1}((-\infty, b])$ is a cap on $\sqrt{k}S^{k-1}$, by Theorem 4.11 (the isoperimetric inequality on the spheres) we have

$$\sigma_{k-1}^{\sqrt{k}}\left(\Pi_{k,d}^{-1}(A_r)\right) \geq \sigma_{k-1}^{\sqrt{k}}\left(\left(\Pi_{k,d}^{-1}(A)\right)_r\right) \geq \sigma_{k-1}^{\sqrt{k}}\left(\left(\Pi_{k,1}^{-1}((-\infty, b])\right)_r\right)\,.$$

Now $\left(\Pi_{k,1}^{-1}((-\infty, b])\right)_r = \Pi_{k,1}^{-1}((-\infty, b + r(k)])$ for some $r(k) \geq 0$ satisfying $\lim_{k\to\infty} r(k) = r$. Therefore, taking the limit $k \to \infty$, we see $\mu_d(A_r) \geq \Phi(b + r)$. Letting $b \uparrow \Phi^{-1}(\mu_d(A))$ yields $\mu_d(A_r) \geq \Phi(\Phi^{-1}(\mu_d(A)) + r) = \Phi(\Phi^{-1}(\mu_d(H)) + r) = \Phi(\lambda + r) = \mu_d(H_r)$. □

The extension of the above theorem to infinite dimension is also straightforward.

Theorem 4.13. *Let $(\mathbb{B}, \mathbb{H}, \mathbb{P})$ be an abstract Wiener space and let A be any measurable subset of $\mathbb{B}$ with $\mu(A) = \Phi(a)$, where $a \in [-\infty, \infty]$. Then*

$$\mathbb{P}(A + tB_{\mathbb{H}}(1)) \geq \Phi(a + t) \quad \forall\, t \geq 0\,, \tag{4.4.4}$$

where $B_{\mathbb{H}}(r) = \{h \in \mathbb{H}\,,\ \|h\| \leq r\}$ is the closed ball of center 0 and radius r in $\mathbb{H}$. In particular, if $\mathbb{P}(A) \geq 1/2$, then

$$\mathbb{P}(A + tB_{\mathbb{H}}(1)) \geq \Phi(t) \quad \forall\, t \geq 0\,, \tag{4.4.5}$$

Let $(\mathbb{B}, \mathbb{H}, \mathbb{P})$ be an abstract Wiener space. Let X be a Gaussian random variable with values in the Banach space $\mathbb{B}$. Assume that the mean of X is 0 and variance is given by (the structure of) $\mathbb{H}$. This means that for any $\ell \in \mathbb{B}' \subseteq \mathbb{H}$, $\langle X, \ell\rangle$ is a mean zero Gaussian random variable with variance $|\ell|_{\mathbb{H}}^2$. The median $M(X)(> 0)$ of $\|X\|$ is a number such that

$$\mathbb{P}(\|X\| \leq M(X)) = \mathbb{P}(\|X\| \geq M(X)) = \frac{1}{2}\,.$$

Define

$$\sigma(X) := \sup_{\ell\in\mathbb{B}':|\ell|_{\mathbb{H}}\le 1} \left\{\mathbb{E}\left(\langle X,\ell\rangle\right)^2\right\}^{1/2} . \tag{4.4.6}$$

Lemma 4.6. *Let X be a Gaussian random variable with values in $\mathbb{B}$. Then, we have*

$$\mathbb{P}\left(|\|X\|-M(X)|>t\right)\le \Psi\left(\frac{t}{\sigma(X)}\right)\le \exp\left(-\frac{t^2}{2\sigma^2(X)}\right) \quad \forall\, t>0 . \tag{4.4.7}$$

Proof. Using a density argument, we can reduce the proof of the theorem to finite dimensional case. We assume that $\mathbb{B}=\mathbb{R}^d$ and $X=(X_1,\cdots,X_d)$ is a Gaussian vector with values in $\mathbb{R}^d$. The covariance matrix Σ of X is a symmetric positive definite matrix. We can write $\Sigma=P^TDP$, where P is an orthogonal matrix, $P^T(=P^{-1})$ is its transpose and D is a diagonal matrix with elements $0<\sigma_1^2\le\cdots\le\sigma_d^2<\infty$. Define $Y=PX$. Then, $\|X\|=\|Y\|$ and the covariance matrix of Y is D. Namely, $Y_1,\cdots,Y_d$ are independent with variance σ_1^2, $\cdots$, σ_d^2. In other words, $Z_1=Y_1/\sigma_1$, $\cdots$, $Z_d=Y_d/\sigma_d$ are d-dimensional standard Gaussian. It is straightforward to see that $\sigma(X)=\sigma_d$, which is denoted by σ. Denote $M=M(X)$ and

$$A=\{Y:=(Y_1,\cdots,Y_d)\,;\ |Y_d|\le M\}=\{(Z_1,\cdots,Z_d)\,,|Z_d|\le M/\sigma\} .$$

Then, $\mu(A)\ge 1/2$. If we set

$$A_{t/\sigma}=\left\{(Z_1,\cdots,Z_d)\,,|Z_d|\le \frac{M}{\sigma}+\frac{t}{\sigma}\right\} ,$$

then

$$A_{t/\sigma}^c=\{|Y_d|\ge M+t\}\supseteq\{\|X\|\ge\|Y\|\ge M+t\} .$$

Thus, from the isoperimetric inequality (4.4.3) (in fact, we only need the one dimensional case, which is much easier to prove), we have

$$\mu\{\ \|X\|\ge M+t\}\le 1-\mu(A_{t/\sigma})\le \frac{1}{2}e^{-\frac{t^2}{2\sigma^2}} .$$

Applying the same argument to $A_1:=\{(Y_1,\cdots,Y_d)\,;|Y_1|\ge M\}$ with σ replaced by σ_1 yield

$$\mu\{\|X\|\ge M-t\}\le\frac{1}{2}e^{-\frac{t^2}{2\sigma_1^2}}\le\frac{1}{2}e^{-\frac{t^2}{2\sigma^2}}$$

since $\sigma_1\le\sigma$. Noticing $\sigma=\sigma(X)$ proves the lemma. □

Theorem 4.14. *Let X be a Gaussian random variable in the Banach space $\mathbb{B}$ with the corresponding $\sigma(X)$. Then*

$$\lim_{t\to\infty}\frac{1}{t^2}\log\mathbb{P}\left(\|X\|>t\right)=-\frac{1}{2\sigma^2(X)}. \tag{4.4.8}$$

Or equivalently,

$$\mathbb{E}\exp\left\{\lambda\|X\|^2\right\}<\infty \quad \textit{if and only if} \quad \lambda<\frac{1}{2\sigma^2(X)}. \tag{4.4.9}$$

Proof. Since X is Radon, the median M of $\|X\|$ exists and is finite. Thus

$$\mathbb{P}\left(\|X\|>t\right)\le\mathbb{P}\left(\|X-M\|>t-M\right)=e^{-\frac{(t-M)^2}{2\sigma^2(X)}}.$$

This proves $\limsup\limits_{t\to\infty}\frac{1}{t^2}\log\mathbb{P}\left(\|X\|>t\right)\le-\frac{1}{2\sigma^2(X)}$.

From the definition of $\sigma(X)$, for any $\varepsilon>0$, there is an $\ell_\varepsilon\in\mathbb{B}'$ with $|\ell_\varepsilon|_{\mathbb{H}}=1$ such that the variance of $\langle X,\ell_\varepsilon\rangle$ is greater than or equal to $\sigma_\varepsilon^2=(\sigma(X)-\varepsilon)^2$. Without loss of generality, we can assume that its mean is 0. Thus

$$\mathbb{P}\left(\|X\|>t\right)\ge\mathbb{P}\left(\langle X,\ell_\varepsilon\rangle>t\right)=2\Psi\left(\frac{t}{\sigma_\varepsilon}\right).$$

Combining this with a passage of $\varepsilon\downarrow 0$ yields

$$\limsup_{t\to\infty}\log\mathbb{P}\left(\|X\|>t\right)\ge-\frac{1}{2\sigma^2(X)}.$$

This shows the theorem. The equivalence between (4.4.8) and (4.4.9) is obvious. □

Remark 4.4. Inequality (4.4.9) is also called the Fernique inequality (without the precise constant), obtained first by [Fernique (1970)]. It is interesting to give a different proof.

Theorem 4.15. *Let $(\mathbb{B},\mathbb{H},\mathbb{P})$ be an abstract Wiener space. There is an $\gamma_0>0$ such that*

$$\mathbb{E}\left(e^{\gamma_0\|x\|^2}\right)=\int_{\mathbb{B}}e^{\gamma_0\|x\|^2}\mathbb{P}(dx)<\infty. \tag{4.4.10}$$

Proof. Let $X:(\Omega,\mathcal{F},P)\to\mathbb{B}$ and $Y:(\Omega,\mathcal{F},P)\to\mathbb{B}$ be two independent $\mathbb{B}$-valued random variables with the law $\mathbb{P}$. It is easy to verify that $\frac{X+Y}{\sqrt{2}}$

and $\frac{X-Y}{\sqrt{2}}$ are independent with law $\mathbb{P}$. Let $t > s$. We have

$$\begin{aligned}
\mathbb{P}(\|x\| \le s)\,\mathbb{P}(\|x\| > t) &= P(\|\frac{X+Y}{\sqrt{2}}\| \le s) P\left(\|\frac{X-Y}{\sqrt{2}}\| > t\right) \\
&= P\left(\|\frac{X+Y}{\sqrt{2}}\| \le s\,,\ \|\frac{X-Y}{\sqrt{2}}\| > t\right) \\
&\le P\left(|\|X\| - \|Y\|| \le \sqrt{2}s\,,\ \|X\| + \|Y\| > \sqrt{2}t\right) \\
&\le P\left(\|X\| \ge \frac{t-s}{\sqrt{2}}\,,\ \|Y\| > \frac{t-s}{\sqrt{2}}\right) \\
&= \left[P\left(\|X\| \ge \frac{t-s}{\sqrt{2}}\right)\right]^2,
\end{aligned}$$

where the above last inequality follows from the fact that

$$\left\{(x,y)\,, x\,, y \ge 0\,, |x-y| \le \sqrt{2}s\,, |x+y| \ge \sqrt{2}t\right\} \subseteq \left\{(x,y)\,, x\,, y \ge \frac{t-s}{\sqrt{2}}\right\}.$$

Thus, we have

$$\mathbb{P}(\|x\| \ge t) \le \frac{\left[\mathbb{P}\left(\|x\| \ge \frac{t-s}{\sqrt{2}}\right)\right]^2}{\mathbb{P}(\|x\| \le s)}. \tag{4.4.11}$$

Define

$$t_0 = s > 0\,, \quad t_{n+1} = s + \sqrt{2}t_n\,, \quad \alpha_n = \frac{\left[\mathbb{P}\left(\|x\| \ge \frac{t_n - s}{\sqrt{2}}\right)\right]^2}{P(\|x\| \le s)}.$$

Then, from (4.4.11) it follows

$$\begin{aligned}
t_n &= (1 + \sqrt{2} + \cdots + (\sqrt{2})^n)s = \frac{2^{\frac{n+1}{2}} - 1}{\sqrt{2}-1}s \le \frac{2^{\frac{n+1}{2}}}{\sqrt{2}-1}s\,, \\
\mathbb{P}(\|x\| \ge t_n) &= \alpha_n \le \alpha_{n-1}^2 \le \alpha_0^{2^n} = e^{2^n \log \alpha_0}\,,
\end{aligned}$$

where

$$\alpha_0 = \frac{\mathbb{P}(\|x\| > s)}{\mathbb{P}(\|x\| \le s)} = \frac{\mathbb{P}(\|x\| > s)}{1 - \mathbb{P}(\|x\| > s)}. \tag{4.4.12}$$

For any $\gamma_0 > 0$, we have

$$\begin{aligned}\mathbb{E}\left[e^{\gamma_0\|x\|^2}\right] &= \mathbb{E}\left[e^{\gamma_0\|x\|^2}\chi_{\{\|x\|<t_0\}}\right] + \sum_{n=0}^{\infty}\mathbb{E}\left[e^{\gamma_0\|x\|^2}\chi_{\{t_n\le\|x\|<t_{n+1}\}}\right] \\ &\le e^{\gamma_0 t_0^2} + \sum_{n=0}^{\infty} e^{\gamma_0 t_{n+1}^2}\mathbb{P}\{t_n \le \|x\| < t_{n+1}\} \\ &\le e^{\gamma_0 s^2} + \sum_{n=0}^{\infty} e^{\gamma_0 t_{n+1}^2}\mathbb{P}\{\|x\| \ge t_n\} \\ &\le e^{\gamma_0 s^2} + \sum_{n=0}^{\infty} e^{\gamma_0 t_{n+1}^2 + 2^n\log(\alpha_0)}\,.\end{aligned}$$

From the definition of t_n, we have that as $n \to \infty$

$$\text{The leading term of } \gamma_0 t_{n+1}^2 + 2^n \log(\alpha_0) = \left(\frac{4\gamma_0 s^2}{(\sqrt{2}-1)^2} + \log(\alpha_0)\right)\cdot 2^n\,.$$

Since $\|\cdot\|$ is measurable and $\mathbb{P}\{\|x\| \ge s\} \downarrow 0$ as $s \uparrow \infty$. One can choose s sufficiently large such that $\alpha_0 < 1$. Thus $\log(\alpha_0) < 0$. Fix one such s, one can choose γ_0 sufficiently small such that we still have $\frac{4\gamma_0 s^2}{(\sqrt{2}-1)^2} + \log(\alpha_0) < 0$. This shows that (4.4.13) is finite. □

An immediate consequence is

Corollary 4.3. *Let $(\mathbb{B}, \mathbb{H}, \mathbb{P})$ be an abstract Wiener space. Then, for any $\alpha \in \mathbb{R}$,*

$$\mathbb{E}\left(e^{\alpha\|x\|}\right) < \infty\,. \tag{4.4.13}$$

Proof. Let $\lambda > 0$ satisfy (4.4.9). For any $\alpha \in \mathbb{R}$, there is a positive constant $C_{\alpha,\lambda}$ such that

$$\alpha x \le \lambda x^2 + C_{\alpha,\lambda}\,.$$

This immediately implies (4.4.13). □

4.5 Small Ball (Left Tail) Estimate

Definition 4.8. Let (E, d) be a metric space and let A be a compact subset of E. The d-metric entropy of A is defined as

$$H(\varepsilon, A) = \log N(\varepsilon, A)\,, \tag{4.5.1}$$

where

$$N(\varepsilon, A) = \min\Big\{n \geq 1\,;\ \text{there are points } a_1, \cdots, a_n \in E \\ \text{such that } \bigcup_{i=1}^{n} B(a_i, \varepsilon) \supset A\Big\}$$

and $B(a_i,\varepsilon) = \{x \in E\,;\ d(x,a_i) < \varepsilon\}$ is the open ball (in E) of radius ε centered at a_i.

Lemma 4.7. *Let $(\mathbb{B},\mathbb{H},\mathbb{P})$ be an abstract Wiener space and let $B_\varepsilon(h) = \{x \in \mathbb{B}\,;\ \|x-h\| < \varepsilon\}$. Let $K = \{h \in \mathbb{H}\,; |h|_{\mathbb{H}} \leq 1\}$ be the unit ball in $\mathbb{H}$. For any $\lambda > 0$ and $\varepsilon > 0$, we have*

$$H(2\varepsilon, \lambda K) \leq \lambda^2/2 - \log \mathbb{P}(B_\varepsilon(0)) \tag{4.5.2}$$

and

$$H(\varepsilon, \lambda K) + \log \mathbb{P}(B_{2\varepsilon}(0)) \geq \log \Phi(\lambda + \alpha_\varepsilon)\,, \tag{4.5.3}$$

where $\Phi(t) = \frac{1}{\sqrt{2\pi}}\int_{-\infty}^{t} e^{-\frac{u^2}{2}}du$ and $\Phi(\alpha_\varepsilon) = \mathbb{P}(B_\varepsilon(0))$.

Proof. Define

$$\begin{aligned} M(\varepsilon,\lambda K) &= \max\Big\{n \geq 1\,;\ \exists\ h_1,\cdots,h_n \in \lambda K\,, |h_i - h_j| \geq 2\varepsilon \quad \forall i \neq j\Big\}\,;\\ N(\varepsilon,\lambda K) &= \min\left\{k \geq 1\,;\ \exists\ h_1,\cdots,h_k \in \lambda K\,, \bigcup_{j=1}^{k} B_\varepsilon(h_j) \supseteq \lambda K\right\}. \end{aligned} \tag{4.5.4}$$

Let $E(\varepsilon,\lambda K)$ and $F(\varepsilon,\lambda K)$ be two such finite sets with the cardinalities $M(\varepsilon,\lambda K)$ and $N(\varepsilon,\lambda K)$. Namely,

$$\#E(\varepsilon,\lambda K) = M(\varepsilon,\lambda K)\,, \quad \#F(\varepsilon,\lambda K) = N(\varepsilon,\lambda K)\,.$$

Thus, $E(\varepsilon,\lambda K)$ and $F(\varepsilon,\lambda K)$ have the following properties: for any $g,h \in E(\varepsilon,\lambda K)$, $\|g-h\| \geq 2\varepsilon$ and $\bigcup_{h\in F(\varepsilon,\lambda K)} B_\varepsilon(h) \supseteq \lambda K$. From the Cameron-Martin formula and then from the Jensen's inequality, it follows that for any $h \in \lambda K$,

$$\begin{aligned} \mathbb{P}(B_\varepsilon(h)) &= \int_{B_\varepsilon(0)} e^{-\langle w,h\rangle - \frac{|h|^2}{2}}\mathbb{P}(dw)\\ &= e^{-\frac{|h|^2}{2}}\mathbb{P}(B_\varepsilon(0))\frac{1}{\mathbb{P}(B_\varepsilon(0))}\int_{B_\varepsilon(0)} e^{-\langle w,h\rangle}\mathbb{P}(dw)\\ &\geq e^{-\frac{\lambda^2}{2}}\mathbb{P}(B_\varepsilon(0))\exp\left\{-\frac{1}{\mathbb{P}(B_\varepsilon(0)}\int_{B_\varepsilon(0)}\langle w,h\rangle\mathbb{P}(dw)\right\}\\ &= e^{-\frac{\lambda^2}{2}}\mathbb{P}(B_\varepsilon(0))\,. \end{aligned} \tag{4.5.5}$$

From the definition of $M(\varepsilon, \lambda K)$ and $E(\varepsilon, \lambda K)$, we have

$$M(\varepsilon, \lambda K) \min_{h \in E(\varepsilon, \lambda K)} \mathbb{P}(B_\varepsilon(h)) \le 1 . \tag{4.5.6}$$

This combined with (4.5.5) gives

$$\log M(\varepsilon, \lambda K) - \frac{\lambda^2}{2} + \log(\mathbb{P}(B_\varepsilon(0)) \le 0 . \tag{4.5.7}$$

We can choose $E(\varepsilon, \lambda K)$ to be maximal which implies

$$\bigcup_{h \in E(\varepsilon, \lambda K)} B_{2\varepsilon}(h) \supseteq \lambda K . \tag{4.5.8}$$

By the definition of the entropy this means

$$H(2\varepsilon, \lambda K) \le \log(\# E(\varepsilon, \lambda K)) = \log M(\varepsilon, \lambda K) . \tag{4.5.9}$$

This together with (4.5.7) yields (4.5.2).

From $\bigcup_{h \in F(\varepsilon, \lambda K)} B_\varepsilon(h) \supseteq \lambda K$ it follows that for any $x \in \lambda K$, there is an $h_x \in F(\varepsilon, \lambda K)$ such that $\|h_x - x\| < \varepsilon$. Thus, for any element $b \in B_\varepsilon(0)$, we have

$$\|h_x - (x + b)\| \le \|h_x - x\| + \|b\| < 2\varepsilon .$$

This means that

$$\bigcup_{h \in F(\varepsilon, \lambda K)} B_{2\varepsilon}(h) \supseteq \lambda K + B_\varepsilon(0) .$$

Hence,

$$N(\varepsilon, \lambda K) \max_{h \in F(\varepsilon, \lambda K)} \mathbb{P}(B_{2\varepsilon}(h)) \ge \mathbb{P}(\lambda K + B_\varepsilon(0)) \ge \Phi(\lambda + \alpha_\varepsilon) , \tag{4.5.10}$$

where the above second inequality follows from the concentration inequality (4.4.4). From the Anderson's inequality (3.3.23), we have for any $h \in \mathbb{H}$, $\mathbb{P}(B_{2\varepsilon}(h)) \le \mathbb{P}(B_{2\varepsilon}(0))$. Hence, (4.5.10) gives

$$N(\varepsilon, \lambda K) \mathbb{P}(B_{2\varepsilon}(0)) \ge \Phi(\lambda + \alpha_\varepsilon)$$

which yields (4.5.3). □

We shall use $A \preccurlyeq B$ (or $B \succcurlyeq$) to represent that there is a constant C such that $A \le CB$. We denote $A \approx B$ if $A \preccurlyeq B$ and $B \preccurlyeq A$. The following theorem is from [Kuelbs and Li (1993)] and [Li and Linde (1999)].

Theorem 4.16. *Let $(\mathbb{B}, \mathbb{H}, \mathbb{P})$ be an abstract Wiener space and let $K = \{h \in \mathbb{H} ; |h|_{\mathbb{H}} \le 1\} \subseteq \mathbb{B}$ be the unit ball in $\mathbb{H}$. Let $f(1/x)$ and $g(1/x)$ be two regularly varying functions at ∞ and let J be a slowly varying function at infinity such that $J(x) \approx J(x^\rho)$ as $x \to \infty$ for each $\rho > 0$. Define*

$$\varphi(\varepsilon) = - \log \mathbb{P}(B_\varepsilon(0)) . \tag{4.5.11}$$

Then, we have the following statements.

(i) It holds

$$H(\varepsilon/\sqrt{2\varphi(\varepsilon)}, K) \succcurlyeq \varphi(2\varepsilon) \quad as \quad \varepsilon \downarrow 0 . \tag{4.5.12}$$

In particular, if $\varphi(\varepsilon) \approx \varepsilon^{-\alpha} J(\varepsilon^{-1})$ *for some* $\alpha > 0$, *then*

$$H(\varepsilon, K) \succcurlyeq \left[\varepsilon^{-\alpha} J(\varepsilon^{-1})\right]^{2/(2+\alpha)} . \tag{4.5.13}$$

(ii) If $\varphi(x) \preccurlyeq f(x)$, *then*

$$H(\varepsilon/\sqrt{f(\varepsilon)}, K) \preccurlyeq f(\varepsilon) . \tag{4.5.14}$$

In particular, if $\varphi(x) \preccurlyeq \varepsilon^{-\alpha} J(\varepsilon^{-1})$, *then*

$$H(\varepsilon, K) \preccurlyeq \left[\varepsilon^{-\alpha} J(\varepsilon^{-1})\right]^{2/(2+\alpha)} . \tag{4.5.15}$$

(iii) If $H(\varepsilon, K) \succcurlyeq g(\varepsilon)$, *then*

$$\varphi(\varepsilon) \succcurlyeq g(\varepsilon/\sqrt{\varphi(\varepsilon)}) . \tag{4.5.16}$$

In particular, if $H(\varepsilon, K) \succcurlyeq \varepsilon^{-\alpha} J(\varepsilon^{-1})$ *with* $0 < \alpha < 2$, *then*

$$\varphi(\varepsilon) \succcurlyeq \left[\varepsilon^{-\alpha} J(\varepsilon^{-1})\right]^{2/(2-\alpha)} . \tag{4.5.17}$$

(iv) If $H(\varepsilon, K) \preccurlyeq g(\varepsilon)$, *then*

$$\varphi(2\varepsilon) \preccurlyeq g(\varepsilon/\sqrt{\varphi(\varepsilon)}) . \tag{4.5.18}$$

In particular, if $H(\varepsilon, K) \preccurlyeq \varepsilon^{-\alpha} J(\varepsilon^{-1})$ *with* $0 < \alpha < 2$, *then*

$$\varphi(\varepsilon) \preccurlyeq \left[\varepsilon^{-\alpha} J(\varepsilon^{-1})\right]^{2/(2-\alpha)} . \tag{4.5.19}$$

Proof. It is easy to check by the definition of the metric entropy that $H(\varepsilon, \lambda K) = H(\varepsilon/\lambda, K)$. Inequality (4.5.3) can be rewritten as

$$H(\varepsilon, \lambda K) \geq \varphi(2\varepsilon) + \log \Phi(\lambda + \alpha_\varepsilon) .$$

With the choice $\lambda = -\alpha_\varepsilon$ this gives

$$H(\varepsilon(-\alpha_\varepsilon)^{-1}, K) \geq \varphi(2\varepsilon) + \log \Phi(0) = \varphi(2\varepsilon) + \log(1/2) . \tag{4.5.20}$$

Recall the definition of α_ε: $\Phi(\alpha_\varepsilon) = \mathbb{P}(B_\varepsilon(0))$. We see that $\alpha_\varepsilon \to -\infty$ as $\varepsilon \to 0$. From the property of Φ we see that

$$-\frac{\alpha_\varepsilon^2}{2} \approx -\log \mathbb{P}(B_\varepsilon(0)) = \varphi(\varepsilon) .$$

Hence,

$$-\alpha_\varepsilon \approx \sqrt{2\varphi(\varepsilon)} . \tag{4.5.21}$$

Thus, (4.5.20) can be written as

$$H(\varepsilon/\sqrt{2\varphi(\varepsilon)}, K) \geq \varphi(2\varepsilon) + \log(1/2) \geq \varphi(2\varepsilon)\,.$$

This implies (4.5.12).

With the choice $\lambda = 2\sqrt{\varphi(\varepsilon)}$ inequality (4.5.2) implies

$$H(\varepsilon/\sqrt{\varphi(\varepsilon)}, K) \leq 3\varphi(\varepsilon) \preccurlyeq f(\varepsilon)\,.$$

This is (4.5.14).

If $H(\varepsilon, K) \succcurlyeq g(\varepsilon)$, then (4.5.2) implies

$$g(2\varepsilon/\lambda) \preccurlyeq \frac{\lambda^2}{2} + \varphi(\varepsilon)\,.$$

Letting $\lambda = 2\sqrt{\varphi(\varepsilon)}$ yields $g(\varepsilon/\sqrt{\varphi(\varepsilon)}) \preccurlyeq 3\varphi(\varepsilon)$. This is (4.5.16).

If $H(\varepsilon, K) \preccurlyeq g(\varepsilon)$, then (4.5.3) implies

$$g(\varepsilon/\lambda) - \varphi(2\varepsilon) \succcurlyeq \log\Phi(\lambda + \alpha_\varepsilon)\,.$$

Take $\lambda = -\alpha_\varepsilon$ to obtain

$$g(-\varepsilon/\alpha_\varepsilon) \succcurlyeq \varphi(2\varepsilon) + \log\Phi(0) = \varphi(2\varepsilon) + \log(1/2) \geq \varphi(2\varepsilon)\,.$$

Inequality (4.5.18) is proved by (4.5.21).

Let $\delta = \varepsilon/\sqrt{2\varphi(\varepsilon)}$ and define the function $\varepsilon = h(\delta)$ by $\varepsilon = h(\delta) = h(\varepsilon/\sqrt{2\varphi(\varepsilon)})$. If (4.5.12) is satisfied, then

$$H(\delta, K) \succcurlyeq \varepsilon^{-\alpha} J(\varepsilon^{-1}) \succcurlyeq (h^{-1}(\delta))^{-\alpha} J(\frac{1}{h^{-1}(\delta)})\,, \tag{4.5.22}$$

where h^{-1} is the inverse function of h.

Now if $\varphi(x) \approx \varepsilon^{-\alpha} J(\varepsilon^{-1})$ then $\delta \approx \varepsilon^{1+\alpha/2}(J(\varepsilon^{-1})^{-1/2}$. This implies $\varepsilon = h^{-1}(\delta) \approx (\delta\sqrt{J(\delta^{-1})})^{\frac{2}{2+\alpha}}$ since J is slowly varying at infinity and $J(x) \approx J(x^\rho)$ as $x \to \infty$ for each $\rho > 0$. Substituting this asymptotics into (4.5.22) we have

$$H(\delta, K) \succcurlyeq \left[\delta^{-\alpha} J(\delta^{-1})\right]^{2/(2+\alpha)}\,.$$

This is (4.5.13) since $\delta \to 0$.

Inequality (4.5.15) can be proved in a similar way. Inequality (4.5.17) is a straightforward consequence of (4.5.16).

We shall prove (4.5.19) under the condition that there is a $\gamma < \infty$ such that $\varphi(\varepsilon) \preccurlyeq \varepsilon^{-\gamma}$. This additional condition can be proved under our assumption (see e.g. [Li and Linde (1999), Proposition 2.4]). Under the condition of statement of (iv) of the theorem, we can take $g(\varepsilon) = \varepsilon^{-\alpha} J(\varepsilon^{-1})$ in inequality (4.5.18) to obtain

$$\varphi(2\varepsilon) \preccurlyeq \varepsilon^{-\alpha} \varphi(\varepsilon)^{\alpha/2} J\left(\sqrt{\varphi(\varepsilon)}/\varepsilon\right)\,.$$

Or

$$\varphi(\varepsilon) \preccurlyeq \varepsilon^{-\alpha}\varphi(\varepsilon/2)^{\alpha/2} J\left(2\sqrt{\varphi(\varepsilon/2)}/\varepsilon\right) . \tag{4.5.23}$$

From the assumption that $1 \le \varphi(x) \preccurlyeq x^{-\varepsilon}$ and from (A.20), we see

$$J(4\sqrt{\varphi(\varepsilon/2)}/\varepsilon) \le C \sup_{1\le\rho\le\gamma'} J(1/\varepsilon^{\rho}) \le CJ(1/\varepsilon) \tag{4.5.24}$$

for some $\gamma' > 1 + \gamma/2$. Substituting this inequality (4.5.24) into (4.5.23) we have

$$\log \varphi(\varepsilon) \le \frac{\alpha}{2} \log \varphi(\varepsilon/2) + \log \psi(\varepsilon) , \tag{4.5.25}$$

where $\psi(\varepsilon) = C\varepsilon^{-\alpha} J(1/\varepsilon)$. Iterating the above inequality (4.5.25) we obtain

$$\begin{aligned}\log \varphi(\varepsilon) &\le \left(\frac{\alpha}{2}\right)^{n} \log \varphi(\varepsilon/2^{n}) + \sum_{j=0}^{n-1} \left(\frac{\alpha}{2}\right)^{j} \log \psi(\varepsilon/2^{j}) \\ &\le \left(\frac{\alpha}{2}\right)^{n} \log(C2^{n\gamma}/\varepsilon^{\gamma}) + \sum_{j=0}^{n-1} \left(\frac{\alpha}{2}\right)^{j} \log \psi(\varepsilon/2^{j}) . \end{aligned} \tag{4.5.26}$$

In the above inequality we used $\varphi(\varepsilon) \preccurlyeq \varepsilon^{-\gamma}$. Since $\alpha < 2$, the first summand in (4.5.26) goes to zero when $n \to \infty$. Thus, we have

$$\begin{aligned}\log \varphi(\varepsilon) &\le \sum_{j=0}^{\infty} \left(\frac{\alpha}{2}\right)^{j} \log \psi(\varepsilon/2^{j}) \\ &= \frac{2}{2-\alpha} \log \psi(\varepsilon) + \sum_{j=0}^{\infty} \left(\frac{\alpha}{2}\right)^{j} \log \frac{\psi(\varepsilon/2^{j})}{\psi(\varepsilon)} .\end{aligned}$$

Application of the inequality (A.19) to the slowly varying function J yields $J(2^{j}\varepsilon^{-1})/J(\varepsilon^{-1}) \le 2^{j+1}$. Thus, we have

$$\frac{\psi(\varepsilon/2^{j})}{\psi(\varepsilon)} = 2^{\alpha j} J(2^{j}\varepsilon^{-1})/J(\varepsilon^{-1}) \le 2^{\alpha j + j + 1} .$$

Consequently, we have

$$\sum_{j=0}^{\infty} \left(\frac{\alpha}{2}\right)^{j} \log \frac{\psi(\varepsilon/2^{j})}{\psi(\varepsilon)} \le \sum_{j=0}^{\infty} \left(\frac{\alpha}{2}\right)^{j} \log 2^{\alpha j + j + 1} < \infty .$$

Hence,

$$\log \varphi(\varepsilon) \le \frac{2}{2-\alpha} \log \psi(\varepsilon) + c .$$

This implies

$$\varphi(\varepsilon) \preccurlyeq \psi(\varepsilon)^{\frac{2}{2-\alpha}} \preccurlyeq \left(\varepsilon^{-\alpha} J(\varepsilon^{-1})\right)^{\frac{2}{2-\alpha}}$$

which is (4.5.19). □

Chapter 5

Nonlinear Functionals on Abstract Wiener Space

5.1 Fock Space and Chaos Expansion

On an abstract Wiener space $(\mathbb{B}, \mathbb{H}, \mathbb{P})$, we see that $\tilde{h}_1, \cdots \tilde{h}_d$ are joint centered Gaussian variables of mean 0 and covariance $(\langle h_i, h_j\rangle)_{1\le i,j\le d}$. Clearly, for any measurable function $f : \mathbb{R}^d \to \mathbb{R}$, $f(\tilde{h}_1, \cdots, \tilde{h}_d)$ is also a random variable on $\mathbb{B}$. We shall see that all random variables in $L^2(\mathbb{B})$ can be described by the Fock space on $\mathbb{H}$. To define the Fock space, first we introduce the tensor product of two Hilbert spaces.

Let $\mathbb{H}_1$ and $\mathbb{H}_2$ be two Hilbert spaces with scalar products $\langle \cdot, \cdot\rangle_1$ and $\langle \cdot, \cdot\rangle_2$. For any elements $\varphi \in \mathbb{H}_1$ and $\psi \in \mathbb{H}_2$, we define the tensor product of φ and ψ as a bilinear form on $\mathbb{H}_1 \times \mathbb{H}_2$, given by

$$\varphi \otimes \psi(\xi_1, \xi_2) = \langle \varphi, \xi_1\rangle_1 \langle \psi, \xi_1\rangle_2 \quad \forall\, \xi_1 \in \mathbb{H}_1\,,\ \xi_2 \in \mathbb{H}_2\,. \tag{5.1.1}$$

Denote the vector space generated by the linear span of such tensor product by

$$\mathbb{H}_1 \otimes_{\text{pre}} \mathbb{H}_2 = \left\{ \sum_{j=1}^{n} \varphi_j \otimes \psi_j\,, \quad \varphi_j \in \mathbb{H}_1\,,\ \psi_j \in \mathbb{H}_2\,, n \in \mathbb{N} \right\}.$$

If $\varphi_1, \varphi_2 \in \mathbb{H}_1$ and $\psi_1, \psi_2 \in \mathbb{H}_2$, we define

$$\langle \varphi_1 \otimes \psi_1\,, \varphi_2 \otimes \psi_2\rangle_{\mathbb{H}_1\otimes\mathbb{H}_2} := \langle \varphi_1\,, \varphi_2\rangle_{\mathbb{H}_1} \langle \psi_1\,, \psi_2\rangle_{\mathbb{H}_2}\,.$$

By (bi)linearity, we can extend the definition of $\langle a, b\rangle_{\mathbb{H}_1\otimes\mathbb{H}_2}$ for all elements of $a, b \in \mathbb{H}_1 \otimes_{\text{pre}} \mathbb{H}_2$.

Proposition 5.1. *$\mathbb{H}_1 \otimes_{\text{pre}} \mathbb{H}_2$ with $\langle \cdot, \cdot\rangle_{\mathbb{H}_1\otimes\mathbb{H}_2}$ is a well-defined scalar product space.*

Proof. If an element $F \in \mathbb{H}_1 \otimes \mathbb{H}_2$ has the following two expressions

$$F = \sum_{j=1}^{n} \varphi_j \otimes \psi_j = \sum_{k=1}^{m} \varphi'_k \otimes \psi'_k ,$$

for some $\varphi_j, \varphi'_k \in \mathbb{H}_1, \psi_j, \psi'_k \in \mathbb{H}_2$, then for any $\varphi \in \mathbb{H}_1$ and $\psi \in \mathbb{H}_2$, we have

$$\sum_{j=1}^{n} \langle \varphi_j, \varphi \rangle \langle \psi_j , \psi \rangle = \sum_{k=1}^{m} \langle \varphi'_j, \varphi \rangle \langle \psi'_j , \psi \rangle = F(\varphi, \psi) .$$

This means that $\langle F, \varphi \otimes \psi \rangle_{\mathbb{H}_1 \otimes \mathbb{H}_2}$ is independent of the form of F. By the same argument, we see that $\langle F, G \rangle_{\mathbb{H}_1 \otimes \mathbb{H}_2}$ is independent of the form of G. Thus $\langle F, G \rangle_{\mathbb{H}_1 \otimes \mathbb{H}_2}$ is well-defined for all $F, G \in \mathbb{H}_1 \otimes_{\text{pre}} \mathbb{H}_2$.

Now assume $F = \sum_{j=1}^{n} \varphi_j \otimes \psi_j$ is a nonzero element in $\mathbb{H}_1 \otimes_{\text{pre}} \mathbb{H}_2$. Let $\{e_1, \cdots, e_\ell)$ be orthonormal set of the subspace of $\mathbb{H}_1$ spanned by $\{\varphi_1, \cdots, \varphi_n\}$. Express φ_j as linear combinations of $\{e_1, \cdots, e_\ell)$ and substitute them into the expression of F. Then we can rewrite $F = \sum_{j=1}^{\ell} e_j \otimes f_j$ for some $f_1, \cdots, f_\ell \in \mathbb{H}_2$, not all zero. Thus,

$$\langle F, F \rangle_{\mathbb{H}_1 \otimes \mathbb{H}_2} = \sum_{j,k=1}^{\ell} \langle e_j, e_k \rangle_{\mathbb{H}_1} \langle f_j, f_k \rangle_{\mathbb{H}_2} = \sum_{j=1}^{\ell} \|f_j\|_{\mathbb{H}_2}^2 > 0 ,$$

proving the proposition. □

Definition 5.1. The Hilbert space obtained by completion of $\mathbb{H}_1 \otimes_{\text{pre}} \mathbb{H}_2$ with respect to the above defined scalar product is called the (Hilbertian) tensor product of $\mathbb{H}_1$ and $\mathbb{H}_2$ and is denoted by $\mathbb{H}_1 \otimes \mathbb{H}_2$.

Proposition 5.2. *If $\{e_1, e_2, \cdots\}$ is an orthonormal basis of $\mathbb{H}_1$ and if $\{\varepsilon_1, \varepsilon_2, \cdots\}$ is an orthonormal basis of $\mathbb{H}_2$, then $\{e_j \otimes \varepsilon_k, j, k = 1, 2, \cdots\}$ is an orthonormal basis of $\mathbb{H}_1 \otimes \mathbb{H}_2$.*

Proof. The orthonormality is obvious. To show the completeness, it suffices to show that $\mathbb{H}_1 \otimes_{\text{pre}} \mathbb{H}_2$ is contained in the closed subspace spanned by $\{e_j \otimes \varepsilon_k, j, k = 1, 2, \cdots\}$. Since an element in $\mathbb{H}_1 \otimes_{\text{pre}} \mathbb{H}_2$ has the form $F = \sum_{j=1}^{n} \varphi_j \otimes \psi_j$ it suffices to show that for any element $\varphi \in \mathbb{H}_1$ and $\psi \in \mathbb{H}_2$, $\varphi \otimes \psi$ is in the closed subspace spanned by $\{e_j \otimes \varepsilon_k, j, k = 1, 2, \cdots\}$. To this end we express $\varphi = \sum_{j=1}^{\infty} a_j e_j$ and $\psi = \sum_{j=1}^{\infty} b_j e_j$. Then it is easy to verify that $f_n = \sum_{j,k=1}^{n} a_j b_k e_j \otimes \varepsilon_k$ will converge to $\varphi \otimes \psi$ in $\mathbb{H}_1 \otimes \mathbb{H}_2$. The proposition is proved. □

If $\{e_1, e_2, \cdots\}$ is an orthonormal basis of $\mathbb{H}_1$ and if $\{\varepsilon_1, \varepsilon_2, \cdots\}$ is an orthonormal basis of $\mathbb{H}_2$, then from this proposition we see that any element F in the tensor product $\mathbb{H}_1 \otimes \mathbb{H}_2$ can be written as

$$F = \sum_{j,k=1}^{\infty} f_{jk} e_j \otimes \varepsilon_k \tag{5.1.2}$$

with the tensor product norm

$$\|F\|^2_{\mathbb{H}_1 \otimes \mathbb{H}_2} = \sum_{j,k=1}^{\infty} |f_{jk}|^2 . \tag{5.1.3}$$

Example 5.1. If $\mathbb{H} = \mathbb{R}^d$, then $\mathbb{R}^d \otimes \mathbb{R}^d = \{M\,; M = (a_{ij})_{1 \le i,j \le d}\}$ is the space of all $d \times d$ matrices.

Let $(X, \mathcal{X}, \mu)$ be a measure space and let $\mathbb{H}$ be a Hilbert space. Let $L^p(X, \mathbb{H}, \mu)$ be the space of all measurable functions $f : X \to \mathbb{H}$ such that norm

$$\|f\|_p = \left(\int_X |f(x)|^p_{\mathbb{H}} \mu(dx) \right)^{1/p} < \infty .$$

Then $L^p(X, \mathbb{H}, \mu)$ is a Banach space with the above norm. When $p = 2$, $L^2(X, \mathbb{H}, \mu)$ is the Hilbert space with the following scalar product

$$\langle f \,, g \rangle = \int_X \langle f(x) \,, g(x) \rangle_{\mathbb{H}} \mu(dx) .$$

If $\mathbb{H} = \mathbb{C}$, then we denote $L^p(X, \mathbb{C}, \mu)$ by $L^p(X, \mu)$.

Example 5.2. Let $(X, \mathcal{X}, \mu)$ and $(Y, \mathcal{Y}, \mu)$ be two measure spaces such that $L^2(X, \mu)$ and $L^2(Y, \nu)$ are separable. Then

$$L^2(X, \mu) \otimes L^2(Y, \nu) \cong L^2(X \times Y, \mu \times \nu) .$$

In fact, if $\{e_j(x)\}$ and $\{\varepsilon_k(y)\}$ are orthonormal bases of $L^2(X, \mu)$ and $L^2(Y, \nu)$, respectively, then we can identify $e_j \otimes \varepsilon_k \in L^2(X, \mu) \otimes L^2(Y, \nu)$ with $e_j(x)\varepsilon_k(y) \in L^2(X \times Y, \mu \times \nu)$ and obtain a mapping from a subspace of $L^2(X, \mu) \otimes L^2(Y, \nu)$ spanned by finite linear combination of $\{e_j(x)\}$ to $L^2(X \times Y, \mu \times \nu)$. It is easy to see that this mapping is one to one and norm preserving. This shows $L^2(X, \mu) \otimes L^2(Y, \nu) \cong L^2(X \times Y, \mu \times \nu)$.

Let $\mathbb{H}_1, \cdots, \mathbb{H}_n$ be Hilbert spaces. Then we can define the (Hilbertian) tensor product space $\mathbb{H}_1 \otimes \cdots \otimes \mathbb{H}_n$ in the same way as in the case $n = 2$.

If $\left\{e_k^j\right\}_{k\in\mathbb{N}}$ is an orthonormal basis of $\mathbb{H}_j$, $j=1,\cdots,n$, then an element $F\in\mathbb{H}_1\otimes\cdots\otimes\mathbb{H}_n$ can be written as

$$F=\sum_{k_1,\cdots,k_n=1}^{\infty} f_{k_1,\cdots,k_n} e_{k_1}^1\otimes\cdots\otimes e_{k_n}^n\,. \tag{5.1.4}$$

The norm of this vector is

$$\|F\|^2_{\mathbb{H}_1\otimes\cdots\otimes\mathbb{H}_n}=\sum_{k_1,\cdots,k_n=1}^{\infty}|f_{k_1,\cdots,k_n}|^2\,. \tag{5.1.5}$$

When $\mathbb{H}_1=\cdots=\mathbb{H}_n=\mathbb{H}$ we denote $\mathbb{H}^{\otimes n}=\overbrace{\mathbb{H}\otimes\cdots\otimes\mathbb{H}}^{n}$.

In general, we have

Proposition 5.3. *If $F\in\mathbb{H}_1\otimes\cdots\otimes\mathbb{H}_n$, then*

$$\|F\|^2_{\mathbb{H}_1\otimes\cdots\otimes\mathbb{H}_n}=\sum_{k_1,\cdots,k_n=1}^{\infty}|\langle F,e_{k_1}^1\otimes\cdots e_{k_n}^n\rangle_{\mathbb{H}_1\otimes\cdots\otimes\mathbb{H}_n}|^2\,, \tag{5.1.6}$$

where $\left\{e_k^i,k=1,2,\cdots\right\}$ is an orthonormal basis of $\mathbb{H}_i$, $i=1,\cdots,n$.

When $F\in\mathbb{H}^{\otimes 2}$, the norm $\|F\|^2_{\mathbb{H}^{\otimes 2}}$ is called the Hilbert-Schmidt norm of F.

Definition 5.2. If $F\in\mathbb{H}^{\otimes 2}$, then we define its trace by

$$\mathrm{Tr}(F)=\sum_{k=1}^{\infty}\langle F\,,e_k\otimes e_k\rangle_{\mathbb{H}^{\otimes 2}}\,, \tag{5.1.7}$$

where $\{e_k,k=1,2,\cdots\}$ is an orthonormal basis of $\mathbb{H}$. It can be proved that the above definition of trace is independent of choice of the basis.

Example 5.3. If $\mathbb{H}=\mathbb{R}^d$, then

$$\left(\mathbb{R}^d\right)^{\otimes m}=\left\{M\,;\;M=(M_{i_1,\cdots,i_m})_{1\le i_1,\cdots,i_m\le d}\right\}$$

is the set of all tensors M of the above form with the norm

$$\|M\|^2=\sum_{i_1,\cdots,i_m=1}^{d}M^2_{i_1,\cdots,i_m}\,.$$

Example 5.4. If $\mathbb{H}=L^2(X,\mu)$, then $\left(L^2(X,\mu)\right)^{\otimes m}=L^2(X^m,\mu^{\otimes m})$ is the space of all square integrable functions (with respect to the measure $\mu^{\otimes m}$) from $\overbrace{X\times\cdots\times X}^{m}$ to $\mathbb{R}$. In particular, we have $\left(L^2([0,T],dx)\right)^{\otimes m}=L^2([0,T]^m,dx_1\cdots dx_m)$.

If $\varphi_1, \cdots, \varphi_n \in \mathbb{H}$, then we define the *symmetrization* of $\varphi_1 \otimes \cdots \otimes \varphi_n$ by

$$Ⓢ(\varphi_1 \otimes \cdots \otimes \varphi_n) = \frac{1}{n!} \sum_{\sigma \in \Sigma_n} \varphi_{\sigma(1)} \otimes \cdots \otimes \varphi_{\sigma(n)}, \tag{5.1.8}$$

where Σ_n is the set of all permutations σ of $\{1, \cdots, n\}$. We also denote the symmetrization of $\varphi_1 \otimes \cdots \otimes \varphi_n$ by

$$\varphi_1 \hat{\otimes} \cdots \hat{\otimes} \varphi_n = Ⓢ(\varphi_1 \otimes \cdots \otimes \varphi_n).$$

The symmetrization of a tensor product $\varphi_1 \otimes \cdots \otimes \varphi_n$ does not depend on the choice of its factors. Namely, if $\varphi_1 \otimes \cdots \otimes \varphi_n = \varphi'_1 \otimes \cdots \otimes \varphi'_n$, then for any $\psi_1, \cdots, \psi_n \in \mathbb{H}$, we have

$$Ⓢ(\varphi_1 \otimes \cdots \otimes \varphi_n)(\psi_1, \cdots, \psi_n) = Ⓢ(\varphi'_1 \otimes \cdots \otimes \varphi'_n)(\psi_1, \cdots, \psi_n).$$

This can be seen immediately from (5.1.8). The symmetrization of $\mathbb{H}_{\text{pre}}^{\otimes k}$ is denoted by $\mathbb{H}_{\text{pre}}^{\hat{\otimes} k}$ and the symmetrization of $\mathbb{H}^{\otimes k}$ is denoted by $\mathbb{H}^{\hat{\otimes} k}$.

Let $\mathbb{H}$ be a Hilbert space. The infinite direct sum of the tensor products of the Hilbert $\mathbb{H}$ will be the Fock space. More precisely, we have the following definition.

Definition 5.3. Let $\mathbb{H}$ be a Hilbert space. Then

$$\Phi(\mathbb{H}) = \bigoplus_{n=0}^{\infty} \mathbb{H}^{\otimes n} = \left\{ a = (a_0, a_1, a_2, \cdots), a_n \in \mathbb{H}^{\otimes n} \quad \text{and} \quad \|a\|^2_{\Phi(\mathbb{H})} := \sum_{n=0}^{\infty} n! \|a_n\|^2_{\mathbb{H}^{\otimes n}} \right\}$$

is called the Fock space over $\mathbb{H}$ and

$$\hat{\Phi}(\mathbb{H}) = \bigoplus_{n=0}^{\infty} \mathbb{H}^{\hat{\otimes} n} = \left\{ a = (a_0, a_1, a_2, \cdots) \in \Phi(\mathbb{H}) \quad \text{and} \quad a_n \in \mathbb{H}^{\hat{\otimes} n} \right\}$$

is called the symmetric Fock space over $\mathbb{H}$. For a Hilbert space $\mathbb{H}$, the Fock space $\Phi(\mathbb{H})$ is also a Hilbert space with the following scalar product:

$$\langle a, b \rangle = \sum_{n=0}^{\infty} n! \langle a_n, b_n \rangle_{\mathbb{H}^{\otimes n}}.$$

Example 5.5. If $e \in \mathbb{H}$, then

$$\mathcal{E}(e) = (1, e, \frac{1}{2!} e^{\otimes 2}, \frac{1}{3!} e^{\otimes 3}, \cdots) \in \mathbb{H}.$$

It is called an exponential vector in the Fock space $\Phi(\mathbb{H})$.

Let $(\mathbb{B}, \mathbb{H}, \mathbb{P})$ be an abstract Wiener space. For any element $h \in \mathbb{H}$, we denote $\tilde{h}$ the centered Gaussian random variable $\tilde{h} : \mathbb{B} \to \mathbb{R}$.

Let $\{e_k, k = 1, 2, \cdots\}$ be an orthonormal basis of $\mathbb{H}$. We consider the set of all f_n in $\mathbb{H}^{\hat{\otimes} n}$ which can be written as

$$f = \sum_{\text{finite}} f_{j_1,\cdots,j_m} e_{j_1}^{\hat{\otimes} k_1} \hat{\otimes} \cdots \hat{\otimes} e_{j_m}^{\hat{\otimes} k_m}, \quad f_{j_1,\cdots,j_m} \in \mathbb{R}\,, \tag{5.1.9}$$

where $j_1, \cdots, j_m$ are different. For this $f \in \mathbb{H}^{\hat{\otimes} n}$ the multiple Wiener-Itô integral is defined as

$$I_n(f) := \sum_{\text{finite}} f_{j_1,\cdots,j_m} H_{j_1}(\tilde{e}_{j_1}) \cdots H_{j_m}(\tilde{e}_{j_m})\,, \tag{5.1.10}$$

where $n = n_1 + \cdots + n_k$ and H is the Hermite polynomial introduced in Section 3.4.

Example 5.6. If $h \in \mathbb{H}$ with $|h| = 1$, then we have

$$I_n(h^{\otimes n}) = H_n(\tilde{h}) \tag{5.1.11}$$

and for $h \in \mathbb{H}$, we have

$$I_n(h^{\otimes n}) = |h|^n H_n(|h|^{-1}\tilde{h})\,. \tag{5.1.12}$$

Example 5.7. If $h_1, \cdots, h_d$ are orthonormal, then

$$I_{n_1+\cdots+n_d}(h_1^{\otimes n_1} \otimes \cdots \otimes h_d^{\otimes n_d}) = H_{n_1}(\tilde{h}_1) \cdots H_{n_d}(\tilde{h}_d)\,. \tag{5.1.13}$$

Example 5.8. It is interesting to note that for general $h \in \mathbb{H}$, $I_n(h^{\otimes n}) \neq H_n(\tilde{h})$. In fact, by (3.4.21), we have

$$\begin{aligned} H_n(\tilde{h}) &= H_n\left(|h|\left[\tilde{h}/|h|\right]\right) \\ &= \sum_{k \le n/2} \frac{n!(|h|^2-1)^k}{2^k k!(n-2k)!} |h|^{n-2k} H_{n-2k}\left(\tilde{h}/|h|\right) \\ &= \sum_{k \le n/2} \frac{n!(|h|^2-1)^k}{2^k k!(n-2k)!} I_{n-2k}\left(h^{\otimes(n-2k)}\right)\,, \end{aligned} \tag{5.1.14}$$

where the last identity follows from (5.1.12).

For general element f_n in $\mathbb{H}^{\hat{\otimes} n}$ we can define the multiple integral $I_n(f_n)$ by the L^2 convergence. It is straightforward to obtain the following Itô isometry equality

$$\mathbb{E}\left[I_n(f_n) I_m(g_m)\right] = \begin{cases} n!\langle f_n, g_n\rangle_{\mathbb{H}^{\otimes n}} & \text{if } m = n \\ 0 & \text{if } m \neq n\,. \end{cases} \tag{5.1.15}$$

Example 5.9. Let $(\mathbb{B}, \mathbb{H}, \mathbb{P})$ be an abstract Wiener space. If $h \in \mathbb{H}$, we denote

$$\mathcal{E}(h) = \mathcal{E}(\tilde{h}) := \exp\left\{\tilde{h} - \frac{1}{2}|h|^2\right\}.$$

From the moment generating formula of the Hermite polynomials, we have

$$\begin{aligned} \mathcal{E}(h) &= \exp\left\{|h|\,\frac{\tilde{h}}{|h|} - \frac{1}{2}|h|^2\right\} \\ &= \sum_{n=0}^{\infty} \frac{|h|^n}{n!} H_n\left(\frac{\tilde{h}}{|h|}\right) = \sum_{n=0}^{\infty} \frac{1}{n!} I_n(h^{\otimes n}). \end{aligned} \tag{5.1.16}$$

We introduce the following sets

$$\begin{aligned} \mathcal{E} &= \left\{F = \sum_{j=1}^{d} a_j \mathcal{E}(h_j)\,;\ d \geq 1,\ h_j \in \mathbb{H},\ \text{and } a_j \in \mathbb{R}\right\}; \\ \mathcal{P} &= \left\{F = f(\tilde{h}_1, \cdots, \tilde{h}_d)\,;\ d \geq 1,\ h_j \in \mathbb{H},\ \text{and } f : \mathbb{R}^d \to \mathbb{R} \text{ is polynomial}\right\}; \\ \mathcal{S} &= \Big\{F = f(\tilde{h}_1, \cdots, \tilde{h}_d)\,;\ d \geq 1,\ h_j \in \mathbb{H}, \\ &\qquad \text{and } f : \mathbb{R}^d \to \mathbb{R} \text{ is smooth and of compact support}\Big\}. \end{aligned} \tag{5.1.17}$$

Lemma 5.1. *Let p be a non-negative integer and let $\mathcal{C}^p(\mathbb{R}^d)$ denote the space of all continuously differentiable functions up to order p. Let μ be a finite measure on $(\mathbb{R}^d, \mathcal{B}(\mathbb{R}^d))$ and let $G \in \mathcal{C}^p(\mathbb{R}^d)$ be such that $\partial^\alpha G \in L^r(\mu)$ for some $r \geq 1$ and any multi-index α such that $|\alpha| := \sum_{j=1}^d \alpha_j \leq p$. Then, there exists a sequence $(G_n)_{n\in\mathbb{N}}$ such that*

(1) *Each G_n is a trigonometric polynomial of several variables, that is*

$$G_n(x_1, \ldots, x_d) = \sum_{\text{finite}} a^n_{l_1,\ldots,l_d} \exp\left\{\imath\xi^n_{l_1} x_1 + \cdots + \imath\xi^n_{l_d} x_d\right\},$$

and where $a^n_{l_1,\ldots,l_d}$ and $\xi^n_{l_j}$ are real numbers.

(2) *We have $\lim_{n\to\infty} \partial^\alpha G_n = \partial^\alpha G$ in $L^r(\mu)$ for any α such that $|\alpha| \leq p$.*

Proof. We sketch the main ideas of its proof. First, given $G \in \mathcal{C}^p(\mathbb{R}^d)$, there exists a sequence of $\mathcal{C}^\infty(\mathbb{R}^d)$ functions with compact support that converge, jointly with their derivatives, to G in $L^r(\mu)$. So, one only needs to approximate a function $G \in \mathcal{C}^\infty(\mathbb{R}^d)$ with support contained in a rectangle of $\mathbb{R}^d$, say K. Moreover, given $\varepsilon > 0$, we can suppose that $\mu(K^c) < \varepsilon$. For a such function, consider its Fourier partial sums on the rectangle K that converge uniformly, jointly with their derivatives to G and its derivatives. Since these partial sums are periodic functions with the same period, their sup-norm on all $\mathbb{R}^d$ is the same that the sup-norm on the compact K. With these ingredients, the result is easily obtained. □

From the above lemma it is easy to deduce the following.

Proposition 5.4. *Each of the three sets $\mathcal{E}$, $\mathcal{P}$ and $\mathcal{S}$ are dense in $L^p(\mathbb{B})$ for any $1 \le p < \infty$.*

The following theorem is called the chaos expansion theorem for non-linear Gaussian functionals.

Theorem 5.1 (Chaos expansion theorem). *Any functional F in $L^2(\mathbb{B})$ admits the following expansion*

$$F = \sum_{n=0}^{\infty} I_n(f_n), \tag{5.1.18}$$

where $f_n \in \mathbb{H}^{\hat{\otimes} n}$ and $\sum_{n=0}^{\infty} n! \|f_n\|_{\mathbb{H}^{\otimes n}}^2 < \infty$. Moreover, the following isometry identity holds.

$$\mathbb{E}(F^2) = \sum_{n=0}^{\infty} n! \|f_n\|_{\mathbb{H}^{\otimes n}}^2 . \tag{5.1.19}$$

Proof. Since $\mathcal{P}$ is dense in $L^2(\mathbb{B})$ it suffices to show that any element $F = f(\tilde{h}_1, \cdots, \tilde{h}_d) \in \mathcal{P}$ can be written as (5.1.18). Let $\ell_1, \cdots, \ell_m$ be an orthonormal basis of the vector space spanned by $h_1, \cdots, h_d$. Then there is a polynomial g such that $F = g(\tilde{\ell}_1, \cdots, \tilde{\ell}_m)$. Write the polynomial g as finite sum of monomial $x_1^{n_1} \cdots x_m^{n_m}$. It suffices for us to show that $\tilde{\ell}_1^{n_1} \cdots \tilde{\ell}_m^{n_m}$ can have an expansion of the form (5.1.18). This is easy since each factor $\tilde{\ell}_i^{n_i}$ can be expressed as linear combination of $H_{k_i}(\tilde{\ell}_i)$ and then $\tilde{\ell}_1^{n_1} \cdots \tilde{\ell}_m^{n_m}$ can be written as linear combination of

$$H_{k_1}(\tilde{\ell}_1) \cdots H_{k_m}(\tilde{\ell}_m) = I_{k_1 + \cdots k_m} \left(\ell_1^{\otimes k_1} \otimes \cdots \otimes \ell_m^{\otimes k_m} \right) .$$

The theorem is then proved. □

Example 5.10. Assume that $h_1, \cdots, h_d$ are orthonormal vectors in $\mathbb{H}$. Let $f : \mathbb{R}^d \to \mathbb{R}$ be measurable and let its Fourier transform $\hat{f}$ exist such that

$f(x) = \int_{\mathbb{R}^d} e^{-\iota\langle x,\xi\rangle} \hat{f}(\xi)d\xi$. Thus, by (5.1.16) and then by (5.1.13), we can find the explicit chaos expansion of of $F = f(\tilde{h}_1, \cdots, \tilde{h}_d)$ as follows

$$\begin{aligned}
f(\tilde{h}_1, \cdots, \tilde{h}_d) &= \int_{\mathbb{R}^d} e^{-\iota\sum_{j=1}^d \xi_j \tilde{h}_j} \hat{f}(\xi)d\xi \\
&= \int_{\mathbb{R}^d} e^{-\frac{1}{2}|\xi|^2} e^{-\iota\sum_{j=1}^d \xi_j \tilde{h}_j + \frac{1}{2}|\xi|^2} \hat{f}(\xi)d\xi \\
&= \sum_{n_1,\cdots,n_d=0}^{\infty} \frac{(-\iota)^{n_1+\cdots+n_d}}{n_1!\cdots n_d!} \int_{\mathbb{R}^d} \xi_1^{n_1}\cdots\xi_d^{n_d} \hat{f}(\xi) e^{-\frac{1}{2}|\xi|^2} d\xi \\
&\qquad H_{n_1}(\tilde{h}_1)\cdots H_{n_d}(\tilde{h}_d) \\
&= \sum_{n=0}^{\infty} I_n(f_n)\,, \qquad (5.1.20)
\end{aligned}$$

where

$$f_n = \sum_{n_1+\cdots+n_d=n}^{\infty} \frac{(-\iota)^{n_1+\cdots+n_d}}{n_1!\cdots n_d!} h_1^{\otimes n_1}\otimes\cdots\otimes h_d^{\otimes n_d} \int_{\mathbb{R}^d} \xi_1^{n_1}\cdots\xi_d^{n_d} \hat{f}(\xi) e^{-\frac{1}{2}|\xi|^2} d\xi\,. \qquad (5.1.21)$$

Example 5.11. If $\mathbb{H} = \mathbb{B} = \mathbb{R}$, then the corresponding abstract Wiener space is $(\mathbb{R}, \mu(dx))$, where $\mu(dx) = (2\pi)^{-1/2}\exp\left\{-\frac{x^2}{2}\right\}dx$ is the standard one dimensional Gaussian measure. It is easy to see that in this case $\mathbb{H}^{\hat{\otimes} n} = \mathbb{R}$ and the Fock space is

$$\Phi(\mathbb{R}) = \{(a_0, a_1, \cdots, a_n, \cdots); a_i \in \mathbb{R}\}\,.$$

The unit vector $e = 1$ in $\mathbb{R}$ constructs the basis of $\mathbb{R}$, and $e^{\otimes n} = e = 1$ is the basis $\mathbb{R}^{\hat{\otimes} n}$. The multiple integral is then $I_n(e^{\otimes n}) = H_n(x)$. The chaos expansion theorem (Theorem 5.1) is exactly the Hermite expansion of a square integrable function (see Proposition 3.7). Namely, for any $f \in L^2(\mathbb{R}, \mu(dx))$,

$$f(x) = \sum_{n=0}^{\infty} a_n H_n(x)\,, \quad \text{for some } a_n \in \mathbb{R}\,.$$

Example 5.12. If $\mathbb{H} = \mathbb{B} = \mathbb{R}^d$ with the Euclidean norm, then the corresponding abstract Wiener space is $(\mathbb{R}^d, \mu(dx))$, where $\mu(dx) = (2\pi)^{-d/2}\exp\left\{-\frac{|x|^2}{2}\right\}dx$ is the standard d-dimensional Gaussian measure. It is easy to see that in this case $\mathbb{H}^{\hat{\otimes} n} = \mathbb{R}^{dn}$ and the Fock space is

$$\Phi(\mathbb{R}^d) = \left\{(a_0, a_1, \cdots, a_n, \cdots); a_n \in \mathbb{R}^{dn}\right\}\,,$$

where we can express

$$a_n = \left(a^{(n)}_{i_1,\cdots i_n}\right)_{1\le i_1,\cdots,i_n\le d}.$$

The unit vectors $e_i = (0,\cdots,0,1,0,\cdots,0)^T$, $i=1,\cdots,d$, in $\mathbb{R}^d$ construct a basis of $\mathbb{R}^d$ and $\tilde{e}_i = x_i$. A basis of $\mathbb{H}^{\hat{\otimes} n} = \mathbb{R}^{dn}$ is the set of all tensor products of the form $e_{j_1}\hat{\otimes}\cdots\hat{\otimes}e_{j_n}$, $1\le j_1,\cdots,j_n\le d$, which can be written as

$$e_1^{\otimes n_1}\hat{\otimes}\cdots\hat{\otimes}e_d^{\otimes n_d},$$

where $n_1,\cdots,n_d\ge 0$, $n_1+\cdots+n_\kappa = n$. The corresponding multiple integral is then

$$I_n(e_1^{\otimes n_1}\hat{\otimes}\cdots\hat{\otimes}e_d^{\otimes n_d}) = H_{n_1}(x_1)\cdots H_{n_d}(x_d).$$

The chaos expansion theorem (Theorem 5.1) is then the Hermite expansion of a square integrable function of d-variables. Namely, for any $f\in L^2(\mathbb{R}^d,\mu(dx))$,

$$f(x) = a_0 + \sum_{n=1}^{\infty}\sum_{n_1+\cdots+n_d=n,n_1,\cdots,n_d\ge 0} a^{(n)}_{n_1,\cdots,n_d}H_{n_1}(x_1)\cdots H_{n_d}(x_d),$$

where $a^{(n)}_{n_1,\cdots,n_d}\in\mathbb{R}$.

We refer to [Meyer (1993)] and the references therein for further details on Fock space and chaos expansion.

5.2 Polarization

Let $\mathbb{H}$ be any vector space (In this section we shall use $\mathbb{H}$ and $\mathbb{B}$ to denote any vector spaces) and let f be an element in $\overbrace{\mathbb{H}\otimes\cdots\otimes\mathbb{H}}^{d}$. Namely, $f(\varphi_1,\cdots,\varphi_d)$ be a d-linear form (see definition below). Associated with f we have a homogeneous polynomial of degree d

$$\bar{f}(\varphi) = f(\varphi,\cdots,\varphi),\ \varphi\in\mathbb{H}.$$

On the other hand, if $f\in\mathbb{H}^{\hat{\otimes} d}$ is symmetric, we can recover f from $\bar{f}$ by the following *polarization* technique:

$$f(\varphi_1,\cdots,\varphi_d) = \frac{1}{2^d d!}\sum_{\varepsilon_j=\pm1,1\le j\le d}\varepsilon_1\cdots\varepsilon_d\bar{f}\left(\sum_{j=1}^d\varepsilon_j\varphi_j\right)^{\otimes d}.$$

This identity is very useful when we discuss multiple integrals and we shall prove it in this section.

Definition 5.4. Let $\mathbb{B}$ and $\mathbb{H}$ be two vector spaces. A mapping $f : \mathbb{B}^d \to \mathbb{H}$ is called a d-linear form if

$$f(\cdots, ax_i + by_i, \cdots) = af(\cdots, x_i, \cdots) + bf(\cdots, y_i, \cdots), \qquad \forall\, x_i, y_i \in \mathbb{B} \text{ and } a, b \in \mathbb{R}.$$

The d-linear form f is called symmetric if

$$f(x_{\sigma(1)}, \cdots, x_{\sigma(d)}) = f(x_1, \cdots, x_d)$$

for any permutation σ of $\{1, \cdots, d\}$. If $f : \mathbb{B}^d \to \mathbb{H}$ is a d-linear form, then the functional $\bar{f} : \mathbb{B} \to \mathbb{H}$ defined by $\bar{f}(x) = f(x, \cdots, x)$ is called a homogeneous polynomial of degree d.

Theorem 5.2 (Polarization formula). *If $f : \mathbb{B}^d \to \mathbb{H}$ is a symmetric d-linear form and if $\bar{f}$ is the associated homogeneous polynomial, then*

$$f(x_1, \cdots, x_d) = \frac{1}{2^d d!} \sum_{\varepsilon_1, \cdots, \varepsilon_d = \pm 1} (-1)^{\frac{d-\varepsilon_1-\cdots-\varepsilon_d}{2}} \bar{f}(x_0 + \varepsilon_1 x_1 + \cdots + \varepsilon_d x_d) \tag{5.2.1}$$

for all $x_1, \cdots, x_d \in \mathbb{B}$, where x_0 is an arbitrarily fixed point in $\mathbb{B}$.

Remark 5.1. We call the right hand side of (5.2.1) the polarization of $\bar{f}$ at point $(x_1, \cdots, x_d) \in \mathbb{B}^d$.

Proof. From the linearity and the symmetry of f, we have

$$\begin{aligned}
&\bar{f}(x_0 + \varepsilon_1 x_1 + \cdots + \varepsilon_d x_d) \\
&\quad = f(x_0 + \varepsilon_1 x_1 + \cdots + \varepsilon_d x_d, \cdots, x_0 + \varepsilon_1 x_1 + \cdots + \varepsilon_d x_d) \\
&\quad = \sum_{j_0 + j_1 + \cdots + j_d = d} \frac{d!}{j_0! j_1! \cdots j_d!} \varepsilon_1^{j_1} \cdots \varepsilon_d^{j_d} \\
&\qquad\quad f(\overbrace{x_0, \cdots, x_0}^{j_0}, \overbrace{x_1, \cdots, x_1}^{j_1} \cdots, \overbrace{x_d, \cdots, x_d}^{j_d})
\end{aligned}$$

Denote the right hand side of (5.2.1) by $g(x_1, \cdots, x_d)$. Then by the above computation we have

$$\begin{aligned}
g(x_1, \cdots, x_d) = &\sum_{j_0 + j_1 + \cdots + j_d = d} \frac{A_{j_0, \cdots, j_d}}{j_0! j_1! \cdots j_d!} \\
&\qquad f(\overbrace{x_0, \cdots, x_0}^{j_0}, \overbrace{x_1, \cdots, x_1}^{j_1} \cdots, \overbrace{x_d, \cdots, x_d}^{j_d}),
\end{aligned}$$

where

$$A_{j_0,\cdots,j_d} = \frac{1}{2^d} \sum_{\varepsilon_1,\cdots,\varepsilon_d=\pm 1} (-1)^{\frac{d-\varepsilon_1-\cdots-\varepsilon_d}{2}} \varepsilon_1^{j_1} \cdots \varepsilon_d^{j_d} .$$

Suppose $j_k = 0$ for some $k = 1, \cdots, d$, say $k = 1$. Then

$$A_{j_0,\cdots,j_d} = \frac{1}{2^d} \sum_{\varepsilon_2,\cdots,\varepsilon_d=\pm 1} \left[(-1)^{\frac{d-1-\varepsilon_2\cdots-\varepsilon_d}{2}} + (-1)^{\frac{d+1-\varepsilon_2\cdots-\varepsilon_d}{2}} \right] \varepsilon_1^{j_2} \cdots \varepsilon_d^{j_d} = 0 .$$

Thus all $j_k = 1$ for $k = 1, \cdots, d$. This also implies $j_0 = 0$. In this case, we have

$$\begin{aligned} A_{0,1,\cdots,1} &= \frac{1}{2^d} \sum_{\varepsilon_1,\cdots,\varepsilon_d=\pm 1} (-1)^{\frac{d-\varepsilon_1-\cdots-\varepsilon_d}{2}} \varepsilon_1 \cdots \varepsilon_d \\ &= \frac{1}{2^d} \sum_{\varepsilon_1,\cdots,\varepsilon_d=\pm 1} (-1)^{\frac{1-\varepsilon_1}{2}} \varepsilon_1 \cdots (-1)^{\frac{1-\varepsilon_d}{2}} \varepsilon_d = 1 \end{aligned}$$

since $(-1)^{\frac{1-\varepsilon_j}{2}} \varepsilon_j = 1$ for all $j = 1, \cdots, d$ and there are exactly 2^d terms in the sum. Thus $g(x_1, \cdots, x_d) = f(x_1, \cdots, x_d)$, proving the theorem. □

Example 5.13. If $d = 2$ and $f(\varphi_1, \varphi_2) = \frac{1}{2}[\varphi_1 \otimes \varphi_2 + \varphi_2 \otimes \varphi_1]$, then $\bar{f}(\varphi) = \varphi^{\otimes 2}$. It is easy to see that the right hand side of (5.2.1) is

$$\begin{aligned} &\frac{1}{8} \left[(\varphi_0 + \varphi_1 + \varphi_2)^{\otimes 2} - (\varphi_0 - \varphi_1 + \varphi_2)^{\otimes 2} - (\varphi_0 + \varphi_1 - \varphi_2)^{\otimes 2} \right. \\ &\qquad \left. + (\varphi_0 - \varphi_1 - \varphi_2)^{\otimes 2} \right] = \frac{1}{2} [\varphi_1 \otimes \varphi_2 + \varphi_2 \otimes \varphi_1] = f(\varphi_1, \varphi_2) . \end{aligned}$$

A generic element in $\mathbb{B}^d$ will be denoted by $\mathbf{x} = (x_1, \cdots, x_d)$. For a function $f : \mathbb{B}^d \to \mathbb{H}$, we define its symmetrization as

$$\circledS_{\mathbf{x}} f = \frac{1}{d!} \sum_{\sigma \in \Sigma} f(x_{\sigma(1)}, \cdots, x_{\sigma(d)}) ,$$

where Σ denotes the set of all permutations of $\{1, \cdots, d\}$.

For any d-homogeneous polynomial $f : \mathbb{B} \to \mathbb{H}$, we denote its polarization by

$$\begin{aligned} \mathrm{Pol}_{\mathbf{x}} f &= \mathrm{Pol}_{x_1,\cdots,x_d} f \\ &= \frac{1}{2^d d!} \sum_{\varepsilon_1,\cdots,\varepsilon_d=\pm 1} (-1)^{\frac{d-\varepsilon_1-\cdots-\varepsilon_d}{2}} f(x_0 + \varepsilon_1 x_1 + \cdots + \varepsilon_d x_d) , \end{aligned}$$

where $\mathbf{x} = (x_1, \cdots, x_d) \in \mathbb{B}^d$. It is clear that the polarization $\mathrm{Pol}_{\mathbf{x}} f$ is independent of the choice of x_0.

An immediate consequence of Theorem 5.2 is the following proposition.

Proposition 5.5. *Let $\mathbb{B}$ and $\mathbb{H}$ be two vector spaces. Let $f : \mathbb{B}^d \to \mathbb{H}$ be a multilinear form, namely, f is linear with respect to all its variables. Let $\bar{f}(x) = f(\overbrace{x, \cdots, x}^{d})$ be the associated homogeneous polynomial of degree d. Then*

$$Ⓢ_{\mathbf{x}} f = \mathrm{Pol}_{\mathbf{x}} \bar{f} \,. \tag{5.2.2}$$

Let $\mathbb{B}$ be any vector space and $x_1, \cdots, x_d \in \mathbb{B}$. Then the symmetric tensor product $x_1 \hat{\otimes} \cdots \hat{\otimes} x_d$ can be considered as a symmetric d-linear form from $\mathbb{B}^d$ to $\mathbb{B}^{\otimes d}$. For this form, the polarization formula becomes

Theorem 5.3. *Let $\mathbb{B}$ be any vector space. Then*

$$\begin{aligned} x_1 \hat{\otimes} \cdots \hat{\otimes} x_d &= \mathrm{Pol}_{x_1, \cdots, x_d} \left(x^{\otimes d} \right) \qquad (5.2.3) \\ &:= \frac{1}{2^d d!} \sum_{\varepsilon_1, \cdots, \varepsilon_d = \pm 1} (-1)^{\frac{d - \varepsilon_1 - \cdots - \varepsilon_d}{2}} (x_0 + \varepsilon_1 x_1 + \cdots + \varepsilon_d x_d)^{\otimes d} \,, \\ &\qquad \forall \; x_1, \cdots, x_d \in \mathbb{B}, \end{aligned}$$

where x_0 is an arbitrarily fixed element in $\mathbb{B}$.

The following consequence of the above theorem will be useful.

Proposition 5.6. *Let $\mathbb{B}$ and $\mathbb{H}$ be two Banach spaces. Let $F, G : \mathbb{B}^{\otimes d} \to \mathbb{H}$ (or $F, G : \mathbb{B}^{\hat{\otimes} d} \to \mathbb{H}$) be continuous multilinear forms. If $F(x^{\otimes d}) = G(x^{\otimes d})$ for all $x \in \mathbb{B}$, then $F(\mathbf{x}) = G(\mathbf{x})$ for all $\mathbf{x} \in \mathbb{B}^{\hat{\otimes} d}$.*

Proof. Since F and ⓈF induce the same homogeneous polynomial, it suffices to show the case that $F, G : \mathbb{B}^{\hat{\otimes} d} \to \mathbb{H}$.

Since $F(x^{\otimes d}) = G(x^{\otimes d})$ for all $x \in \mathbb{B}$ and since F and G are linear, we have by (5.2.3)

$$F(x_1 \hat{\otimes} \cdots \hat{\otimes} x_d) = G(x_1 \hat{\otimes} \cdots \hat{\otimes} x_d) \quad \forall \; x_1, \cdots, x_d \in \mathbb{B} \,.$$

By the linearity again we have

$$F\Big(\sum_{\text{finite}} a_{i_1, \cdots, i_d} x_{i_1} \hat{\otimes} \cdots \hat{\otimes} x_{i_d}\Big) = G\Big(\sum_{\text{finite}} a_{i_1, \cdots, i_d} x_{i_1} \hat{\otimes} \cdots \hat{\otimes} x_{i_d}\Big) \,.$$

Since the set of all elements of the form $\left\{ \sum_{\text{finite}} a_{i_1, \cdots, i_d} x_{i_1} \hat{\otimes} \cdots \hat{\otimes} x_{i_d} \right\}$ is dense in $\mathbb{B}^{\hat{\otimes} d}$, the proposition is proved by using the continuity. □

We are going to extend the above results to a function which is symmetric among some groups of variables.

Let $\mathbb{B}_1, \cdots, \mathbb{B}_k$ and $\mathbb{H}$ be vector spaces. A point in $\mathbb{B}_i^{d_i}$ will be denoted by $\mathbf{x}_i = (x_{i,1}, \cdots, x_{i,d_i})$. If σ_i is any permutation of $(1, \cdots, d_i)$, we denote $\mathbf{x}_{i,\sigma_i} = (x_{i,\sigma(1)}, \cdots, x_{i,\sigma(d_i)})$.

For a mapping $f : \mathbb{B}_1^{d_1} \oplus \cdots \oplus \mathbb{B}_k^{d_k} \to \mathbb{H}$, we define

$$\circledS_{\mathbf{x}_i} f(\mathbf{x}_1, \cdots, \mathbf{x}_k) = \frac{1}{i_d!} \sum_{\sigma_i} f(\mathbf{x}_1, \cdots, \mathbf{x}_{i-1}, \mathbf{x}_{i,\sigma_i}, \mathbf{x}_{i+1}, \cdots, \mathbf{x}_k)$$
$$= \frac{1}{i_d!} \sum_{\sigma_i} f(\mathbf{x}_1, \cdots, x_{i,\sigma(1)}, \cdots, x_{i,\sigma(i_d)}, \cdots, \mathbf{x}_k),$$

where σ_i runs all permutation of $\{1, \cdots, d_i\}$.

Definition 5.5. Let $f : \mathbb{B}_1^{d_1} \oplus \cdots \oplus \mathbb{B}_k^{d_k} \to \mathbb{H}$ be a multilinear form. Then

$$\bar{f}(x_1, \cdots, x_d) = \bar{f}(\overbrace{x_1, \cdots, x_1}^{d_1}, \cdots, \overbrace{x_k, \cdots, x_k}^{d_k}), \quad \forall\, x_1 \in \mathbb{B}_1, \cdots, x_k \in \mathbb{B}_k$$

is called a $d_1 \times \cdots \times d_k$ homogeneous polynomial.

The following statement is a consequence of Theorem 5.2

Theorem 5.4 (Multilinear-polarization). *Let $\mathbb{B}_1, \cdots, \mathbb{B}_k$ and $\mathbb{H}$ be vector spaces and let $f : \mathbb{B}_1^{d_1} \oplus \cdots \oplus \mathbb{B}_k^{d_k} \to \mathbb{H}$ be a multilinear form and let $\bar{f}$ be the associated $d_1 \times \cdots \times d_k$ homogeneous polynomial. Then*

$$\circledS_{\mathbf{x}_1} \cdots \circledS_{\mathbf{x}_k} f(\mathbf{x}_1, \cdots, \mathbf{x}_k) = \mathrm{Pol}_{\mathbf{x}_1} \cdots \mathrm{Pol}_{\mathbf{x}_k} \bar{f}, \tag{5.2.4}$$

where $\mathbf{x}_1 \in \mathbb{B}_1^{d_1}$, $\cdots$, $\mathbf{x}_k \in \mathbb{B}_k^{d_k}$. In particular, if $\bar{f}$ and $\bar{g}$ are two homogeneous polynomials of k-variables such that

$$\bar{f}(x_1, \cdots, x_k) = \bar{g}(x_1, \cdots, x_k) \quad \forall \;\; x_1 \in \mathbb{B}_1, \cdots, x_k \in \mathbb{B}_k,$$

then their inducing $d_1 \times \cdots \times d_k$-multilinear symmetric forms f and g are also the same. Namely, we have

$$f(\mathbf{x}_1, \cdots, \mathbf{x}_k) = g(\mathbf{x}_1, \cdots, \mathbf{x}_k) \quad \forall\; \mathbf{x}_1 \in \mathbb{B}_1^{d_1}, \cdots, \; \mathbf{x}_k \in \mathbb{B}_k^{d_k}.$$

Proof. $\circledS_{\mathbf{x}_1} \cdots \circledS_{\mathbf{x}_k} f(\mathbf{x}_1, \cdots, \mathbf{x}_k)$ is a symmetric function of $\mathbf{x}_k$. Thus, (5.2.1) gives

$$\circledS_{\mathbf{x}_1} \cdots \circledS_{\mathbf{x}_k} f(\mathbf{x}_1, \cdots, \mathbf{x}_k) = \mathrm{Pol}_{\mathbf{x}_k} (\circledS_{\mathbf{x}_1} \cdots \circledS_{\mathbf{x}_{k-1}} f)(\mathbf{x}_1, \cdots, \mathbf{x}_{k-1}, \overbrace{x_k, \cdots, x_k}^{d_k}),$$

where $\mathrm{Pol}_{\mathbf{x}_k}$ means the polarization with respect to the variable $\mathbf{x}_k$ while other variables are kept fixed. Now,

$$\mathrm{Pol}_{\mathbf{x}_k} (\circledS_{\mathbf{x}_1} \cdots \circledS_{\mathbf{x}_{k-1}} f)(\mathbf{x}_1, \cdots, \mathbf{x}_{k-1}, \overbrace{x_k, \cdots, x_k}^{d_k})$$

is a symmetric function of $\mathbf{x}_{k-1}$ and hence,

$$\begin{aligned}
&\mathrm{Pol}_{\mathbf{x}_k}(\circledS_{\mathbf{x}_1}\cdots\circledS_{\mathbf{x}_{k-1}}f)(\mathbf{x}_1,\cdots,\mathbf{x}_{k-1},\overbrace{x_k,\cdots,x_k}^{d_k})\\
&=\mathrm{Pol}_{\mathbf{x}_{k-1}}\mathrm{Pol}_{\mathbf{x}_k}(\circledS_{\mathbf{x}_1}\cdots\circledS_{\mathbf{x}_{k-2}}f)(\mathbf{x}_1,\cdots,\mathbf{x}_{k-2},\overbrace{x_{k-1},\cdots,x_{k-1}}^{d_{k-1}},\overbrace{x_k,\cdots,x_k}^{d_k})\,.
\end{aligned}$$

Continuing this way yields (5.2.4). □

Similar to Theorem 5.3, we have

Theorem 5.5. *We have the following polarization formula for tensor product which is symmetric among some groups.*

$$\begin{aligned}
&(x_{11}\hat{\otimes}\cdots\hat{\otimes}x_{1d_1})\otimes\cdots\otimes(x_{k1}\hat{\otimes}\cdots\hat{\otimes}x_{kd_k})\\
&\quad=\mathrm{Pol}_{(x_{11},\cdots,x_{1d_1})}(x_1)\cdots\mathrm{Pol}_{(x_{k1},\cdots,x_{kd_k})}(x_k)\left(x_1^{\otimes d_1}\otimes\cdots\otimes x_k^{\otimes d_k}\right)\,,
\end{aligned}\tag{5.2.5}$$

where $\mathrm{Pol}_{(x_1,\cdots,x_d)}(\mathbf{x})$ *denotes the symmetrization with respect to the variable* $\mathbf{x}$ *at point* $(x_1,\cdots,x_d)$.

Since the linear combinations of the elements of the form of the left hand of (5.2.5) is dense in $\mathbb{B}_1^{\hat{\otimes}d_1}\otimes\cdots\otimes\mathbb{B}_k^{\hat{\otimes}d_k}$, we have

Proposition 5.7. *Let* $\mathbb{B}_1,\cdots,\mathbb{B}_k$ *and* $\mathbb{H}$ *be Banach spaces. Let* $F,G:\mathbb{B}_1^{\otimes d_1}\otimes\cdots\otimes\mathbb{B}_k^{\otimes d_k}\to\mathbb{H}$ *(or* $F,G:\mathbb{B}_1^{\hat{\otimes}d_1}\otimes\cdots\otimes\mathbb{B}_k^{\hat{\otimes}d_k}\to\mathbb{H}$*) be multilinear and continuous. If* $F(x_1^{\otimes d}\otimes\cdots\otimes x_k^{\otimes d_k})=G(x_1^{\otimes d}\otimes\cdots\otimes x_k^{\otimes d_k})$ *for all* $x_1\in\mathbb{B}_1,\cdots,x_k\in\mathbb{B}_k$*, then* $F(\mathbf{x})=G(\mathbf{x})$ *for all* $\mathbf{x}\in\mathbb{B}_1^{\hat{\otimes}d_1}\otimes\cdots\otimes\mathbb{B}_k^{\hat{\otimes}d_k}$.

5.3 Multiple Wiener-Itô Integrals

We introduced multiple Wiener-Itô integral in Section 5.1. If $f_n\in\mathbb{H}^{\hat{\otimes}n}$, then we can define $I_n(f_n)$. In fact, for any $f\in\mathbb{H}^{\otimes n}$ we can define

$$I_n(f_n)=I_n(\circledS f_n)\,.$$

The important chaos expansion theorem says that any square integrable functional $F:\mathbb{B}\to\mathbb{R}$ can be represented as linear combination of such multiple integrals. The study of a nonlinear functional on $\mathbb{B}$ (random variable) is then led to the study of each multiple integral. In this section, we shall further study these integrals.

Let $I_n(f_n)$ and $I_m(g_m)$ be two multiple integrals, where $f_n\in\mathbb{H}^{\hat{\otimes}n}$ and $g_m\in\mathbb{H}^{\hat{\otimes}m}$. They are in $L^p(\mathbb{B})$ for any $p\in[1,\infty)$. Thus $I_n(f_n)\cdot I_m(g_m)$ is also $L^2(\mathbb{B})$. To give the chaos expansion of this product, we introduce the following contractions of $f_n \in \mathbb{H}^{\hat{\otimes} n}$ and $g_m \in \mathbb{H}^{\hat{\otimes} m}$. Let $\{e_k, k = 1, 2, \cdots\}$ be an orthonormal basis of $\mathbb{H}$.

Definition 5.6. The k-th contraction of f and g, denoted by $f \otimes_k g$ is defined as

$$f\hat{\otimes}_k g = \sum_{n_1,\cdots,n_k=1}^{\infty} \langle f, e_{n_1} \otimes \cdots \otimes e_{n_k} \rangle_{\mathbb{H}^{\otimes(n-k)}} \hat{\otimes} \langle g, e_{n_1} \otimes \cdots \otimes e_{n_k} \rangle_{\mathbb{H}^{\otimes(m-k)}} . \tag{5.3.1}$$

Theorem 5.6. *If $f_n \in \mathbb{H}^{\hat{\otimes} n}$ and $g_m \in \mathbb{H}^{\hat{\otimes} m}$, then*

$$I_n(f_n) \cdot I_m(g_m) = \sum_{k=0}^{n\wedge m} \frac{n!m!}{k!(n-k)!(m-k)!} I_{n+m-2k}(f\hat{\otimes}_k g) . \tag{5.3.2}$$

Proof. Let $f, g \in \mathbb{H}$. Then , for any $s, t \in \mathbb{R}$, we have

$$\begin{aligned}
\mathcal{E}(t\tilde{f})\mathcal{E}(s\tilde{g}) &= \mathcal{E}(\widetilde{tf + sg})e^{st\langle f,g\rangle} = \sum_{\ell=0}^{\infty} \frac{1}{\ell!} I_\ell\left((tf+sg)^{\otimes \ell}\right) \sum_{k=0}^{\infty} \frac{1}{k!} s^k t^k \langle f, g\rangle^k \\
&= \sum_{\ell=0}^{\infty} \sum_{j=0}^{\ell} \sum_{k=0}^{\infty} \frac{1}{j!(\ell-j)!k!} t^j s^{\ell-j} t^k s^k \langle f, g\rangle^k I_\ell(f^{\otimes j} \otimes g^{\otimes(\ell-j)}) \\
&= \sum_{m,n=0}^{\infty} \sum_{k=0}^{m\wedge n} \frac{1}{(m-k)!(n-k)!k!} t^n s^m \langle f, g\rangle^k \\
&\qquad \cdot I_{n+m-2k}(f^{\otimes(m-k)} \otimes g^{\otimes(n-k)}) .
\end{aligned} \tag{5.3.3}$$

The coefficient of $s^n t^m$ of $\mathcal{E}(t\tilde{f})\mathcal{E}(s\tilde{g})$ is $\frac{1}{n!m!} I_n(f^{\otimes n}) I_m(g^{\otimes m})$. Compared it with the coefficient of $s^n t^m$ in (5.3.3), we have

$$\begin{aligned}
I_n(f^{\otimes n}) I_m(g^{\otimes m}) = \sum_{k=0}^{m\wedge n} &\frac{m!n!}{(m-k)!(n-k)!k!} \langle f, g\rangle^k \\
&I_{n+m-2k}(f^{\otimes(m-k)} \otimes g^{\otimes(n-k)}) .
\end{aligned} \tag{5.3.4}$$

Let

$$\begin{aligned}
F(f_n, g_m) &:= I_n(f_n) I_m(g_m) \quad \text{and} \\
G(f_n, g_m) &:= \sum_{k=0}^{n\wedge m} \frac{n!m!}{k!(n-k)!(m-k)!} I_{n+m-2k}(f_n \hat{\otimes}_k g_m)
\end{aligned}$$

Both F and G can be considered as $n\times m$ multilinear forms on $\mathbb{H}^{\hat{\otimes} n}\otimes\mathbb{H}^{\hat{\otimes} m}$. Equation (5.3.4) shows that the induced homogeneous polynomials of F and G are the same:

$$F(f^{\otimes n}\otimes g^{\otimes m}) = G(f^{\otimes n}\otimes g^{\otimes m})\,.$$

By the polarization technique (Proposition 5.7) we know that F and G coincide on $\mathbb{H}^{\hat{\otimes} n}\otimes\mathbb{H}^{\hat{\otimes} m}$, proving (5.3.2). □

To obtain a product formula for m multiple Wiener-Itô integrals $I_{n_1}(f_1)$, $\cdots$, $I_{n_m}(f_m)$, it is more convenient to introduce the Feynman diagram.

Definition 5.7. Given m positive integers $n_1,\cdots,n_m$, denote the set of all vertices by

$$V=\{(p,r),\ 1\le p\le m\,,\ 1\le r\le n_p\}\,.$$

Any pair of vertices $[(p_1,r_1),(p_2,r_2)]$ is called an edge, where $(p_1,r_1),(p_2,r_2)\in V$. An admissible edge is a pair of vertices (an edge) $[(p_1,r_1),(p_2,r_2)]$ satisfying $p_1<p_2$. A set of admissible edges

$$E=\Big\{\ [(p_1,r_1),(p_2,r_2)]\ ,\ \text{where } p_1,p_2,r_1,r_2 \text{ are certain (not necessarily all) numbers satisfying } 1\le p_1<p_2\le m\,, 1\le r_l\le n_{p_1}, 1\le r_2\le n_{p_2}\Big\}$$

is called admissible if each vertex (p,r) appears in at most one edge. By convention we assume that empty set is admissible. Until the end of this section, all edges or all sets of edges that we mention will always be assumed to be admissible. A diagram γ is a pair $(V,E(\gamma))$, where $E(\gamma)$ is a set of (admissible) edges. The set of all diagrams is denoted by $\Gamma(n_1,\cdots,n_m)$.

Let $f_1\in\mathbb{H}^{\hat{\otimes} n_1}$, $\cdots$, $f_m\in\mathbb{H}^{\hat{\otimes} n_m}$ and let $\{e_1,e_2,\cdots\}$ be an orthonormal basis of $\mathbb{H}$. For any edge $[(p_1,r_1),(p_2,r_2)]$ we associate a contraction operation on $f_1\otimes\cdots\otimes f_n$ by contracting between f_{p_1} and f_{p_2}.

$$\begin{aligned}&\text{II}_{[(p_1,r_1),(p_2,r_2)]}(f_1\otimes\cdots\otimes f_n)\\ &=\sum_{k=1}^{\infty} f_1\otimes\cdots\otimes f_{p_1-1}\otimes\langle f_{p_1},e_k\rangle_{r_1}\otimes\langle f_{p_2},e_k\rangle_{r_2}\otimes\cdots\otimes f_{p_2+1}\otimes\cdots\otimes f_m\,,\end{aligned}$$

where $\langle f_{p_1},e_k\rangle_{r_1}$ denotes the scalar product with respect to the r_1-th variables of f_{p_1} and $\langle f_{p_2},e_k\rangle$ denotes the scalar product with respect to the r_2-th variable of f_{p_1}. Namely, if $f_{p_1}=\sum a_{i_1,\cdots,i_{n_{p_1}}} e_{i_1}\otimes\cdots\otimes e_{i_{n_{p_1}}}$, then

$$\langle f_{p_1},e_k\rangle_{r_1}=\sum a_{i_1,\cdots,i_{n_{p_1}}} e_{i_1}\otimes\cdots\otimes e_{i_{r_1-1}}\otimes e_{i_{r_1+1}}\otimes\cdots\otimes e_{i_{n_{p_1}}}\langle e_{i_{r_1}},e_k\rangle\,.$$

But if f_{p_1} is symmetric, this does not matter here. The appearance of r_1 and r_2 is only for counting purpose in our current situation.

It is obvious that $\amalg_{[(p_1,r_1),(p_2,r_2)]}(f_1 \otimes \cdots \otimes f_n)$ is an element of $\mathbb{H}^{\otimes(n_1+\cdots+n_m-2)}$. If E is a set of admissible edges, for each edge of E we apply the contraction operation $\amalg$. The result will be denoted by $\amalg_E$. Thus

$$\amalg_E(f_1 \otimes \cdots \otimes f_n) = \left(\prod_{[(p_1,r_1),(p_2,r_2)] \in E} \amalg_{[(p_1,r_1),(p_2,r_2)]} \right) (f_1 \otimes \cdots \otimes f_n) \,.$$

Note that one f_i may be used for contraction many times but can be used at most n_i times. If $|E|$ denotes the cardinality (the total number of elements) of the set E, then $\amalg_E(f_1 \otimes \cdots \otimes f_n) \in \mathbb{H}^{\otimes(n_1+\cdots+n_m-2|E|)}$.

Now we can state our formula for the product of m multiple Wiener-Itô integrals.

Theorem 5.7. *Let $f_i \in \mathbb{H}^{\hat{\otimes} n_i}$, $i = 1, \cdots, m$. For any diagram $\gamma \in \Gamma(n_1, \cdots, n_m)$, denote by $E(\gamma)$ its associated set of edges. Then*

$$\prod_{i=1}^m I_{n_i}(f_i) = \sum_{\gamma \in \Gamma(n_1,\cdots,n_m)} I_{|\gamma|}\left(\amalg_{E(\gamma)}(f_1 \otimes \cdots \otimes f_m)\right) \,, \tag{5.3.5}$$

where $|\gamma| := n_1 + \cdots + n_m - 2|E|$.

Proof. First, let us prove this identity when $m = 2$. In this case we can decompose $\Gamma(n_1, n_2)$ into disjoint subsets Γ_k, where Γ_k is the set of diagram γ such that $|E(\gamma)| = k$. When $\gamma \in \Gamma_k$,

$$I_{|\gamma|}\left(\left[\amalg_{E(\gamma)}(f_1 \otimes \cdots \otimes f_m)\right]\right) = I_{n_1+n_2-2k}(f_1 \hat{\otimes}_k f_2) \,.$$

Let us compute the cardinality of Γ_k. A diagram in Γ_k has the edges of following form $\{[(1, r_{11}), (2, r_{21})], \cdots, [(1, r_{1k}), (2, r_{2k})]\}$. There are $\binom{n_1}{k}$ ways to choose $r_{11}, \cdots, r_{1k}$ and there are $\binom{n_2}{k}$ ways to choose $r_{21}, \cdots, r_{2k}$. Once these $r_{1,i}, r_{2i}$'s are chosen, all permutations

$$\left\{ \left[(1, r_{11}), (2, r_{2\sigma(1)})\right], \cdots, \left[(1, r_{1k}), (2, r_{2\sigma(k)})\right] \right\} \,,$$

where σ is a permutation of $\{1, \cdots, k\}$, are different set of edges. There is a contribution of $k!$ from these permutations. Thus the total number of elements in Γ_k is $k!\binom{n_2}{k}\binom{n_2}{k}$. Thus, the right hand side of (5.3.5) is

$$\sum_{k \le n_1 \wedge n_2} k! \binom{n_2}{k}\binom{n_2}{k} I_{n_1+n_2-2k}(f_1 \hat{\otimes}_k f_2)$$

By (5.3.2) we see (5.3.5) holds true when $m = 2$.

To show (5.3.5) for general m, we use induction. Assume the identity is true for $m-1$. Then

$$\begin{aligned}
\prod_{i=1}^{m-1} I_{n_i}(f_i) &= \left(\prod_{i=1}^{m-1} I_{n_i}(f_i)\right) I_{n_m}(f_m) \\
&= \sum_{\gamma\in\Gamma(n_1,\cdots,n_{m-1})} I_{|\gamma|}\left(\left[\Pi_{E(\gamma)}(f_1\otimes\cdots\otimes f_{m-1})\right]\right) I_{n_m}(f_m) \\
&= \sum_{\gamma\in\Gamma(n_1,\cdots,n_{m-1})} \sum_{\gamma'\in\Gamma(|\gamma|,n_m)} \\
&\qquad I_{|\gamma'|}\left(\Pi_{E(\gamma')}\left[\Pi_{E(\gamma)}(f_1\otimes\cdots\otimes f_{m-1})\right]\otimes f_m\right) .
\end{aligned} \tag{5.3.6}$$

A diagram $\gamma \in \Gamma(n_1, \cdots, n_{m-1})$ and a diagram $\gamma' \in \Gamma(|\gamma|, n_m)$ can be associated with a diagram $\gamma'' = (\gamma, \gamma')$ in $\Gamma(n_1, \cdots, n_m)$ in the following natural way. Let us consider the set $V_m = \{(p, r_p)\}$ of all vertices and the set $V_{m-1} = \{(p, r_p)\,,\ p \le m-1\}$. The edges in a diagram $\Gamma \in \Gamma(n_1, \cdots, n_{m-1})$ involve only the vertices in V_{m-1}. We call an vertex (p, r_p) a free vertex with respect to the diagram γ if it does not belong to any end point of any edge in $E(\gamma)$. The total number of free vertices in V_{m-1} is $|\gamma|$. An edge $\gamma' \in \Gamma(|\gamma|, n_m)$ can be identified as the form $[(p, r_p), (m, r_m)]$, where (p, r_p) is a free vertex in V_{m-1} with respect to γ. If

$$E(\gamma) = \{[(p_{11}, r_{p_{11}}), (p_{21}, r_{p_{21}})]\ , \cdots , [(p_{1\ell}, r_{p_{1\ell}}), (p_{2\ell}, r_{p_{2\ell}})]\}\ ,$$

then we can associate (γ, γ') with an element $\Gamma(n_1, \cdots, n_m)$ in the following way:

$$\begin{aligned}
&(\gamma, \gamma') \leftrightarrow \\
&\{[(p_{11}, r_{p_{11}}), (p_{21}, r_{p_{21}})]\ , \cdots , [(p_{1\ell}, r_{p_{1\ell}}), (p_{2\ell}, r_{p_{2\ell}})]\, , [(p, r_p), (m, r_m)]\}\ .
\end{aligned}$$

This mapping is obviously a bijection. This combined with (5.3.6) shows (5.3.5) for m. The theorem is then proved by induction. □

As indicated in the previous proof in the case $m = 2$, the formula (5.3.2) is more explicit than (5.3.5). Next theorem will be an analogy to the formula (5.3.2) for the product of m multiple integrals.

Presumably, one can use the Feynman diagram formula to obtain a more explicit formula for the product of m multiple Wiener-Itô integrals by counting all terms which have the same form. But it seems that this combinatorial is complex. We use the idea of polarization. Assume that $h_1, \cdots, h_m \in \mathbb{H}$. Then for any real numbers $t_1, \cdots, t_m \in \mathbb{R}$, we have

$$\begin{aligned}
\mathcal{E}(t_1 h_1) \cdots \mathcal{E}(t_m h_m) &= \mathcal{E}(t_1 h_1 + \cdots + t_m h_m) \prod_{1 \le i < j \le m} \exp\{t_i t_j \langle h_i, h_j \rangle\} \\
&= \sum_{\ell, k_{12}, \cdots, k_{m-1,m}=0}^{\infty} \frac{1}{\ell!} I_\ell\left((t_1 h_1 + \cdots + t_m h_m)^{\otimes \ell}\right) \prod_{1 \le i < j \le m} \frac{(t_i t_j)^{k_{ij}}}{k_{ij}!} \langle h_i, h_j \rangle^{k_{ij}} \\
&= \sum_{\ell_1, \cdots, \ell_m, k_{12}, \cdots, k_{m-1,m}=0}^{\infty} \frac{t_1^{\ell_1} \cdots t_m^{\ell_m}}{\ell_1! \cdots \ell_m!} \prod_{1 \le i < j \le m} \frac{(t_i t_j)^{k_{ij}}}{k_{ij}!} \langle h_i, h_j \rangle^{k_{ij}} \\
&\quad I_{\ell_1 + \cdots + \ell_m}\left(h_1^{\otimes \ell_1} \otimes h_m^{\otimes \ell_m}\right).
\end{aligned}$$

For any set of non-negative integers $\kappa = \{k_{ij}, 1 \le i < j \le m\}$, where $k_{ij} \in \mathbb{Z}_+$, we denote

$$\kappa_i = \sum_{j=i+1}^{m} k_{i,j} + \sum_{j=1}^{i-1} k_{j,i} \tag{5.3.7}$$

$$|\kappa| = \sum_{1 \le i < j \le m} k_{ij} = \frac{1}{2}\left[\kappa_1 + \cdots + \kappa_m\right]; \tag{5.3.8}$$

$$\kappa! = \prod_{1 \le i < j \le m} k_{ij}!. \tag{5.3.9}$$

With these notations, we have

$$\begin{aligned}
\mathcal{E}(t_1 h_1) \cdots \mathcal{E}(t_m h_m) &= \sum_{\ell_1, \cdots, \ell_m, k_{12}, \cdots, k_{m-1,m}=0}^{\infty} \frac{t_1^{\ell_1 + \kappa_1} \cdots t_m^{\ell_m + \kappa_m}}{\ell_1! \cdots \ell_m! \kappa!} \\
&\quad \prod_{1 \le i < j \le m} \langle h_i, h_j \rangle^{k_{ij}} I_{\ell_1 + \cdots + \ell_m}\left(h_1^{\otimes \ell_1} \otimes h_m^{\otimes \ell_m}\right).
\end{aligned} \tag{5.3.10}$$

Let $f_1 \in \mathbb{H}^{\hat{\otimes} n_1}$, $\cdots$, $f_m \in \mathbb{H}^{\hat{\otimes} n_m}$ and let $\{e_1, e_2, \cdots\}$ be an orthonormal basis of $\mathbb{H}$. We denote by $\amalg_{i,j}$ the contraction between f_i and f_j, in the expression $f_1 \otimes \cdots \otimes f_m$, namely,

$$\begin{aligned}
\amalg_{i,j}(f_1 \otimes \cdots \otimes f_m) &= \sum_{k=1}^{\infty} f_1 \otimes \cdots \otimes f_{i-1} \otimes \langle f_i, e_k \rangle \otimes f_{i+1} \otimes \cdots \otimes f_{j-1} \\
&\quad \otimes \langle f_j, e_k \rangle \otimes f_{j+1} \otimes \cdots \otimes f_m
\end{aligned}$$

For any $\kappa = \{k_{ij}, 1 \le i < j \le m\}$, we denote

$$\amalg_\kappa = \prod_{1 \le i < j \le m} \amalg_{ij}. \tag{5.3.11}$$

Thus, we deduce from (5.3.10) that

$$\mathcal{E}(t_1h_1)\cdots\mathcal{E}(t_mh_m) = \sum_{\kappa_1\le n_1,\cdots,\kappa_m\le n_m} \frac{t_1^{n_1}\cdots t_m^{n_m}}{(n_1-k_1)!\cdots(n_m-k_m)!\kappa!} I_{n_1+\cdots+n_m-2|\kappa|}\left(\amalg_\kappa\left[h_1^{\otimes n_1}\otimes\cdots\otimes h_m^{\otimes n_m}\right]\right) . \tag{5.3.12}$$

Comparing the coefficients of $t_1^{n_1}\cdots t_m^{n_m}$ we see that

$$\prod_{i=1}^m I_{n_i}(h_i^{\otimes n_i}) = \sum_{\kappa_1\le n_1,\cdots,\kappa_m\le n_m} \frac{n_1!\cdots n_m!}{(n_1-\kappa_1)!\cdots(n_m-\kappa_m)!\kappa!} I_{n_1+\cdots+n_m-2|\kappa|}\left(\amalg_\kappa\left[h_1^{\otimes n_1}\otimes\cdots\otimes h_m^{\otimes n_m}\right]\right) . \tag{5.3.13}$$

We can generalize the above formula to a more general one.

Theorem 5.8. *Let* $f_1 \in \mathbb{H}^{\hat{\otimes} n_1}$, $\cdots$, $f_m \in \mathbb{H}^{\hat{\otimes} n_m}$. *With the notations* (5.3.7)-(5.3.9) *and* (5.3.11), *we have*

$$\prod_{i=1}^m I_{n_i}(f_i) = \sum_{\kappa\in\mathbb{K},\kappa_1\le n_1,\cdots,\kappa_m\le n_m} \frac{n_1!\cdots n_m!}{(n_1-\kappa_1)!\cdots(n_m-\kappa_m)!\kappa!} I_{n_1+\cdots+n_m-2|\kappa|}\left(\amalg_\kappa\left[f_1\otimes\cdots\otimes f_m\right]\right) , \tag{5.3.14}$$

where $\mathbb{K} = \{k_{ij}\,, k_{ij}\in\mathbb{Z}_+\,, 1\le i<j\le n\}$.

Proof. Equation (5.3.13) we have obtained implies that equation (5.3.14) holds true when $f_1 = h_1^{\otimes n_1}$, $\cdots$, $f_m = h_m^{\otimes n_m}$. The polarization technique (Proposition 5.7) yields the theorem. □

When $n_1 = \cdots = n_m = 1$ (5.3.14) reduces to

Corollary 5.1. *Let* $f_1\cdots, f_m \in \mathbb{H}$. *Then*

$$\prod_{i=1}^m I_1(f_i) = \sum_{i_1<j_1,\cdots,i_k<j_k} \prod_{\ell=1}^k \langle f_{i_\ell}, f_{j_\ell}\rangle I_{m-2k}\left(\Lambda_{i_1,j_1,\cdots,i_k,j_k} f_1\otimes\cdots\otimes f_m\right) , \tag{5.3.15}$$

where the set of distinct elements $\{i_1, j_1, \cdots, i_k, j_k\}$ *is a subset of* $\{1,\cdots,m\}$ *and* $\Lambda_{i_1,j_1,\cdots,i_k,j_k} f_1\otimes\cdots\otimes f_m$ *means that* $f_{i_1}, f_{j_1}, \cdots, f_{i_k}, f_{j_k}$

are removed from $f_1 \otimes \cdots \otimes f_m$. In particular, we have the following formula. If $X_1, \cdots, X_m$ are centered Gaussians, then the following Feynman diagram formula holds true.

$$\mathbb{E}(X_1 \cdots X_m) = \begin{cases} \sum_{i_1<j_1,\cdots,i_k<j_k} \mathbb{E}(X_{i_1} X_{j_1}) \cdots \mathbb{E}(X_{i_k} X_{j_k}) & \text{if } m = 2k \\ 0 & \text{if } m \text{ is odd}, \end{cases} \tag{5.3.16}$$

where $\{i_1, j_1, \cdots, i_k, j_k\} = \{1, \cdots, m\}$.

Definition 5.8. Let $f_n \in \mathbb{H}^{\hat{\otimes} n}$. f_n is called *negative definite* on $\mathbb{H}^{\hat{\otimes} n}$ if

$$\left\langle f_n, h^{\hat{\otimes} n} \right\rangle_{\mathbb{H}^{\hat{\otimes} n}} \le 0 \qquad \forall\ h \in \mathbb{H}\,.$$

Example 5.14. The following statements are proved in [Hu and Kallianpur (1998)].

(i) If $f_2 \in \mathbb{H}^{\hat{\otimes} 2}$, then there is an $\alpha > 0$ such that $\mathbb{E}\exp[\alpha I_2(f_2)] < \infty$. And $\mathbb{E}\exp[\alpha I_2(f_2)] < \infty$ is true for all $\alpha > 0$ iff f_2 is negative definite (see [Hu and Kallianpur (1998)], Theorem 5.1).

(ii) For any nonzero f_{2n+1} in $\mathbb{H}^{\hat{\otimes} 2n+1}$ ($n \ge 1$) and for any $\lambda \in \mathbb{R}$, $\mathbb{E}\left(\exp\{\lambda I_{2n+1}(f_{2n+1})\}\right) = \infty$. (see [Hu and Kallianpur (1998)], Theorem 5.2).

It is conjectured (see [Hu and Kallianpur (1998)]) that

(iii) If n is even, then $\mathbb{E}\exp[I_n(f_n)] < \infty$ iff f_n is negative definite on $\mathbb{H}^{\hat{\otimes} n}$.

This is useful in the so-called Euclidean quantum field theory (see [Glimm and Jaffe (1987)], [Hu and Kallianpur (1998)]).

The following theorem is due to [Üstünel and Zakai (1989)]. A simpler proof is given by [Kallenberg (1991)].

Theorem 5.9. *Let $f \in \mathbb{H}^{\hat{\otimes} n}$ and $g \in \mathbb{H}^{\hat{\otimes} m}$. Then $I_n(f)$ and $I_m(g)$ are independent if and only if*

$$f \hat{\otimes}_{\mathbb{H}} g = 0\,, \tag{5.3.17}$$

where $f \hat{\otimes}_{\mathbb{H}} g \in \mathbb{H}^{\hat{\otimes} n+m-2}$ defined by

$$f \hat{\otimes}_{\mathbb{H}} g = \sum_{n=1}^{\infty} \langle f\,, e_n \rangle_{\mathbb{H}} \hat{\otimes} \langle g\,, e_n \rangle_{\mathbb{H}}\,.$$

Proof. From (5.3.2) we see

$$\mathbb{E}\left[I_n(f_n)I_m(g_m)\right]^2 = \sum_{k=0}^{n\wedge m}\left(\frac{n!m!}{k!(n-k)!(m-k)!}\right)^2 (m+n-2k)!\|f\hat{\otimes}_k g\|^2$$
$$\geq (m+n)!\|f\hat{\otimes}g\|^2 . \tag{5.3.18}$$

It is easy to see that

$$(m+n)!\|f\hat{\otimes}g\|^2 = n!m!\|f\|^2\|g\|^2 + \sum_{k=1}^{m\wedge n}\binom{n}{k}\binom{m}{k}n!m!\|f\hat{\otimes}_k g\|^2 .$$

If $I_n(f)$ and $I_m(g)$ are independent, then

$$\mathbb{E}\left[I_n(f_n)I_m(g_m)\right]^2 = \mathbb{E}\left[I_n(f_n)\right]^2\mathbb{E}\left[I_m(g_m)\right]^2 = m!n!\|f\|^2\|g\|^2 .$$

Thus

$$n!m!\|f\|^2\|g\|^2 \geq (m+n)!\|f\hat{\otimes}g\|^2$$
$$= n!m!\|f\|^2\|g\|^2 + \sum_{k=1}^{m\wedge n}\binom{n}{k}\binom{m}{k}n!m!\|f\hat{\otimes}_k g\|^2 .$$

This implies that $f\hat{\otimes}_k g = 0$ for $k = 1, \cdots, n\wedge m$. In particular, we have $f\hat{\otimes}_1 g = 0$. This proves the necessity of the condition (5.3.17).

Now we prove the sufficiency. Define $\mathbb{H}_f$ as the closed subspace of $\mathbb{H}$ spanned by

$$\left\{\langle f, g_1\otimes\cdots\otimes g_{n-1}\rangle_{\mathbb{H}^{\otimes(n-1)}}, g_1, \cdots, g_{n-1}\in\mathbb{H}\right\} .$$

We can also define $\mathbb{H}_g$ in a similar way. Let $\{\xi_1, \cdots, \xi_k, \cdots\}$ be an orthonormal basis of $\mathbb{H}_f$ and let $\{\eta_1, \cdots, \eta_k, \cdots\}$ be an orthonormal basis of $\mathbb{H}_g$. From the condition $f\otimes_1 g = 0$, we see that $\langle\xi_i, \eta_j\rangle = 0$ for all i, j. For an orthonormal basis $\{e_1, e_2, \cdots, \}$ of $\mathbb{H}$, we write

$$f = \sum f_{i_1,\cdots,i_n} e_{i_1}\hat{\otimes}e_{i_2}\hat{\otimes}\cdots\hat{\otimes}e_{i_n} ,$$

$\langle f, e_{i_2}\hat{\otimes}\cdots\hat{\otimes}e_{i_n}\rangle \in \mathbb{H}_f$ implies that $\sum_{i_1} f_{i_1,\cdots,i_n}e_{i_1} = \sum_{i_1} h_{i_1,\cdots,i_n}\xi_{i_1}$. Thus,

$$f = \sum h_{i_1,\cdots,i_n}\xi_{i_1}\hat{\otimes}e_{i_2}\hat{\otimes}\cdots\hat{\otimes}e_{i_n} = \sum h_{i_1,\cdots,i_n}e_{i_2}\hat{\otimes}\cdots\hat{\otimes}e_{i_n}\hat{\otimes}\xi_{i_1} .$$

Continue this way to write

$$f = \sum \tilde{f}_{i_1,\cdots,i_n}\xi_{i_1}\hat{\otimes}\xi_{i_2}\hat{\otimes}\cdots\hat{\otimes}\xi_{i_n} .$$

Similarly, we have

$$g = \sum \tilde{g}_{i_1,\cdots,i_m}\eta_{i_1}\hat{\otimes}\eta_{i_2}\hat{\otimes}\cdots\hat{\otimes}\eta_{i_m} .$$

Thus, $I_n(f)$ is a function of $\tilde{\xi}_1, \cdots, \tilde{\xi}_k, \cdots$ and $I_m(g)$ is a function of $\tilde{\eta}_1, \cdots, \tilde{\eta}_k, \cdots$. Since $\left\{\tilde{\xi}_1, \cdots, \tilde{\xi}_k, \cdots\right\}$ and $\{\tilde{\eta}_1, \cdots, \tilde{\eta}_k, \cdots\}$ are independent, $I_n(f)$ and $I_m(g)$ are independent. □

5.4 Multiple Stratonovich Integrals

In general the multiple Wiener-Itô integral $I_p(f) = I_p(f)(x)$ (p is an positive integer) is defined only for almost sure $x \in \mathbb{B}$ and it is not a continuous functional of x with respect to the topology of $\mathbb{B}$. Under some additional condition it may be continuous in x. Even in this continuous case, $I_p(f)(\sigma x) \neq \sigma^p I_p(x)$. This means that the multiple Wiener-Itô integrals are not homogeneous as a functional on $\mathbb{B}$. To make it more clear, let us consider $h \in \mathbb{B}'$, then $I_1(h)(x) = \langle x, h\rangle = \tilde{h}$. For this functional, we do have $I_1(h)(\sigma x) = \sigma I_1(h)(x)$. For any $n \geq 1$, we have

$$I_p(h^{\otimes p}) = \sum_{k\leq p/2} \frac{(-1)^k p!}{2^k k!(p-2k)!} |h|^{2k} \left[I_1(h)\right]^{p-2k} .$$

From here we have

$$I_p(h^{\otimes p})(\sigma x) = \sum_{k\leq p/2} \frac{(-1)^k p!}{2^k k!(p-2k)!} |h|^{2k} \sigma^{p-2k} \left[I_1(h)\right]^{p-2k} .$$

We see clearly that $I_p(h^{\otimes p})(\sigma x) \neq \sigma^p I_p(h^{\otimes p})(x)$ for $p \geq 2$. We will introduce the multiple Stratonovich integrals which will satisfy the homogeneous property.

Let $\{e_1, e_2, \dots\}$ be an orthonormal basis of $\mathbb{H}$. Let $f \in \mathbb{H}^{\hat{\otimes} p}$ and consider the following random variable.

$$S_p^n(f) = \sum_{k_1, \cdots, k_p=1}^{n} \langle f\,, e_{k_1} \hat{\otimes} \cdots \hat{\otimes} e_{k_p}\rangle_{\mathbb{H}^{\hat{\otimes} n}} \tilde{e}_{k_1} \cdots \tilde{e}_{k_p} . \tag{5.4.1}$$

Definition 5.9. If as $n \to \infty$, $\left\{S_p^n(f)\,, n = 1, 2, \cdots\right\}$ is convergent in probability, then we say that the multiple Stratonovich integral of f exists. The limit is called the multiple Stratonovich integral of f and is denoted by $S_p(f)$.

Definition 5.10. Denote

$$\begin{aligned} &\mathrm{Tr}^{k,n} f \\ &= \sum_{i_1, \cdots, i_{p-k}=1}^{p} \langle f\,, e_{i_1} \hat{\otimes} e_{i_1} \hat{\otimes} \cdots \hat{\otimes} e_{i_k} \hat{\otimes} e_{i_k} \hat{\otimes} e_{i_{k+1}} \hat{\otimes} \cdots \hat{\otimes} e_{i_{p-2k}}\rangle e_{i_{k+1}} \hat{\otimes} \cdots \hat{\otimes} e_{i_{p-2k}} \end{aligned}$$

which is considered as an element in $\mathbb{H}^{\hat{\otimes}(p-2k)}$. If as $n \to \infty$, $\mathrm{Tr}^{k,n} f$ converges in $\mathbb{H}^{\hat{\otimes}(p-2k)}$, then we say that f possesses k-th order trace. The limit is called the k-th order trace of f and is denoted by

$$\mathrm{Tr}^k f = \sum_{i_1, \cdots, i_k=1}^{\infty} \langle f\,, e_{i_1} \hat{\otimes} e_{i_1} \hat{\otimes} \cdots \hat{\otimes} e_{i_k} \hat{\otimes} e_{i_k}\rangle_{\mathbb{H}^{\hat{\otimes} 2k}} . \tag{5.4.2}$$

Theorem 5.10. *Let $f \in \mathbb{H}^{\hat{\otimes} p}$. If the traces of order k of f exist for all $k \le p/2$, then the multiple Stratonovich integral of f, namely $S_p(f)$, exists and the following Hu-Meyer formulaes hold*

$$\begin{cases} S_p(f) = \displaystyle\sum_{k \le p/2} \frac{p!}{2^k k!(p-2k)!} I_{p-2k}(\mathrm{Tr}^k f)\,; & (5.4.3) \\ I_p(f) = \displaystyle\sum_{k \le p/2} \frac{(-1)^k p!}{2^k k!(p-2k)!} S_{p-2k}(\mathrm{Tr}^k f)\,. & (5.4.4) \end{cases}$$

Proof. Using the Feynman diagram formula (5.3.15) for the product $\tilde{e}_{\ell_1} \cdots \tilde{e}_{\ell_p}$ and using the notation there we have

$$S_p^n(f) = \sum_{\ell_1, \cdots, \ell_p = 1}^{n} \langle f, e_{\ell_1} \hat{\otimes} \cdots \hat{\otimes} e_{\ell_p} \rangle_{\mathbb{H}^{\hat{\otimes} p}} \sum_{k \le p/2} \sum_{i_1 < j_1, \cdots, i_k < j_k} \langle e_{\ell_{i_1}}, e_{\ell_{j_1}} \rangle$$
$$\cdots \langle e_{\ell_{i_k}}, e_{\ell_{j_k}} \rangle I_{p-2k} \left(\Lambda_{\ell_{i_1}, \cdots, \ell_{j_k}} e_{\ell_1} \hat{\otimes} \cdots \hat{\otimes} e_{\ell_p} \right).$$

By the symmetry we can replace the above product inside the third summation for $i_1, j_1, \cdots, i_k, j_k$ by the product with $i_1, j_1, \cdots, i_k, j_k$ being $1, 2, \cdots, 2k-1, 2k$. The total number to sum for the above third summation is the number of combinations of n objects taking $2k$ and then from these $2k$ objects taking p for $i_1, \cdots, i_k$. The remaining p objects can be paired with $i_1, \cdots, i_k$ in $k!$ ways. Finally, because of the condition $i_1 < j_i, \cdots, i_k < j_k$, we need to divide the total by $1/2^k$. Thus, the total number to sum for the above third summation is

$$\binom{n}{2k} \binom{2k}{k} k! \frac{1}{2^k} = \frac{n!}{2^k k!(p-2k)!}.$$

Thus we have

$$S_p^n(f) = \sum_{k \le p/2} \frac{p!}{2^k k!(p-2k)!} \sum_{\ell_1, \cdots, \ell_p = 1}^{n} \langle f, e_{\ell_1} \hat{\otimes} \cdots \hat{\otimes} e_{\ell_p} \rangle_{\mathbb{H}^{\hat{\otimes} n}}$$
$$\langle e_{\ell_1}, e_{\ell_2} \rangle \cdots \langle e_{\ell_{2k-1}}, e_{\ell_{2k}} \rangle I_{p-2k} \left(e_{\ell_{2k+1}} \hat{\otimes} \cdots \hat{\otimes} e_{\ell_p} \right)$$
$$= \sum_{k \le p/2} \frac{p!}{2^k k!(p-2k)!} I_{p-2k} \Bigg(\sum_{\ell_1, \cdots, \ell_{p-k} = 1}^{n} \langle f, e_{\ell_1} \hat{\otimes} e_{\ell_1} \hat{\otimes} \cdots \hat{\otimes} e_{\ell_k} \hat{\otimes} e_{\ell_k}$$
$$\hat{\otimes} e_{\ell_{k+1}} \hat{\otimes} \cdots \hat{\otimes} e_{\ell_{p-k}} \rangle e_{\ell_{k+1}} \hat{\otimes} \cdots \hat{\otimes} e_{\ell_{p-k}} \Bigg)$$
$$= \sum_{k \le p/2} \frac{p!}{2^k k!(p-2k)!} I_{p-2k} \left(\mathrm{Tr}^{k,n} f \right),$$

where in the above second equality we used the fact that $\{e_k\}$ is orthonormal. By the assumption $\mathrm{Tr}^{k,n} f$ converges to $\mathrm{Tr}^k f$ in $\mathbb{H}^{\hat{\otimes}(p-2k)}$, we see that $I_{p-2k}\left(\mathrm{Tr}^{k,n} f\right)$ converges to $I_{p-2k}\left(\mathrm{Tr}^k f\right)$ in $L^2(\mathbb{B})$. This proves that the multiple Stratonovich integral exist and (5.4.3) holds true.

Clearly, $S_{p-2k}(\mathrm{Tr}^k f)$ is well-defined by the above argument. To show (5.4.4), it suffices to show the identity holds true when f is of the form $f = f_1 \hat{\otimes} \cdots \hat{\otimes} f_p$, where $f_1, \cdots, f_p \in \mathbb{H}$. By the polarization technique (5.2.1), it suffices to show (5.4.4) when $f = h^{\otimes p}$ for any $h \in \mathbb{H}$.

If $f = h^{\otimes p}$ for any $h \in \mathbb{H}$, then all its traces exist and by the definition of the multiple Stratonovich integral we have

$$S_p(h^{\otimes p}) = \tilde{h}^p \,.$$

Now for any $t \in \mathbb{R}$,

$$\begin{aligned}
\mathcal{E}(t\tilde{h}) = e^{t\tilde{h}} e^{-\frac{t^2|h|^2}{2}} &= \sum_{\ell,k=0}^{\infty} \frac{t^\ell \tilde{h}^\ell}{\ell!} \frac{(-1)^k |h|^{2k} t^{2k}}{2^k k!} \\
&= \sum_{p=0}^{\infty} t^p \sum_{k \le p/2} \frac{(-1)^k}{2^k k!(p-2k)!} |h|^{2k} \tilde{h}^{p-2k} \\
&= \sum_{p=0}^{\infty} t^p \sum_{k \le p/2} \frac{(-1)^k}{2^k k!(p-2k)!} S_{p-2k}(\mathrm{Tr}^k(h^{\otimes p})) \,, \qquad (5.4.5)
\end{aligned}$$

where in the above second equation we made a substitution $\ell + 2k \to p$. But the coefficient of t^p is $\frac{1}{p!} I_p(h^{\otimes p})$ when expanding $\varepsilon(t\tilde{h})$. Comparing the coefficients of t^p in (5.4.5) yields (5.4.4) for $f = h^{\otimes p}$. The identity (5.4.4) is proved by the polarization technique. □

Remark 5.2. As for the multiple Wiener-Itô integral the multiple Stratonovich integral $S_p(f)$ can be defined directly for $f \in \mathbb{H}^{\otimes p}$, which is not necessarily symmetric. If it exists then

$$S_p(f) = S_p(\mathbb{S}(f)) \,.$$

Example 5.15. If $f_1, \cdots, f_p \in \mathbb{H}$, then by the definition of multiple Stratonovich integral we have immediately

$$S_p(f_1 \hat{\otimes} f_2 \hat{\otimes} \cdots \hat{\otimes} f_p) = \tilde{f}_1 \tilde{f}_2 \cdots \tilde{f}_p \,. \qquad (5.4.6)$$

For this $f = f_1 \otimes f_2 \otimes \cdots \otimes f_p$ we have

$$S_p(f)(\sigma x) = \tilde{f}_1(\sigma x) \tilde{f}_2(\sigma x) \cdots \tilde{f}_p(\sigma x) = \sigma^p \tilde{f}_1(x) \tilde{f}_2(x) \cdots \tilde{f}_p(x) = \sigma^p S_p(f) \,.$$

Thus S_p is (infinite dimensional) homogeneous "polynomial" of degree p.

The trace operator $f \to \mathrm{Tr}^k f$ is a linear operator. But it is not closable (see Example 6.1). There is no maximal extension. Different approximation will induce different traces and hence different approximation theorems. There have been several other approximations of the multiple Stratonovich integrals. Here we present one approximation different than the one by (5.4.1), following the idea of [Hu and Meyer (1993)] (see also the references therein).

Let $\{\varphi_n\,,\ n = 1, 2, \cdots\}$ be a sequence of elements in $\mathbb{H}^{\hat{\otimes}2}$. We identify φ_n as a bounded linear operator from $\mathbb{H}$ to $\mathbb{H}$. Assume that it is an approximation of identity of $\mathbb{H}$. Namely, for any $h \in \mathbb{H}$, $\langle h, \varphi_n\rangle = \varphi_n(h)$ converges to h in $\mathbb{H}$ as $n \to \infty$. The linear operator $\varphi_n : \mathbb{H} \to \mathbb{H}$ is bounded with respect to the norm of $\mathbb{H}$. We also assume that for each n it is also bounded with respect to the norm of $\mathbb{B}$. Thus, φ_n can be extended to $x \in \mathbb{B}$ and for each n we can consider $\langle x, \varphi_n\rangle$ as an $\mathbb{H}$-valued random variable. $\langle x, \varphi_n\rangle^{\otimes p}$ is then a $\mathbb{H}^{\otimes p}$-valued random variable. Thus, for any $f \in \mathbb{H}^{\hat{\otimes}p}$, the following is a well-defined random variable

$$S_p^n(f)(x) = \langle f, \langle x, \varphi_n\rangle^{\otimes p}\rangle\,. \tag{5.4.7}$$

As $n \to \infty$, $S_p^n(f)(x)$ converges to $\langle f, x^{\otimes p}\rangle$ if $x \in \mathbb{H}$ since ϕ_n converges to identity. Usually the above convergence does not hold for every fixed $x \in \mathbb{B}$. In fact, $\langle f, x^{\otimes p}\rangle$ is even ill-defined. We want to study under what condition sequence (5.4.7) does have a limit (in mean square sense) as random variables.

First, notice that we can extend the previous computations to Hilbert space valued multiple integrals with the product replaced by the tensor product. We shall use them freely. We can write by the Hu-Meyer formula (5.4.3)

$$\begin{aligned}\langle x, \varphi_n\rangle^{\otimes p} &= S_p(\varphi_n^{\otimes p})\\ &= \sum_{k\le p/2} \frac{p!}{2^k k!(p-2k)!}\langle \varphi_n, \varphi_n\rangle^{\otimes k} \otimes I_{p-2k}(\varphi_n^{\otimes(p-2k)})\,,\end{aligned}$$

where $\langle \varphi_n, \varphi_n\rangle^{\otimes k}$ is an element in $\mathbb{H}^{\otimes(2k)}$ and $I_{p-2k}(\varphi_n^{\otimes(p-2k)})$ is an $\mathbb{H}^{\otimes(p-2k)}$-valued random variable. Hence, we have the following expression for (5.4.7)

$$\begin{aligned}S_p^n(f) &= \sum_{k\le p/2} \frac{p!}{2^k k!(p-2k)!}\langle f, \langle \varphi_n, \varphi_n\rangle^{\otimes k} \otimes I_{p-2k}(\varphi_n^{\otimes(p-2k)})\rangle\\ &= \sum_{k\le p/2} \frac{p!}{2^k k!(p-2k)!} I_{p-2k}\left(\langle f, \langle \varphi_n, \varphi_n\rangle^{\otimes k} \otimes \varphi_n^{\otimes(p-2k)}\rangle\right),\end{aligned}$$

where $\langle f, \langle \varphi_n, \varphi_n \rangle^{\otimes k} \otimes \varphi_n^{\otimes(p-2k)} \rangle$ is an element in $\mathbb{H}^{\otimes(p-2k)}$. Since φ_n is an approximation of identity, we have that $\langle f, \varphi_n^{\otimes(p-2k)} \rangle$ will converge to f. On the other hand, if $g = g_1 \otimes g_2 \in \mathbb{H}^{\otimes 2}$, then by the symmetry of φ_n,

$$\begin{aligned} \langle g\,, \langle \varphi_n\,, \varphi_n \rangle \rangle &= \langle g_1 \otimes g_2\,, \langle \varphi_n\,, \varphi_n \rangle \rangle \\ &= \langle \varphi_n(g_1)\,, \varphi_n(g_2) \rangle \\ &\to \langle g_1, g_2 \rangle\,. \end{aligned}$$

More generally, if $f = \sum_{finite} a_{ij} e_i \otimes e_j$ for an orthonormal basis $\{e_i\}$, then

$$\langle f, \langle \varphi_n\,, \varphi_n \rangle \rangle \to \sum_{finite} a_{ij} \langle e_i, e_j \rangle = \sum a_{ii}\,.$$

Thus if convergent, then $\langle f, \langle \varphi_n\,, \varphi_n \rangle \rangle$ will converge to the trace $\mathrm{Tr} f$ of f. This gives the following theorem.

Theorem 5.11. *Assume that $\{\varphi_n\,,\ n = 1, 2, \cdots\}$ is a sequence of elements in $\mathbb{H}^{\hat{\otimes} 2}$ which converges to the identity and assume that $\varphi_n(x)$ can be extended to $x \in \mathbb{B}$. Suppose that for any $k \le p/2$, $\langle f, \langle \varphi_n, \varphi_n \rangle^{\otimes k} \otimes \varphi_n^{\otimes(p-2k)} \rangle$ converges in $\mathbb{H}^{\otimes(p-2k)}$ as $n \to \infty$, whose limit is denoted by $\mathrm{Tr}^k f$. Then $S_p^n(f) = \langle f, \langle x, \varphi_n \rangle^{\otimes p} \rangle$ converges in $L^2(\mathbb{B})$ and its limit is*

$$S_p(f) = \sum_{k \le p/2} \frac{p!}{2^k k! (p-2k)!} I_{p-2k}\left(\mathrm{Tr}^k f\right)\,.$$

Example 5.16. If $\mathbb{H}$ is a Hilbert space and $\{e_i\}$ is an orthonormal basis of $\mathbb{H}$. Then one choice of φ_n is

$$\varphi_n = \sum_{i=1}^{n} e_i \otimes e_i\,. \tag{5.4.8}$$

This is an approximation of identity since for any $f \in \mathbb{H}$,

$$\langle \varphi_n, f \rangle = \sum_{i=1}^{n} \langle f, e_n \rangle e_n \to f \quad \text{in } \mathbb{H}\,.$$

It is also obvious that φ_n can be extended to $\mathbb{B}$ as a random variable with values in $\mathbb{H}$:

$$\varphi_n(x) = \sum_{i=1}^{n} \tilde{e}_n(x) e_i\,.$$

This choice of φ_n coincides with the approximation (5.4.1). This approximation was extensively studied by [Budhiraja and Kallianpur (1997)], [Hu and Meyer (1993)], [Johnson and Kallianpur (1990)] and the references therein.

Fixed $T > 0$. Let $(\mathbb{B}, \mathbb{H}, \mathbb{P})$ be the canonical Wiener space on $[0, T]$. Namely, $\mathbb{B} = C_0([0, T], \mathbb{R})$ and $\mathbb{P}$ is the canonical Wiener measure.

$$\mathbb{H} = \left\{ h \in \mathbb{B}, h \text{ is absolutely continuous and } \dot{h} \in L^2([0, T]) \right\} ,$$

with the product $\langle h_1, h_2 \rangle = \int_0^T \dot{h}_1(t)\dot{h}_2(t)dt$, is the Cameron-Martin space. For notational simplicity, we shall identify $\mathbb{H}$ with $L^2([0, T])$ by

$$L^2([0, T]) \ni g \leftrightarrow \int_0^{\cdot} g(t)dt \in \mathbb{H}$$

with

$$\langle \int_0^{\cdot} g_1(t)dt, \int_0^{\cdot} g_2(t)dt \rangle_{\mathbb{H}} = \int_0^T g_1(t)g_2(t)dt =: \langle g_1, g_2 \rangle$$

for any $g_1, g_2 \in L^2 = L^2([0, T])$. Similarly, we can also identify $\mathbb{H}^{\otimes p}$ with $L^2([0, T]^p)$. For example, for any element φ in $L^2([0, T]^2)$, $\int_0^{\cdot} \int_0^{\cdot} \varphi(s, t)dsdt$ is the corresponding element in $\mathbb{H}^{\otimes 2}$.

For an element $f \in L^2([0, T]^p)$ we can define its multiple Wiener-Itô integral. Its multiple Stratonovich integral exists if the corresponding traces exist. We shall give more examples in this particular canonical Wiener space case.

Example 5.17. Let $(\mathbb{B}, \mathbb{H}, \mathbb{P})$ be the canonical Wiener space on $[0, T]$. Let $\pi : 0 = t_0 < t_1 < \cdots < t_{n-1} < t_n = T$ be a partition of interval $[0, T]$. A polygonal approximation of the Brownian motion is

$$W^\pi(t) = W(t_i) + \frac{W(t_{i+1}) - W(t_i)}{t_{i+1} - t_i}(t - t_i) \quad \text{for} \quad t_i \le t < t_{i+1} .$$

This approximation can be considered as a special form of the general one in Theorem 5.11 by the following choice of φ:

$$\varphi^\pi(s, t) = \sum_{i=0}^{n-1} \frac{1}{t_{i+1} - t_i} I_{[t_i, t_{i+1})}(s) I_{[t_i, t_{i+1})}(t) . \tag{5.4.9}$$

For each π, φ^π is an element in $L^2([0, T]^2)$. For any element $h \in L^2([0, T])$,

$$\langle h, \varphi^\pi \rangle = \int_0^T \varphi^\pi(s, t)h(t)dt = \sum_{i=0}^{n-1} \frac{1}{t_{i+1} - t_i} \int_{t_i}^{t_{i+1}} h(t)dt \, I_{[t_i, t_{i+1})}(s)$$

which converges to h, where and in what follows we use $\langle \cdot, \cdot \rangle = \langle \cdot, \cdot \rangle_{L^2}$. To see that this approximation yields the polygonal approximation we do the following computation:

$$\begin{aligned} \langle W, \varphi \rangle_{\mathbb{H}}(t) &= \int_0^t \int_0^T \varphi^\pi(s, r)\dot{W}(r)drds \\ &= W(t_i) + \frac{W(t_{i+1}) - W(t_i)}{t_{i+1} - t_i}(t - t_i) \quad \text{for} \quad t_i \le t < t_{i+1} . \end{aligned}$$

It is obvious that

$$\langle \varphi^\pi , \varphi^\pi \rangle = \varphi^\pi(s,t)\,.$$

Denote $T_i = [t_i, t_{i+1})$ and $\Delta_i = t_{i+1} - t_i$. The k-trace of any function $f(t_1, \cdots, t_p) \in \mathbb{H}^{\hat{\otimes} p}$ is the limit in $L^2([0,T]^{p-2k})$ of

$$\mathrm{Tr}^{k,\pi} f = \sum_{i_1,\cdots,i_k=0}^{n-1} \frac{1}{\Delta_{i_1}\cdots\Delta_{i_k}} \int_{T_{i_1}^2\times\cdots\times T_{i_k}^2} f(t_1,\cdots,t_{2k},s_1,\cdots,s_{p-2k})dt_1\cdots dt_{2k}\,. \tag{5.4.10}$$

Theorem 5.11 gives

Corollary 5.2. *If* $\mathrm{Tr}^{k,\pi} f$ *given by* (5.4.10) *converges to* $\mathrm{Tr}^{k,pol} f$ *in* $L^2([0,T]^{p-2k})$ *as* $\pi \to 0$, *then as* $\pi \to 0$

$$S_p^\pi(f) = \int_{[0,T]^p} f(t_1,\cdots,t_p)\dot{W}^\pi(t_1)\cdots\dot{W}^\pi(t_p)dt_1\cdots dt_p$$

converges to

$$S_p(f) = \sum_{k\le p/2} \frac{p!}{2^k k!(p-2k)!} I_{p-2k}\left(\mathrm{Tr}^{k,pol} f\right)\,.$$

Example 5.18. Still let $(\mathbb{B}, \mathbb{H}, \mathbb{P})$ be the canonical Wiener space over the interval $[0,T]$ and consider the following

$$\varphi^\varepsilon(s,t) = \frac{1}{2\varepsilon} I_{[(s-\varepsilon)^+, s+\varepsilon]}(t) \tag{5.4.11}$$

with the convention that when s,t are outside $[0,T]^2$, $\varphi(s,t) = 0$.

The approximation of the Brownian motion is

$$\begin{aligned} W^\varepsilon(t) &= \int_0^t \int_0^T \varphi^\varepsilon(s,r)dW(r)ds = \frac{1}{2\varepsilon}\int_0^t\int_0^T I_{[(s-\varepsilon)^+,s+\varepsilon]}(r)dW(r) \\ &= \frac{1}{2\varepsilon}\int_0^t (W(s+\varepsilon) - W((s-\varepsilon)^+))ds\,. \end{aligned}$$

The k-th trace for $f \in L^2([0,T]^p)$ is the limit in $L^2([0,T]^{p-2k})$, if the limit exists, of

$$\begin{aligned} \mathrm{Tr}^{k,\varepsilon} f = \frac{1}{(2\varepsilon)^k} \int_{[0,T]^{2k}} & I_{[(s_1-\varepsilon)^+, s_1+\varepsilon]}(t_1) I_{[(s_k-\varepsilon)^+, s_k+\varepsilon]}(t_k) \\ & f(s_1,t_1,\cdots,s_k,t_k,r_1,\cdots,r_{p-2k})ds_1dt_1\cdots ds_kdt_k\,. \end{aligned} \tag{5.4.12}$$

Corollary 5.3. *If the limit of* (5.4.12) *exists in* $L^2([0,T]^{p-2k})$ *for all* $k \le p/2$*, denoted by* $\mathrm{Tr}^{k,sym} f$*, then*

$$\int_{[0,T]^p} f(t_1, \cdots, t_p)\dot{W}^\varepsilon(t_1)\cdots \dot{W}^\varepsilon(t_p)dt_1\cdots dt_p$$

has a limit in $L^2(\mathbb{B})$ *given by*

$$S_p(f) = \sum_{k\le p/2} \frac{p!}{2^k k!(p-2k)!} I_{p-2k}\left(\mathrm{Tr}^{k,sym} f\right).$$

One may also use the following approximation of identity:

$$\varphi^\varepsilon(s,t) = \frac{1}{\varepsilon} I_{[(s-\varepsilon)^+,s]}(t) \tag{5.4.13}$$

to approximate the Brownian motion.

5.5 Right Tail Estimate for Homogeneous Chaos

Let $\mathbb{B}$ and $\mathbb{H}$ be two linear spaces. A function $f : \mathbb{B} \to \mathbb{H}$ is called a homogeneous polynomial of degree d if there is a symmetric d-linear mapping $\bar{f} : \mathbb{B}^d \to \mathbb{H}$ such that

$$f(x) = \bar{f}(x, x, \cdots, x) \quad \forall\, x \in \mathbb{B}.$$

If we define $\pi : \mathbb{B} \to \mathbb{B}^d$ by $\pi(x) = (x, \cdots, x)$, then we can write the above as $f(x) = \bar{f} \circ \pi(x)$.

Let $(\mathbb{B}, \mathbb{H}, \mathbb{P})$ be an abstract Wiener space and let B be a separable Banach space. Let q be a positive integer. If $f \in \mathbb{H}^{\hat{\otimes} q}$, then the multiple Wiener-Itô integral $I_q(f)$ is well defined. But we know that it is not homogeneous. If the traces of f exist, then the multiple Stratonovich $S_q(f)$ exists and $S_q(f)(\alpha x) = \alpha^q S_q(f)$. Now, let $\{f_k\}$ be a sequence of elements in $\mathbb{H}^{\hat{\otimes} q}$ such that $S_q(f_k)$ is well-defined in the sense of definition 5.9. If $e_1, \cdots, e_n$ are elements in B, then

$$F_n = \sum_{k=1}^n S_q(f_k)e_k$$

is an element such that $|||F||| = \mathbb{E}\|F_n\|_B < \infty$. The completion of such F_n's with respect to the norm $|||\cdot|||$ is denoted by $H_q(B)$. An element in $H_q(B)$ is called a B-valued homogeneous chaos of degree q.

Proposition 5.8. *Let* B *be a linear topological space with a continuous seminorm* $\|\cdot\| = \|\cdot\|_B$*. Let* $F : \mathbb{B} \to B$ *be a* B*-valued homogeneous* *chaos of degree q. For any $h \in \mathbb{H}$, $F(x+th)$ is well-defined and there are homogeneous chaos $G_k(x,h)$ of degree $q-k$, $k=0,1,\cdots,q$ (in x) which is also a homogeneous polynomial of degree k in h such that*

$$F(x+th) = \sum_{k=0}^{q} \binom{q}{k} t^k G_k(x,h) \tag{5.5.1}$$

and

$$\sup_{\|h\|_{\mathbb{H}}\le 1} \|G_k(x,h)\|_B < \infty \quad \text{almost surely for all} \quad x \in \mathbb{B}\,. \tag{5.5.2}$$

Proof. Let $F_n = \sum_{k=1}^{n} S_q(f_k)e_k$ (f_k and e_k may depend on n and we omit this explicit dependence on n to alleviate notation) such that F_n converges to F almost surely. Define

$$A = \left\{ x \in \mathbb{B}\,,\ \lim_{n\to\infty} F_n(x) \text{ exists in the norm of } B \right\}$$

and

$$G(x) = \begin{cases} \lim_{n\to\infty} F_n(x) & \text{if } x \in A \\ 0 & \text{if } x \notin A\,. \end{cases}$$

Then $F = G$ almost surely. It is easy to see that $\lambda A = A$ for all $\lambda \in \mathbb{R}$ and $\lambda \neq 0$. For any $r_1, \cdots, r_q \in \mathbb{R}$, we have

$$\begin{aligned} 1 &= \mathbb{P}((r_1^2 + \cdots + r_q^2 + 1)^{1/2} A) \\ &= \mathbb{P} \times \cdots \times \mathbb{P} \times \mathbb{P}\left(\{x_1, \cdots, x_q, y \in A\,,\ r_1x_1 + \cdots + r_qx_q + y \in A\}\right) \\ &= \int \mathbb{P}(A - r_1x_1 - \cdots - r_qx_q)\mathbb{P}(dx_1)\cdots\mathbb{P}(dx_q)\,. \end{aligned}$$

Thus, we can find a measurable subset $C_i \subseteq \mathbb{B}$ with $\mathbb{P}(C_i) = 1$, $i = 1, \cdots, q$ such that

$$\mathbb{P}(A - r_1x_1 - \cdots - r_qx_q) = 1$$

for all $x_i \in C_i\,,\ r_i \in \mathbb{Z}\,, i = 1, \cdots, q$. Take $C = \cap_{i=1}^{q} C_i$. We have $\mathbb{P}(C) = 1$ and

$$\begin{gathered} \mathbb{P}\left(\cap_{r_1,\cdots,r_q\in\mathbb{Z}} \{A - r_1x_1 - \cdots - r_qx_q\}\right) = 1 \\ \forall\, x_1, \cdots, x_q \in C\,,\ r_1, \cdots, r_q \in \mathbb{Z}\,. \end{gathered}$$

Since $\mathbb{P}(\cdot + h)$ is equivalent to $\mathbb{P}(\cdot)$ for any $h \in \mathbb{H}$ we can find a measurable subset still denoted by C such that

$$\mathbb{P}\left(\cap_{r_1,\cdots,r_q\in\mathbb{Z}} \{A - r_1x_1 - \cdots - r_qx_q - s_1h_1 - \cdots - s_qh_q\}\right) = 1 \tag{5.5.3}$$

for all $x_i \in C_i, h_i \in \mathbb{H}$, $r_i, s_i \in \mathbb{Z}, i = 1, \cdots, q$. Hence, we can assume that $\mathbb{H} \subseteq C$ such that (5.5.3) holds for all $x_i \in C$, $r_i \in \mathbb{Z}, i = 1, \cdots, q$. Let $x_1, \cdots, x_q \in C$ and choose a $x_0 \in \cap_{r_1, \cdots, r_q \in \mathbb{Z}} \{A - r_1 x_1 - \cdots - r_q x_q\}$. Then $x_0 + r_1 x_1 + \cdots + r_q x_q \in A$ for all $r_i \in \mathbb{Z}$. Hence from the identity (5.2.1)

$$\bar{G}(x_1, \cdots, , x_q) = \lim_{n \to \infty} \bar{F}_n(x_1, \cdots, x_q)$$

is well-defined for all $x_i \in C \supseteq \mathbb{H}$. Moreover, we have $\|\bar{G}(x_1, \cdots, x_q)\| < \infty$. $\bar{G}$ can be considered as a q-linear form from $\mathbb{B}^q$ to F. Thus, for any $x \in \mathbb{B}$, $h_1, \cdots, h_{k-1} \in C$,

$$C \ni h \to \tilde{G}(x) = \bar{G}(\overbrace{x, \cdots, x}^{q-k}, h_1, \cdots, h_{k-1}, h)$$

is measurable from $\mathbb{B}$ to B. Thus, from Theorem 4.9, we know that restricted to $\mathbb{H}$ the mapping $\tilde{G}$ is bounded. Thus

$$\phi(x, x_1, \cdots, x_{k-1}) := \sup_{h \in \mathbb{H}, \|h\|_{\mathbb{H}} \le 1} \|\bar{G}(\overbrace{x, \cdots, x}^{q-k}, x_1, \cdots, x_{k-1}, h)\|_B$$

is finite and sub-additive in x_{k-1}. In the same argument, we can show that

$$\sup_{h_i \in \mathbb{H}, \|h_i\|_{\mathbb{H}} \le 1, \ i=1,\cdots,k} \|\bar{G}(\overbrace{x, \cdots, x}^{q-k}, h_1, \cdots, h_k)\|_B < \infty .$$

Therefore, we have

$$\sup_{h \in \mathbb{H}, \|h\|_{\mathbb{H}} \le 1} \|G_k(x, h)\|_B < \infty \quad \forall x \in \mathbb{B},$$

where we define

$$G_k(x, h) = \bar{G}(\overbrace{x, \cdots, x}^{q-k}, \overbrace{h, \cdots, h}^{k}) .$$

In particular, we have

$$\sup_{h \in \mathbb{H}, \|h\|_{\mathbb{H}} \le 1} \|G_q(h)\|_B < \infty .$$

It is easy to see that (5.5.1) holds. □

From the above proof, we can also identify $G_q(h)$ as the following almost sure limit

$$G_q(h) = \lim_{t \to \infty} \frac{1}{t^q} F(x + th) . \tag{5.5.4}$$

Now we define

$$\sigma(F) = \sup_{h\in\mathbb{H}\,,\|h\|_{\mathbb{H}}\le 1} \|G_q(h)\|_B\,. \tag{5.5.5}$$

Theorem 5.12. *Let B be a Banach space with norm $\|\cdot\|$ and let F be a B-valued homogeneous Wiener chaos of degree q. If $\sigma(F)$ is defined by (5.5.5), then*

$$\lim_{t\to\infty} \frac{1}{t^{2/q}} \log \mathbb{P}(\|F\| > t) = -\frac{1}{2\sigma(F)^{2/q}}\,. \tag{5.5.6}$$

As a consequence, we have that for any $c < \frac{1}{2\sigma(F)^{2/q}}$

$$\mathbb{E}\exp\left\{c\|F\|^{2/q}\right\} < \infty\,, \tag{5.5.7}$$

and for any $c > \frac{1}{2\sigma(F)^{2/q}}$

$$\mathbb{E}\exp\left\{c\|F\|^{2/q}\right\} = \infty\,. \tag{5.5.8}$$

Proof. For $k = 0, 1, \cdots, q-1$ and for $M_k > 0$, define

$$A_k = \left\{x\,, \binom{q}{k} \sup_{\|h\|_{\mathbb{H}}\le 1} \|G_k(x,h)\|_B \le M_k\right\} \tag{5.5.9}$$

and $A = \cap_{k=0}^{q-1} A_k$. From (5.5.2), we see that there are positive constants M_k, $k = 0, 1, \cdots, q-1$ such that $P(A_k) > 1 - 1/(2q)$. This implies $P(A) > 1/2$. Thus from the isoperimetric inequality (4.4.5), we have

$$\mathbb{P}(A + tB_{\mathbb{H}}(1)) \ge \Phi(t)\,.$$

On the other hand, if $x \in A$ and $h \in \mathbb{B}_{\mathbb{H}}(1)$, then

$$\begin{aligned} \|F(x+th)\|_B &\le \sum_{k=0}^{q} \binom{q}{k} t^k \|G_k(x,h)\|_B \\ &\le \sum_{k=0}^{q-1} M_k t^k + \sigma(F) t^q\,. \end{aligned}$$

This means

$$A + tB_{\mathbb{H}}(1) \subseteq \left\{x \in \mathbb{B}, \|F(x)\|_B \le \sum_{k=0}^{q-1} M_k t^k + \sigma(F)t^q\right\}.$$

Hence,

$$\mathbb{P}\left\{x \in \mathbb{B}, \|F(x)\|_B \le \sum_{k=0}^{q-1} M_k t^k + \sigma(F)t^q\right\} \ge \Phi(t)\,,$$

where we recall $\Phi(t) = \frac{1}{\sqrt{2\pi}} \int_{-\infty}^t e^{-\frac{u^2}{2}} du$. Or

$$\mathbb{P}\left\{x \in \mathbb{B}, \|F(x)\|_B > \sum_{k=0}^{q-1} M_k t^k + \sigma(F) t^q\right\} \le \Psi(t)\,.$$

Let $s = \sum_{k=0}^{q-1} M_k t^k + \sigma(F) t^q$ to obtain $t = \rho(s)$ which is the largest root of this equation. Then,

$$\mathbb{P}\left\{x \in \mathbb{B}, \|F(x)\|_B > t\right\} \le \Psi(\rho(t)) := 1 - \Phi(t)\,.$$

It is easy to see that $\rho(t) \approx \left(\frac{t}{\sigma(F)}\right)^{1/q}$ as $t \to \infty$. This implies

$$\limsup_{t\to\infty} \frac{1}{t^{2/q}} \mathbb{P}\left\{x \in \mathbb{B}, \|F(x)\|_B > t\right\} \le -\frac{1}{2\sigma(F)^{2/q}}\,.$$

Now it remains to show the converse of the above inequality. By Definition (5.5.5), for any $\varepsilon > 0$, there is a $\ell_\varepsilon \in \mathbb{B}' \subseteq \mathbb{H}$ with $\|\ell_\varepsilon\| = 1$ such that

$$\sigma(F) < \|G_q(\ell_\varepsilon)\|_B + \varepsilon\,. \tag{5.5.10}$$

For any $x \in \mathbb{B}$, we can write $x = y_x + z_x$, where $y_x = \langle x, \ell_\varepsilon\rangle \ell_\varepsilon$ and $z_x = x - \langle x, \ell_\varepsilon\rangle \ell_\varepsilon$ are independent. Therefore, we can expand

$$F(x) = \langle x, \ell_\varepsilon\rangle^q G_q(\ell_\varepsilon) + \sum_{k=1}^q \langle x, \ell_\varepsilon\rangle^{q-k} G_k(z_x, \ell_\varepsilon)\,.$$

As in the first part of the proof we can find positive number M, such that

$$\mathbb{P}(A) \ge 1/2\,, \text{where } A = \cap_{k=1}^q \left\{x \in \mathbb{B}\,;\ \|G_k(x, \ell_\varepsilon)\| \le M\,, k = 1, \cdots q\right\}\,.$$

On the other hand, when $x \in A$,

$$\|F(x)\|_B \ge |\langle x, \ell_\varepsilon\rangle|^q \,\|G_q(\ell_\varepsilon)\|_B - M \sum_{k=1}^q |\langle x, \ell_\varepsilon\rangle|^{q-k}\,.$$

Thus

$$A \cap \left\{ |\langle x, \ell_\varepsilon\rangle|^q \,\|G_q(\ell_\varepsilon)\|_B - M \sum_{k=1}^q |\langle x, \ell_\varepsilon\rangle|^{q-k} > t \right\} \subseteq \{F(x) > t\}\,.$$

From the independence of y_x and z_x, we have

$$\begin{aligned}
\mathbb{P}\{F(x) > t\} &\ge \mathbb{P}(A)\mathbb{P}\left\{ |\langle x, \ell_\varepsilon\rangle|^q \,\|G_q(\ell_\varepsilon)\|_B - M \sum_{k=1}^q |\langle x, \ell_\varepsilon\rangle|^{q-k} > t \right\} \\
&\succcurlyeq \frac{1}{2}\mathbb{P}\left\{|\langle x, \ell_\varepsilon\rangle|^q \,\|G_q(\ell_\varepsilon)\|_B > t\right\} \quad \text{as } t \to \infty \\
&= \frac{1}{2}\mathbb{P}\left\{|\langle x, \ell_\varepsilon\rangle| > (t/\|G_q(\ell_\varepsilon)\|_B)^{1/q}\right\} \\
&= \Psi\left(\left(\frac{t}{\|G_q(\ell_\varepsilon)\|_B}\right)^{1/q}\right)\,.
\end{aligned}$$

This implies that

$$\liminf_{t\to\infty}\frac{1}{t^{2/q}}\mathbb{P}\left\{x\in\mathbb{B}, \|F(x)\|_B>t\right\}\geq -\frac{1}{2G_q(\ell_\varepsilon)^{2/q}}\geq -\frac{1}{2(\sigma(F)-\varepsilon)^{2/q}}\,.$$

The proof of the theorem follows immediately since $\varepsilon>0$ is arbitrary. □

One can define the Banach space B valued multiple Wiener-Itô integral F as the limit of $\sum_{k=1}^n I_q(f_k(n))e_k^{(n)}$, where $f_k(n)\in\mathbb{H}^{\hat{\otimes}q}$ and $e_k^{(n)}\in B$:

$$\lim_{n\to\infty}\mathbb{E}\left[\|F-\sum_{k=1}^n I_q(f_k(n))e_k^{(n)}\|\right]\,.$$

It is possible to consider the exponential integrability of the Banach space valued multiple integral.

Corollary 5.4. *Let B be a Banach space with norm $\|\cdot\|$ and let F be a B-valued homogeneous Wiener chaos of degree q which has the form*

$$F=\sum_{k=1}^\infty I_q(f_k)e_k\,.$$

(The above convergence means $\lim_{n\to\infty}\mathbb{E}\|F-\sum_{k=1}^n I_q(f_k)e_k\|_B=0$.) Define

$$G_q(h)=\sum_{k=1}^\infty\langle f_k,h^{\otimes q}\rangle_{\mathbb{H}^{\otimes q}}e_k\,.$$

If $\sigma(F)$ (defined by (5.5.5)) is positive, then

$$\lim_{t\to\infty}\frac{1}{t^{2/q}}\log\mathbb{P}(\|F\|>t)=-\frac{1}{2\sigma(F)^{2/q}}\,. \tag{5.5.11}$$

As a consequence, we have for any $c<\frac{1}{2\sigma(F)^{2/q}}$

$$\mathbb{E}\exp\left\{c\|F\|^{2/q}\right\}<\infty \tag{5.5.12}$$

and for any $c>\frac{1}{2\sigma(F)^{2/q}}$

$$\mathbb{E}\exp\left\{c\|F\|^{2/q}\right\}=\infty\,. \tag{5.5.13}$$

Remark 5.3. If $B = \mathbb{R}$ and if $F = I_q(f_q)$ is a Wiener chaos of order q, then from the hypercontractivity, we have for any $p \geq 1$,

$$\begin{aligned}\|I_q(f_q)\|_p &= (p-1)^{q/2}\|\Gamma\left(\frac{1}{\sqrt{p-1}}\right) I_q(f_q)\|_p \\ &\leq (p-1)^{q/2}\|I_q(f_q)\|_2 = (p-1)^{q/2}\|f_q\|_{\mathbb{H}^q}\,.\end{aligned}$$

Namely,

$$\mathbb{E}\left(|I_q(f_q)|^{\frac{2}{q}}\right)^p \leq C^p p^p \|f_q\|_{\mathbb{H}^q}^{\frac{2}{q}p}\,.$$

By Stirling formula $p! \approx \sqrt{2\pi} p^{p+1/2} e^{-p}$ this implies that there is a constant $c_0 > 0$ such that for all $c < c_0$,

$$\mathbb{E}\exp\left\{c_0\,|I_q(f_q)|^{\frac{2}{q}}\right\} < \infty\,. \tag{5.5.14}$$

Thus, the hypercontractivity can also be used to obtain the sharp exponent $2/q$ for the exponential integrability. But the constant c_0 may not be the best as in Corollary 5.4.

5.6 Chaos Expansion of Exit Time and Skeleton of Wiener Functional

Let $(W_t, t \geq 0)$ be a d-dimensional Brownian motion and let D be a domain of $\mathbb{R}^d$.

$$T_D = \inf\{t > 0, W_t^x \in D\}$$

is called the hitting time of D and

$$\tau_D = \inf\{t > 0, W_t^x \notin D\}$$

is called the first exit time of D. Both T_D and τ_D are stooping time. Define the shift operator

$$\theta_t\omega(s) = \omega(s+t)\,.$$

It is easy to see that for any stopping time s,

$$\tau_D = s + \tau_D \circ \theta_s \quad \text{on } \{s < \tau_D\}\,.$$

This identity implies that for any $t, s \geq 0$,

$$\begin{aligned}I_{\{\tau_D > t+s\}} &= I_{\{\tau_D > t\}} I_{\{\tau_D - t > s\}} = I_{\{\tau_D > t\}} I_{\{\tau_D\circ\theta_t > s\}} \\ &= I_{\{\tau_D > t\}} I_{\{\tau_D > s\}} \circ \theta_t\,. \end{aligned}\tag{5.6.1}$$

This means that $M_t = I_{\{\tau_D > t\}}$ is a multiplicative functional (in the theory of Markov processes).

Adjoin a point ∂ to the domain D and the process defined by

$$W_t^D = \begin{cases} W_t & \text{if } t < \tau_D \\ \partial & \text{if } t \geq \tau_D \,. \end{cases}$$

is called the Brownian motion killed outside of the domain D. The state space of this process is $D_\partial = D \cup \{\partial\}$.

Definition 5.11. The boundary of D is denoted by $\partial D = \bar{D} \cap \overline{D^c}$. A point z is called a regular boundary point of D if $z \in \partial D$ and $P^z(\tau_D = 0) = 1$. The set of regular boundary points of D is denoted by $(\partial D)_r$ and D is said to be regular if $(\partial D)_r = \partial D$.

It is well-known (see [Chung and Zhao (1995), Theorem 2.4] or [Meyer (1962)] for general theory of Markov processes defined by multiplicative functionals) that W_t^D is Markov process whose transition probability has a density $p^D(t,x,y)$ such that

$$\begin{aligned} P_t^D f(x) &= \int_D p^D(t,x,y) f(y) dy = \mathbb{E}^x \left[f(W_t) I_{\{\tau_D > t\}} \right] \\ &= \mathbb{E}^x \left[f(W_t^D) \right] , \quad \forall\, x \in D \,, \end{aligned} \tag{5.6.2}$$

where we use the convention $f(\partial) = 0$ and $\mathbb{E}^x$ means that when one evaluates the expectation, the Brownian motion W is considered as a stochastic process starting at x. Moreover, we have if D is regular domain, then

$$\begin{cases} \frac{\partial}{\partial t} p^D(t,x,y) = \frac{1}{2} \Delta_x p^D(t,x,y) \\ \lim_{t \to 0} p^D(t,x,y) = \delta(x-y) \\ \lim_{x \to z, x \in D} p^D(t,x,y) = 0 \,, \quad \forall\, z \in \partial D \,, \end{cases} \tag{5.6.3}$$

where $\Delta_x = \sum_{i=1}^d \frac{\partial^2}{\partial x_i^2}$ is the Laplacian with respect to x-variable.

Remark 5.4. Here we used ∂ to denote the killing point, partial derivative and boundary without ambiguity.

Example 5.19. Consider one dimensional Brownian motion $\{W_t\,, t \geq 0\}$. Let $\varepsilon > 0$ and let the domain D be the symmetric interval $[-\varepsilon, \varepsilon]$. The exit time is denoted by $\tau_\varepsilon = \inf\{t \geq 0\,, |W_t| > \varepsilon\}$. The transition density of the Brownian motion killed outside of $[-\varepsilon, \varepsilon]$ is

$$P_t^\varepsilon f(x) = \mathbb{E}^x \left[f(W_t) I_{\{\tau_\varepsilon > t\}} \right] = \int_{-\varepsilon}^{\varepsilon} p^\varepsilon(t,x,y) f(y) dy \,, \quad \forall\, x \in (-\varepsilon, \varepsilon) \,,$$

where

$$p^\varepsilon(t,x,y) = \frac{1}{\varepsilon} \sum_{k=1}^\infty \exp\left[-\frac{k^2 \pi^2 t}{8 \varepsilon^2} \right] \sin \frac{k\pi}{2} \left(\frac{x}{\varepsilon} + 1 \right) \sin \frac{k\pi}{2} \left(\frac{y}{\varepsilon} + 1 \right) .$$

If D is a bounded domain of $\mathbb{R}^d$, then from the Markov property, we have

$$\begin{aligned}\mathbb{E}^x\left[f(W_t)I_{\{\tau_D>t\}}\middle|\mathcal{F}_s\right] &= I_{\{\tau_D>s\}}P^D_{t-s}f(W_s)\\ &= I_{\{\tau_D>s\}}\left[P^D_t f(x) + \int_0^s \left[\nabla P^D_{t-r}f(W_r)\right]\cdot dW_r\right] =: \Theta\,,\end{aligned}$$

where the last identity follows from the Itô formula. We claim that

$$\Theta = P^D_t f(x) + \int_0^s I_{\{\tau_D>r\}}\left[\nabla P^D_{t-r}f(W_r)\right]\cdot dW_r\,. \tag{5.6.4}$$

In fact, if $\tau_D > s$, then $\tau_D > r$, it is easy to see that (5.6.4) is true. If $\tau_D \le s$, then $\Theta = 0$. By Itô formula, the right hand side of (5.6.4) is

$$P^D_t f(x) + \int_0^{\tau_D} \left[\nabla P^D_{t-r}f(W_r)\right]\cdot dW_r = P^D_t f(W_{\tau_D}) = 0$$

since $W_{\tau_D} \in \partial D$ and the domain D is regular. Hence, we have

$$\mathbb{E}^x\left[f(W_t)I_{\{\tau_D>t\}}\middle|\mathcal{F}_s\right] = P^D_t f(0) + \int_0^s I_{\{\tau_D>r\}}\left[\nabla P^D_{t-r}f(W_r)\right]\cdot dW_r\,.$$

Letting $s \uparrow t$, we have

Proposition 5.9. *Let D be a bounded regular domain of $\mathbb{R}^d$ and let f be twice continuously differentiable on D and continuous on $\bar{D}$. Then for any $x \in D$,*

$$\begin{aligned}f(W^x_t)I_{\{\tau_D>t\}} &= P^D_t f(x) + \int_0^t I_{\{\tau_D>r\}}\left[\nabla P^D_{t-r}f(W^x_r)\right]\cdot dW_r\\ &= \mathbb{E}^x\left[f(W_t)I_{\{\tau_D>t\}}\right] + \int_0^t I_{\{\tau_D>r\}}\left[\nabla P^D_{t-r}f(W^x_r)\right]\cdot dW_r\,.\end{aligned} \tag{5.6.5}$$

The above proposition gives the Clark derivative of the exit time τ_D of a domain $D \subseteq \mathbb{R}^d$. If $g : [0,\infty) \to \mathbb{R}$ is continuously differentiable, then $g(\tau_D) = \int_0^\infty I_{\{\tau_D>t\}}dg(t) + g(0)$. Thus we have

$$g(\tau_D) = \mathbb{E}\left[g(\tau_D)\right] + \int_0^\infty \left\{\int_0^\infty I_{\{t\wedge\tau_D>r\}}\left[\nabla P^D_{t-r}f(W_r)\right]dg(t)\right\}\cdot dW_r\,. \tag{5.6.6}$$

Applying equation (5.6.5) to the integrand $I_{\{\tau_D>r\}}\nabla P^D_{t-r}f(W^x_r)$, we have

$$\begin{aligned}f(W^x_t)I_{\{\tau_D>t\}} &= P^D_t f(x) + \int_0^t P^D_{s_1}\left[\nabla P^D_{t-s_1}f(W^x_{s_1})\right]\cdot dW_{s_1}\\ &\quad + \int_{0<s_1<s_2<t} I_{\{\tau_D>s_1\}}\left[\nabla P^D_{s_1}\right]\otimes\left[\nabla P^D_{t-s_2}f(W^x_{s_1})\right]\cdot dW_{s_1}\otimes dW_{s_2}\,,\end{aligned}$$

where

$$\begin{aligned}&\left[\nabla P^D_{s_1}\right]\otimes\left[\nabla P^D_{t-s_2}f(W^x_{s_1})\right]\,dW_{s_1}\otimes dW_{s_2}\\&=\sum_{i,j=1}^d\left\{\partial_j P^D_{s_1}\left[\partial_i P^D_{t-s_2}f\right]\right\}(W^x_{s_1})\,dW^j_{s_1}dW^i_{s_2}\,.\end{aligned}$$

Similar notations will be used for higher orders. Continue this iteration and we obtain the chaos expansion

Proposition 5.10. *Let D be a regular domain and let τ_D be the first exit time of D. Then we have the following chaos expansion concerning the first exit time:*

$$\begin{aligned}f(W^x_t)I_{\{\tau_D>t\}}=\mathbb{E}\left[f(W^x_t)I_{\{\tau_D>t\}}\right]+\sum_{n=1}^\infty\int_{0<s_1<\cdots<s_n<t}P^D_{s_1}\left[\nabla P^D_{s_2-s_1}\right]\\\otimes\cdots\otimes\left[\nabla P^D_{t-s_n}f(x)\right]\cdot dW_{s_1}\otimes\cdots\otimes dW_{s_n}\end{aligned}\tag{5.6.7}$$

In particular, if g is continuously differentiable such that $\mathbb{E}\left[g(\tau_D)\right]<\infty$, then we have

$$\begin{aligned}g(\tau_D)=\mathbb{E}\left[g(\tau_D)\right]+\sum_{n=1}^\infty\int_{0<s_1<\cdots<s_n<\infty}P^D_{s_1}\left[\nabla P^D_{s_2-s_1}\right]\\\otimes\cdots\otimes\left[\int_{s_n}^\infty\nabla\left(P^D_{t-s_n}1\right)(x)dg(t)\right]\cdot dW_{s_1}\otimes\cdots\otimes dW_{s_n}\,.\end{aligned}\tag{5.6.8}$$

Now let us assume that $\{W_t, 0\le t\le 1\}$ is a one dimensional Brownian motion on the canonical Wiener space $(\mathbb{B},\mathbb{H},\mathbb{P})$. Let $F:\mathbb{B}\to\mathbb{R}$ be a random variable. Usually F is defined almost surely on $\mathbb{B}$. One may arbitrarily modify its values on any set of probability zero. Since $\mathbb{H}$ is subset of $\mathbb{B}$ with measure zero, there is no meaning to talk about the values of F on $\mathbb{H}$ in general. However, we can give the following definition.

Definition 5.12. Let F be a random variable. We define the value of F at $h\in\mathbb{H}$ by the following limit if it exists

$$F(h)=\lim_{\varepsilon\downarrow 0}\mathbb{E}[F|\sup_{0\le t\le 1}|W_t-h_t|<\varepsilon]=\lim_{\varepsilon\downarrow 0}\frac{\mathbb{E}\left[FI_{\{\sup_{0\le t\le 1}|W_t-h_t|<\varepsilon\}}\right]}{\mathbb{P}\left(\sup_{0\le t\le 1}|W_t-h_t|<\varepsilon\right)}\,.\tag{5.6.9}$$

It is the value (skeleton) at h of F.

In the study of Feynman path integral, Hu and Meyer gave (in [Hu and Meyer (1988a)], [Hu and Meyer (1988b)]) a definition of the value at $h \in \mathbb{H}$ for multiple Itô-Wiener integrals (given by the now so-called Hu-Meyer formula). To introduce this definition we formally define the traces of a symmetric function $f_n : [0,1]^n \to \mathbb{R}$,

$$\mathrm{Tr}^k f_n(s_1, \cdots, s_{n-2k}) = \int_{[0,1]^k} f_n(s_1, \cdots, s_{n-2k}, t_1, t_1, \cdots, t_k, t_k) dt_1 \cdots dt_k \,. \tag{5.6.10}$$

With this notion of traces, the value of $F = I_n(f_n)$ at h is defined by

$$I_n(f_n)(h) = \sum_{k \le n/2} \frac{(-1)^k n!}{2^k k!(n-2k)!} \langle \mathrm{Tr}^k f_n, h^{\otimes(n-2k)} \rangle_{L^2([0,1]^{n-2k})} \,. \tag{5.6.11}$$

In the study of Feynman integral by analytic continuation, one has also to define the value of an Itô-Wiener functional at $W^\sigma = \sigma W$ (notice that the measures defined by W^σ and W are mutually singular when $\sigma \ne 1$). One can define the multiple Iô-Wiener integral $I_n^\sigma(W^\sigma)$ with respect to Brownian motion W^σ in the same way as in the case $\sigma = 1$:

$$I_n^\sigma(f_n) = \int_{[0,t]^n} f_n(t_1, \cdots, t_n) dW_{t_1}^\sigma \cdots dW_{t_n}^\sigma \,.$$

[We use the usual convention that when $\sigma = 1$ we omit the dependence on σ: $I_n^1(f_n) = I_n(f_n)$). Then the value of $I_n^\sigma(f_n)$ at $W^\tau = \tau W$ is given by the following Hu-Meyer formula ([Hu and Meyer (1988a)], [Hu and Meyer (1988b)]):

$$I_n^\sigma(f_n)(W^\tau) = \sum_{k \le n/2} \frac{(\tau^2 - \sigma^2)^k n!}{2^k k!(n-2k)!} I_{n-2k}^\tau \left(\mathrm{Tr}^k f_n\right)(W^\tau) \,. \tag{5.6.12}$$

It is interesting to note that (5.6.11) can be considered as a particular case of (5.6.12) when $\sigma = 1$ and $\tau = 0$ with the convention that when variance $\tau = 0$, $W^\tau = h$ is a differential function and $I_n^0(f_n)(h) = \langle f_n, h^{\otimes n} \rangle$.

It is equally interesting to see the meaning of the case when $\sigma = 0$ and $\tau = 1$. In this case the identity (5.6.12) becomes

$$I_n^0(f_n)(W) = \sum_{k \le n/2} \frac{n!}{2^k k!(n-2k)!} I_{n-2k} \left(\mathrm{Tr}^k f_n\right)(W) \,. \tag{5.6.13}$$

This identity is understood in the following way. First, $I_n^0(W)$ is defined as the limit of

$$I_n^0(W_\varepsilon) = \langle f_n \,, W_\varepsilon \rangle, \quad \text{where } W_\varepsilon \text{ is a differentiable approximation of } W \,.$$

And then, we take the limit as $\varepsilon \to \infty$:

$$
\begin{aligned}
I_n^0(f_n)(W) &= \lim_{\varepsilon} I_n^0(W_\varepsilon) \\
&= \sum_{k\le n/2} \frac{n!}{2^k k!(n-2k)!} I_{n-2k}\left(\mathrm{Tr}^k f_n\right)(W)\,. \qquad (5.6.14)
\end{aligned}
$$

Let us point out that the above identity (5.6.14) has been established by various authors under various approximations and various conditions on f_n (see [Budhiraja and Kallianpur (1997)], [Farré, Jolis and Utzet (2010)], [Delgado and Jolis (1990)], [Delgado and Sanz (1992)], [Hu and Meyer (1993)], [Johnson and Kallianpur (1990)], [Solé and Utzet (1994)], [Sugita (1989)]), [Toh and Chew (2005)]). In Section 5.4, we have studied some of these approximations that appeared in [Hu and Meyer (1993)].

In this section we shall present one rigorous definition of (5.6.11). This work was originally studied in [Carmona and Nualart (1992)]. Our presentation follows [Hu (1992)].

If $F = I_n(f_n)$ is a multiple Itô-Wiener integral, where $f_n \in L^2([0,1]^n)$, we want to evaluate $F(h)$, $h \in \mathbb{H}$. We will only compute $F(0)$ and the general $F(h)$ can be computed by using Girsanov transformation (see Chapter 6.5).

Theorem 5.13. *Let*

$$
F = I_n(f_n) = \int_{0\le s_1,\cdots,s_n\le 1} f_n(s_1,\cdots,s_n)dW(s_1)\cdots dW(s_n)
$$

be a multiple Itô-Wiener integral, where $n = 2m$ is even and $f_n \in L^{1+\frac{n}{2}}([0,1]^n)$. If the m-trace of f_n exists in the following sense

$$
\begin{aligned}
\mathrm{Tr}^m(f_n) = \lim_{\varepsilon\downarrow 0}\frac{1}{\varepsilon^m}\int_{0<t_1<t_2<1,\cdots,0<t_{2m-1}<t_{2m}<1} & f_n(t_1,\ldots,t_n) \\
& \psi((t_2-t_1)/\varepsilon)\ldots\psi((t_n-t_{n-1})/\varepsilon)dt_1\cdots dt_n\,, \qquad (5.6.15)
\end{aligned}
$$

where $\psi(t) = e^{-t}$, then the following limit exists and is given by the following identity.

$$
F(0) = \lim_{\varepsilon\downarrow 0}\frac{\mathbb{E}\left[FI_{\{\sup_{0\le t\le 1}|W_t|<\varepsilon\}}\right]}{\mathbb{P}\left(\sup_{0\le t\le 1}|W_t|<\varepsilon\right)} = \frac{(-1)^m(2m)!}{2^m m!}\mathrm{Tr}^m f_n\,. \qquad (5.6.16)
$$

Proof. *Step 1: Denominator.* We want to compute the limit

$$
F_\varepsilon := \frac{\mathbb{E}\left[FI_{\{\sup_{0\le t\le 1}|W_t|<\varepsilon\}}\right]}{\mathbb{P}\left(\sup_{0\le t\le 1}|W_t|<\varepsilon\right)}\,, \qquad (5.6.17)
$$

where $\varepsilon \downarrow 0$. Let $\tau_\varepsilon = \inf\{t > 0\,;\ |W_t| > \varepsilon\}$ be first exit time of W out of the interval $[-\varepsilon, +\varepsilon]$. We know that $M_t^\varepsilon = I_{\{\tau_\varepsilon > t\}}$ is a multiplicative functional. Denote by $Q_t^\varepsilon f(x) = \mathbb{E}^x[f(W_t)M_t^\varepsilon]$ the associated killing Markov semigroup. From Example 5.19, we see that this semigroup has transition density $q_{t,\varepsilon}(x,y)$, namely,

$$Q_t^\varepsilon f(x) = \mathbb{E}^x[f(W_t)M_t^\varepsilon] = \int_{-\varepsilon}^{\varepsilon} q_{t,\varepsilon}(x,y) f(y) dy\,.$$

Moreover, the density is given by

$$q_{t,\varepsilon}(x,y) = \frac{1}{\varepsilon}\sum_{k=1}^\infty e^{-\lambda_k t/\varepsilon^2} r_k\left(\frac{x}{\varepsilon}\right) r_k\left(\frac{y}{\varepsilon}\right) = \frac{1}{\varepsilon}\, q_t(x/\varepsilon\,, y/\,\varepsilon)\,, \tag{5.6.18}$$

where

$$\begin{cases} \lambda_k = k^2\pi^2/8\,, \quad \text{and} \quad r_k(x) = \sin\dfrac{k\pi}{2}(x+1)\,; \\ q_t(x,y) = \displaystyle\sum_{k=1}^\infty e^{-\lambda_k t} r_k(x)\, r_k(y)\,. \end{cases} \tag{5.6.19}$$

It is well-known that the denominator in (5.6.17) is

$$\mathbb{P}\left(\sup_{0\le t\le 1} |W_t| < \varepsilon\right) \sim \frac{4}{\pi} e^{-\lambda_1/\varepsilon^2}\,. \tag{5.6.20}$$

This can also be obtained by

$$\mathbb{P}\left(\sup_{0\le t\le 1} |W_t| < \varepsilon\right) = \mathbb{E}[M_1^\varepsilon] = \int_{-\varepsilon}^{\varepsilon} q_{t,\varepsilon}(x,y) dy$$

with the explicit expression for $q_{t,\varepsilon}(x,y)$.

Step 2: Numerator. By (5.6.7), we see that

$$\begin{aligned} I_{\left\{\sup_{0\le t\le 1}|W_t|<\varepsilon\right\}} &= I_{\{\tau_\varepsilon > 1\}} \\ &= \mathbb{P}(\tau_\varepsilon > 1) + \sum_{n=1}^\infty \int_{\Sigma_n} g_\varepsilon(t_1,\dots,t_n) dW_{s_1}\cdots dW_{s_n}\,, \end{aligned}$$

where

$$\Sigma_n = \{(t_1,\cdots,t_n)\in\mathbb{R}^n\,;\ 0 < t_1 < \cdots < t_n < 1\}$$

and

$$g_\varepsilon(t_1,\dots,t_n) = q_{t_1,\varepsilon} * q'_{t_2-t_1,\varepsilon} * \dots * q'_{1-t_n,\varepsilon}(x,y) * 1(0)\,. \tag{5.6.21}$$

Here and in what follows we use the notation $p*q(x,y) = \int_{\mathbb{R}} p(x,z)q(z,y)dz$ and $p*q(x) = \int_{\mathbb{R}} p(x,z)q(z)dz$.

Therefore, the numerator in (5.6.17) is given by the following expression:

$$\mathbb{E}\left[I_n(f_n)I_{\{\sup_{0\le t\le 1}|W_t|<\varepsilon\}}\right] = n!\int_{\Sigma_n} f(t_1,\dots,t_n)\,g_\varepsilon(t_1,\dots,t_n)\,dt_1\dots dt_n\,. \tag{5.6.22}$$

As a consequence, we have $F_\varepsilon = n!\tilde{F}_\varepsilon$, where

$$\tilde{F}_\varepsilon = \frac{\pi}{4}e^{\lambda_1/\varepsilon^2}\int_{\Sigma_n} f(t_1,\dots,t_n)\,g_\varepsilon(t_1,\dots,t_n)\,dt_1\dots dt_n\,. \tag{5.6.23}$$

Step 3: Decomposition. Now we want to study the above limit. The idea is to separate the principal terms from the negligible ones. We shall consider the principal term in this step. To this end, let us consider the following two orthogonal systems (which are not complete) of $L^2(\mathbb{R})$ (we omit the explicit dependence on ε to simplify notation),

$$\rho_k(x) = (1/\sqrt{\varepsilon})\,r_k(x/\varepsilon)\,,\quad k\ge 1\,,\qquad \tilde\rho_k(x) = (1/\sqrt{\varepsilon})\,\tilde r_k(x/\varepsilon)\,,\quad k\ge 0\,,$$

where

$$r_k(x) = I_{\{|x|<1\}}\sin\frac{k\pi}{2}(x+1)\,,\qquad \tilde r_k(x) = I_{\{|x|<1\}}\cos\frac{k\pi}{2}(x+1)\,.$$

Notice

$$\int \tilde\rho_k(x)\,\rho_\ell(x)\,dx = \begin{cases} 0 & \text{if } k\equiv \ell \pmod 2,\\ 4\ell/\pi(\ell^2-k^2) & \text{otherwise}. \end{cases} \tag{5.6.24}$$

With these notations, one can write

$$q_{t,\varepsilon} = \sum_{k=1}^\infty e^{-\lambda_k t/\varepsilon^2}\,\rho_k(x)\,\rho_k(y)$$

$$q'_{t,\varepsilon} = \sum_{k=1}^\infty \frac{k\pi}{2\varepsilon}e^{-\lambda_k t/\varepsilon^2}\,\tilde\rho_k(x)\,\rho_k(y)\,.$$

To find the principal term we decompose $q_{t,\varepsilon}$ and $q'_{t,\varepsilon}$ into two parts,

$$q_{t,\varepsilon} = e^{-\lambda_1 t/\varepsilon^2}(a+b_t)\,,\quad q'_{t,\varepsilon} = e^{-\lambda_1 t/\varepsilon^2}(a'+b'_t)\,, \tag{5.6.25}$$

where a, a' are the first terms of the corresponding series for q and q', respectively (which do not depend on t), and while b_t or b'_t denotes the remaining parts. More precisely, we have

$$a(x,y) = \rho_1(x)\,\rho_1(y)\,,\quad a'(x,y) = \frac{\pi}{2\varepsilon}\,\tilde\rho_1(x)\,\rho_1(y)\,,$$

$$b_t(x,y) = \sum_{k\ge 2} e^{-(\lambda_k-\lambda_1)t/\varepsilon^2}\,\rho_k(x)\,\rho_k(y)\,,$$

$$b'_t(x,y) = \sum_{k\ge 2} e^{-(\lambda_k-\lambda_1)t/\varepsilon^2}\,\frac{k\pi}{2\varepsilon}\,\tilde\rho_k(x)\,\rho_k(y)\,.$$

In the expression (5.6.23), we compute $g_\varepsilon(t_1,\dots,t_n)$ using the expression (5.6.21) by replacing each factor q and each factor q' according to (5.6.25).

Now we consider the substitution of q and q' in the expression (5.6.21) by (5.6.25). First, we notice that the factor $e^{-\lambda_1/\varepsilon^2}$ will cancel the factor $e^{\lambda_1/\varepsilon^2}$ in (5.6.23). Using the fact that

$$a * a' = a' * a' = 0$$

and the fact that b'_{t/ε^2} starts with a term $e^{-(\lambda_2-\lambda_1)t/\varepsilon^2}$, it is natural to think that to find the principal term one should place as many terms a or a' as possible, but not consecutive, in the expression (5.6.23). Thus, we see that the principal term in (5.6.23) when $n = 2m$ is even, will be contributed by the value at 0 of the function

$$e^{-\lambda_1/\varepsilon^2} a * b'_{(t_2-t_1)/\varepsilon^2} * a' \dots * b'_{(t_n-t_{n-1})/\varepsilon^2} * a' * 1\,. \tag{5.6.26}$$

Noticing that

$$a' * 1 = \frac{4}{\pi}\sqrt{\varepsilon}\,\tilde\rho_1\,,$$

we see that the factor $\pi/4$ will also disappear. Thus, we are led to the decomposition

$$\begin{aligned}
\tilde F_\varepsilon &= \int_{\Sigma_n} f_n(t_1,\dots,t_n)\,(a + b_{t_1/\varepsilon^2}) * (a' + b'_{(t_2-t_1)/\varepsilon^2}) * \dots \\
&\qquad * (a' + b'_{(t_n-t_{n-1})/\varepsilon^2}) * (a' + b'_{(1-t)/\varepsilon^2}) * 1(0)dt \\
&= \int_{\Sigma_n} f_n(t_1,\dots,t_n)\, j_\varepsilon(t_1,\dots,t_n)\, dt_1\dots dt_n \\
&\qquad + \int_{\Sigma_n} f_n(t_1,\dots,t_n)\, h_\varepsilon(t_1,\dots,t_n)\, dt_1\dots dt_n \\
&=: I_1 + I_2\,,
\end{aligned}$$

where

$$\begin{aligned}
j_\varepsilon(t_1,\dots,t_n) = (\frac{\pi}{2\varepsilon})^m \sqrt{\varepsilon}\,\rho_1(0) \int_{\mathbb{R}^n} \rho_1(x_1)\, b'_{(t_2-t_1)/\varepsilon^2}(x_1,x_2)\,\tilde\rho_1(x_2)\,\rho_1(x_2) \\
b'_{(t_4-t_3)/\varepsilon^2}\dots b'_{(t_n-t_{n-1})/\varepsilon^2}(x_{n-1},x_n)\,\tilde\rho(x_n)\,dx_1\dots dx_n
\end{aligned}$$

and h_ε is the remaining terms of the expansion (the other terms except the term given by (5.6.26)) in the expansion of

$$(a+b_{t_1/\varepsilon^2}) * (a'+b'_{(t_2-t_1)/\varepsilon^2}) * \dots * (a'+b'_{(t_n-t_{n-1})/\varepsilon^2}) * (a'+b'_{(1-t)/\varepsilon^2}) * 1(0)\,.$$

Step 4: Principal term. On interval $[0,1]$, denote

$$\beta'_t(x,y) = \sum_{k\ge 2} e^{-(\lambda_k-\lambda_1)t}\tilde r_k(x)\, r_k(y)$$

$$\phi(t) = -\pi \int r_1(x)\,\beta'_t(x,y)\,\tilde r_1(y)\,dxdy\,.$$

Then by using the explicit expression of a, b, a' and b', we have

$$j_\varepsilon(t_1,\ldots,t_n)=\frac{(-1)^m}{2^m\varepsilon^{2m}}\phi((t_2-t_1)/\varepsilon^2)\ldots\phi((t_n-t_{n-1})/\varepsilon^2)\,,$$

where

$$\phi(t)=\sum_{\substack{k\geq 2\\ k\ \text{pair}}}\frac{8k^2}{(k^2-1)^2}\,e^{-(k^2-1)\pi^2t/8}\,.$$

Moreover, we have that $\int_0^\infty \phi(t)\,dt=1$. Therefore, the principal term of F_ε is computed as follows

$$\begin{aligned}
n!I_1 &= \int_{\Sigma_n} f_n(t_1,\ldots,t_n)j_\varepsilon(t_1,\ldots,t_n)dt\\
&= \frac{(-1)^m(2m)!}{2^m\varepsilon^{2m}}\int_{\Sigma_n} f_n(t_1,\ldots,t_n)\phi((t_2-t_1)/\varepsilon^2)\ldots\phi((t_n-t_{n-1})/\varepsilon^2)dt\\
&= \frac{(-1)^m(2m)!}{2^m\varepsilon^{2m}m!}\int_{0<t_1<t_2<1,\cdots,0<t_{2m-1}<t_{2m}<1} f_n(t_1,\ldots,t_n)\\
&\qquad \phi((t_2-t_1)/\varepsilon^2)\ldots\phi((t_n-t_{n-1})/\varepsilon^2)dt\\
&\to \frac{(-1)^m(2m)!}{2^m m!}\mathrm{Tr} f_n\,,
\end{aligned} \tag{5.6.27}$$

where the above convergence follows from the assumption on the existence of m-trace of f_n.

Step 5: The remaining terms. From (5.6.27), we see that to show (5.6.16) it suffices to show that I_2 converges to 0. To this end, we shall use the following facts whose proofs are straightforward.

Let $\{\rho_k\}$ and $\{\tilde\rho_k\}$ be two orthogonal systems in $L^2([0,1])$ and let G be the operator given by the kernel

$$g(x,y)=\sum_{k=1}^\infty a_k\,\tilde\rho_k(x)\,\rho_k(y)\,.$$

Then the norm $\|\,G\,\|_{22}$ as an operator from $L^2([0,1])$ to $L^2([0,1])$ is bounded by $\sup_k|a_k|$, and the norm $\|\,G\,\|_{2\infty}$ as an operator from $L^2([0,1])$ to $L^\infty([0,1])$ is bounded by $\sup_x(\sum_{k=1}^\infty|\,a_k\tilde\rho_k(x)\,|^2)^{1/2}$.

We need to show that all terms in

$$I_2=\int_{\Sigma_n} f_n(t_1,\ldots,t_n)\,h_\varepsilon(t_1,\ldots,t_n)\,dt_1\ldots dt_n\,, \tag{5.6.28}$$

converge to 0. We are going to show that h_ε converges to 0 in $L^p(\Sigma_n)$ for certain $p<2$, This will require that $f_n\in L^q(\Sigma_n)$ (we will see that q can be chosen as $q=1+n/2$ as given in the statement of the theorem).

First, we notice that each term in I_2 can be written as $u_0 * u_1 * \dots u_n * 1(0)$, which is bounded by

$$\| u_0 \|_{2\infty} \| u_1 \|_{22} \dots \| u_n \|_{22} \| 1 \|_2 ,$$

where u_j is a, a', b or b'.

We need to bound all of above factors. The last one is equal to $\| 1 \|_2 = \sqrt{2\varepsilon}$. Now let us start to bound $\| u_j \|_{22}$ and $\| u_j \|_{2\infty}$. We shall use C to denote a generic constant which may vary from place to place. For the bounds for a, we have easily

$$\| a \|_{2\infty} \leq \sup_x |\rho_1(x)| = 1/\sqrt{\varepsilon}, \quad \| a' \|_{22} \leq \frac{1}{\varepsilon}. \tag{5.6.29}$$

Next, for all $\alpha \in [0,1]$,

$$\begin{aligned} \| b \|_{2\infty} &\leq \frac{1}{\sqrt{\varepsilon}} (\sum_{k=1}^{\infty} e^{-Ck^2 t/\varepsilon^2})^{1/2} \leq \frac{1}{\sqrt{\varepsilon}} (\sum_{k=1}^{\infty} k^{\alpha} e^{-Ck^2 t/\varepsilon^2})^{1/2} \\ &\leq C \frac{1}{\sqrt{\varepsilon}} \Big(\int_0^{\infty} x^{\alpha} e^{-Ctx^2/\varepsilon^2} dx \Big)^{1/2} \leq C\varepsilon^{\alpha/2} t^{-(1+\alpha)/4} . \end{aligned} \tag{5.6.30}$$

As for the derivative of b, we have

$$\| b'_{t/\varepsilon^2} \|_{22} \leq \frac{C}{\varepsilon} \sup_{k \geq 2} k e^{-Ck^2 t/\varepsilon^2} .$$

Using the inequality $e^{-x} \leq Cx^{\alpha-1}$ for all $\alpha \in [0,1]$ and for all $x \geq 0$, we obtain

$$\| b'_{t/\varepsilon^2} \|_{22} \leq \frac{C}{\varepsilon} (\frac{\varepsilon^2}{t})^{1-\alpha} . \tag{5.6.31}$$

After substituting the above bounds into the generic term $u_0 * u_1 * \dots u_n * 1(0)$ of I_2, we will see a multiple integral of the form $J_m(t,\alpha) := \int_{\{0<r_1<\dots<r_m<t\}} \prod_{i=1}^m (r_i - r_{i-1})^{\alpha_i} dr$, where by convention, $r_0 = 0$. This integral can be bounded easily by

$$J_m(t,\alpha) = \int_{\{0<r_1<\dots<r_m<t\}} \prod_{i=1}^m (r_i - r_{i-1})^{\alpha_i} dr \leq \frac{\kappa^m t^{\sum_{i=1}^m \alpha_i + m}}{\Gamma(\sum_{i=1}^m \alpha_i + m + 1)}, \tag{5.6.32}$$

where κ is a universal constant and Γ is the Gamma function.

To explain our idea more clearly, let us consider only the second chaos case ($n = 2$). If we omit the principal term and the zero terms, the remaining terms are

$$J_1 = a * b'_{(t-s)/\varepsilon^2} * b'_{(1-t)/\varepsilon^2} * 1(0) , \quad J_2 = b_{s/\varepsilon^2} * a' * b'_{(1-t)/\varepsilon^2} * 1(0) ,$$

$$J_3 = b_{s/\varepsilon^2} * b'_{(t-s)/\varepsilon^2} * a' * 1(0) , \quad J_4 = b_{s/\varepsilon^2} * b'_{(t-s)/\varepsilon^2} * b'_{(1-t)/\varepsilon^2} * 1(0) .$$

Let us explain the typical term J_1 (which contains a) and J_3 (which contains b). J_1 can be bounded by

$$|J_1| \le \frac{C}{\sqrt{\varepsilon}}\frac{1}{\varepsilon}(\frac{\varepsilon^2}{t-s})^{1-\alpha}(\frac{\varepsilon^2}{1-s})^{1-\alpha}\sqrt{\varepsilon} = C\varepsilon^{4(1-\alpha)-2}(t-s)^{1-\alpha}(1-s)^{1-\alpha}.$$

The integral of $\int_{0<s<t<1}|J_1|^p(s,t)dsdt$ is finite if α satisfies $-p(1-\alpha) > -1$ and it converges to 0 as $\varepsilon \to 0$ if $4(1-\alpha)-2>0$. One can find that α satisfies the above two conditions if $p<2$.

As for J_3 we choose two constants $\beta, \alpha > 0$ and apply the above bounds for a', b and b' to obtain

$$\begin{aligned}|J_3| &\le C\,\frac{\varepsilon^{\beta/2}}{s^{(1+\beta)/4}}\,\frac{1}{\varepsilon}\,\frac{1}{\varepsilon}\,(\frac{\varepsilon^2}{t-s})^{1-\alpha}\sqrt{\varepsilon}\\ &= C\,\varepsilon^{\beta/2+2(1-\alpha)-3/2}\,s^{-(1+\beta)/4}\,(1-t)^{-(1-\alpha)}.\end{aligned}$$

The condition we need to impose so that the integral of

$$\int_{0<s<t<1}|J_3|^p(s,t)dsdt$$

is finite and converges to 0 as $\varepsilon \to 0$ is

$$\beta/2 - 2\alpha + 1/2 > 0\,, \quad -p(1+\beta)/4 > -1\,, \quad -p(1-\alpha) > -1\,.$$

This is possible if $p<2$, with the choice $\beta \sim 1$ and $\alpha \sim 1/2$.

The general case can be done in a similar way. We can distinguish the following two types of terms.

(i) The term J_1^{generic} of first type starts with a, which can also be considered as a term obtained from the principal term by replacing k (among the m) factors a' by b'.
(ii) The term J_2^{generic} of second type starts with b.

The study of the first type is easy. We use (5.6.29) to bound a' and (5.6.31) to bound b'. Thus we have that a generic term of first type can be bounded by

$$|J_1^{\text{generic}}| \le \frac{C}{\sqrt{\varepsilon}}\,\frac{1}{\varepsilon^{m-1}}\,\frac{1}{\varepsilon^{m+1}}\,\frac{\varepsilon^{2(m+1)(1-\lambda)}}{\prod_i \Delta_i^{1-\alpha}}\sqrt{\varepsilon}\,,$$

where the index i correspond to $m+1$ terms b' under the consideration and $\Delta_i = t_{i+1}-t_i$. Raise to the power p and we obtain (as $n=2m$)

$$|J_1^{\text{generic}}| \le C\varepsilon^{p((n+2)(1-\alpha)-n)}\prod_i \Delta_i^{-p(1-\alpha)}$$

and the integral $\int_{\Sigma_n} |J_1^{\text{generic}}|dt$ is finite and converges to 0 as $\varepsilon \downarrow 0$, if

$$p(1-\alpha) < 1\,, \quad (n+2)(1-\alpha) > n\,,$$

which is implied by $p < \frac{n+1}{n}$. This means if $f \in L^q$, where $q = 1 + \frac{n}{2}$, then $\int_{\Sigma_n} J_1^{\text{generic}} f_n(t_1, \cdots, t_n)dt$ converges to 0 as $\varepsilon \downarrow 0$.

The estimate of the term J_2^{generic} of second type can be done in a similar way. However, we need to use (5.6.30) for b and (5.6.31) for b'. This imposes the use of two different exponents β and α (like what we did above for J_3). Since the procedure is very similar, we omit the details. □

Chapter 6

Analysis of Nonlinear Wiener Functionals

6.1 Gross-Sobolev Derivatives

First, let us recall some results from functional analysis (see for example [Yosida (1980)]).

Definition 6.1. Let $\mathbb{B}_1$ and $\mathbb{B}_2$ be two Banach spaces and let T be an (unbounded) linear operator from $\mathrm{Dom}(T) \subseteq \mathbb{B}_1$ to $\mathbb{B}_2$. The graph $G(T)$ of the operator T is the set $\{(x, Tx) : x \in \mathrm{Dom}(T)\} \subseteq \mathbb{B}_1 \times \mathbb{B}_2$. T is called a closed linear operator if the graph $G(T)$ constitutes a closed linear subspace of $\mathbb{B}_1 \times \mathbb{B}_2$.

From this definition it is easy to see that T is closed if and only if the following condition is satisfied:

The conditions $\{x_n\} \subseteq \mathrm{Dom}(T)$,

$$\lim_{n\to\infty} \|x_n - x\|_{\mathbb{B}_1} = 0, \quad \text{and} \quad \lim_{n\to\infty} \|Tx_n - y\|_{\mathbb{B}_2} = 0$$

implies that $x \in \mathrm{Dom}(T)$ and $Tx = y$.

Example 6.1. The trace operator Tr defined by (5.4.10), when $k = 1$ and $p = 2$, is not a closed operator. In fact, take $\mathbb{B}_1 = L^2([0, T]^2)$ and $\mathbb{B}_2 = \mathbb{R}$. Let $f_n(s,t) = I_{\{|s-t|\le \frac{1}{n}\}}$. Then, it is easy to see that $\int_0^T \int_0^T f_n^2(s,t)dsdt \to 0$ and $\mathrm{Tr} f_n \to T$. This means $f_n \to 0 \in \mathrm{Dom}(\mathrm{Tr})$. But $\mathrm{Tr}(0) = 0 \ne T$.

Definition 6.2. A linear operator $T : \mathrm{Dom}(T) \to H$ from its domain $\mathrm{Dom}(T) \subseteq \mathbb{B}_1$ to $\mathbb{B}_2$ is called closable if it has a closed extension $\tilde{T} : D(\tilde{T}) \subseteq \mathbb{B}_1 \to \mathbb{B}_2$ where $D(T)$ is a subset $D(\tilde{T})$. The extension B of A is sometimes called the closure of A.

Proposition 6.1. *If $\mathbb{B}_1$ and $\mathbb{B}_2$ are two Banach spaces, then T is closable if and only if the following condition is satisfied:*

$\{x_n\} \subseteq \mathrm{Dom}(T)$, $\lim_{n\to\infty} \|x_n\|_{\mathbb{B}_1} = 0$ and $\lim_{n\to\infty} \|Tx_n - y\|_{\mathbb{B}_2} = 0$ implies that $y = 0$.

Proof. First, we prove the "only if" part. Since $\tilde{T}$ is the extension of T, we have that $\lim_{n\to\infty} \|Tx_n - y\|_{\mathbb{B}_2} = 0$ implies $\lim_{n\to\infty} \|\tilde{T}x_n - y\|_{\mathbb{B}_2} = 0$. Since $\tilde{T}$ is closed. The condition in the theorem implies $y = \tilde{T}0 = 0$.

Now we turn to the "if" part. Let us define the smallest closed extension $\tilde{T}$ of T as follows:

$x \in D(\tilde{T})$ if and only if there exists a sequence of x_n in $\mathrm{Dom}(T)$ such that $\lim_{n\to\infty} \|x_n - x\|_{\mathbb{B}_1} = 0$ and $\lim_{n\to\infty} \|Tx_n - y\|_{\mathbb{B}_2} = 0$ exists; and we define $\tilde{x} = y$.

It is easy to check that $\tilde{T}$ is well-defined closed operator and $\mathrm{Dom}(T) \subseteq D(\tilde{T})$. □

Definition 6.3. Let $(\mathbb{B}, \mathbb{H}, \mathbb{P})$ be an abstract Wiener space. If $F \in \mathcal{S}_{\exp}$, namely, there are $\ell_1, \cdots, \ell_d \in \mathbb{H}$, such that $F(x) = f(\langle x, \ell_1\rangle, \cdots, \langle x, \ell_d\rangle)$ for some smooth function f of at most exponential growth: there are two constants c and C such that

$$|f(x)| \le C \exp\{c|x|\}\,, \quad \forall\, x \in \mathbb{R}^d\,,$$

then we define its derivative at $x \in \mathbb{B}$ along the direction $h \in \mathbb{H}$ as

$$D_h F(x) = \lim_{\varepsilon\to 0} \frac{F(x+\varepsilon h) - F(x)}{\varepsilon}\,,$$

where the above limit is in probability.

It is easy to see that for such a functional, we have

$$\begin{aligned} D_h F(x) &= \sum_{k=1}^d \partial_k f(\langle x, \ell_1\rangle, \cdots, \langle x, \ell_d\rangle)\langle h, \ell_k\rangle \\ &= \langle h, DF(x)\rangle\,, \end{aligned}$$

where

$$DF(x) = \sum_{k=1}^d \partial_k f(\langle x, \ell_1\rangle, \cdots, \langle x, \ell_d\rangle)\ell_k\,.$$

Fixed $x \in \mathbb{B}$, the derivative $DF(x)$ is identified as an element in $\mathbb{H}$. The operator D can be considered as a linear (unbounded) operator from $\mathcal{S}_{\exp}$ to $L^p(\mathbb{B}, \mathbb{H})$.

If $F(x) = f(\langle x, \ell_1 \rangle, \cdots, \langle x, \ell_d \rangle) \in \mathcal{S}_{\exp}$, then we can also define its higher order derivative $D^k F$ for a positive integer k as

$$D^k_{h_1, \cdots, h_k} F(x) = \sum_{i_1, \cdots, i_k = 1}^{d} \partial^k_{i_1, \cdots, i_k} f(\langle x, \ell_1 \rangle, \cdots, \langle x, \ell_d \rangle) \langle h_1, \ell_{i_1} \rangle \cdots \langle h_k, \ell_{i_k} \rangle \tag{6.1.1}$$

and

$$D^k F(x) = \sum_{i_1, \cdots, i_k = 1}^{d} \partial^k_{i_1, \cdots, i_k} f(\langle x, \ell_1 \rangle, \cdots, \langle x, \ell_d \rangle) \ell_{i_1} \otimes \cdots \otimes \ell_{i_k} \tag{6.1.2}$$

where we use $\partial^k_{i_1, \cdots, i_k} = \dfrac{\partial^k}{\partial x_{i_1} \cdots \partial x_{i_k}}$. The operator D^k can be considered as a linear (unbounded) operator from $\mathcal{S}_{\exp}$ to $L^p(\mathbb{B}; \mathbb{H}^{\otimes k})$

Example 6.2. Let $h \in \mathbb{H}$. Then

$$\mathcal{E}(h) = \mathcal{E}(\tilde{h}) := \sum_{n=0}^{\infty} \frac{1}{n!} I_n(h^{\otimes n}) = \exp\left(\tilde{h} - \frac{1}{2}\|h\|^2_{\mathbb{H}}\right)$$

is called an *exponential vector* (in $L^2(\mathbb{B}, \mathbb{H}, \mathbb{P})$).

For an exponential vector $\mathcal{E}(h)$ and a $g \in \mathbb{H}$, we have

$$D_g \mathcal{E}(h) = \mathcal{E}(h) \langle h, g \rangle\,, \quad D\mathcal{E}(h) = \mathcal{E}(h) h\,.$$

From now on we denote

$$\mathcal{E} = \left\{ \sum_{finite} a_i \mathcal{E}(h_i)\,, \quad a_i \in \mathbb{R}\,,\ h_i \in \mathbb{H} \right\}.$$

We also denote

$$\mathcal{S}_0 = \Big\{ F(x) = f(\langle x, \ell_1 \rangle, \cdots, \langle x, \ell_d \rangle)\,, d \in \mathbb{N}\,,\ \ell_1, \cdots, \ell_d \in \mathbb{H}\,,$$
$$f \text{ is smooth with compact support} \Big\}\,;$$

$$\mathcal{P} = \Big\{ F(x) = f(\langle x, \ell_1 \rangle, \cdots, \langle x, \ell_d \rangle)\,, d \in \mathbb{N}\,,\ \ell_1, \cdots, \ell_d \in \mathbb{H}\,,$$
$$f \text{ is a polynomial of } d \text{ variables} \Big\}\,.$$

We see that all $\mathcal{E}$, $\mathcal{S}_0$, and $\mathcal{P}$ are dense subsets of $L^p(\mathbb{B}, \mathbb{P})$ for any $p \in [1, \infty)$.

The first thing we want to show is the closability of the derivative operators D^k so that we can extend them to general functionals.

Theorem 6.1. *For any $p \geq 1$ and for any positive integer k, D^k is closable from $L^p(\mathbb{B})$ to $L^p(\mathbb{B}; \mathbb{H}^{\otimes k})$.*

Proof. We shall prove the case $k = 1$. General case is similar.
Step 1: First, we claim that for any $F \in \mathcal{E}$ and $h \in \mathbb{H}$,

$$\mathbb{E}\left(\langle DF, h\rangle\right) = \mathbb{E}(F\tilde{h})\,, \quad \text{where we recall } \tilde{h} = \langle x, h\rangle\,. \tag{6.1.3}$$

By linearity, it suffices to show (6.1.3) for $F(x) = \exp\left\{\langle x, \ell\rangle\right\}$, where $\ell \in \mathbb{H}$. First, we have for any $t \in \mathbb{R}$,

$$\mathbb{E}\left(F \exp\left[t\tilde{h}\right]\right) = \mathbb{E}\exp\left[\langle x, \ell + th\rangle\right] = \exp\left[|\ell + th|^2/2\right]\,.$$

Differentiating both sides with respect to t (at $t = 0$), we have

$$\mathbb{E}(F\tilde{h}) = \exp\left[|\ell|^2/2\right]\langle \ell, h\rangle = \mathbb{E}\exp\left[\langle x, \ell\rangle\right]\langle \ell, h\rangle = \mathbb{E}\left\langle DF, h\rangle\right]$$

which is (6.1.3).
Step 2: Applying (6.1.3) to FG and noticing the product rule $D(FG) = FDG + GDF$, we see that if $F, G \in \mathcal{E}$ and if $h \in \mathbb{H}$, then

$$\mathbb{E}(G\langle DF, h\rangle) = -\mathbb{E}(F\langle DG, h\rangle) + \mathbb{E}(FG\tilde{h})\,. \tag{6.1.4}$$

By a density argument, we can show the above identity for all $F, G \in \mathcal{S}_{\text{exp}}$.
Step 3: Let F_n, $n = 1, 2, \cdots$ be a sequence of functionals such that F_n converges to 0 in $L^p(\mathbb{B})$ and DF_n converges to η in $L^p(\mathbb{B}; \mathbb{H})$. We want to show $\eta = 0$. For any $\varepsilon > 0$, any $h \in \mathbb{H}$ and any $F \in \mathcal{S}_0$, we have by (6.1.4),

$$\begin{aligned}
\mathbb{E}\left(\langle \eta, Fhe^{-\varepsilon\tilde{h}^2}\rangle\right) &= \lim_{n\to\infty} \mathbb{E}\left(\langle DF_n, Fhe^{-\varepsilon\tilde{h}^2}\rangle\right) \\
&= -\lim_{n\to\infty} \mathbb{E}\left(F_n\langle D(Fe^{-\varepsilon\tilde{h}^2}), h\rangle\right) + \lim_{n\to\infty} \mathbb{E}\left[F_n F\tilde{h}e^{-\varepsilon\tilde{h}^2}\right] = 0\,.
\end{aligned}$$

However, the linear combinations of $\mathbb{H}$-valued functionals of the above form $Fhe^{-\varepsilon\tilde{h}^2}$ is dense in $L^p(\mathbb{B}; \mathbb{H})$. Hence $\eta = 0$. The theorem is then proved. □

Remark 6.1. Using the divergence operator δ that we shall introduce later, we shall see that equation (6.1.4) means

$$\delta(Gh) = G\tilde{h} - \langle DG, h\rangle\,.$$

For any $p \geq 1$, we denote by $\mathbb{D}_{k,p} = \mathbb{D}_{k,p}(\mathbb{B})$ the closure of $\mathcal{E}$ (or $\mathcal{S}_{\exp}$) with respect to the norm

$$\|F\|_{k,p} = \left[\mathbb{E}(|F|^p) + \sum_{j=1}^{k} \mathbb{E}(|D^j F|^p_{\mathbb{H}^{\otimes j}}) \right]^{1/p} , \tag{6.1.5}$$

where $|D^j F|_{\mathbb{H}^{\otimes j}}$ is defined as a tensor product norm. Since D^k is closable, we have

Theorem 6.2. *For any $p \geq 1$ and any positive integer k, D^k can be extended to a linear bounded operator from $\mathbb{D}_{k,p}(\mathbb{B})$ to $L^p(\mathbb{B})$.*

When $p = 2$, then $\mathbb{D}_{k,2}(\mathbb{B})$ is a Hilbert space with the scalar product

$$\langle F, G \rangle_{\mathbb{D}_{k,2}} = \mathbb{E}(FG) + \sum_{j=1}^{k} \mathbb{E}(\langle D^j F \, , D^j G \rangle_{\mathbb{H}^{\otimes j}}) \, .$$

For any $t \in \mathbb{R}$ and $\ell \in \mathbb{H}$ with $|\ell| = 1$, we have

$$D_h \mathcal{E}(t\tilde{\ell}) = t\mathcal{E}(t\tilde{\ell}) \langle \ell, h \rangle \, .$$

Compare the coefficient of t^n, we have

$$D_h I_n(\ell^{\otimes n}) = n I_{n-1}(\ell^{\otimes (n-1)}) \langle \ell, h \rangle = n I_{n-1}(\langle \ell^{\otimes n}, h \rangle) \, .$$

By polarization technique, we see that $D_h I_n(f_n) = n I_{n-1} \left(\langle f_n \, , h \rangle \right)$ for $f_n = \ell_1 \hat{\otimes} \cdots \hat{\otimes} \ell_n$ and hence for linear combinations of such functionals. Since the subspace of such linear combinations is dense in $\mathbb{H}^{\hat{\otimes} n}$, we have the following identity: $D_h I_n(f_n) = n I_{n-1} \left(\langle f_n \, , h \rangle \right)$ for $f_n = \mathbb{H}^{\hat{\otimes} n}$. Similar argument can also be applied to higher derivatives. We write them as

Proposition 6.2. *If $f_n \in \mathbb{H}^{\hat{\otimes} n}$ and $h \in \mathbb{H}$, then $\langle f_n \, , h \rangle$ is an element in $H^{\hat{\otimes} (n-1)}$ and we have*

$$D_h I_n(f_n) = n I_{n-1} \left(\langle f_n \, , h \rangle \right) \tag{6.1.6}$$

and similarly we have

$$D^k_{h_1, \cdots, h_k} I_n(f_n) = \begin{cases} \frac{n!}{(n-k)!} I_{n-k} \left(\langle f_n \, , h_1 \otimes \cdots \otimes h_k \rangle \right) & \text{if } k \leq n \\ 0 & \text{if } k > n \, . \end{cases} \tag{6.1.7}$$

For an element $f_n \in \mathbb{H}^{\hat{\otimes} n}$, it can be considered an element in $\mathbb{H}^{\hat{\otimes} m}$ but with value in $\mathbb{H}^{\hat{\otimes} n-m}$. Thus, multiple integral $I_m(f_n)$ is well-defined as an random variable (nonlinear functional) with values in $\mathbb{H}^{\hat{\otimes} n-m}$. With this identification we can write

$$D^k I_n(f_n) = \begin{cases} \frac{n!}{(n-k)!} I_{n-k}(f_n) & \text{if } k \le n \\ 0 & \text{if } k > n\,, \end{cases} \tag{6.1.8}$$

where $I_{n-k}(f_n)$ is considered as a multiple integral with values in $\mathbb{H}^{\hat{\otimes} k}$.

If $f_n \in \mathbb{H}^{\hat{\otimes} n}$, then

$$DI_n(f_n) = nI_{n-1}(f_n) = n\sum_{j=1}^{\infty} I_{n-1}(\langle f_n, \ell_j\rangle)\ell_j\,,$$

where $\{\ell_j\}$ is an orthonormal basis of $\mathbb{H}$. Thus

$$\begin{aligned}\mathbb{E}\left[|DI_n(f_n)|^2_{\mathbb{H}}\right] &= n^2\mathbb{E}|I_{n-1}(f_n)|^2_{\mathbb{H}} = n^2\sum_{j=1}^{\infty}\mathbb{E}\left(I_{n-1}(\langle f_n, \ell_j\rangle)\right)^2 \\ &= n^2(n-1)!\sum_{j=1}^{\infty}|\langle f_n, \ell_j\rangle|^2_{\mathbb{H}^{\otimes(n-1)}} = n^2(n-1)!|f_n|^2_{\mathbb{H}^{\otimes n}} \\ &= n\mathbb{E}(I_n(f_n)^2)\,. \end{aligned} \tag{6.1.9}$$

The above computation gives

Proposition 6.3. *Let $F = \sum_{n=0}^{\infty} J_n F \in L^2(\mathbb{B})$ be the chaos expansion of F, where J_n is the projection operator to the n-th chaos. Then*

$$F \in \mathbb{D}_{1,2} \iff \sum_{n=0}^{\infty} n\mathbb{E}|J_n F|^2 < \infty \tag{6.1.10}$$

and

$$\mathbb{E}\left[|DF|^2\right] = \sum_{n=0}^{\infty} n\mathbb{E}|J_n F|^2\,. \tag{6.1.11}$$

The following proposition also follows from the above computation.

Proposition 6.4. *If $F \in \mathbb{D}_{1,2}$, then*

$$DJ_n(F) = J_{n-1}(DF) \tag{6.1.12}$$

The following proposition is about the chain rule of derivative operator.

Proposition 6.5. *Let $\phi : \mathbb{R}^m \to \mathbb{R}$ be a continuously differentiable function with bounded partial derivatives and let $F = (F_1, \cdots, F_m) \in \mathbb{D}_{1,p}$ for some $p \ge 1$. Then, $\phi(F) \in \mathbb{D}_{1,p}$ and*

$$D\phi(F) = \sum_{j=1}^{m} \partial_j \phi(F) DF_j\,. \tag{6.1.13}$$

Proof. We only need to approximate F by elements in $\mathcal{S}_{\exp}$. □

Lemma 6.1. *Let $F_n \in \mathbb{D}_{1,2}$, $n = 1, 2, \cdots$. Assume that $F_n \to F$ in $L^2(\mathbb{B})$ and*

$$\sup_{n\geq 1} \mathbb{E}(|DF_n|^2) < \infty\,. \tag{6.1.14}$$

Then $F \in \mathbb{D}_{1,2}$ and DF_n converges to DF weakly in $L^2(\mathbb{B};\mathbb{H})$.

Proof. From (6.1.14) it follows that there is a subsequence and an element G in $L^2(\mathbb{B};\mathbb{H})$ such that DF_{n_k} converges to G weakly in $L^2(\mathbb{B};\mathbb{H})$. Write the chaos expansion: $F_{n_k} = \sum_{m=0}^{\infty} J_m F_{n_k}$. Then

$$DF_{n_k} = \sum_{m=1}^{\infty} DJ_m F_{n_k} = \sum_{m=0}^{\infty} J_m DF_{n_k}\,.$$

Thus, $J_m DF_{n_k}$ converges to $J_m G$ weakly in $L^2(\mathbb{B};\mathbb{H})$. But $J_m DF_{n_k} = DJ_{m+1}F_{n_k}$ which implies then $DJ_{m+1}F = J_m G$. This implies that F is in $\mathbb{D}_{1,2}$ and $DF = G$. On the other hand, we can show similarly that for any weakly convergent subsequence the limit must be DF. This means that the whole sequence converges weakly to DF. □

Proposition 6.6. *Let φ be Lipschitzian, namely, there is a constant K such that*

$$|\varphi(x) - \varphi(y)| \leq K|x - y|\,, \quad \forall\, x, y \in \mathbb{R}^d\,.$$

Let $F = (F_1, \cdots, F_d) \in \mathbb{D}_{1,2}$. Then, $\varphi(F) \in \mathbb{D}_{1,2}$ and there is a random vector $G = (G_1, \cdots, G_d)$ uniformly bounded by K ($|G| \leq K$) such that

$$D\varphi(F) = \sum_{k=1}^{d} G_k DF_k\,. \tag{6.1.15}$$

Proof. Let φ_n be a sequence of smooth functions such that φ_n converges to ϕ almost surely and $|\nabla\varphi_n| \leq K$. By the chain rule (6.1.13), we have

$$D\varphi_n(F) = \sum_{k=1}^{d} \partial_k\varphi_n(F) DF_k\,. \tag{6.1.16}$$

It is easy to verify that $\varphi_n(F)$ converges in $L^2(\mathbb{B})$ to $\varphi(F)$ and the condition (6.1.14) is satisfied for $\varphi_n(F)$. Thus, $\varphi(F) \in \mathbb{D}_{1,2}$ and $D\varphi_n(F)$ converges to $D\varphi(F)$ weakly in $L^2(\mathbb{B};\mathbb{H})$. On the other hand, $\nabla\varphi_n(F)$ is bounded by K and hence there is a subsequence $\nabla\varphi_{n_k}(F)$ converging to $(G_1, \cdots, G_m)$ weakly in $L^2(\mathbb{B})$. Taking the weak limit in (6.1.16), we obtain (6.1.15). It is also easy to see that $|G| \leq K$. □

Proposition 6.7. *Let $F \in \mathbb{D}_{1,1}$. If $DF = 0$, then $F = \mathbb{E}(F)$ is a constant.*

Proof. If $F \in \mathbb{D}_{1,2}$, then $DF = 0$ means that $DJ_n(F) = J_{n-1}(DF) = 0$ for all $n \geq 1$. This implies $J_n(F) = 0$ for all $n \geq 1$. Thus, $F = J_0(F) = \mathbb{E}(F)$. Let ϕ_n be a smooth function such that $\phi_n(x) = 1$ when $|x| \leq n$ and $\phi_n(x) = 0$ when $|x| \geq n+1$. Then, $F_n := \phi_n(F) \in \mathbb{D}_{1,2}$ and $DF_n = \phi_n'(F)DF = 0$. Thus, $F_n = \mathbb{E}(F_n)$. Clearly, F_n converges to F almost surely and in L^1. Taking the limit in $F_n = \mathbb{E}(F_n)$ yields the proposition. □

Proposition 6.8. *Let $A \subset \mathbb{B}$ be a measurable set. Then, the indicator function I_A of A belongs to $\mathbb{D}_{1,1}$ if and only if $P(A)$ is equal to 0 or 1.*

Proof. If I_A is in $\mathbb{D}_{1,1}$, then by chain rule, we have

$$DI_A = DI_A^2 = 2I_A DI_A\,.$$

Thus, DI_A is zero on A^c. When $\omega \in A$, then $DI_A = 2DI_A$, this also implies that $DI_A = 0$. Thus, $DI_A \equiv 0$. The previous proposition implies that I_A is a constant function, proving the proposition. □

Let $(\mathbb{B}, \mathbb{H}, \mathbb{P})$ be an abstract Wiener space and let $h \in \mathbb{H}$. Consider the mapping

$$\tau_h : \mathbb{B} \ni x \to \tau_h(x) = x + h \in \mathbb{B}\,.$$

Theorem 6.3 (Cameron-Martin). *The mapping $\tau_h : \mathbb{B} \to \mathbb{B}$ is absolutely continuous. Namely, if $A \in \mathcal{B}(\mathbb{B})$ is such that $\mathbb{P}(A) = 0$, then $\mathbb{P}(\{x \in \mathbb{B}, x+h \in A\}) = 0$. Moreover, For any $F \in L^p(\mathbb{B}), p > 1$, we have the following Cameron-Martin formula:*

$$\mathbb{E}(F(\cdot + h)) = \mathbb{E}\left(F \exp\left\{\tilde{h} - \frac{1}{2}|h|_{\mathbb{H}}^2\right\}\right)\,. \tag{6.1.17}$$

Proof. Since $\mathcal{E}$ is dense in $L^p(\mathbb{B})$, it suffices to show (6.1.17) for $F(x) = \exp\{\langle x, g\rangle_{\mathbb{H}}\}$ for any $g \in \mathbb{H}$. When F is of this form, the left hand side of (6.1.17) is

$$\mathbb{E}(F(\cdot + h)) = \mathbb{E}\exp\{\tilde{g} + \langle g, h\rangle_{\mathbb{H}}\} = \exp\left\{\frac{1}{2}|g|_{\mathbb{H}}^2 + \langle g, h\rangle_{\mathbb{H}}\right\}\,.$$

The right hand side of (6.1.17) is

$$\begin{aligned}\mathbb{E}\left(F \exp\left\{\tilde{h} - \frac{1}{2}|h|_{\mathbb{H}}^2\right\}\right) &= \mathbb{E}\left(\exp\left\{\tilde{g} + \tilde{h} - \frac{1}{2}|h|_{\mathbb{H}}^2\right\}\right) \\ &= \exp\left\{\frac{1}{2}|g|_{\mathbb{H}}^2 + \langle g, h\rangle_{\mathbb{H}}\right\}\,.\end{aligned}$$

This proves (6.1.17) for $F(x) = \exp\{\langle x, g\rangle_{\mathbb{H}}\}$. The linearity argument and the limiting argument can be routinely used to show (6.1.17) for all $F \in L^p$. □

Let $F : \mathbb{B} \to \mathbb{R}$ be a measurable function. Then, F is only almost surely defined. The functional $F(x_1 + x), x_1, x \in \mathbb{B}$, may not be well-defined random variable as a function of x (see [Hu and Meyer (1988a,b)] and reference therein for a discussion). But from the above theorem, we can define $\tau_h F$ by $\tau_h F(x) = F(\tau_h x) = F(x + h)$ for any $h \in \mathbb{H}$.

Proposition 6.9. *If $F \in L^p$ for some $p > 1$, then for any fixed $h \in \mathbb{H}$, $\tau_h F \in L^q$ for any $1 < q < p$.*

Proof. The formula (6.1.17) gives

$$\mathbb{E}|(F(\cdot + h))|^q = \mathbb{E}\left(|F|^q \exp\left\{\tilde{h} - \frac{1}{2}|h|_{\mathbb{H}}^2\right\}\right).$$

An application of Hölder inequality implies the corollary. □

Lemma 6.2. *Let $F \in \mathbb{D}_{1,p}$ for some $p > 1$ and let $h \in \mathbb{H}$. Then, for any $a < b$,*

$$F(x+bh) - F(x+ah) = \int_a^b \langle DF(x+th), h\rangle_{\mathbb{H}} dt\,, \quad \text{for } x \in \mathbb{B} \text{ almost surely.} \tag{6.1.18}$$

Proof. Let $F_k \in \mathcal{E}$ and $F_k \to F$ in $\mathbb{D}_{1,p}$ as $k \to \infty$. Then, it is easy to see that

$$F_k(x + bh) - F_k(x + ah) = \int_a^b \langle DF_k(x + th), h\rangle_{\mathbb{H}} dt\,. \tag{6.1.19}$$

It is easy to see that $\tau_{bh}F_k$ and $\tau_{ah}F_k$ converge to $\tau_{bh}F$ and $\tau_{ah}F$ in L^1 by Proposition 6.9. We also have

$$\begin{aligned}
&\mathbb{E}\left|\int_a^b \langle DF_k(\cdot + th), h\rangle_{\mathbb{H}} dt - \int_a^b \langle DF(\cdot + th), h\rangle_{\mathbb{H}} dt\right| \\
&\le \|h\|_{\mathbb{H}} \int_a^b \mathbb{E}\left|DF_k(\cdot + th) - DF(\cdot + th)\right|_{\mathbb{H}} dt \\
&= \|h\|_{\mathbb{H}} \int_a^b \mathbb{E}\left\{|DF_k - DF|_{\mathbb{H}} \exp\left[t\tilde{h} - \frac{t^2|h|_{\mathbb{H}}^2}{2}\right]\right\} dt \\
&= \|h\|_{\mathbb{H}}\|DF_k - DF\|_p \int_a^b \exp\left[\frac{t^2|h|_{\mathbb{H}}^2}{2(p-1)}\right] dt \to 0\,.
\end{aligned}$$

Letting $k \to \infty$ in (6.1.19) yields (6.1.18). □

Proposition 6.10. *Let $F \in \mathbb{D}_{1,p}$ for some $p > 1$ and let $h \in \mathbb{H}$. Then for almost all $x \in \mathbb{B}$, we have*

$$\lim_{\varepsilon \downarrow 0} \frac{1}{\varepsilon} [F(x + \varepsilon h) - F(x)] = \langle DF(x), h \rangle_{\mathbb{H}} . \tag{6.1.20}$$

Proof. By Lemma 6.2 we see that

$$\frac{1}{\varepsilon} [F(x + \delta h + \varepsilon h) - F(x + \delta h)] = \frac{1}{\varepsilon} \int_0^{\varepsilon} \langle DF(x + \delta h + th), h \rangle_{\mathbb{H}} dt$$

converges to $\langle DF(x + \delta h), h \rangle_{\mathbb{H}}$ for almost all $(x, \delta) \in \mathbb{B} \times \mathbb{R}$. This proves (6.1.20) for almost all $x \in \mathbb{B}$ with respect to the measure Q, where $\frac{dQ}{dP} = \exp\left[-\delta\tilde{h} - \delta^2 |h|^2_{\mathbb{H}}/2\right]$. But since Q and P are equivalent, the proposition is proved. □

Now let us consider the classical Wiener space case. Denote $D_t^{(j)}$ the Malliavin derivative with respect to the j-th Brownian motion B^j. It is easy to verify (see also (6.2.11)) that

$$D_r^{(j)} \int_0^T f(s) dW^i(s) = \int_r^T D_r^{(j)} f(s) dW^i(s) + f(r)\delta_{ij} , \tag{6.1.21}$$

for all f adapted such that $\mathbb{E} \int_r^T |D_r^{(j)} f(s)|^2 ds < \infty$, where δ_{ij} is the Kronecker symbol.

Proposition 6.11. *Let $F \in \mathbb{D}_{1,2}$ be a Wiener functional of d-dimensional Brownian motion $W = \left\{(W^1(t), \cdots, W^d(t)), 0 \le t \le T\right\}$. Then, we have the following Clark-Ocone formula*

$$F = \mathbb{E}(F) + \sum_{j=1}^{d} \int_0^T \mathbb{E}\left[D_t^{(j)} F \middle| \mathcal{F}_t\right] dW^j(t) , \tag{6.1.22}$$

where $\mathcal{F}_t$ is the σ-algebra generated by $\left\{(B^1(s), \cdots, W^d(s)), 0 \le s \le t\right\}$.

Proof. First, we assume that $F \in \mathcal{P}$. From the chaos expansion theorem we see that there are adapted processes $\left\{f_t^j, 0 \le t \le T, j = 1, \cdots, d\right\}$ such that

$$F = \mathbb{E}(F) + \sum_{i=1}^{d} \int_0^T f_s^i dW^i(s) .$$

Hence, from (6.1.21)

$$D_t^{(j)} F = \sum_{i=1}^{d} \int_t^T D_t^{(j)} f_s^i dW^i(s) + f_t^j$$

and then

$$\mathbb{E}\left[D_t^{(j)} F \middle| \mathcal{F}_t\right] = f_t^j .$$

This proves (6.1.22) for $F \in \mathcal{P}$. Since $\mathcal{P}$ is dense in $\mathbb{D}_{1,2}$, the general case follows easily from a limiting argument. □

6.2 Divergence Operator

Definition 6.4. Let $F : \mathbb{B} \to \mathbb{H}$ be measurable. We say that $F \in \mathrm{Dom}_p(\delta)$ ($1 < p < \infty$) if for any $G \in \mathbb{D}_{1,q}$, where q is the conjugate number of p (namely, $1/p + 1/q = 1$), we have

$$\mathbb{E}\left[\langle F, DG\rangle\right] \le C_{p,q}(F)\|G\|_q \tag{6.2.1}$$

for some constant $C_{p,q}(F)$, independent of G. By the Riesz representation theorem (see [Yosida (1980)]) there is an element in L^p, denoted by $\delta(F)$, such that

$$\mathbb{E}\left[G\delta(F)\right] = \mathbb{E}\left[\langle F, DG\rangle\right], \quad \forall\, G \in \mathbb{D}_{1,q}\,. \tag{6.2.2}$$

This means that the divergence operator δ is the adjoint operator of the derivative operator D.

From the definition, it is easy to see that

Lemma 6.3. *If $F \in \mathrm{Dom}_p(\delta)$, then $F \in L^p(\mathbb{H})$ and $\delta(F) \in L^p$.*

Here are some properties of divergence operator δ.

Proposition 6.12.

(i) If $F \in \mathrm{Dom}_p(\delta)$, then

$$\mathbb{E}(\delta(F)) = 0\,. \tag{6.2.3}$$

(ii) If $h \in \mathbb{H}$, then $h \in \mathrm{Dom}_p(\delta)$ for any $p > 1$ and

$$\delta(h) = \langle \cdot, h\rangle_{\mathbb{H}} = \tilde{h}\,.$$

(iii) If $G \in \mathbb{D}_{1,r}$, and $F \in \mathrm{Dom}_p(\delta)$ with $\frac{1}{r} + \frac{1}{p} < 1$, then $GF \in \mathrm{Dom}_{p'}(\delta)$, where $\frac{1}{r} + \frac{1}{p} = \frac{1}{p'}$, and

$$\delta(GF) = G\delta(F) - \langle DG, F\rangle_{\mathbb{H}}\,. \tag{6.2.4}$$

Proof. (i): Letting $G = 1$ in (6.2.2) yields (6.2.3).

(ii): For any $h' \in \mathbb{H}$, let $\mathcal{E}(h') = \exp\left\{\langle x, h'\rangle - \frac{|h'|^2}{2}\right\}$. We see $D\mathcal{E}(h') = \mathcal{E}(h')h'$ and then

$$\begin{aligned}\mathbb{E}\langle h, D\mathcal{E}(h')\rangle &= \langle h, h'\rangle \mathbb{E}\mathcal{E}(h') = \langle h, h'\rangle \\ &= \mathbb{E}\left[\langle x, h\rangle \mathcal{E}(h')\right]\,.\end{aligned}$$

Thus, by a linearity argument, we have $\mathbb{E}\langle h, D\phi\rangle = \mathbb{E}\left[\langle x, h\rangle\phi\right]$ for all $\phi \in \mathcal{E}$. Since $\langle x, h\rangle$ is in any L^p and $\mathcal{E}$ is dense in any L^q, we have that $h \in \mathrm{Dom}_p(\delta)$ and $\delta(h) = \tilde{h}$.

(iii): For any $\phi \in \mathbb{D}_{1,q}$, we have

$$\begin{aligned}\mathbb{E}\left[\langle GF\,, D\phi\rangle_{\mathbb{H}}\right] &= \mathbb{E}\left[\langle F\,, GD\phi\rangle_{\mathbb{H}}\right] = \mathbb{E}\left[\langle F\,, D(G\phi) - \phi DG\rangle_{\mathbb{H}}\right] \\ &= \mathbb{E}\left[\langle F\,, D(G\phi)\rangle\right] - \mathbb{E}\left[\langle F\,, \phi DG\rangle_{\mathbb{H}}\right] \\ &= \mathbb{E}\left[G\delta(F)\phi\right] - \mathbb{E}\left[\langle F\,, DG\rangle_{\mathbb{H}}\phi\right] \\ &= \mathbb{E}\left[\left(G\delta(F) - \langle F\,, DG\rangle_{\mathbb{H}}\right)\phi\right]\,. \end{aligned} \tag{6.2.5}$$

It is easy to see that $G\delta(F) - \langle F\,, DG\rangle_{\mathbb{H}} \in L^{p'}$. The equation (6.2.4) follows easily from (6.2.5). □

Theorem 6.4. *If $F : \mathbb{B} \to \mathbb{H}$ is in $\mathbb{D}_{1,2}(\mathbb{H})$, then $F \in \mathrm{Dom}_2(\delta)$ and for any orthonormal basis $\{h_1, h_2, \cdots\}$ of $\mathbb{H}$, we have*

$$\begin{aligned}\delta(F) &= \sum_{i=1}^{\infty} \langle F\,, h_i\rangle_{\mathbb{H}}\delta(h_i) - \langle DF\,, h_i \otimes h_i\rangle_{\mathbb{H}^{\otimes 2}} \\ &= \sum_{i=1}^{\infty} \langle F\,, h_i\rangle_{\mathbb{H}}\delta(h_i) - \mathrm{Tr}(DF)\,. \end{aligned} \tag{6.2.6}$$

Proof. Since $\{h_1, h_2, \cdots\}$ is an orthonormal basis of $\mathbb{H}$ we can write $F = \sum_{i=1}^{\infty} \langle F\,, h_i\rangle_{\mathbb{H}} h_i$. Thus, from (6.2.4),

$$\begin{aligned}\delta(F) &= \sum_{i=1}^{\infty} \delta\left(\langle F\,, h_i\rangle_{\mathbb{H}} h_i\right) \\ &= \sum_{i=1}^{\infty} \langle F\,, h_i\rangle_{\mathbb{H}}\delta\left(h_i\right) - \sum_{i=1}^{\infty} \langle D\langle F\,, h_i\rangle_{\mathbb{H}}\,, h_i\rangle_{\mathbb{H}}\,. \end{aligned}$$

The condition $F \in \mathbb{D}_{1,2}(\mathbb{H})$ assures the legitimacy of the above identities. □

Next, we demonstrate the locality of derivative operator D.

Lemma 6.4. *Let $F \in \mathbb{D}_{1,2}(\mathbb{H})$. Let $\phi : \mathbb{R} \to \mathbb{R}$ be a smooth function with compact support such that $\phi(x) \geq 0$ and $\phi(0) = 1$ and let $\phi_\varepsilon(x) = \phi(\frac{x}{\varepsilon})$. Then, for any fixed $\varepsilon > 0$, $\phi_\varepsilon(|F|_{\mathbb{H}}^2)$ is also in $\mathbb{D}_{1,2}$.*

Proof. Set $\tilde{F}_\varepsilon = \phi_\varepsilon(|F|_{\mathbb{H}}^2)$. Then

$$D.\tilde{F}_\varepsilon = 2\phi'_\varepsilon(|F|_{\mathbb{H}}^2)\langle D.F, F\rangle_{\mathbb{H}}\,.$$

Since ϕ'_ε is bounded, we have

$$\begin{aligned}\||D.\tilde{F}_\varepsilon|_{\mathbb{H}}\|_2 &\leq C_\varepsilon \||\langle D.F, F\rangle_{\mathbb{H}}|_{\mathbb{H}}\|_2 \leq C_\varepsilon \||DF|_{\mathbb{H}^{\otimes 2}}^{1/2} |F|_{\mathbb{H}}^{1/2}\|_2 \\ &\leq C_\varepsilon \||DF|_{\mathbb{H}^{\otimes 2}}\|_2^{1/2} \||F|_{\mathbb{H}}^{1/2}\|_2^{1/2} \end{aligned}$$

which is finite since $F \in \mathbb{D}_{1,2}(\mathbb{H})$. □

Proposition 6.13. *Let $F \in \mathbb{D}_{1,2}(\mathbb{H})$ and $A \in \mathcal{F}$ satisfying that $F = 0$ on A. Then $\delta(F) = 0$ a.s. on A.*

Proof. We follow the notation of Lemma 6.4. Let G be an arbitrary element in $\mathcal{S}$. Consider $G_\varepsilon = G\phi_\varepsilon(|F|_{\mathbb{H}}^2)$. Then, it is easy to see from Lemma 6.4 that $G_\varepsilon \in \mathbb{D}_{1,2}$. By the duality relation we have

$$\mathbb{E}\left[\delta(F)\phi_\varepsilon(|F|_{\mathbb{H}}^2)G\right] = \mathbb{E}\left[\langle F\,, D[G\phi_\varepsilon(|F|_{\mathbb{H}}^2)]\rangle_{\mathbb{H}}\right] = \mathbb{E}[V_\varepsilon]\,,$$

where

$$V_\varepsilon := \phi_\varepsilon(|F|_{\mathbb{H}}^2)\langle F\,, DG\rangle_{\mathbb{H}} + 2G\phi_\varepsilon'(|F|_{\mathbb{H}}^2)\left[\langle F\otimes F\,, DF\rangle_{\mathbb{H}^{\otimes 2}}\right]\,.$$

From this expression it is obvious that if $|F|_{\mathbb{H}} = 0$, then $V_\varepsilon = 0$. However, when $|F|_{\mathbb{H}} \neq 0$, then $\phi_\varepsilon(|F|_{\mathbb{H}}^2) \to 0$ a.s. On the other hand, we have

$$\begin{aligned}|V_\varepsilon| &\le \phi_\varepsilon(|F|_{\mathbb{H}}^2)|F|_{\mathbb{H}}|DG|_{\mathbb{H}} + \phi_\varepsilon'(|F|_{\mathbb{H}}^2)|F|_{\mathbb{H}}^2|DF|_{\mathbb{H}^{\otimes 2}}\\ &\le C(|DG|_{\mathbb{H}} + |DF|_{\mathbb{H}^{\otimes 2}})\end{aligned}$$

since $\phi_\varepsilon(x)\sqrt{x}$ and $x\phi_\varepsilon'(x)$ are bounded with respect to ε. By dominated convergence theorem, we have that

$$\mathbb{E}\left[\delta(F)\phi_\varepsilon(|F|_{\mathbb{H}}^2)G\right] = \mathbb{E}[V_\varepsilon] = \mathbb{E}[V_\varepsilon I_{\{|F|_{\mathbb{H}}\neq 0\}}]$$

converges to 0. On the other hand it is also easy to show that $\mathbb{E}\left[\delta(F)\phi_\varepsilon(|F|_{\mathbb{H}}^2)G\right]$ converges to $\mathbb{E}\left[\delta(F)I_{\{|F|_{\mathbb{H}}=0\}}G\right]$, which is then also 0. This implies

$$\delta(F)I_{\{|F|_{\mathbb{H}}=0\}} = 0 \quad \text{a.s.}\,,$$

proving the proposition. □

Proposition 6.14. *Let $F \in \mathbb{D}_{1,1}$ satisfy $F = c$ a.s. on some set $A \in \mathcal{F}$ and for some constant c. Then, $DF = 0$ a.s. on A.*

Proof. First, we can assume $F \in \mathbb{D}_{1,1}\cap L^\infty$, replacing F, say, by $\arctan(F)$. Let ϕ and ϕ_ε be defined as in Lemma 6.4 but with $\phi(c) = 1$ instead of $\phi(0) = 1$. Denote $\psi_\varepsilon(x) = \int_{-\infty}^x \phi_\varepsilon(u)du$. Then, $D\psi_\varepsilon(F) = \phi_\varepsilon(F)DF$. For any $G = \sum_{i=1}^d G_i h_i$, where $G_i \in \mathcal{S}_0$ and $h_i \in \mathbb{H}$, we have

$$\begin{aligned}\mathbb{E}\left[\phi_\varepsilon(F)\langle DF, G\rangle_{\mathbb{H}}\right] &= \mathbb{E}\left[\langle D\psi_\varepsilon(F), G\rangle_{\mathbb{H}}\right]\\ &= \mathbb{E}\left[\psi_\varepsilon(F)\delta(G)\right]\,,\end{aligned}$$

where $\delta(G) = \sum_{i=1}^d \left[G_i\langle x, h_i\rangle_{\mathbb{H}} - \langle DG_i, h_i\rangle_{\mathbb{H}}\right]$. Letting $\varepsilon \downarrow 0$ and using the dominated convergence theorem, we have

$$0 = \mathbb{E}\left[I_{\{F=c\}}\langle DF, G\rangle_{\mathbb{H}}\right] = \mathbb{E}\left[\langle I_{\{F=c\}}DF, G\rangle_{\mathbb{H}}\right]\,.$$

Since the G's of the above form is dense in $L^1(\mathbb{H})$ we have $I_{\{F=c\}}DF = 0$ a.s., proving the proposition. □

Now we can extend D and δ to more general nonlinear functionals without assuming the global integrability.

Let $\{B_n\,, n=1,2,\cdots\}$ be a sequence of measurable subsets of $\mathbb{B}$ and let F and $\{F_n\,, n=1,2,\cdots\}$ be measurable from $\mathbb{B}$ to $\mathbb{R}$ (or to $\mathbb{H}$). Assume that

(i) $B_n \uparrow \mathbb{B}$ a.s.
(ii) $F_n = F$ on B_n a.s.

Definition 6.5. If $F_n \in \mathbb{D}_{k,p}$ (or $F_n \in \mathbb{D}_{1,2}(\mathbb{H})$), then we say $F \in \mathbb{D}_{k,p}^{\text{loc}}$ (or $F \in \text{Dom}_{\text{loc}}(\delta)$, respectively). We also define

$$D^j F\big|_{B_n} = D^j F_n\big|_{B_n}\,, \quad j = 1,\cdots,k\,; \tag{6.2.7}$$

$$\delta(F)\big|_{B_n} = \delta(F_n)\big|_{B_n}\,. \tag{6.2.8}$$

Remark 6.2. By the lemmas 6.13 and 6.14, we see that $D^j F$ and $\delta(F)$ is independent of the choices of B_n.

Example 6.3. Let $h_1,\cdots,h_d \in \mathbb{B}'$. If $f:\mathbb{R}^d \to \mathbb{R}$ is smooth then $F = f(\langle x,h_1\rangle_{\mathbb{H}}\,,\cdots,\langle x,h_d\rangle_{\mathbb{H}})$ is in $\mathbb{D}_{k,p}^{\text{loc}}$ for any $p>1$ and integer $k \geq 1$.

In fact, without loss of generality, we can assume $\|h_i\|_{\mathbb{B}'} = 1$. We use $B_n = \{x \in \mathbb{B} : \|x\|_{\mathbb{B}} \leq n\}$. Then,

$$|\langle x,h_i\rangle_{\mathbb{H}}| \leq \|x\|_{\mathbb{B}}\|h_i\|_{\mathbb{B}'} \leq n \quad \forall x \in B_n\,.$$

We can construct f_n such that $f_n \in C_0^\infty(\mathbb{R}^d,\mathbb{R})$ and when $\sup_{1\leq i\leq d}|x_i| \leq n$, $f_n = f$. It is clear that $F_n = f_n(\langle x,h_1\rangle_{\mathbb{H}}\,,\cdots,\langle x,h_d\rangle_{\mathbb{H}}) \in \mathbb{D}_{k,p}$ for any k and p and $F_n|_{B_n} = F|_{B_n}$.

Proposition 6.15. *Let $f \in \mathbb{H}^{\hat{\otimes} d}$. Then, $F(\cdot) = I_{d-1}(f(\cdot))$ is an $\mathbb{H}$-valued random variable. We have*

$$\delta(I_{d-1}(f(\cdot))) = I_d(f)\,. \tag{6.2.9}$$

Proof. For any $h \in \mathbb{H}$, we have

$$\begin{aligned}\mathbb{E}\left[\delta(I_{d-1}(f(\cdot)))\mathcal{E}(h)\right] &= \mathbb{E}\left[\langle I_{d-1}(f(\cdot))\,, D\mathcal{E}(h)\rangle_{\mathbb{H}}\right]\\ &= \mathbb{E}\left[\langle I_{d-1}(f(\cdot))\,, \mathcal{E}(h)h\rangle_{\mathbb{H}}\right]\\ &= \mathbb{E}\left[I_{d-1}(\langle f\,, h\rangle_{\mathbb{H}})\mathcal{E}(h)\right] = \langle f\,, h^{\otimes d}\rangle_{\mathbb{H}^{\hat{\otimes} d}} = \mathbb{E}\left[I_d(f)\mathcal{E}(h)\right]\,.\end{aligned}$$

By linearity, we have $\mathbb{E}\left[\delta(I_{d-1}(f(\cdot)))G\right] = \mathbb{E}\left[I_d(f)G\right]$ for any $G \in \mathcal{E}$. Thus, we have $\delta(I_{d-1}(f(\cdot))) = I_d(f)$. □

The operator $L = -\delta D$ is called Ornstein-Uhlenbeck operator or the number operator. For any $f \in \mathbb{H}^{\hat{\otimes} n}$, we have

$$LI_n(f) = \delta DI_n(f) = d\delta I_{n-1}(f(\cdot)) = nI_n(f)\,. \tag{6.2.10}$$

We can also compute $D\delta(F)$.

Proposition 6.16. *If $F : \mathbb{B} \to \mathbb{H}$ is in $\mathbb{D}_{1,2}(\mathbb{H})$, then*

$$D\delta(F) = F + \delta.(DF.)\,. \tag{6.2.11}$$

Proof. For any $h, h' \in \mathbb{H}$, we have

$$\begin{aligned}
\mathbb{E}\left[\langle \mathcal{E}(h)h'\,, D\delta(F)\rangle_{\mathbb{H}}\right] &= \mathbb{E}\left[\delta(\mathcal{E}(h)h')\,, \delta(F))_{\mathbb{H}}\right] \\
&= \mathbb{E}\left[(\mathcal{E}(h)\delta(h') - \mathcal{E}(h)\langle h, h'\rangle_{\mathbb{H}})\,\delta(F)\right] \\
&= \mathbb{E}\left[\langle D\left(\mathcal{E}(h)\delta(h') - \mathcal{E}(h)\langle h, h'\rangle_{\mathbb{H}}\right)\,, F\rangle_{\mathbb{H}}\right] \\
&= \mathbb{E}\left[\mathcal{E}(h)\delta(h')\langle h, F\rangle_{\mathbb{H}} - \mathcal{E}(h)\langle h, h'\rangle_{\mathbb{H}}\langle h\,, F\rangle_{\mathbb{H}}\right] \\
&\quad + \mathbb{E}\left[\mathcal{E}(h)\langle h', F\rangle_{\mathbb{H}}\right]\,.
\end{aligned} \tag{6.2.12}$$

On the other hand, we have

$$\begin{aligned}
\mathbb{E}\left[\langle \mathcal{E}(h)h'\,, \delta.(DF.)\rangle_{\mathbb{H}}\right] &= \mathbb{E}\left[\langle \mathbb{D}.\mathcal{E}(h)h'\,, DF.\rangle_{\mathbb{H}}\right] \\
= \mathbb{E}\left[\langle h.\mathcal{E}(h)h'\,, DF.\rangle_{\mathbb{H}}\right] &= \mathbb{E}\left[\langle \mathcal{E}(h)h'\,, D\langle h\,, F\rangle_{\mathbb{H}}\rangle_{\mathbb{H}}\right] \\
&= \mathbb{E}\left[\delta(\mathcal{E}(h)h')\langle h\,, F\rangle_{\mathbb{H}}\right] \\
&= \mathbb{E}\left[\mathcal{E}(h)\delta(h')\langle h, F\rangle_{\mathbb{H}} - \mathcal{E}(h)\langle h, h'\rangle_{\mathbb{H}}\langle h\,, F\rangle_{\mathbb{H}}\right]\,.
\end{aligned} \tag{6.2.13}$$

Comparing (6.2.12) and (6.2.13), we have

$$\mathbb{E}\left[\langle \mathcal{E}(h)h'\,, D\delta(F)\rangle_{\mathbb{H}}\right] = \mathbb{E}\left[\langle \mathcal{E}(h)h'\,, F + \delta.(DF.)\rangle_{\mathbb{H}}\right]\,.$$

Since the elements of the form $\sum_{finite} \mathcal{E}(h_i)h_i'$ is dense in $\mathbb{D}_{1,2}(\mathbb{H})$, we have $D\delta(F) = F + \delta.(DF.)$. □

From the above computation we immediately have

Proposition 6.17. *If $F \in \mathbb{D}_{1,2}(\mathbb{H})$, then*

$$\mathbb{E}\left[\delta(F)\right]^2 = \mathbb{E}\left[|F|_{\mathbb{H}}^2 + \langle D.F_*\,, D_*F.\rangle_{\mathbb{H}^{\otimes 2}}\right]\,. \tag{6.2.14}$$

Remark 6.3. $\langle D.F_*\,, D_*F.\rangle_{\mathbb{H}^{\otimes 2}}$ can also be expressed as the trace of $(DF)^*DF$ and its exact meaning is explained as follows.

$$\begin{aligned}
\langle D.F_*\,, D_*F.\rangle_{\mathbb{H}^{\otimes 2}} &= \mathrm{Tr}\left((DF)^*DF\right) \\
&= \sum_{i,j=1}^{\infty} \left(D_{e_i}\langle F\,, e_j\rangle\right) D_{e_j}\langle F, e_i\rangle
\end{aligned}$$

if $\{e_j\,, j = 1, 2, \cdots\}$ is an orthonormal basis.

Proof. Using (6.2.11) we have

$$\begin{aligned}\mathbb{E}\left[\delta(F)\right]^2 &= \mathbb{E}\left[\langle D\delta(F)\,, F\rangle_{\mathbb{H}}\right] = \mathbb{E}\left[\langle F + \delta_{\cdot}(DF_{\cdot})\,, F\rangle_{\mathbb{H}}\right] \\ &= \mathbb{E}\left[|F|_{\mathbb{H}}^2\right] + \mathbb{E}\left[\langle \delta_{\cdot}(DF_{\cdot})\,, F\rangle_{\mathbb{H}}\right] \\ &= \mathbb{E}\left[|F|_{\mathbb{H}}^2\right] + \mathbb{E}\left[\langle \mathbb{D}_* F_{\cdot}\,, \mathbb{D}_{\cdot} F_*\rangle_{\mathbb{H}^{\otimes 2}}\right],\end{aligned}$$

proving the proposition. □

Proposition 6.18. *Let $F \in \mathbb{D}_{1,2}$ and $u \in \mathrm{Dom}(\delta)$ such that $Fu \in L^2(\mathbb{B}, \mathbb{H})$. Then, $Fu \in \mathrm{Dom}(\delta)$ and*

$$\delta\left(Fu\right) = F\delta(u) - \langle DF, u\rangle_{\mathbb{H}}\,, \tag{6.2.15}$$

provided the right-hand side is square integrable.

Proof. For any $G \in \mathbb{D}_\infty$, we have

$$\begin{aligned}\mathbb{E}\left[G\delta\left(Fu\right)\right] &= \mathbb{E}\left[\langle DG\,, Fu\rangle\right] = \mathbb{E}\left[\langle FDG\,, u\rangle\right] \\ &= \mathbb{E}\left[\langle D(FG)\,, u\rangle\right] - \mathbb{E}\left[\langle GDF\,, u\rangle\right] \\ &= \mathbb{E}\left[GF\delta(u)\rangle\right] - \mathbb{E}\left[G\langle DF\,, u\rangle\right] \\ &= \mathbb{E}\left[G\left[F\delta(u) - \langle DF\,, u\rangle\right]\right].\end{aligned}$$

This proves the identity since G is arbitrary. □

Example 6.4. Let $F = I_n(f_n)$ for some $f_n \in \mathbb{H}^{\hat{\otimes} n}$. From $DI_n(f_n) = nI_{n-1}(f_n)$ and (5.3.2) we have

$$\begin{aligned}|DF|_{\mathbb{H}}^2 &= n^2|I_{n-1}(f_n)|_{\mathbb{H}}^2 = n^2\sum_{r=0}^{n-1} r!\binom{n-1}{r}^2 I_{2n-2-2r}(f_n\hat{\otimes}_r f_n) \\ &= n^2\sum_{r=1}^{n}(r-1)!\binom{n-1}{r-1}^2 I_{2n-2r}(f_n\hat{\otimes}_r f_n) \\ &= nn!|f|_{\mathbb{H}^{\otimes n}}^2 + n^2\sum_{r=1}^{n-1}(r-1)!\binom{n-1}{r-1}^2 I_{2n-2r}(f_n\hat{\otimes}_r f_n) \\ &= n\mathbb{E}(F^2) + n^2\sum_{r=1}^{n-1}(r-1)!\binom{n-1}{r-1}^2 I_{2n-2r}(f_n\hat{\otimes}_r f_n)\,. \end{aligned}\tag{6.2.16}$$

Thus, we have

$$\begin{aligned}\mathrm{Var}(|DF|_{\mathbb{H}}^2) &= n^4\sum_{r=1}^{n-1}(2n-2r)!\,[(r-1)!]^2\binom{n-1}{r-1}^4 |f_n\hat{\otimes}_r f_n|_{\mathbb{H}^{\otimes(2n-2r)}}^2 \\ &= \sum_{r=1}^{n-1} r^2(2n-2r)!\,[(r-1)!]^2\binom{n}{r}^4 |f_n\hat{\otimes}_r f_n|_{\mathbb{H}^{\otimes(2n-2r)}}^2\,.\end{aligned}\tag{6.2.17}$$

Using $L = \delta D$, and $LF = nF$, we have

$$\begin{aligned}\mathbb{E}(F^4) &= \frac{1}{n}\mathbb{E}\left[(\delta DF)\, F^3\right] = \frac{1}{n}\mathbb{E}\left[\langle DF\,, D(F^3)\rangle\right] \\ &= \frac{3}{n}\mathbb{E}\left[F^2|DF|_{\mathbb{H}}^2\right]\,. \end{aligned} \tag{6.2.18}$$

Denoting the summation in (6.2.16) by G, then

$$\begin{aligned}\mathbb{E}(F^4) - 3\left[\mathbb{E}(F^2)\right]^2 &= \frac{3}{n}\mathbb{E}\left[F^2|DF|_{\mathbb{H}}^2\right] \\ &= \frac{3}{n}\mathbb{E}\left[F^2\left\{n\mathbb{E}(F^2) + G\right\}\right] - 3\left[\mathbb{E}(F^2)\right]^2 = \frac{3}{n}\mathbb{E}\left[F^2 G\right]\,. \end{aligned}$$

Now we can use (5.3.14) to obtain the following

$$\begin{aligned}\mathbb{E}(F^4) - 3\left(\mathbb{E}(F^2)\right)^2 &= \frac{3}{n}\sum_{r=1}^{n-1} rr!^2\binom{n}{r}^4(2n-2r)!\|f_n\hat{\otimes}_r f_n|_{\mathbb{H}\otimes(2n-2r)}^2 \\ &= \sum_{r=1}^{n-1} n!^2\binom{n}{r}^2\left\{|f_n\otimes_r f_n|_{\mathbb{H}\otimes 2n-2r}^2 + \binom{2n-2r}{n-r}|f_n\hat{\otimes}_r f_n|_{\mathbb{H}^{2n-2r}}^2\right\}\,. \end{aligned} \tag{6.2.19}$$

Comparing the above identity with (6.2.17), we have

$$\mathrm{Var}\left(|DF|_{\mathbb{H}}^2\right) \le \frac{n(n-1)}{3}\left[\mathbb{E}(F^4) - 3\left[\mathbb{E}(F^2)\right]^2\right] \le 3n\mathrm{Var}\left(|DF|_{\mathbb{H}}^2\right)\,. \tag{6.2.20}$$

6.3 Regularity of Density of Wiener Functional

Let $(\mathbb{B}, \mathbb{H}, \mathbb{P})$ be an abstract Wiener space. Given a (vector-valued) random variable $F : \mathbb{B} \to \mathbb{R}^d$, its (probability) law $\mathbb{P}\circ F^{-1}$ is

$$\mu_F(A) = \mathbb{P}\circ F^{-1}(A) = \mathbb{P}\left(\{x\in\mathbb{B}\,,\ F(x)\in A\}\right)\,, \quad \forall \text{ Borel subset } A \text{ of } \mathbb{R}^d\,.$$

Definition 6.6. We say that F has a density (with respect to the Lebesgue measure) if there is a probability density function $p_F(x) \ge 0$ such that

$$\mu_F(A) = \mathbb{P}\circ F^{-1}(A) = \int_A p_F(x)dx\,, \quad \forall \text{ Borel subset } A \text{ of } \mathbb{R}^d\,.$$

In the remaining part of this book, when we say density it is always with respect to the Lebesgue measure.

Theorem 6.5. *Let $F : \mathbb{B} \to \mathbb{R}^d$ be measurable and satisfy*

(i) $F \in \mathbb{D}_{1,p}^{\mathrm{loc}}$ *for some* $p > 1$.
(ii) *The Malliavin covariance matrix* $\Gamma_F = (\langle DF_k, DF_j\rangle_{\mathbb{H}})_{1\le i,j\le d}$ *is invertible almost surely.*

Then, the law of F *is absolutely continuous with respect to the Lebesgue measure.*

Proof. By a localization argument, one can assume that $F = (F_1, \cdots, F_d) \in \mathbb{D}_{1,p}$. We shall use the operator $\tau_h F(x) = F(x+h)$. Fix a complete orthonormal basis $\{e_i, i = 1, 2, \cdots\}$ and define $\phi_n = (\phi_{1,n}, \cdots, \phi_{d,n})$, where

$$\phi_{k,n}(t_1, \cdots, t_n) = \tau_{t_1e_1+\cdots+t_ne_n} F_k\,.$$

From the assumption that $F \in \mathbb{D}_{1,p}$, we see that $\phi_{k,n}$ is approximately differentiable in $t_1, \cdots, t_n$ for almost all $x \in \mathbb{B}$. Moreover, we have

$$\mathrm{ap}\,\partial_i\phi_{k,n}(t) = \langle \tau_{t_1e_1+\cdots+t_ne_n} DF_k, e_i\rangle_{\mathbb{H}}\,.$$

Consequently, we have

$$\begin{aligned} J_{kj}^{(n)}(t) &:= \langle \mathrm{ap}\,\partial\phi_{k,n}(t), \mathrm{ap}\,\partial\phi_{j,n}(t)\rangle \\ &= \sum_{\ell=1}^n \langle \tau_{t_1e_1+\cdots+t_ne_n} DF_k, e_\ell\rangle_{\mathbb{H}} \cdot \langle \tau_{t_1e_1+\cdots+t_ne_n} DF_j, e_\ell\rangle_{\mathbb{H}}\,. \end{aligned}$$

Application of Corollary A.2 to ϕ_n yields

$$\int_{\phi_n^{-1}(B)} \det(J^{(n)}) dt_1 \cdots dt_n = 0\,, \quad \text{where} \quad J^{(n)}(t) = \left(J_{kj}^{(n)}(t)\right)_{1\le k,j\le d}\,.$$

Taking the expectation we have

$$\begin{aligned} 0 &= \int_{\mathbb{R}^n} \mathbb{E}\Bigg[\det\Bigg(\sum_{\ell=1}^n \langle \tau_{t_1e_1+\cdots+t_ne_n} DF_k, e_\ell\rangle_{\mathbb{H}} \cdot \\ &\qquad \langle \tau_{t_1e_1+\cdots+t_ne_n} DF_j, e_\ell\rangle_{\mathbb{H}}\Bigg) I_{\phi_n^{-1}(B)}\Bigg] dt_1\cdots dt_n \\ &= \int_{\mathbb{R}^n} \mathbb{E}\Bigg[\tau_{t_1e_1+\cdots+t_ne_n}\Bigg\{\det\Bigg(\sum_{\ell=1}^n \langle DF_k, e_\ell\rangle_{\mathbb{H}} \cdot \\ &\qquad \langle DF_j, e_\ell\rangle_{\mathbb{H}}\Bigg) I_{F^{-1}(B)}\Bigg\}\Bigg] dt_1\cdots dt_n \\ &= \int_{\mathbb{R}^n} \mathbb{E}\Bigg[\det\Bigg(\sum_{\ell=1}^n \langle DF_k, e_\ell\rangle_{\mathbb{H}} \cdot \langle DF_j, e_\ell\rangle_{\mathbb{H}}\Bigg) \\ &\qquad I_{F^{-1}(B)} \exp\Bigg\{\sum_{\ell=1}^n (t_\ell\tilde{e}_\ell - t_\ell^2/2)\Bigg\}\Bigg] dt_1\cdots dt_n\,. \end{aligned}$$

Thus, we have

$$\det\left(\sum_{\ell=1}^{n}\langle DF_k, e_\ell\rangle_{\mathbb{H}}\cdot\langle DF_j, e_\ell\rangle_{\mathbb{H}}\right) I_{F^{-1}(B)} = 0 \quad \text{almost surely}\,.$$

Letting $n\to\infty$ and noticing

$$\det\left(\sum_{\ell=1}^{n}\langle DF_k, e_\ell\rangle_{\mathbb{H}}\cdot\langle DF_j, e_\ell\rangle_{\mathbb{H}}\right) \to \det(\langle DF_k, DF_j\rangle_{\mathbb{H}})$$

in probability, we have

$$\det\left(\langle DF_k, DF_j\rangle_{\mathbb{H}}\right) I_{F^{-1}(B)} = 0 \quad \text{almost surely}\,.$$

This combined with Condition (ii) of the theorem implies that $P(F^{-1}(B)) = 0$. The theorem is proved. □

Proposition 6.19. *Let $F\in\mathbb{D}_{1,p}$ be a real valued random variable. Assume there is an element $u\in L^r(\mathbb{B};\mathbb{H})$ for some $r>1$ such that $\langle DF, u\rangle\neq 0$ a.s. and $\frac{u}{\langle DF,u\rangle}\in \mathrm{Dom}_q(\delta)$. Then F has a density and the density is given by*

$$p_F(x) = \mathbb{E}\left[I_{\{F>x\}}\,\delta\left(\frac{u}{\langle DF,u\rangle}\right)\right]. \tag{6.3.1}$$

Proof. Denote the right hand side of (6.3.1) by $f(x)$. For any smooth function $g\in C_0^\infty(\mathbb{R})$, set $G(x)=\int_\infty^x g(y)dy$. Then, $G(F)\in\mathbb{D}_{1,p}$. We have

$$\begin{aligned}
\int_{\mathbb{R}} g(x)f(x)dx &= \mathbb{E}\left[\int_{\mathbb{R}} I_{\{F>x\}}g(x)dx\;\delta\left(\frac{u}{\langle DF,u\rangle}\right)\right]\\
&= \mathbb{E}\left[G(F)\delta\left(\frac{u}{\langle DF,u\rangle}\right)\right] = \mathbb{E}\left[\left\langle DG(F)\,,\frac{u}{\langle DF,u\rangle}\right\rangle\right]\\
&= \mathbb{E}\left[g(F)\left\langle DF\,,\frac{u}{\langle DF,u\rangle}\right\rangle\right] = \mathbb{E}\left[g(F)\right].
\end{aligned}$$

This shows that $f(x)$ is the density of F (with respect to the Lebesgue measure). □

Definition 6.7. Given a random vector $F=(F_1,\cdots,F_d)\in\mathbb{D}_{1,\mathrm{loc}}$, its Malliavin covariance matrix is

$$\gamma_F = \left(\langle DF_i, DF_j\rangle_{\mathbb{H}}\right)_{1\le i,j\le d}\,.$$

F is called *non-degenerated* if $\det(\gamma_F)^{-1}\in\cap_{p\ge1}L^p(\mathbb{B})$.

Let A be an open set of $\mathbb{R}^d$. F is called *locally non-degenerated* on A if there is a $u=u_A=(u_1,\cdots,u_d)\in\mathbb{D}_{1,\mathrm{loc}}(\mathbb{B},\mathbb{H})$ such that $\left(\langle u_i, DF_j\rangle I_{\{F\in A\}}\right)_{1\le i,j\le d}$ can be extended to a random matrix γ_u (namely, $\gamma_u^{ij}=\langle u_i, DF_j\rangle$ on $\{F\in A\}$) such that $\det(\gamma_u)^{-1}\in\cap_{p\ge1}L^p(\mathbb{B})$.

Remark 6.4. In the above definition the element u depends on A and it may not be unique.

Lemma 6.5. *Suppose that $\gamma \in \mathbb{D}_{1,\mathrm{loc}}$ is an $m \times m$ random matrix such that γ is invertible a.s. Then, $\gamma^{-1} \in \mathbb{D}_{1,\mathrm{loc}}$ and*

$$D\gamma^{-1} = -\gamma^{-1}(D\gamma)\gamma^{-1}, \tag{6.3.2}$$

Proof. For any $h \in \mathbb{H}$, $x \in \mathbb{B}$, and $\varepsilon > 0$, we have

$$\gamma^{-1}(x+\varepsilon h) - \gamma^{-1}(x) = -\gamma^{-1}(x+\varepsilon h)\left[\gamma(x+\varepsilon h) - \gamma(x)\right]\gamma^{-1}(x).$$

The lemma then follows. □

We refer to Section 6.8 for more detailed discussion on how to differentiate a function of matrix valued nonlinear Wiener functional. For any $\alpha = (\alpha_1, \cdots, \alpha_k) \in \{1, \cdots, d\}^k$, denote $\partial_\alpha = \frac{\partial^k}{\partial x_{\alpha_1} \cdots \partial x_{\alpha_k}}$.

Theorem 6.6. *Let $F = (F_1, \cdots, F_d) \in \mathbb{D}_{k,p}$ be a locally nondegenerate on an open set A of $\mathbb{R}^d$ and let $\varphi \in C_p^k(\mathbb{R}^d, \mathbb{R})$. Let $G \in \mathbb{D}_{k,q}$ satisfy that $G = 0$ on the set $\{F \notin A\}$. Then, for any $\alpha = (\alpha_1, \cdots, \alpha_k) \in \{1, \cdots, d\}^k$, there is an $H_\alpha(u, G)$ such that the following integration by parts formula holds*

$$\mathbb{E}\left[\partial_\alpha \varphi(F) G\right] = \mathbb{E}\left[\varphi(F) H_\alpha(u, G)\right]. \tag{6.3.3}$$

Moreover, $H_\alpha(u, G)$ is recursively given as follows.

$$\begin{cases} H_{(i)}(u,G) = \sum_{j=1}^d \delta\left(G(\gamma_u^{-1})^{ij} u_j\right) \\ \qquad = \sum_{j=1}^d G(\gamma_u^{-1})^{ij} \delta(u_j) + \sum_{j=1}^d (\gamma_u^{-1})^{ij} \langle DG, u_j \rangle_{\mathbb{H}} \\ \qquad\quad + \sum_{j=1}^d G \langle D(\gamma_u^{-1})^{ij}, u_j \rangle_{\mathbb{H}}, \\ H_\alpha(u,G) = H_{(\alpha_k)}(u, H_{(\alpha_1, \cdots, \alpha_{k-1})}(u,G)). \end{cases}$$

Furthermore, for any $1 \le p < q < \infty$, we have

$$\|H_\alpha(G)\|_p \le C_{p,q} \|\gamma_u^{-1}\|_{k,2^{k-1}r}^k \|G\|_{k,p} \tag{6.3.4}$$

where $\frac{1}{p} = \frac{1}{q} + \frac{1}{r}$.

Proof. First, we assume that $k = 1$ and $\alpha = (i)$. In this case, the right hand side of (6.3.3) can be computed as follows.

$$\begin{aligned}
\mathbb{E}\left[\varphi(F)H_{(i)}(u,G)\right] &= \sum_{j=1}^{d} \mathbb{E}\left[\varphi(F)\delta\left(G(\gamma_u^{-1})^{ij}u_j\right)\right] \\
&= \sum_{j=1}^{d} \mathbb{E}\left\langle D\varphi(F)\,, G(\gamma_u^{-1})^{ij}u_j\right\rangle \\
&= \sum_{j,\ell=1}^{d} \mathbb{E}(\partial_\ell\varphi)(F)(\gamma_u^{-1})^{ij}G\left\langle DF_\ell\,, u_j\right\rangle \\
&= \sum_{j,\ell=1}^{d} \mathbb{E}(\partial_\ell\varphi)(F)(\gamma_u^{-1})^{ij}G\gamma_u^{j,\ell} \\
&= \sum_{j,\ell=1}^{d} \mathbb{E}(\partial_\ell\varphi)(F)G\delta_{i\ell} = \mathbb{E}\left[(\partial_i\varphi)(F)G\right]\,,
\end{aligned}$$

which is the left hand side of (6.3.3). Where in the second last equality, we use the fact that G is 0 outside $\{F \in A\}$ and on $\{F \in A\}$, $\langle DF_\ell\,, u_j\rangle = \gamma_u^{j,\ell}$. The general case can be shown by induction on the length $|\alpha|$ of the multi-index α.

Since the divergence operator δ is bounded from $\mathbb{D}_{1,p}$ to L^p, we have

$$\|H_{(\alpha_1,\cdots,\alpha_k)}\|_p \le c_p \sum_{j=1}^{d} \|H_{(\alpha_1,\cdots,\alpha_{k-1})}(\gamma_u^{-1})^{\alpha_k,j}u_{\alpha_j}\|\,.$$

This implies (6.3.4) for $k = 1$. The general case follows from the above inequality together with an induction on k. □

Remark 6.5. In applying (6.3.1) or (6.3.3) to study the existence or smoothness of the density the most difficult task may be to show the locally non-degeneracy of F. Usually in practice, one chooses $u = DF$ to prove $\det(\langle DF, DF\rangle)^{-1} \in \cap_{p\ge 1}L^p(\mathbb{B})$. Another popular choice is $u = DL^{-1/2}F$. There are many other choices of u.

Example 6.5. Let $F = \exp\left\{-\lambda B_1^2\right\}$, where $\lambda > 1/2$ is a positive number. Then, $D_rF = -2\lambda F B_1 I_{[0,1]}(r)$. Thus

$$\langle DF, DF\rangle = 4\lambda^2F^2B_1^2\,.$$

It is easy to see that $\langle DF, DF\rangle$ has no negative moments. But it is easy to see that F has the density given by

$$f_F(y) = \sqrt{\frac{1}{-8\lambda\pi\ln y}}y^{2\lambda-1}\,, \quad 0 < y < 1\,.$$

Definition 6.8. Let $F : \mathbb{B} \to \mathbb{R}^d$ be a random vector. Recall that the law μ_F is a probability measure on $(\mathbb{R}^d, \mathcal{B}(\mathbb{R}^d))$ defined by

$$\mu_F(A) = \mathbb{P}(\omega \in \mathbb{B}, F(\omega) \in A).$$

The topological support of the law of F is defined as

$$\mathrm{Supp}(\mu_F) = \left\{x \in \mathbb{R}^d, \quad \mathbb{P}(|x - F| < \varepsilon) > 0 \quad \forall\, \varepsilon > 0\right\}. \tag{6.3.5}$$

Proposition 6.20. *Let* $F = (F_1, \cdots, F_d) \in \mathbb{D}_{1,p}$ *for some* $p > 1$*. Then,* $\mathrm{Supp}(\mu_F)$ *is a closed and connected subset of* $\mathbb{R}^d$.

Proof. If x is a limiting point of $\mathrm{Supp}(\mu_F)$, then for any $\varepsilon > 0$, there is a $x_\varepsilon \in \mathrm{Supp}(\mu_F)$ such that $|x - x_\varepsilon| < \varepsilon/2$ and $\mathbb{P}(|x_\varepsilon - F| < \varepsilon/2) > 0$. Thus, $\mathbb{P}(|x - F| < \varepsilon) > \mathbb{P}(|x_\varepsilon - F| < \varepsilon/2) > 0$. Hence, $\mathrm{Supp}(\mu_F)$ is closed.

If $\mathrm{Supp}(\mu_F)$ is not connected, then it can be decomposed into two disjoint closed subsets A and B. By definition of the support this implies that $\mathbb{P}(F \in A) > 0$ and $\mathbb{P}(F \in B) > 0$ which in turn implies that $0 < \mathbb{P}(A) < 1$.

Let ψ_R be a smooth function which is 1 when $|x| \le R$ and 0 when $|x| \ge R + 1$, and $\sup_{x,R} |\nabla \psi_R(x)| < \infty$.

Now we show that $\mathrm{Supp}(\mu_F)$ is connected. Suppose not. Let A and B two disjoint connected components of $\mathrm{Supp}(\mu_F)$. Let ϕ_R be a smooth function which is 1 when x is in a neighborhood of $A \cap B(0, R)$ and 0 when x is in a neighborhood of $B \cap B(0, R)$, where $B(0, R)$ is the ball of center 0 with radius R. Moreover, we can assume that $\phi_R \in [0, 1]$. By the chain rule we have

$$\begin{aligned} D(\phi_R \psi_R)(F) &= \sum_{j=1}^{d} \left[(\phi_R \partial_i \psi_R)(F) + (\psi_R \partial_i \phi_R)(F)\right] DF_i \\ &= \sum_{j=1}^{d} (\phi_R \partial_i \psi_R)(F) DF_i . \end{aligned}$$

This identity yields

$$\sup_R \|D(\phi_R \psi_R)(F)\|_{\mathbb{H}} = \sum_{j=1}^{d} \sup_R |\partial_i \psi_R|_\infty \|DF_i\|_{\mathbb{H}} \in L^p .$$

It is easy to see that $(\phi_R \psi_R)(F)$ converges to $I_{\{F \in A\}}$ in L^p as $R \to \infty$. Theorem 7.6 says that $I_{\{F \in A\}}$ is in $\mathbb{D}_{1,p}$. Then, Proposition 6.8 implies that $I_{\{F \in A\}}$ is either 0 or 1, which is contradictory to the fact $0 < \mathbb{P}(F \in A) < 1$. $\square$

Proposition 6.21. *Let $F = (F_1, \cdots, F_d) \in \mathbb{D}_{1,p}$ for some $p > 2$. Assume that F has a density $p(x)$ and the density $p(x)$ is a locally Lipschitz continuous in $x \in \mathrm{Supp}(\mu_F)$. Then, $p(a) > 0$ for all interior point $a \in \mathrm{Supp}(\mu_F)$.*

Proof. Suppose that a is an interior point of $\mathrm{Supp}(\mu_F)$ and $p(a) = 0$. Let $r = \frac{2p}{p+2} > 1$. Since $0 < \mathbb{P}(F > a) < 1$, we see that $I_{\{F>a\}} \notin \mathbb{D}_{1,r}$.

Fix $\varepsilon > 0$ and define

$$\phi_\varepsilon(x) = \frac{1}{2\varepsilon} \int_{-\infty}^{x} I_{[a-\varepsilon, a+\varepsilon]}(y) dy\,.$$

First, it is easy to see that $\phi_\varepsilon(F)$ converges to $I_{\{F>a\}}$ in L^r. Furthermore, we have $\phi_\varepsilon(F) \in \mathbb{D}_{1,r}$ and

$$D\phi_\varepsilon(F) = \frac{1}{2\varepsilon} I_{[a-\varepsilon, a+\varepsilon]}(F) DF\,. \tag{6.3.6}$$

The local Lipschitz condition imposed on p implies that $p(x) = p(x) - p(a) \le K|x - a|$. This together with (6.3.6) yields

$$\begin{aligned}\mathbb{E}\left[\|D\phi_\varepsilon(F)\|_{\mathbb{H}}^r\right] &\le \left(\mathbb{E}(|DF|_{\mathbb{H}}^p\right)^{\frac{2}{2+p}} \left(\frac{1}{(2\varepsilon)^2} \int_{a-\varepsilon}^{a+\varepsilon} p(x)dx\right)^{\frac{p}{p+2}} \\ &\le 2^{-\frac{2p}{p+2}} K^{\frac{p}{p+2}} \left(\mathbb{E}(|DF|_{\mathbb{H}}^p\right)^{\frac{2}{2+p}}\end{aligned}$$

which is bounded uniformly on ε. Theorem 7.6 implies that $I_{\{F>a\}}$ is in $\mathbb{D}_{1,r}$, which is contradictory to $I_{\{F>a\}} \notin \mathbb{D}_{1,r}$. The theorem is proved. □

Let

$$\begin{aligned}\mathcal{S}(\mathbb{R}^d) &= \Big\{ f \in C^\infty(\mathbb{R}^d, \mathbb{R})\,;\ \sup_{x\in\mathbb{R}^d} (1 + |x|^2)^m |f^{(k)}(x)| < \infty\,, \\ &\qquad \text{for all } k, m \ge 1 \Big\}\end{aligned}$$

be the Schwartz space of rapid decreasing functions and let $\mathcal{S}'(\mathbb{R}^d)$ be the Schwartz space of the tempered distributions (the space of continuous linear functionals on $\mathcal{S}(\mathbb{R}^d)$.

From Proposition 3.18, we see that the operator $(1 + x^2 - \Delta)$ has the discrete spectrum $2(n+1)$ with eigenfunctions $\mathfrak{H}_n(x)$, $n = 0, 1, \cdots$. Thus, $(1 + x^2 - \Delta)^s$ is well-defined for any $s \in \mathbb{R}$. By interpolation inequality, we see that $(1 + x^2 - \Delta)^s$ maps $\mathcal{S}(\mathbb{R}^d)$ to $\mathcal{S}(\mathbb{R}^d)$. For every $\varphi \in \mathcal{S}(\mathbb{R}^d)$, we denote

$$\|\phi\|_s = \sup_{x\in\mathbb{R}^d} |(1 + x^2 - \Delta)^{s/2} \phi(x)|\,.$$

We define a family of Banach subspaces of $\mathcal{S}_s(\mathbb{R}^d)$ to be the completion of $\mathcal{S}(\mathbb{R}^d)$ with respect to the norm $\|\cdot\|_s$ so that

$$\mathcal{S}(\mathbb{R}^d) \subset \cdots \subset \mathcal{S}_1 \subset \mathcal{S}_0 = \hat{C}(\mathbb{R}^d) \subset \mathcal{S}_{-1} \subset \cdots \subset \mathcal{S}'(\mathbb{R}^d)$$

and

$$\mathcal{S}(\mathbb{R}^d) = \cap_{n=1}^\infty \mathcal{S}_n, \quad \mathcal{S}'(\mathbb{R}^d) = \cup_{n=1}^\infty \mathcal{S}_{-n}.$$

Definition 6.9. Given a d-dimensional Wiener functional $F : W \to \mathbb{R}^d$ and $T \in \mathcal{S}'(\mathbb{R}^d)$, we say that $T \circ F$ is defined in $\mathbb{D}_{-k,p}$ and $T \circ F = \Phi$ for some $\Phi \in \mathbb{D}_{-k,p}$ if there exist $n \in \mathbb{Z}^+$ (the set of non-negative integers), $1 < p < \infty$, $\alpha > 0$ such that $T \in \mathcal{S}_{-2n}$, $\Phi \in \mathbb{D}_{-\alpha,p}$ and the following holds: For every sequence $\{\phi_k : k = 1, \ , \cdots\} \subset \mathcal{S}(\mathbb{R}^d)$ such that $\|\phi_k - T\|_{\mathcal{S}_{-2n}} \to 0$ as $k \to \infty$, it holds that $\|\phi_k \circ F - \Phi\|_{-\alpha,p} \to 0$ as $k \to \infty$ (Note that $\phi_k \circ F \in L^\infty \subset \mathbb{D}_{-\alpha,p}$ for all k).

Clearly, the above Φ, if exists, is uniquely determined by F and T.

Theorem 6.7. *Let $F = (F_1, \cdots, F_d)$ be a nondegenerate random vector. Assume that $\varphi \in \mathcal{S}_{-2k}$ for some $k \in \mathbb{N}$. Then, for any $p > 1$, there exists a constant $c(p,k,F) \in (0,\infty)$ such that*

$$\|\varphi(F)\|_{-2k,p} \le c(p,k,F)\|\varphi\|_{-2k}, \quad \forall \varphi \in \mathcal{S}_{-2k}(\mathbb{R}^d). \tag{6.3.7}$$

Proof. Let $\psi = (1 + x^2 - \Delta)^{-k}\phi \in \mathcal{S}(\mathbb{R}^d)$. By Proposition 6.6 we see that for any $G \in \mathbb{D}_\infty$,

$$\begin{aligned}\mathbb{E}\left[\phi(F)G\right] &= \mathbb{E}\left[(1+x^2-\Delta)^k\psi(F)G\right] \\ &= \mathbb{E}\left[(1+x^2-\Delta)^{k-1}\psi(F)G\right] + \mathbb{E}\left[F^2(1+x^2-\Delta)^{k-1}\psi(F)G\right] \\ &\qquad -\mathbb{E}\left(\Delta\left[(1+x^2-\Delta)^{k-1}\psi\right)(F)G\right] \\ &= \mathbb{E}\left[(1+x^2-\Delta)^{k-1}\psi(F)\left\{G + F^2G + \sum_{i=1}^d H_{(ii)}(G)\right\}\right],\end{aligned}$$

where $H_{(ii)}$ is defined in Proposition 6.6. By induction, one can show that there is a random variable $R_k(G) \in \mathbb{D}_\infty$ which depends linearly on the derivatives of G up to order $2k$ such that

$$\mathbb{E}\left[\phi(F)G\right] = \mathbb{E}(\psi(F)R_k(G)).$$

Let q be the conjugate number of p: $1/p + 1/q = 1$. We have

$$|\mathbb{E}\left[\phi(F)G\right]| \le \|\psi\|_\infty \mathbb{E}|R_k(G)| \le c(p,k,F)\|G\|_{2k,q}\|\phi\|_{-2k}.$$

This implies that

$$\|\phi(F)\|_{2k} \le c(p,k,F)\|\phi\|_{-2k}.$$

The theorem is proved. □

6.4 Girsanov Transformation: Finite Dimension

Definition 6.10. Let $\mathbb{H}$ be a separable Hilbert space and let $\{e_n, n = 1, 2, \cdots\}$ be a complete orthonormal system. Let $K : \mathbb{H} \to \mathbb{H}$ be a bounded linear operator. The operator norm of K is defined as

$$\|K\| = \sup_{h\in\mathbb{H}, |H|_{\mathbb{H}}=1} |Kh|_{\mathbb{H}} . \tag{6.4.1}$$

We say that K is a Hilbert-Schmidt operator if

$$\|K\|_2^2 = \sum_{i,j=1}^{\infty} \langle Ae_i, e_j\rangle^2 < \infty \tag{6.4.2}$$

and $\|K\|_2$ is called the Hilbert-Schmidt norm of K. We say that K is a trace class operator if $\sum_{i=1}^{\infty} |\langle Ae_i, e_i\rangle| < \infty$ and we define its trace as

$$\mathrm{Tr}(K) = \sum_{i=1}^{\infty} \langle Ae_i, e_i\rangle . \tag{6.4.3}$$

If $K : \mathbb{H} \to \mathbb{H}$ has eigenvalues $\lambda_1, \lambda_2, \cdots$, then we have

$$\mathrm{Tr}(K) = \sum_{i=1}^{\infty} \lambda_i , \quad \|K\|_2^2 = \sum_{i=1}^{\infty} \lambda_i^2 , \quad \|K\| = \sup_{i\ge 1} |\lambda_i| . \tag{6.4.4}$$

We also define the $\|K\|_{\mathrm{Tr}} = \sum_{i=1}^{\infty} |\lambda_i|$.

Definition 6.11. If $K : \mathbb{H} \to \mathbb{H}$ has eigenvalues $\lambda_1, \lambda_2, \cdots$, then we define the following determinants of $I + K$

$$\det(I + K) = \prod_{i=1}^{\infty} (1 + \lambda_i) , \quad \text{and} \quad \det{}_2(I + K) = \prod_{i=1}^{\infty} (1 + \lambda_i) \exp(-\lambda_i) \tag{6.4.5}$$

if the limits exist.

It is easy to see from (6.4.4) that if K is of trace class, then it is Hilbert-Schmidt. In fact, we have $\|K\|_2 \le \sqrt{\|K\| \, \|K\|_{\mathrm{Tr}}}$. If $(I + K)$ is of trace class, then $\det(I + K)$ exists and if K is Hilbert-Schmidt, then $\det_2(I + K)$ exists. Thus, $\det_2$ is defined for a larger class of linear operators. It is also easy to see

$$\begin{cases} \det(I + K) = \exp[\mathrm{Tr}\log(I + K)] , \\ \det_2(I + K) = \exp[\mathrm{Tr}\log(I + K) - \mathrm{Tr}(K)] . \end{cases} \tag{6.4.6}$$

Lemma 6.6. *If* $(I+K)$ *is of trace class, then* $\det(I+K)$ *exists and if* K *is Hilbert-Schmidtian, then* $\det_2(I+K)$ *exists. Moreover,*

$$\begin{cases} |\det(I+K)| \le e^{\mathrm{Tr}(K)+\frac{1}{2}\|K\|_2^2}, & \text{and} \\ |\det_2(I+K)| \le e^{\frac{1}{2}\|K\|_2^2}. \end{cases} \tag{6.4.7}$$

Proof. We shall use the following inequality:

$$\log|1+\beta| \le \beta + \frac{\beta^2}{2}, \quad \forall\, \beta \in \mathbb{R}. \tag{6.4.8}$$

This inequality can be proved easily. Since the function te^{-t+1} attains its maximum at $t=1$, we have $te^{-t+1} \le 1$ or $t \le e^{t-1}$. Substituting $t=(1+\beta)^2$, we have $(1+\beta)^2 \le e^{2\beta+\beta^2} \le 1$. taking the logarithm yields (6.4.8).

Let $\lambda_1, \lambda_2, \cdots$ be the eigenvalues of K. Then,

$$\begin{aligned} |\det{}_2(I+K)| &\le \prod_{i=1}^{\infty} |1+\lambda_i| e^{-\lambda_i} \le \exp\left\{\sum_{i=1}^{\infty}[\log|1+\lambda_i| - \lambda_i]\right\} \\ &\le \exp\left\{\frac{1}{2}\sum_{i=1}^{\infty}|\lambda_i|^2\right\} = \exp\left\{\frac{1}{2}\|K\|_2^2\right\}. \end{aligned}$$

This proves the second equality in (6.4.7). The first inequality can be proved similarly. □

Remark 6.6. We refer to [Gohberg, Goldberg and Krupnik (2000); Simon (2005)] for more readings on determinants.

Definition 6.12. Let $T:(\Omega_1,\mathcal{F}_1) \to (\Omega_2,\mathcal{F}_2,\mu)$ be a measurable mapping. We define the probability measure $\mu\circ T^{-1}$ on the measurable space $(\Omega_1,\mathcal{F}_1)$ by

$$\mu\circ T^{-1}(A) = \mu\left(x : T(x) \in A\right), \quad \forall\, A \in \mathcal{F}_1.$$

Let $\mu_d(dx) = (2\pi)^{-d/2} e^{-\frac{|x|^2}{2}} dx$ be the standard Gaussian measure on $\mathbb{R}^d$. If $u:\mathbb{R}^d \to \mathbb{R}^d$ is a continuously differentiable mapping such that $T = I+u$ is diffeomorphism of $\mathbb{R}^d$, then for any smooth function $F:\mathbb{R}^d \to \mathbb{R}^d$ of polynomial growth we have from the usual change of variable formula that

$$\begin{aligned}
&\int_{\mathbb{R}^d} F(x)\mu_d(dx) \\
&= (2\pi)^{-d/2} \int_{\mathbb{R}^d} F(x+u(x))\,|\det(I+\nabla u(x))| \exp\left\{-\frac{|x+u(x)|^2}{2}\right\} dx \\
&= \int_{\mathbb{R}^d} F(x+u(x))\,|\det(I+\nabla u(x))| \exp\left\{-\langle u(x), x\rangle - \frac{1}{2}|u(x)|^2\right\} \mu_d(dx) .
\end{aligned} \tag{6.4.9}$$

The extension of this identity to infinite dimension will be the Girsanov (Girsanov-Ramer) theorem. First, we state more precisely the finite dimensional case.

Proposition 6.22. *Let* $\mathbb{R}^d \ni x \to u(x) \in \mathbb{R}^d$ *satisfy* $\sup_{x\in\mathbb{R}^d} \|\nabla u(x)\| \leq c_1 < 1$ *and* $\sup_{x\in\mathbb{R}^d} \|\nabla u(x)\|_2 \leq c_2 < \infty$. *Then,*

(i) $\mathbb{R}^d \ni x \to T(x) = x + u(x) \in \mathbb{R}^d$ *is bijective. If* $T^{-1} = I + V$ *is the inverse of* T*, then*

$$\sup_{x\in\mathbb{R}^d} \|\nabla V(x)\| \leq \frac{c_1}{1-c_1} \quad \text{and} \quad \sup_{x\in\mathbb{R}^d} \|\nabla V(x)\|_2 \leq \frac{c_2}{1-c_1} . \tag{6.4.10}$$

(ii) Denote

$$\begin{aligned}
\delta(u(x)) &= \langle u(x), x\rangle - \mathrm{Tr}(\nabla u(x)) \\
&= \langle u(x), x\rangle - \sum_{i=1}^{d} \partial_i u_i(x) ,
\end{aligned} \tag{6.4.11}$$

and

$$\begin{aligned}
\Lambda_u(x) &= |\det(I+\nabla u(x))| \exp\left\{-\langle u(x), x\rangle - \frac{1}{2}|u(x)|^2\right\} \\
&= |\det{}_2(I+\nabla u(x))| \exp\left\{-\delta(u(x)) - \frac{1}{2}|u(x)|^2\right\} .
\end{aligned} \tag{6.4.12}$$

Then, $\int_{\mathbb{R}^d} \Lambda_u(x)\mu_d(dx) = 1$. *If we define the probability measure* ν_d *by*

$$\frac{\nu_d(dx)}{\mu_d(dx)} = \Lambda_u(x) , \tag{6.4.13}$$

then for any measurable function $F : \mathbb{R}^d \to \mathbb{R}$, *we have*

$$\int_{\mathbb{R}^d} F(x)\mu_d(dx) = \int_{\mathbb{R}^d} F(T(x))\Lambda_u(x)\mu_d(dx) \tag{6.4.14}$$

$$= \int_{\mathbb{R}^d} F(T(x))\nu_d(dx) . \tag{6.4.15}$$

This means that $T(x)$ *under the probability measure* $\nu_d(dx)$ *has the same law as* x *under the probability measure* $\mu_d(dx)$.

(iii) The probability measures $\mu_d \circ T^{-1}$, $\mu_d \circ T$, *and* μ_d *are mutually absolutely continuous. Define*

$$\begin{aligned}\Lambda_V(x) &= |\det(I+\nabla V(x))| \exp\left\{-\langle V(x), x\rangle - \frac{1}{2}|V(x)|^2\right\} \\ &= |\det{}_2(I+\nabla V(x))| \exp\left\{-\delta(V(x)) - \frac{1}{2}|V(x)|^2\right\}.\end{aligned} \tag{6.4.16}$$

Then, $\int_{\mathbb{R}^d} \Lambda_V(x)\mu_d(dx) = 1$, *and*

$$\begin{cases} \dfrac{\mu_d \circ T^{-1}(dx)}{\mu_d(dx)} = \Lambda_V(x)\,; & (6.4.17) \\ \dfrac{\mu_d \circ T(dx)}{\mu_d(dx)} = \Lambda_u(x)\,. & (6.4.18) \end{cases}$$

Proof. First, we are going to show that T has inverse, denoted by $I+V$. Clearly, if T has an inverse $I+V$, then V satisfies $x = T(x+V(x)) = x + V(x) + u(x+V(x))$. Or V satisfies $V(x) = -u(x+V(x))$. We shall solve this equation. Define

$$\begin{aligned} V_0(x) &= 0 \\ V_{n+1}(x) &= -u(x+V_n(x))\,, \quad n = 0, 1, \cdots \end{aligned}$$

We have (since $V_1(x) = u(x)$ and $V_0(x) = 0$)

$$|V_{n+1}(x) - V_n(x)| \le c_1 |V_n(x) - V_{n-1}(x)| \le \cdots \le c_1^n |u(x)|\,.$$

This shows that V_n uniformly converges to a limit $V(x) = \sum_{n=0}^\infty (V_{n+1}(x) - V_n(x))$. It is easy to see that $\sup_{x\in\mathbb{R}^d} |V(x)| \le \frac{1}{1-c_1} \sup_{x\in\mathbb{R}^d} |u(x)|$ and $V(x) = -u(x+V(x))$. Hence $I+V$ is the inverse of $T = I+u$. This shows that T is bijective.

By the triangle inequality, we have

$$\begin{aligned} |V_{n+1}(x+h) - V_{n+1}(x)| &\le |u(x+h+V_n(x+h)) - u(x+V_n(x))| \\ &\le c_1|h| + c_1|V_n(x+h) - V_n(x)|\,. \end{aligned}$$

Letting $n \to \infty$ we see that $|V(x+h) - V(x)| \le \frac{c_1}{1-c_1}|h|$. This proves that $\|\nabla V(x)\| \le \frac{c_1}{1-c_1}$. Similarly, we have $\|\nabla V(x)\|_2 \le \frac{c_2}{1-c_1}$.

Equations (6.4.14) and (6.4.15) follow immediately from (6.4.9). Letting $F = 1$ in (6.4.14) yields that $\int_{\mathbb{R}^d} \Lambda_u(x)\mu_d(dx) = 1$.

If $T = I + u : \mathbb{R}^d \to \mathbb{R}^d$ is invertible, then we can define $\Gamma(x) = \Lambda \circ T^{-1}(x)$, or $\Gamma \circ T(x) = \Lambda_u(x)$. If we denote $G(x) = F(x)\Gamma(x)$ and

$\tilde{\Gamma}(x) = \frac{1}{\Gamma(x)}$, then from (6.4.14) we have

$$\int_{\mathbb{R}^d} G \circ T(x)\mu_d(dx) = \int_{\mathbb{R}^d} F(x+u(x))\Lambda_u(x)\mu_d(dx)$$
$$= \int_{\mathbb{R}^d} F(x)\mu_d(dx) = \int_{\mathbb{R}^d} G(x)\tilde{\Gamma}(x)\mu_d(dx)\,.$$

Since G can be arbitrary this means

$$\frac{\mu_d \circ T^{-1}(dx)}{\mu_d(dx)} = \tilde{\Gamma}(x)\,. \qquad (6.4.19)$$

Now we want to find an expression for $\tilde{\Gamma}(x)$. Using $T \circ T^{-1}(x) = x$ and the chain rule for differentiation, we have

$$\left[(I+\nabla u(x)) \circ T^{-1}\right] \cdot (I+\nabla V(x)) = \left[(\nabla T) \circ T^{-1}\right] \cdot \nabla T^{-1} = I\,.$$

This means that

$$\frac{1}{\det(I+\nabla u)} \circ T^{-1}(x) = \det(I+\nabla V(x))\,. \qquad (6.4.20)$$

On the other hand, $x = (x+u(x))|_{x=T^{-1}(x)}$ and $T^{-1}(x) = x + V(x)$ implies that $u \circ T^{-1}(x) = -V(x)$. Then, we have

$$\left[\langle u(x), x\rangle + \frac{|u(x)|^2}{2}\right] \circ T^{-1}(x) = -\langle V(x), x\rangle - \frac{1}{2}|V(x)|^2\,. \quad (6.4.21)$$

By the definitions of Λ and $\tilde{\Gamma}$, we have

$$\tilde{\Gamma}(x) = \frac{1}{\Lambda} \circ T^{-1}(x)$$
$$= \left\{\frac{1}{|\det(I+\nabla u(x))|} \exp\left\{\langle u(x), x\rangle + \frac{1}{2}|u(x)|^2\right\}\right\}\Bigg|_{x=T^{-1}(x)}$$
$$= |\det(I+\nabla V(x))| \exp\left\{-\langle V(x), x\rangle - \frac{1}{2}|V(x)|^2\right\} = \Lambda_V(x)\,.$$

Substituting this into (6.4.19) yields (6.4.17). Exchanging the roles of u and V yields (6.4.18). □

Remark 6.7. Let $h_1, \cdots, h_d$ be an orthonormal vectors in $\mathbb{H}$. If a functional from $\mathbb{B}$ to $\mathbb{H}$ has the form $u(x) = \sum_{i=1}^d u_i(\tilde{h}_1, \cdots, \tilde{h}_d)h_i$, then it can be considered as $u : \mathbb{R}^d \to \mathbb{R}^d$. The quantity $\langle u(x), x\rangle = \sum_{i=1}^d \langle u_i(\tilde{h}_1, \cdots, \tilde{h}_d)h_i, h\rangle$ can be extended to a functional on $\mathbb{B}$ as $\sum_{i=1}^d u_i(\tilde{h}_1, \cdots, \tilde{h}_d)\tilde{h}_i$. Thus,

$$\delta\left(\sum_{i=1}^d u_i(\tilde{h}_1, \cdots, \tilde{h}_d)h_i\right) = \sum_{i=1}^d u_i(\tilde{h}_1, \cdots, \tilde{h}_d)\tilde{h}_i - \sum_{i=1}^d \partial_i u_i(\tilde{h}_1, \cdots, \tilde{h}_d)$$

is the divergence (Skorohod integral) of u. Thus, the operator δ introduced in (6.4.11) is indeed the divergence operator (or Skorohod integral). It is also easy to see that $\langle u(x), x\rangle$ corresponds to the Stratonovich integral.

6.5 Girsanov-Ramer Theorem in Abstract Wiener Space

Lemma 6.7. *If $F \in \mathbb{D}_{1,p}(\mathbb{H})$ for some $p > 1$ satisfies that $\|\nabla F\| \le c_1$ and $\|\nabla F\|_2 \le c_2$, where $\|\cdot\|$, $\|\cdot\|_2$ denote the operator and Hilbert-Schmidt norms from $\mathbb{H}$ to $\mathbb{H}$, then there is a cylindrical functional $F_n \in \mathbb{D}_{1,p}(\mathbb{R}^n)$, namely, $F_n = \sum_{j=1}^n f_{n,j}(\tilde{h}_1, \cdots, \tilde{h}_n)h_j$, where $h_1, \cdots, h_n, \cdots \in \mathbb{H}$, such that $F_n \to F$ in $\mathbb{D}_{1,p}(\mathbb{R}^n)$ and $\|\nabla F_n\| \le c_1$ and $\|\nabla F_n\|_2 \le c_2$.*

Proof. Let $\{h_n, n = 1, 2, \cdots\}$ be an orthonormal basis of $\mathbb{H}$ and let $\mathbb{H}_n$ be the Hilbert space generated by $\{h_1, \cdots, h_n\}$. Let π_n be the projection to $\mathbb{H}_n$ and let V_n be the sigma algebra generated by $\left\{\tilde{h}_1, \cdots, \tilde{h}_n\right\}$. Define

$$F_n = \mathbb{E}\left[\pi_n Q_{1/n} F | V_n\right] ,$$

where $Q_t F(x) = \mathbb{E}^{\tilde{x}} F(e^{-t}x + \sqrt{1 - e^{-2t}}\tilde{x})$ is the Ornstein-Uhlenbeck semigroup. Then, F_n is of the form $F_n = \sum_{j=1}^n f_{n,j}(\tilde{h}_1, \cdots, \tilde{h}_n)h_j$ and $F_n \to F$ in $D_{1,p}$. Now

$$DF_n = e^{-1/n}\mathbb{E}\left[\pi_n \otimes \pi_n Q_{1/n} DF | V_n\right] .$$

Jessen's inequality yields

$$\|DF_n\| \le e^{-1/n}\mathbb{E}\left[\pi_n \otimes \pi_n Q_{1/n} \|DF\| V_n\right] \le c_1 e^{-1/n} \le c_1 .$$

Similarly, we have $\|DF_n\|_2 \le e^{-1/n} c_2 \le c_2$. The lemma is proved. □

The main theorem of this section that we plan to prove is the following

Theorem 6.8. *Let $u : \mathbb{B} \to \mathbb{H}$ be measurable and the derivative Du exists. Let u satisfy*

$$\|Du\| \le c_1 < 1 , \quad \text{and} \quad \|Du\|_2 \le c_2 < \infty . \tag{6.5.1}$$

Then,

(i) The mapping $T(x) = x + u(x)$ from $\mathbb{B}$ to $\mathbb{B}$ is bijective almost surely. The inverse, denoted by $T^{-1}(x) = x + V(x)$, satisfies that $V \in \mathbb{D}_{1,p}$ for any $p > 1$. Moreover,

$$\|DV(x)\| \le \frac{c_1}{1 - c_1} \quad \text{and} \quad \|DV(x)\|_2 \le \frac{c_2}{1 - c_1} \quad \textit{for a.s. } x \in \mathbb{B} . \tag{6.5.2}$$

(ii) Denote

$$\Lambda_u = |\det{}_2(I+Du)| \exp\left\{\delta(u) - \frac{1}{2}|u|_{\mathbb{H}}^2\right\}. \tag{6.5.3}$$

Then, $\mathbb{E}^{\mathbb{P}}(\Lambda_u) = 1$. *If we define the probability measure* $\mathbb{Q}$ *by* $\frac{d\mathbb{Q}}{d\mathbb{P}} = \Lambda_u$, *then*

$$\mathbb{E}^{\mathbb{P}}[F(x)] = \mathbb{E}^{\mathbb{P}}[F(T(x))\Lambda_u(x)] = \mathbb{E}^{\mathbb{Q}}[F(T(x))]. \tag{6.5.4}$$

This means that x *under the probability measure* $\mathbb{P}$ *and* $T(x)$ *under the probability measure* $\mathbb{Q}$ *have the same law.*

(iii) The measure $\mu\circ T^{-1}$, $\mu\circ T$, *and* μ *are mutually absolutely continuous and*

$$\begin{cases} \dfrac{d\mathbb{P}\circ T^{-1}}{d\mathbb{P}} = \Lambda_V(x)\,; & (6.5.5) \\ \dfrac{d\mathbb{P}\circ T}{d\mathbb{P}} = \Lambda_u(x)\,. & (6.5.6) \end{cases}$$

Proof. *Step 1.* From Lemma 6.7 we see that there is a sequence $u_n(x) = \sum_{j=1}^n u_{n,j}(\tilde{h}_1,\cdots,\tilde{h}_n)h_j$ such that u_n converges to u in $\mathbb{D}_{1,p}$ for any $p\ge 1$ and

$$\|Du_n\| \le c_1 < 1\,, \quad \text{and} \quad \|Du_n\|_2 \le c_2 < \infty\,.$$

We can consider u_n as a mapping from $\mathbb{R}^n$ to $\mathbb{R}^n$ satisfying the above condition. From Proposition 6.22, it follows that there are V_n, Λ_{u_n}, Λ_{V_n} such that (6.4.10)-(6.4.18) hold. We shall prove the theorem by letting $n\to\infty$.

Step 2. First, we show the sequence Λ_{u_n}, Λ_{V_n} are uniformly integrable. We will prove that Λ_{u_n} is uniformly integrable and the proof of Λ_{V_n} is similar. By Theorem A.5 it suffices to show that $\sup_{n\ge1}\mathbb{E}[\Lambda_{u_n}|\log\Lambda_{u_n}|]<\infty$. By (6.4.18), this becomes $\sup_{n\ge1}\mathbb{E}\left[|\log\Lambda_{u_n}|\circ T_n^{-1}\right]<\infty$. By definition of Λ_{u_n}, we are led to show

$$\mathbb{E}\left[\left|\det{}_2(I+Du_n\circ T_n^{-1})\right| + \left|\delta(u_n)\circ T_n^{-1}\right| + \frac{1}{2}|u_n\circ T_n^{-1}|^2\right] \le C < \infty\,. \tag{6.5.7}$$

Consider the first term on the left hand side of (6.5.7). From (6.4.7), we see that

$$\left|\det{}_2(I+Du_n\circ T_n^{-1})\right| \le \exp\left\{\frac{1}{2}\|Du_n\circ T_n^{-1}\|_2^2\right\} \le e^{\frac{c_2^2}{2}}\,.$$

For the third term on the left hand side of (6.5.7) we use $u_n \circ T_n^{-1} = -V_n(x)$ to obtain that for some $\xi \in \mathbb{R}^n$, the following inequality holds true.

$$|u_n \circ T^{-1}|^2 = |V_n(x)|^2 = \langle DV_n(\xi), x\rangle^2 \le \frac{c_2^2}{(1-c_1)^2}|x|^2 .$$

This implies that there is an $\alpha > 0$, independent of n, such that $\sup_{n\ge1} \mathbb{E}e^{\alpha|u_n\circ T^{-1}|^2} < \infty$, which implies

$$\sup_{n\ge1} \mathbb{E}\left[\frac{1}{2}|u_n \circ T_n^{-1}|^2\right] < \infty .$$

We use (6.4.21) for the second term on the left hand side of (6.5.7). Equation (6.4.21) can be rewritten as

$$\delta(u_n) \circ T_n^{-1} = -\delta(V_n) - |V_n|^2 + \mathrm{Tr}\left[(Du_n \circ T_n^{-1}) \cdot DV_n\right] .$$

The above first and the second terms have been bounded earlier. The last one is bounded by

$$\begin{aligned}
&\mathbb{E}\left|\mathrm{Tr}\left[(Du_n \circ T_n^{-1}) \cdot DV_n\right]\right| \\
&\quad \le \left(\mathbb{E}\|(Du_n \circ T_n^{-1})\|_2^2\right)^{1/2}\left(\mathbb{E}\|DV_n\|_2^2\right)^{1/2} \le \frac{c_2^2}{1-c_1},
\end{aligned}$$

where we have used (6.5.1) and (6.5.2). Thus, all three terms in (6.5.7) are uniformly bounded.

Step 3. Next, we show that V_n has a limit. We have

$$\begin{aligned}
|V_n - V_m|_{\mathbb{H}} &\le |u_n \circ T_n - u_m \circ T_m|_{\mathbb{H}} \\
&\le |u_n \circ T_n - u_m \circ T_n|_{\mathbb{H}} + |u_m \circ T_n - u_m \circ T_m|_{\mathbb{H}} \\
&\le |u_n \circ T_n - u_m \circ T_n|_{\mathbb{H}} + c_1|V_n - V_m|_{\mathbb{H}} .
\end{aligned}$$

Thus, we have

$$|V_n - V_m|_{\mathbb{H}} \le \frac{1}{1-c_1}|u_n \circ T_n - u_m \circ T_n|_{\mathbb{H}} .$$

Consequently, we have

$$\begin{aligned}
\mathbb{P}\left(|V_n - V_m|_{\mathbb{H}} > K\right) &\le \mathbb{P}\left(\frac{1}{1-c_1}|u_n \circ T_n - u_m \circ T_n|_{\mathbb{H}} > K\right) \\
&\le \frac{1}{(1-c_1)K}\mathbb{E}\left(|u_n \circ T_n - u_m \circ T_n|_{\mathbb{H}}\right) \\
&\le \frac{1}{(1-c_1)K}\mathbb{E}\left(\Lambda_{u_n}|u_n - u_m|_{\mathbb{H}}\right) \to 0
\end{aligned}$$

since Λ_{u_n} is uniformly integrable. Thus, V_n converges to a certain random variable V in probability. By the uniform integrability, the convergence is

also true in L^p for any $p > 1$. Since ∇V_n is bounded, we also have that V_n converges to V in $\mathbb{D}_{1,p}$ for any $p > 1$.
Step 4. We show $T^{-1} = S := I + V$. For any $a > 0$,

$$\begin{aligned}
&\mathbb{P}\left(\|T \circ S(x) - x\|_{\mathbb{B}} > a\right) \\
&\le \mathbb{P}\left(\|T \circ S - T_n \circ S\|_{\mathbb{B}} > a/2\right) + \mathbb{P}\left(\|T_n \circ S - T_n \circ T^{-1}\|_{\mathbb{B}} > a/2\right) \\
&\le \mathbb{E}\left[\Lambda_{u_n} I_{\|T_n - T\|_{\mathbb{B}} > a/2}\right] + \mathbb{P}\left(|u_n(x + V(x)) - u_n(x + V_n(x))| > a/2\right) \\
&\le \mathbb{E}\left[\Lambda_{u_n} I_{\|T_n - T\|_{\mathbb{B}} > a/2}\right] + \mathbb{P}\left(|V - V_n| > a/(2c_1)\right) \\
&\to 0
\end{aligned}$$

as $n \to \infty$. Similarly, we have $\mathbb{P}\left(\|S \circ T(x) - x\|_{\mathbb{B}} > a\right) = 0$. This means that $T^{-1} = S$.
Step 5. Letting $n \to \infty$ in (6.4.10)-(6.4.18) yields (6.5.2)-(6.5.6). □

Remark 6.8. For a more through discussion of Girsanov transformation, we refer to [Buckdahn (1994); Üstünel and Zakai (2000)].

An immediate corollary is the following Cameron-Martin theorem.

Theorem 6.9. *Let $h \in \mathbb{H}$. Consider the mapping $T : \mathbb{B} : x \to x + h \in \mathbb{B}$. Then, we have the following statements.*

(i) Denote

$$\Lambda_h = \exp\left\{\tilde{h} - \frac{1}{2}|h|_{\mathbb{H}}^2\right\} \tag{6.5.8}$$

and define the probability measure $\mathbb{Q}$ by $\frac{d\mathbb{Q}}{d\mathbb{P}} = \Lambda_h$, then

$$\mathbb{E}^{\mathbb{P}}\left[F(x)\right] = \mathbb{E}^{\mathbb{Q}}\left[F(T(x))\right]. \tag{6.5.9}$$

This means that x under the probability measure $\mathbb{P}$ and $T(x) = x + h$ under the probability measure $\mathbb{Q}$ have the same law.

(ii) The measure $\mu \circ T^{-1}$, $\mu \circ T$, and μ are mutually absolutely continuous and.

$$\begin{cases}
\dfrac{d\mathbb{P} \circ T^{-1}}{d\mathbb{P}} = \Lambda_{-h}(x); & (6.5.10) \\[2ex]
\dfrac{d\mathbb{P} \circ T}{d\mathbb{P}} = \Lambda_h(x). & (6.5.11)
\end{cases}$$

Let $K : [0,1] \times [0,1] \to \mathbb{R}$ be a measurable function such that $\int_{[0,1]^2} K^2(s,t)dsdt < \infty$. Associated with this kernel K we define a (Hilbert-Schmidt) operator $Kf(s) = \int_0^1 K(s,t)f(t)dt$ from $L^2([0,1]$ to itself. It is easy to see that $\|K\|_2^2 = \int_0^1 \int_0^1 K^2(s,t)dsdt$. If $K : [0,1] \times [0,1] \to \mathbb{R}$ is a continuous function, then $\mathrm{Tr} K = \int_0^1 K(s,s)ds$. Then, from [Gohberg, Goldberg and Krupnik (1997), 2.15], we have

$$\det{}_2(I + \lambda K) = \exp\left\{\sum_{m=2}^{\infty} \frac{(-1)^{m+1}\lambda^m}{m} \mathrm{Tr}\,(K^m)\right\} \tag{6.5.12}$$

when λ is sufficiently small, where the kernel of the operator K^m is given by

$$K^{(m)}(s,t) = \int_{[0,1]^m} K(s,t_1)K(t_1,t_2)\cdots K(t_{m-1},t)dt_1 \cdots t_{m-1} .$$

If the kernel satisfies $K(s,t) = 0$ for all $0 < s < t < 1$, then we see that $\mathrm{Tr}\,(K^m) = 0$ for all $m \geq 1$. For such a kernel we have by (6.5.12) that $\det_2(I + K) = 1$. In the same way, if the kernel K satisfies $K(s,t) = 0$ for all $0 < t < s < 1$, we also have $\det_2(I + K) = 1$.

Now let $\mathbb{B} = C_0([0,1])$, $\mathbb{H} = \left\{h \in \mathbb{B},\ \int_0^1 |\dot{h}_t|^2 dt < \infty\right\}$ be the canonical Wiener space with the canonical Wiener measure $\mathbb{P}$. Let $W : \mathbb{B} \to \mathbb{R}$ be the Wiener process. From $W_t(x) = x(t)$, $\forall\, x \in \mathbb{B}$, we can identify W with x. Let $u : [0,1] \times \mathbb{B} \to \mathbb{R}$ be an adapted process such that $\int_0^1 \mathbb{E}\left[|u(s)|^2\right] ds < \infty$. Consider the transformation

$$T : \mathbb{B} \to \mathbb{B} \quad \text{defined by} \quad T(W)_\cdot = W_\cdot + \int_0^\cdot u(s,W)ds .$$

It is clear that (denoting $\tilde{u} = \int_0^\cdot u(s,W)ds$)

$$D_h \tilde{u} = \int_0^\cdot \int_0^1 D_r u(s) \dot{h}_r drds = \int_0^\cdot \int_0^s D_r u(s) \dot{h}_r drds , \quad \forall\, h \in \mathbb{H} .$$

If we define $K_{\tilde{u}}(s,r) = D_r u(s) I_{\{r<s\}}$, then $D\tilde{u} : \mathbb{H} \to \mathbb{H}$ is given by kernel $K_{\tilde{u}}$ (notice that we identify $\mathbb{H}$ as $L^2([0,1])$ by $\mathbb{H} \ni h = \int_0^\cdot \dot{h}_t dt \to \dot{h}_\cdot \in L^2([0,1])$). Thus, we have $\det_2(I + D\tilde{u}) - 1$ a.s. On the other hand, from Remark 6.7 and the adaptedness of u, it follows that $\delta(\tilde{u}) = \int_0^1 u(s)dW_s$ (Itô integral). Thus, we have the corollary.

Corollary 6.1. *Let $u : [0,1] \times \mathbb{B} \to \mathbb{R}$ be an adapted process satisfying $\int_0^1 \mathbb{E}\,[u_t]^2\, dt < \infty$. Then, mapping $T(W) = W + \int_0^\cdot u(t,W)dt$ has inverse $T^{-1}(\tilde{W}) = \tilde{W} - \int_0^\cdot u(t,W)dt$, where $\tilde{W} = W + \int_0^\cdot u(t,W)dt$. Denote*

$$\Lambda_u = \exp\left\{\int_0^1 u_t dW_t - \frac{1}{2}\int_0^1 |u_t|^2 dt\right\} \tag{6.5.13}$$

$$\tilde{\Lambda}_u = \exp\left\{-\int_0^1 u_t dW_t - \frac{1}{2}\int_0^1 |u_t|^2 dt\right\} \tag{6.5.14}$$

$$\tag{6.5.15}$$

and assume $\mathbb{E}^{\mathbb{P}}(\Lambda_u) = 1$. Then, we have the following statements.

(i) If we define the probability measure $\mathbb{Q}$ *by* $\frac{d\mathbb{Q}}{d\mathbb{P}} = \Lambda_u$*, then*

$$\mathbb{E}^{\mathbb{P}}[F(W.)] = \mathbb{E}^{\mathbb{P}}[F(T(W.))\Lambda_u(W.)] = \mathbb{E}^{\mathbb{Q}}[F(T(W.))] . \tag{6.5.16}$$

This means that W *under the probability measure* $\mathbb{P}$ *and* $T(W)$ *under the probability measure* $\mathbb{Q}$ *have the same law. In other word,* $\{T(W)_t, t \in [0,1]\}$ *is a Brownian motion under the probability measure* $\mathbb{Q}$.

(ii) The measure $\mu \circ T^{-1}$, $\mu \circ T$, *and* μ *are mutually absolutely continuous and*

$$\begin{cases} \dfrac{d\mathbb{P} \circ T^{-1}}{d\mathbb{P}} = \tilde{\Lambda}_u ; & (6.5.17) \\ \dfrac{d\mathbb{P} \circ T}{d\mathbb{P}} = \Lambda_u . & (6.5.18) \end{cases}$$

Remark 6.9.

(i) Corollary 6.1 is not exactly the consequence of Theorem 6.8 since in Corollary we use the condition $\mathbb{E}^{\mathbb{P}}(\Lambda_u) = 1$. This condition can be verified by the well-known Novikov condition or a less known Kazamaki condition ([Kazamaki (1978)]) (See also [Hu and Kallianpur (1998)] for an application of the Kazamaki condition).

(ii) There are some more elementary ways to obtain the Girsanov formula for adapted shift by using for example the martingale argument.

6.6 Wick Product

Definition 6.13. If $f_n \in \mathbb{H}^{\hat{\otimes} n}$ and $g_m \in \mathbb{H}^{\hat{\otimes} m}$, then the Wick product of $I_n(f_n)$ and $I_m(g_m)$ is defined as

$$I_n(f_n) \diamond I_m(g_m) = I_{n+m}(f_n \hat{\otimes} g_m) , \tag{6.6.1}$$

where $f_n \hat{\otimes} g_m \in \mathbb{H}^{\hat{\otimes}(n+m)}$ denotes the symmetric tensor product of f_n and g_m. Namely, for any vectors $\{e_1, \cdots, e_{n+m}\}$ in $\mathbb{H}$, we have

$$\begin{aligned}&\langle f_n \hat{\otimes} g_m , e_1 \otimes \cdots \otimes e_{n+m} \rangle \\ &\quad = \frac{1}{(n+m)!} \sum_{\sigma \in \Sigma_{n+m}} \langle f_n , e_{\sigma(1)} \otimes \cdots \otimes e_{\sigma(n)} \rangle \langle g_m , e_{\sigma(n+1)} \times \cdots \otimes e_{\sigma(n+m)} \rangle ,\end{aligned}$$

where Σ_{n+m} denotes the set of all permutations σ of $\{1, 2, \cdots, n+m\}$.

Example 6.6. If $\mathbb{H} = L^2([0,T], \mathbb{R})$, then $\mathbb{H}^{\otimes n} = L^2([0,T]^n, \mathbb{R})$.

$$\begin{aligned}\mathbb{H}^{\hat{\otimes} n} &= \hat{L}^2([0,T]^n, \mathbb{R}) \\ &= \{ f \in L^2([0,T]^n, \mathbb{R}) , \quad f \text{ is a symmetric function of } n\text{-variables} \} .\end{aligned}$$

If $f \in \mathbb{H}^{\hat{\otimes} n}$ and $g \in \mathbb{H}^{\hat{\otimes} m}$, then

$$\begin{aligned} & f\hat{\otimes}g(t_1, \cdots, t_{n+m}) \\ & = \frac{1}{(n+m)!} \sum_{\sigma \in \Sigma_{n+m}} f(t_{\sigma(1)}, \cdots, t_{\sigma(n)}) g(t_{\sigma(n+1)}, \cdots, t_{\sigma(n+m)}) . \end{aligned}$$

If $F = \sum_{n=0}^{\infty} I_n(f_n)$ and $G = \sum_{m=0}^{\infty} I_m(g_m)$, then we may define the Wick product of $F \diamond G$ by linearity and limiting argument

$$F \diamond G = \sum_{n=0}^{\infty} \sum_{m=0}^{\infty} I_{n+m}(f_n \hat{\otimes} g_m) ,$$

if the limit on the right hand side exists in some sense.

Remark 6.10. Of course, there are various topologies that one can use. We do not specify any of them here since we will not deal with various test spaces or distribution spaces that have been discussed extensively in literature. We refer to [Meyer (1993); Hu and Kallianpur (1998)] and references therein for some use of Wick product in quantum physics .

An immediate consequence of the above definition is the following proposition.

Proposition 6.23. *If $h_1, \cdots, h_p \in \mathbb{H}$, then*

$$\tilde{h}_1 \diamond \cdots \diamond \tilde{h}_p = I_p(h_1 \otimes \cdots \otimes h_p) . \tag{6.6.2}$$

It is clear that $F, G \in L^2$ does not imply that $F \diamond G$ is a well-defined object in L^2. Now we present a sufficient condition on F and G such that $F \diamond G \in L^2$. To this end, we need to introduce some new norms.

If $F = \sum_{n=0}^{\infty} I_n(f_n)$, we define

$$\|F\|_{(r)}^2 = \|\Gamma(r)F\|^2 = \sum_{n=0}^{\infty} n! r^{2n} \|f_n\|_2^2 ,$$

where we may still consider $\Gamma(r)$ as the second quantization operator. But now r is greater than 1. So $\Gamma(r)F = \sum_{n=0}^{\infty} r^n I_n(f_n)$ may not converge for some $F \in L^2$. If $r = 1$, then $\|F\|_{(r)} = \|F\|_2$.

Proposition 6.24. *Let $\frac{1}{p^2} + \frac{1}{q^2} = \frac{1}{r^2}$, $p, q, r > 0$. Then*

$$\|F \diamond G\|_{(r)} \le \|F\|_{(p)} \|G\|_{(q)} .$$

Proof. Let $F = \sum_{n=0}^{\infty} I_n(f_n)$ and $G = \sum_{n=0}^{\infty} I_n(g_n)$. Denote $h_n = \sum_{k+j=n} f_k \hat{\otimes} g_j$. Let $a = r^{-2}q^2 - 1$. Then, $1 + a^{-1} = r^{-2}p^2$. We have

$$
\begin{aligned}
\sqrt{n!}\|h_n\| \le \sqrt{n!} \sum_{k+j=n} \|f_k\|\|g_j\| &= \sum_{k+j=n} \binom{n}{k}^{1/2} \sqrt{k!j!}\|f_k\|\|g_j\| \\
&\le \left(\sum_{k+j=n} \binom{n}{k} a^k \right)^{1/2} \left(\sum_{k+j=n} a^{-k} k!j!\|f_k\|^2\|g_j\|^2 \right)^{1/2} \\
&= \left(\sum_{k+j=n} (1+a)^n a^{-k} k!j!\|f_k\|^2\|g_j\|^2 \right)^{1/2} \\
&= \left(\sum_{k+j=n} (1+a^{-1})^k k!\|f_k\|^2 (1+a)^j j!\|g_j\|^2 \right)^{1/2},
\end{aligned}
$$

which implies the proposition. □

As a direct consequence of the above proposition we obtain a condition on F and G such that $F \diamond G$ is in L^2.

Corollary 6.2. *If $\Gamma(p)F$, $\Gamma(q)G$ are in L^2 with $\frac{1}{p^2} + \frac{1}{q^2} = 1$, then $F \diamond G$ exists as an element in L^2.*

Let $f \in \mathbb{H}$. Recall that $\tilde{f}$ denotes the centered Gaussian random variable $\langle \cdot, f \rangle_{\mathbb{H}}$ on $(\mathbb{B}, \mathbb{H}, \mathbb{P})$. We recall

$$
\mathcal{E}(f) := \sum_{n=0}^{\infty} \frac{1}{n!} I_n(f^{\otimes n}) = \exp\left\{ \tilde{f} - \frac{\|f\|_{\mathbb{H}}^2}{2} \right\}, \tag{6.6.3}
$$

the exponential vector of f. It is easy to verify

$$
\mathcal{E}(f) \diamond \mathcal{E}(g) = \mathcal{E}(f+g), \quad \forall\, f, g \in \mathbb{H}. \tag{6.6.4}
$$

Proposition 6.25. *If $f_1, f_2 \in \mathbb{H}$ are two unit vectors which are orthogonal, then*

$$
H_n(\tilde{f}_1) \diamond H_m(\tilde{f}_2) = H_n(\tilde{f}_1) H_m(\tilde{f}_2).
$$

Under some suitable conditions on F and G, we have

$$
D_g(F \diamond G) = D_g F \diamond G + F \diamond D_g G \tag{6.6.5}
$$

$$
D(F \diamond G) = DF \diamond G + F \diamond DG \tag{6.6.6}
$$

Proposition 6.26. *If $g \in \mathbb{H}$, $F \in L^2(\mathbb{B}, \mathbb{H}, \mathbb{P})$ and if $D_g F$ exists and is in $L^2(\mathbb{B}, \mathbb{H}, \mathbb{P})$, then $F \diamond \tilde{g}$ exists in $L^1(\mathbb{B}, \mathbb{H}, \mathbb{P})$ and*

$$
F \diamond \tilde{g} = F\tilde{g} - D_g F = F\tilde{g} - \langle DF, g \rangle_{\mathbb{H}}. \tag{6.6.7}
$$

Proof. Recall

$$\mathcal{E} = \left\{ \sum_{finite} a_i e^{\tilde{f}_i}\,, \quad \text{where} \quad a_i \in \mathbb{R}\,, \quad f_i \in H \right\}.$$

Then, $\mathcal{E}$ is dense in L^2. It is easy to see that (6.6.7) is true for all elements in $\mathcal{E}$ by (6.6.4). A density argument shows the proposition. □

Proposition 6.27. *If $g \in \mathbb{H}$, $F \in L^2(\mathbb{B}, \mathbb{H}, \mathbb{P})$ and if D_gF exists and is in $L^2(\mathbb{B}, \mathbb{H}, \mathbb{P})$, then $F \diamond \tilde{g}$ exists in $L^1(\mathbb{B}, \mathbb{H}, \mathbb{P})$ and*

$$F \diamond \tilde{g} = \delta(Fg)\,. \tag{6.6.8}$$

Proof. Let $G \in \mathcal{S}$. Then

$$\begin{aligned}\mathbb{E}\left((F \diamond g)G\right) &= \mathbb{E}\left(FG\tilde{g} - G\langle DF, g\rangle_{\mathbb{H}}\right)\\ &= \mathbb{E}\left(\langle D(FG)\,, g\rangle_{\mathbb{H}} - \langle GDF, g\rangle_{\mathbb{H}}\right)\\ &= \mathbb{E}\left(\langle FDG\,, g\rangle_{\mathbb{H}}\right) = \mathbb{E}\left(\langle DG\,, Fg\rangle_{\mathbb{H}}\right)\,.\end{aligned}$$

Since G is arbitrary, we show the proposition. □

The following proposition is used in [Duncan, Hu and Pasik-Duncan (2000); Hu (2005)].

Proposition 6.28.

$$\mathbb{E}\left[\left(F \diamond \tilde{f}\right)(G \diamond \tilde{g})\right] = \mathbb{E}\left[FG\langle f\,, g\rangle_{\mathbb{H}} + D_gFD_fG\right]\,. \tag{6.6.9}$$

Proof.

$$\begin{aligned}\mathbb{E}\left[\left(F \diamond \tilde{f}\right)(G \diamond \tilde{g})\right] &= \mathbb{E}\left[\delta\left(Ff\right)(G \diamond \tilde{g})\right] = \mathbb{E}\left[FD_f\left(G \diamond \tilde{g}\right)\right]\\ &= \mathbb{E}\left[F\left(D_fG \diamond \tilde{g} + G \diamond D_f\tilde{g}\right)\right]\\ &= \mathbb{E}\left[F\left(D_fG \diamond \tilde{g} + G\langle f, g\rangle_{\mathbb{H}}\right)\right]\\ &= \mathbb{E}\left[F\delta\left(D_fGg\right) + FG\langle f, g\rangle_{\mathbb{H}}\right]\\ &= \mathbb{E}\left[FG\langle f\,, g\rangle_{\mathbb{H}} + D_gFD_fG\right]\,.\end{aligned}$$

This is the proposition. □

Let $F = \exp\{s\tilde{f} - s^2\|f\|^2/2\}$ and $G = \exp\{t\tilde{g} - t^2\|g\|^2/2\}$, where s and t are two arbitrary real numbers (we use $|\cdot|$ to denote $|\cdot|_{\mathbb{H}}$). We have

$$\begin{aligned}F \diamond G &= \exp\{s\tilde{f} + t\tilde{g} - |sf + tg|^2/2\}\\ &= \exp\{s\tilde{f} - s^2|f\|^2/2\}\exp\{t\tilde{g} - t^2|g|^2/2\}\exp\{-st\langle f, g\rangle\}\\ &= \sum_{i,j,p=0}^{\infty} s^i t^j (st)^p \frac{(-1)^p I_i(f^{\hat{\otimes} i}) I_j(g^{\hat{\otimes} j})\langle f, g\rangle^p}{i!j!p!}\\ &= \sum_{m,n=0}^{\infty} s^m t^n \sum_{p \le m \wedge n} \frac{(-1)^p I_{m-p}(f^{\hat{\otimes} m-p}) I_{n-p}(g^{\hat{\otimes} n-p})\langle f, g\rangle^p}{p!(n-p)!(m-p)!}.\end{aligned}$$

On the other hand, we have

$$F \diamond G = \sum_{n,m=1}^{\infty} s^m t^n I_m(f^{\hat{\otimes} m}) \diamond I_n(g^{\hat{\otimes} n})/m!n!\,.$$

Comparing the coefficients of $s^m t^n$, we can write the above formula as

$$I_m(f^{\hat{\otimes} m}) \diamond I_n(g^{\hat{\otimes} n}) = \sum_{p\le m\wedge n} \frac{(-1)^p n!m!\langle f, g\rangle^p}{p!(n-p)!(m-p)!} I_{m-p}(f^{\hat{\otimes} m-p}) I_{n-p}(g^{\hat{\otimes} n-p}).$$

The above formula can be written as

$$I_m(f^{\hat{\otimes} m}) \diamond I_n(g^{\hat{\otimes} n}) = \sum_{p\le m\wedge n} \frac{(-1)^p}{p!} \langle D^p I_m(f^{\hat{\otimes} m}), D^p I_n(g^{\hat{\otimes} n})\rangle_{\mathbb{H}^{\hat{\otimes} p}},$$

where $D^p F$ is identified as a mapping from $\mathbb{B}$ to $\mathbb{H}^{\hat{\otimes} p}$. By the polarization technique, we have if f_m and g_n are continuous symmetric functions of m and n-variables, then

$$I_m(f_m) \diamond I_n(g_n) = \sum_{p\le m\wedge n} \frac{(-1)^p}{p!} \langle D^p I_m(f_m), D^p I_n(g_n)\rangle_{\mathbb{H}^{\hat{\otimes} p}}. \qquad (6.6.10)$$

In the same way we can obtain that

$$I_m(f_m) I_n(g_n) = \sum_{p\le m\wedge n} \frac{1}{p!} \langle D^p I_m(f_m)\,, \diamond D^p I_n(g_n)\rangle_{\mathbb{H}^{\hat{\otimes} p}}\,, \qquad (6.6.11)$$

where we consider $D^p I_m(f_m)$ and $D^p I_n(g_n)$ as two random variables with values in $H^{\hat{\otimes} p}$ and $\langle D^p I_m(f_m)\,, \diamond D^p I_n(g_n)\rangle_{\mathbb{H}^{\hat{\otimes} p}}$ is the *Wick scalar product.* More precisely,

$$\begin{aligned}
&\langle D^p I_m(f_m)\,, \diamond D^p I_n(g_n)\rangle_{\mathbb{H}^{\hat{\otimes} p}} \\
&= \sum_{k_1,\cdots,k_p=1}^{\infty} \langle D^p I_m(f_m)\,, e_{k_1}\hat{\otimes}\cdots\hat{\otimes} e_{k_p}\rangle_{\mathbb{H}^{\hat{\otimes} p}} \diamond \langle D^p I_n(g_n)\,, e_{k_1}\hat{\otimes}\cdots\hat{\otimes} e_{k_p}\rangle_{\mathbb{H}^{\hat{\otimes} p}} \\
&= \sum_{k_1,\cdots,k_p=1}^{\infty} m(m-1)\cdots(m-p+1)n(n-1)\cdots(n-p+1) I_{m+n-2p}(h)
\end{aligned}$$

with

$$h = \sum_{k_1,\cdots,k_p=1}^{\infty} \langle f_n\,, e_{k_1}\hat{\otimes}\cdots\hat{\otimes} e_{k_p}\rangle \hat{\otimes} \langle g_n\,, e_{k_1}\hat{\otimes}\cdots\hat{\otimes} e_{k_p}\rangle\,,$$

where as before, $\{e_1, e_2\,, \cdots\}$ is an orthonormal basis of $\mathbb{H}$.

We can define $\langle D^p G\,, \diamond D^p G\rangle_{\mathbb{H}^{\hat{\otimes} p}}$ in a similar way.

The above two formulae are for single chaos. We can use the linearity and limiting argument to obtain the following theorem.

Theorem 6.10. *If all the derivatives of F and G exists and satisfy for $H = F$ and $H = G$*

$$\sum_{p=1}^{\infty} \frac{1}{p!} \|D^p H\|^2_{\mathbb{H}^{\hat{\otimes} p}} < \infty \,, \tag{6.6.12}$$

then $F \diamond G$ exists as an element in $L^1(\mathbb{B}, \mathbb{P})$ and

$$F \diamond G = \sum_{p=0}^{\infty} \frac{(-1)^p}{p!} \langle D^p F, D^p G \rangle_{\mathbb{H}^{\hat{\otimes} p}} . \tag{6.6.13}$$

We can show that the condition (6.6.12) also implies following identity:

$$FG = \sum_{p=0}^{\infty} \frac{1}{p!} \langle D^p F \,, \, \diamond D^p G \rangle_{\mathbb{H}^{\hat{\otimes} p}} \,.$$

Example 6.7. If $G = \tilde{g}$ for some $g \in H$, then from (6.6.13)

$$F \diamond G = F\tilde{g} - \langle DF, g \rangle \,.$$

Example 6.8. If $G = I_2(g_2)$ for some $g_2 \in \mathbb{H}^{\hat{\otimes} 2}$, then from (6.6.13)

$$F \diamond G = F I_2(g_2) - 2\langle DF, I_1(g_2) \rangle + \langle D^2 F, g_2 \rangle \,.$$

The following proposition states that if F and G are independent and if they are in single chaos form, then their Wick product and the usual product are the same.

Proposition 6.29. *If $F = I_n(f_n)$ and $G = I_m(g_m)$ are independent, where $f_n \in \mathbb{H}^{\hat{\otimes} n}$ and $g_m \in \mathbb{H}^{\hat{\otimes} m}$, then*

$$F \diamond G = FG \,. \tag{6.6.14}$$

Proof. It is a direct consequence of (6.6.10) and Theorem 5.9. □

From this proposition it is natural to conjecture that if F and G are independent, then $F \diamond G = FG$.

Denote by $\tau_\xi F$ the translation operator: $\tau_\xi F(\omega) = F(\omega + \xi)$, $\omega \in B$. We have the following identities.

$$\tau_\xi F = [\mathcal{E}(\xi) F] \diamond \mathcal{E}(\xi) \tag{6.6.15}$$

$$\tau_\xi (F \diamond G) = (\tau_\xi F) \diamond (\tau_\xi G) \,. \tag{6.6.16}$$

Let us consider the multiple Wiener-Itô integral by approximation. In the previous section, we know that for an approximation $\langle \phi_n, x \rangle$, the quantity $\langle f, \langle \phi_n, x \rangle^{\otimes p} \rangle$ converges to the multiple Stratonovich integral

$S_p(f)$ if exists. If we replace $\langle \phi_n, x\rangle^{\otimes p}$ by the Wick product $\langle \phi_n, x\rangle^{\diamond p}$, then the limit will be multiple Wiener-Itô integral. We shall explain more about this point.

Let $\phi_n \in \mathbb{H}^{\otimes 2}$ be an approximation of identity. It is identified as a bounded map from $\mathbb{H}$ to $\mathbb{H}$:

$$\phi_n = \sum_{i=1}^{\infty} e_i \otimes a_{n,i}\,, \quad \phi_n(h) = \sum_{i=1}^{\infty} a_{n,i}(h) e_i\,,$$

where $\{e_i\}_{i=1}^{\infty}$ is an orthonormal basis of $\mathbb{H}$ and $a_{ni} \in \mathbb{H}$. a_{ni} can be considered as a bounded linear form on $\mathbb{H}$: $a_{ni}(h) = \langle a_{ni}, h\rangle$. The assumption that for any $h \in \mathbb{H}$, $\phi_n(h) \to h$ as $n \to \infty$ implies that

$$\langle a_{ni}, h\rangle = \langle e_i, \phi_n(h)\rangle \;\to\; \langle e_i, h\rangle\,.$$

This means that a_{ni} converges to e_i weakly. Assume that for each n, ϕ_n : $\mathbb{H} \to \mathbb{H}$ is also bounded with respect to the norm $\mathbb{B}$, namely, ϕ_n can be extended to a bounded operator from $\mathbb{B}$ to $\mathbb{H}$. Then,

$$\phi_n(x) = \langle \phi_n, x\rangle = \sum_{i=1}^{\infty} \langle a_{ni}, x\rangle e_i$$

can be considered as an $\mathbb{H}$-valued random variable. The Wick power of $\phi_n(x)$ is considered as an $\mathbb{H}^{\otimes p}$ valued random variable in the following way

$$\begin{aligned}
\phi_n(x)^{\diamond p} &= \sum_{i_1,\cdots,i_p=1}^{\infty} \langle a_{ni_1}, x\rangle \diamond \cdots \diamond \langle a_{ni_p}, x\rangle e_{i_1} \otimes \cdots \otimes e_{i_p} \\
&= \sum_{i_1,\cdots,i_p=1}^{\infty} I_p\left(a_{ni_1} \otimes \cdots \otimes a_{ni_p}\right) e_{i_1} \otimes \cdots \otimes e_{i_p}\,, \qquad (6.6.17)
\end{aligned}$$

where the last equation follows from equation (6.6.2).

Now if $f \in \mathbb{H}^{\otimes p}$, then

$$\begin{aligned}
\langle \phi_n(x)^{\diamond p}, f\rangle &= \sum_{i_1,\cdots,i_p=1}^{\infty} I_p\left(a_{ni_1} \otimes \cdots \otimes a_{ni_p}\right) \langle e_{i_1} \otimes \cdots \otimes e_{i_p}, f\rangle \\
&= I_p\left(\sum_{i_1,\cdots,i_p=1}^{\infty} \langle e_{i_1} \otimes \cdots \otimes e_{i_p}, f\rangle a_{ni_1} \otimes \cdots \otimes a_{ni_p}\right) \\
&= I_p\left(\langle \phi_n^{\otimes p}, f\rangle\right)\,.
\end{aligned}$$

Theorem 6.11. *Let $\phi_n \in \mathbb{H}^{\otimes 2}$ be an approximation of identity and let $f \in \mathbb{H}^{\otimes p}$. If $\langle \phi_n^{\otimes p}, f\rangle$ is convergent in $\mathbb{H}^{\otimes p}$, then $\langle f, \phi_n(x)^{\diamond p}\rangle$ converges in $L^2(\mathbb{B})$ to $I_p(f)$.*

Proof. Since a_{ni} converges weakly to e_i, if

$$\Psi_n := \langle \phi_n^{\otimes p}, f\rangle = \sum_{i_1,\cdots,i_p=1}^{\infty} \langle f, e_{i_1}\otimes\cdots\otimes e_{i_p}\rangle a_{ni_1}\otimes\cdots\otimes a_{ni_p}$$

is convergent in $\mathbb{H}^{\otimes p}$, then it converges weakly in $\mathbb{H}^{\otimes p}$ to $\sum_{i_1,\cdots,i_p=1}^{\infty}\langle e_{i_1}\otimes\cdots\otimes e_{i_p}, f\rangle e_{i_1}\otimes\cdots\otimes e_{i_p} = f$. So Ψ_n converges strongly to f in $\mathbb{H}^{\otimes p}$. Theorem is then proved. □

6.7 Wick Renormalization

It is natural to define the Wick power

$$F^{\diamond k} = \overbrace{F\diamond F\diamond\cdots\diamond F}^{k}\,.$$

Example 6.9. If $l\in H$, then

$$\tilde{l}^{\diamond n} = I_n\left(l^{\hat{\otimes} n}\right) = \|l\|^n H_n\left(\frac{\tilde{l}}{\|l\|}\right);$$

$$\exp^{\diamond}\left(\tilde{l}\right) = \exp\left(\tilde{l} - \frac{\|l\|^2}{2}\right).$$

If $f:\mathbb{R}^d\to\mathbb{R}$ is a real valued entire function, then

$$f(x) = \sum_{n=0}^{\infty}\sum_{n_1+\cdots+n_d=n} a_{n_1,\cdots,n_d}x^{n_1}\cdots x_d^{n_d}\,. \tag{6.7.1}$$

Theorem 6.12. *Let $X_1=\tilde{l}_1,\cdots,X_d=\tilde{l}_d$ be independent standard Gaussian random variables (We can assume that $X_1,\cdots,X_n$ are any indepependent Gaussians). If $f:\mathbb{R}^d\to\mathbb{R}$ is real entire function given above such that*

$$\sum_{n=0}^{\infty}\sum_{n_1+\cdots+n_d=n} n_1!\cdots n_d! a_{n_1,\cdots,n_d}\|X_1\|_2^{2n_1}\ldots\|X_d\|_2^{2n_d} < \infty\,, \tag{6.7.2}$$

then

$$f^{\diamond}(X_1,\cdots,X_d) = \sum_{n=0}^{\infty}\sum_{n_1+\cdots+n_d=n} a_{n_1,\cdots,n_d}X_1^{\diamond n_1}\diamond\cdots\diamond X_d^{\diamond n_d} \tag{6.7.3}$$

is well-defined as an element in L^2. Moreover,

$$\begin{aligned}&\mathbb{E}\left[f^{\diamond}(X_1,\cdots,X_d)\right]^2\\ &=\sum_{n=0}^{\infty}\sum_{n_1+\cdots+n_d=n} n_1!\cdots n_d! a^2_{n_1,\cdots,n_d}\|X_1\|_2^{2n_1}\ldots\|X_d\|_2^{2n_d}\,.\end{aligned} \tag{6.7.4}$$

Proof. It is easy to see that

$$\mathbb{E}\,[X_1^{\diamond n_1}\diamond\cdots\diamond X_d^{\diamond n_d}]^2 = n_1!\cdots, n_d!\|X_1\|_2^{2n_1}\ldots\|X_d\|_2^{2n_d}\,.$$

The theorem is proved by the orthogonality of each term in (6.7.3). □

Definition 6.14. Let $X_1\,,\cdots,X_d$ be independent standard Gaussian random variables. We call (6.7.3) the renormalization of $f(X_1,\cdots,X_d)$ and we denote

$$:f(X_1,\cdots,X_d):=\sum_{n=0}^{\infty}\sum_{n_1+\cdots+n_d=n}a_{n_1,\cdots,n_d}X_1^{\diamond n_1}\diamond\cdots\diamond X_d^{\diamond n_d}\,. \tag{6.7.5}$$

A limiting argument can be used to discuss the case $d=\infty$.

Theorem 6.13. *Let $X=(X_1\,,\cdots,X_d)$ be iid standard Gaussians and let f be as in the previous theorem. Let $(Y_1\,,\cdots,Y_d)$ be independent copy of $X_1\,,\cdots,X_d$ ($Y_1\,,\cdots,Y_d$ and $X_1\,,\cdots,X_d$ are independent and have the same joint probability distribution). Then*

$$f^{\diamond}(X_1,\cdots,X_d)=\mathbb{E}\,[f(X_1+\imath Y_1\,,\cdots,X_d+\imath Y_d)|X_1\,,\cdots,X_d]\,. \tag{6.7.6}$$

Proof. We have (denote $\sum=\sum_{n=0}^{\infty}\sum_{n_1+\cdots+n_d=n}$)

$$\begin{aligned}&\mathbb{E}\,[f(X+\imath Y)|X]\\&=\sum a_{n_1,\cdots,n_d}\mathbb{E}\left[(X_1+\imath Y_1)^{n_1}\cdots(X_d+\imath Y_d)^{n_d}\big|X_1\,,\cdots,X_d\right]\\&=\sum a_{n_1,\cdots,n_d}\mathbb{E}\left[(X_1+\imath Y_1)^{n_1}\big|X_1\right]\cdots\mathbb{E}\left[(X_d+\imath Y_d)^{n_d}\big|X_d\right]\,.\end{aligned}$$

Thus, it suffices to show that for every $n\ge 1$,

$$\mathbb{E}\left[(X_1+\imath Y_1)^n\big|X_1\right]=X_1^{\diamond n}\,. \tag{6.7.7}$$

In fact, we have for all $t\in\mathbb{R}$,

$$\begin{aligned}\mathbb{E}\left[e^{t(X_1+\imath Y_1)}\big|X_1\right]&=e^{tX_1-\frac{t^2}{2}\|Y_1\|_2^2}=e^{tX_1-\frac{t^2}{2}\|X_1\|_2^2}\\&=\sum_{n=0}^{\infty}\frac{t^n}{n!}X_1^{\diamond n}\,.\end{aligned}$$

Expanding the left hand side in term of t^n and comparing the coefficients of t^n, we prove (6.7.7) and then the theorem. □

The Wick product can also be defined for any two functionals of Gaussians. If X and Y are two mean zero Gaussians, then (for any $\xi,\eta\in\mathbb{R}$)

$$\mathcal{E}(X):=\exp\left\{X-\frac{\mathbb{E}(X^2)}{2}\right\}\,; \tag{6.7.8}$$

$$:e^{\xi X}:=\mathcal{E}(\xi X)=\exp\left\{\xi X-\frac{\xi^2}{2}\mathbb{E}(X^2)\right\}\,; \tag{6.7.9}$$

$$\mathcal{E}(X)\diamond\mathcal{E}(Y)=\mathcal{E}(X+Y)\,. \tag{6.7.10}$$

This implies

$$e^{\xi X}\diamond e^{\eta Y}=e^{\xi X+\eta Y-\xi\eta\mathbb{E}(XY)}\,. \tag{6.7.11}$$

Let us first use Theorem 6.10 to compute $:e^{\lambda X^2}:$, where X is a standard normal random variable.

Proposition 6.30. *Let X be a standard normal random variable. If $|\lambda|<1$, then the Wick renormalization of $e^{\frac{1}{2}\lambda X^2}$ exists in the sense of Theorem 6.12 and*

$$:e^{\frac{1}{2}\lambda X^2}:=\frac{1}{\sqrt{\lambda+1}}e^{\frac{\lambda}{2(\lambda+1)}X^2}\,. \tag{6.7.12}$$

Proof. First, we have

$$e^{\frac{1}{2}\lambda x^2}=\sum_{n=0}^{\infty}a_nx^n\,,$$

where

$$a_n=\begin{cases}0 & \text{if } n=2k+1\\ \frac{\lambda^k}{2^kk!} & \text{if } n=2k\,.\end{cases}$$

Thus, (6.7.2) is equivalent to $|\lambda|<1$.

Let Y be a standard normal independent of X. From Theorem 6.10, we have

$$\begin{aligned}:e^{\frac{1}{2}\lambda X^2}: &=\mathbb{E}\left[e^{\frac{1}{2}\lambda(X+\imath Y)^2}|X\right]\\ &=\mathbb{E}\left[e^{\frac{1}{2}\lambda X^2+\imath\lambda XY-\frac{1}{2}\lambda Y^2}|X\right]\\ &=e^{\frac{1}{2}\lambda X^2}\frac{1}{\sqrt{2\pi}}\int_{-\infty}^{\infty}e^{\imath\lambda Xy-\frac{\lambda+1}{2}y^2}dy\\ &=\frac{1}{\sqrt{\lambda+1}}e^{\frac{\lambda}{2(\lambda+1)}X^2}\,.\end{aligned}$$

This is the identity (6.7.12). □

Example 6.10. Let $f:\mathbb{R}\to\mathbb{R}$ be a nice function. The Fourier transform of f is given by $\hat{f}(\xi)=\int_{\mathbb{R}}e^{-\imath z\xi}f(z)dz$ and

$$f(x)=\frac{1}{2\pi}\int_{\mathbb{R}}e^{\imath x\xi}\hat{f}(\xi)d\xi=\frac{1}{2\pi}\int_{\mathbb{R}^2}e^{\imath(x-z)\xi}f(z)dzd\xi\,.$$

If X is a mean 0 Gaussian, then

$$\begin{aligned}
f^{\diamond}(X) &= \int_{\mathbb{R}^2} : e^{\imath(X-z)\xi} : f(z)dzd\xi \\
&= \int_{\mathbb{R}^2} e^{\imath(X-z)\xi+\frac{\xi^2}{2}\mathbb{E}(X^2)} f(z)dzd\xi \\
&= \sum_{k=0}^{\infty} \frac{\left[\mathbb{E}(X^2)\right]^k}{2^k k!} \int_{\mathbb{R}^2} e^{\imath(X-z)\xi}\xi^{2k} f(z)dzd\xi \\
&= \sum_{k=0}^{\infty} \frac{(-1)^k\left[\mathbb{E}(X^2)\right]^k}{2^k k!} \frac{\partial^{2k}}{\partial X^{2k}} \int_{\mathbb{R}^2} e^{\imath(X-z)\xi} f(z)dzd\xi \\
&= \sum_{k=0}^{\infty} \frac{(-1)^k\left[\mathbb{E}(X^2)\right]^k}{2^k k!} f^{(2k)}(X) . \qquad (6.7.13)
\end{aligned}$$

In particular, we have for all mean 0 Gaussian X,

$$X^{\diamond n} = \sum_{k\le n/2} \frac{(-1)^k n!}{2^k k!(n-2k)!} \left[\mathbb{E}(X^2)\right]^k X^{n-2k} . \qquad (6.7.14)$$

Example 6.11. We continue the previous example. If g is another nice function and if X and Y are jointly mean 0 Gaussians, then

$$\begin{aligned}
f(X)\diamond g(Y) &= \int_{\mathbb{R}^4} e^{\imath(X-x)\xi} \diamond e^{\imath(Y-y)\eta} f(x)g(y)dxd\xi dyd\eta \\
&= \int_{\mathbb{R}^4} e^{\imath(X-x)\xi+\imath(Y-y)\eta+\xi\eta\mathbb{E}(XY)} f(x)g(y)dxd\xi dyd\eta \\
&= \sum_{k=0}^{\infty} \frac{\xi^k\eta^k\left[\mathbb{E}(XY)\right]^k}{k!} \int_{\mathbb{R}^4} e^{\imath(X-x)\xi+\imath(Y-y)\eta} f(x)g(y)dxd\xi dyd\eta \\
&= \sum_{k=0}^{\infty} \frac{(-1)^k\left[\mathbb{E}(XY)\right]^k}{k!} \\
&\qquad \frac{\partial^{2k}}{\partial X^k \partial Y^k} \int_{\mathbb{R}^4} e^{\imath(X-x)\xi+\imath(Y-y)\eta} f(x)g(y)dxd\xi dyd\eta \\
&= \sum_{k=0}^{\infty} \frac{(-1)^k\left[\mathbb{E}(XY)\right]^k}{k!} f^{(k)}(X)g^{(k)}(Y) . \qquad (6.7.15)
\end{aligned}$$

Example 6.12. One may combine (6.7.13) and (6.7.15) to compute $f(X)\diamond$

$g^\diamond(Y)$ as follows.

$$\begin{aligned} f(X) \diamond g^\diamond(Y) &= \sum_{m=0}^{\infty} \frac{(-1)^m \left[\mathbb{E}(Y^2)\right]^m}{2^m m!} f(X) \diamond g^{(2m)}(Y) \\ &= \sum_{k,m=0}^{\infty} \frac{(-1)^{k+m}}{2^m m! k!} \left[\mathbb{E}(XY)\right]^k \left[\mathbb{E}(Y^2)\right]^m f^{(k)}(X) g^{(k+2m)}(Y) . \end{aligned} \tag{6.7.16}$$

When we take $g(x) = x^p$ in (6.7.16), we obtain immediately the following proposition.

Proposition 6.31. *Let p be a positive integer. Let X and Y be two mean 0, jointly Gaussian random variables and let $G : \mathbb{R} \to \mathbb{R}$ be differentiable up to order p such that $G^{(j)}$ for any $j = 0, \ldots, p$ are of at most exponential growth. Then, the Wick product $G(X) \diamond Y^{\diamond p}$ can be expressed in terms of ordinary products as*

$$\begin{aligned} G(X) \diamond Y^{\diamond p} &= G(X) Y^p \\ &+ \sum_{0<l+2m\le p} \frac{(-1)^{m+l} p!}{2^m m! (p-2m-l)!} G^{(l)}(X) \left[\mathbb{E}(XY)\right]^l \left[\mathbb{E}(Y^2)\right]^m Y^{p-2m-l} . \end{aligned} \tag{6.7.17}$$

Example 6.13. In order to illustrate the kind of correction terms we obtain, let us write formula (6.7.17) for $p = 1, 2, 3$:

$$\begin{aligned} G(X) \diamond Y &= G(X)Y - G'(X)\mathbb{E}(XY) \\ G(X) \diamond Y^{\diamond 2} &= G(X)Y^2 - G(X)\mathbb{E}(Y^2) - 2G'(X)\mathbb{E}(XY)Y \\ &\quad + G''(X) \left[\mathbb{E}(XY)\right]^2 \\ G(X) \diamond Y^{\diamond 3} &= G(X)Y^3 - 3G(X)\mathbb{E}(Y^2)Y \\ &\quad + 3G'(X)\mathbb{E}(XY)\mathbb{E}(Y^2) - 3G'(X)\mathbb{E}(XY)Y^2 \\ &\quad + 3G''(X) \left[\mathbb{E}(XY)\right]^2 Y - G'''(X) \left[\mathbb{E}(XY)\right]^3 . \end{aligned}$$

The following multidimensional analogy can be used similarly and is also used in [Hu, Jolis and Tindel (2013)].

Proposition 6.32. *Let $g_1, \ldots, g_d, h_1 \ldots, h_d \in \mathbb{H}$. Consider the random variables $X_1 = I_1(g_1), \ldots, X_d = I_1(g_d)$, and $Y_1 = I_1(h_1), \ldots, Y_d = I_1(h_d)$. Suppose that $Y_1, \cdots, Y_d$ are independent and also that X_j and Y_k are independent for $k \neq j$. Let $p = (p_1, \ldots, p_d)$ be a multi-index and set $|p| = \sum_{j=1}^d p_j$. Assume that $G \in C^{|p|}(\mathbb{R}^d)$ is such that $\partial^\alpha G \in L^r(\mu_{\bar{g}})$*

for any multi-index $\alpha = (\alpha_1 \dots, \alpha_d)$ *and for some* $r > 2$, *with* $\alpha_k \le p_k$, $k = 1, \dots, d$. *Then* $G(X_1, \cdots, X_d) \diamond Y_1^{\diamond p_1} \diamond \cdots \diamond Y_d^{\diamond p_d}$ *is well defined in* $L^2(\mathbb{B})$ *and the following formula holds:*

$$
\begin{aligned}
& G(X_1, \cdots, X_d) \diamond Y_1^{\diamond p_1} \diamond \cdots \diamond Y_d^{\diamond p_d} \\
&\quad = \sum_{l_1 + 2m_1 \le p_1} \cdots \sum_{l_d + 2m_d \le p_d} \partial^{l_1, \dots, l_d} G(X_1, \cdots, X_d) \\
&\quad \times \prod_{k=1}^{d} \left[\frac{(-1)^{(m_k + l_k)} p_k!}{2^{m_k} m_k! l_k! (p_k - 2m_k - l_k)!} \left(\mathbb{E}(X_k Y_k)\right)^{l_k} \left(\mathbb{E}(Y_k^2)\right)^{m_k} Y_k^{p_k - 2m_k - l_k} \right],
\end{aligned}
\tag{6.7.18}
$$

where $\partial^{j_1, \cdots, j_d}$ *denotes* $\frac{\partial^{j_1 + \cdots + j_d}}{\partial x_1^{j_1} \cdots \partial x_d^{j_d}}$.

For general element in $\mathbb{H}^{\hat{\otimes} n}$ we give the following examples.

Example 6.14. Let $f \in \mathbb{H}^{\hat{\otimes} 2}$ and consider $: \exp\left(\frac{1}{2} I_2(f)\right) :$. It is well-known that there is an orthonormal basis $\{e_1, e_2, \cdots\}$ of $\mathbb{H}$ and $\alpha_n, n = 1, 2, \cdots$ such that $\sum_{n=1}^{\infty} \lambda_n^2 < \infty$ and

$$
f = \sum_{n=1}^{\infty} \lambda_n e_n \hat{\otimes} e_n .
$$

Thus

$$
I_2(f) = \sum_{n=1}^{\infty} \lambda_n (\tilde{e}_n^2 - 1) .
$$

Therefore

$$
\begin{aligned}
: \exp\left(\frac{1}{2} I_2(f)\right) : &= \prod_{n=1}^{\infty} : e^{\frac{1}{2} \lambda_n (\tilde{e}_n^2 - 1)} : \\
&= \prod_{n=1}^{\infty} e^{-\frac{\lambda_n}{2} - \frac{1}{2} \log(\lambda_n + 1) + \frac{\lambda_n}{2(\lambda_n + 1)} \tilde{e}_n^2}
\end{aligned}
$$

Example 6.15. If we use the multiple Stratonovich integral, we have

$$
\begin{aligned}
: S_n(f_1 \hat{\otimes} f_2 \hat{\otimes} \cdots \hat{\otimes} f_n) : &= : \tilde{f}_1 \tilde{f}_2 \cdots \tilde{f}_n : \\
&= \tilde{f}_1 \diamond \tilde{f}_2 \diamond \cdots \diamond \tilde{f}_n \\
&= I_n(f_1 \hat{\otimes} f_2 \hat{\otimes} \cdots \hat{\otimes} f_n) .
\end{aligned}
$$

Let $f \in \mathbb{H}^{\hat{\otimes} n}$ have all traces of order k, $k \le n/2$ and let $S_n^N(f)$ be given by (5.4.1). Then

$$\begin{aligned} : S_n^N(f) : &= \sum_{k_1,\cdots,k_n=1}^{N} \langle f, e_{k_1} \otimes \cdots \otimes e_{k_n} \rangle_{\mathbb{H}^{\hat{\otimes} n}} \tilde{e}_{k_1} \diamond \cdots \diamond \tilde{e}_{k_n} \\ &= I_n \left(\sum_{k_1,\cdots,k_n=1}^{N} \langle f, e_{k_1} \otimes \cdots \otimes e_{k_n} \rangle_{\mathbb{H}^{\hat{\otimes} n}} e_{k_1} \otimes \cdots \otimes e_{k_n} \right) \\ &\to I_n(f) \qquad (\text{in } L^2) . \end{aligned}$$

Thus, we see that if $f \in \mathbb{H}^{\hat{\otimes} n}$ has traces of all order k, $k \le n/2$, then $S_n(f)$ exists as an element of L^2 and

$$: S_n(f) := I_n(f) . \tag{6.7.19}$$

Combining the above with the Hu-Meyer formula we have

Proposition 6.33. *If $f \in \mathbb{H}^{\hat{\otimes} n}$ have all traces of order k, $k \le n/2$, then*

$$: I_n(f) := \sum_{k \le n/2} \frac{(-1)^k n!}{2^k k!(n-2k)!} I_{n-2k} \left(\mathrm{Tr}^k f \right) . \tag{6.7.20}$$

Proof. From the Hu-Meyer formula (5.4.4), we have

$$\begin{aligned} : I_n(f) : &= \sum_{k \le n/2} \frac{(-1)^k n!}{2^k k!(n-2k)!} : S_{n-2k}(\mathrm{Tr}^k f) : \\ &= \sum_{k \le n/2} \frac{(-1)^k n!}{2^k k!(n-2k)!} I_{n-2k}(\mathrm{Tr}^k f) . \end{aligned}$$

□

It is also easy to verify that

$$: FG := (: F :) \diamond (: G :) . \tag{6.7.21}$$

In Euclidean quantum field theory, the existence of interacting field leads to the exponential integrability problem of $S_n(f)$ for some very special f on some specific abstract Wiener space. The Wick renormalization method reduces the problem to exponential integrability problem of $I_n(f)$. For more detailed discussion see [Simon (1974)] (see also [Hu and Kallianpur (1998)]).

6.8 (Noncommutative) Composition of Wiener Functional

Lemma 6.8. *If $A \in \mathbb{H}^{\otimes \ell}$ and $B \in \mathbb{H}^{\otimes m}$, then*

$$\|A \otimes B\|_{\mathbb{H}^{\otimes(\ell+m)}} = \|A\|_{\mathbb{H}^{\otimes \ell}} \|B\|_{\mathbb{H}^{\otimes m}} \,. \tag{6.8.1}$$

Proof. Let $\{e_j, j = 1, 2, \cdots\}$ be an orthonormal basis of $\mathbb{H}$. To simplify the notation, we shall omit the sub-index.

$$\begin{aligned}
\|A \otimes B\|^2 &= \sum_{i_1,\cdots,i_{\ell+m}=1}^{\infty} \langle A \otimes B, e_{i_1} \otimes \cdots \otimes e_{i_{\ell+m}} \rangle^2 \\
&= \sum_{i_1,\cdots,i_{\ell+m}=1}^{\infty} \langle A, e_{i_1} \otimes \cdots \otimes e_{i_\ell} \rangle^2 \langle B, e_{i_{\ell+1}} \otimes \cdots \otimes e_{i_{\ell+m}} \rangle^2 \\
&= \sum_{i_1,\cdots,i_\ell=1}^{\infty} \langle A, e_{i_1} \otimes \cdots \otimes e_{i_\ell} \rangle^2 \sum_{i_{\ell+1},\cdots,i_\ell=1}^{\infty} \langle B, e_{i_{\ell+1}} \otimes \cdots \otimes e_{i_{\ell+m}} \rangle^2 \\
&= \|A\|^2_{\mathbb{H}^{\otimes \ell}} \|B\|^2_{\mathbb{H}^{\otimes m}} \,.
\end{aligned}$$

□

Let $A : \mathbb{B} \to (\mathbb{R}^m)^{\otimes 2}$ be a matrix valued random variable, namely, $A = (a_{ij})_{1 \le i,j \le m}$ with $a_{ij} : \mathbb{B} \to \mathbb{R}$. We say $A \in \mathbb{D}_{1,\text{loc}}$, if all its entries a_{ij} are in $\mathbb{D}_{1,\text{loc}}$ and we define

$$DA = (Da_{ij})_{1 \le i,j \le m} \in \mathbb{H} \otimes (\mathbb{R}^m)^{\otimes 2} \,, \quad \text{a.s.}$$

We can identify DA as a matrix whose elements are $\mathbb{H}$-valued random variables.

The operator D is a derivative operator. This means that it satisfies the chain rule (6.1.13). If f is a differentiable function and if $A \in \mathbb{D}_{1,\text{loc}}$ is a square matrix, then $f(A)$ is a well-defined matrix valued random variable as well. It is interesting to know if $f(A) \in \mathbb{D}_{1,\text{loc}}$ and if yes, how to compute $Df(A)$ in terms of DA.

If $g : \mathbb{R}^m \to \mathbb{R}^n$, then $\partial g(x) = \left(\partial_{x_j} g_i\right)_{1 \le i \le n, 1 \le j \le m}$ is an $n \times m$ matrix. We can also identify $\partial g(x)$ as an element in $\mathbb{R}^{n \times m} = \mathbb{R}^n \otimes \mathbb{R}^m$. If $g : \mathbb{R}^{n \times m} \to \mathbb{R}^{k \times \ell}$, then $\partial g(x) \in \mathbb{R}^{k \times \ell} \otimes \mathbb{R}^{n \times m}$.

If $f : \mathbb{R} \to \mathbb{R}$ is a nice function and $A \in \mathbb{R}^{m \otimes m}$ is a square matrix, then this function can be extended to a function $A = (a_{ij})_{1 \le i,j \le m} \to B = f(A) = (b_{ij})_{1 \le i,j \le m}$ from $\mathbb{R}^{n \otimes n}$ to $\mathbb{R}^{n \otimes n}$. We will show that if f is differentiable, then for every $k, \ell = 1, \cdots, m$, $b_{k\ell}$ is differentiable in $a_{ij}, 1 \le i,j \le m$ and we also want to express these derivatives. Clearly, the derivative $\partial f(A)$ can be considered as an element in $\mathbb{R}^{n \otimes n} \otimes \mathbb{R}^{n \otimes n}$.

To express this derivative it is more convenient to use the matrix notation and the directional derivative. As in [Hu (1995, 1994)] we introduce the following derivative along the direction of a matrix H

$$\partial_H f(A) = \lim_{\varepsilon \to 0} \frac{1}{\varepsilon} \left[f(A + \varepsilon H) - f(A) \right] \tag{6.8.2}$$

if the limit exists entry-wise.

It is obvious to have the following proposition.

Proposition 6.34 (Noncommutative product rule). *Let $f, g : \mathbb{R} \to \mathbb{R}$ be continuous function. If both of $\partial_H f(A)$ and $\partial_H g(A)$ exist, then $\partial_H [f(A)g(A)]$ exists and*

$$\partial_H [f(A)g(A)] = [\partial_H f(A)]\, g(A) + f(A) [\partial_H g(A)] \, . \tag{6.8.3}$$

Proposition 6.35. *If $f(x) = x^{-1}$, then*

$$\partial_H A^{-1} = -A^{-1} H A^{-1} \, , \quad \text{where } A \text{ is invertible}. \tag{6.8.4}$$

If $f(x) = x^d$, then

$$\partial_H A^d = \sum_{k=0}^{d-1} A^k H A^{d-1-k} \, . \tag{6.8.5}$$

Proof. The following identity

$$\frac{1}{\varepsilon} \left[(A + \varepsilon H)^{-1} - A^{-1} \right] = -(A + \varepsilon H)^{-1} \frac{1}{\varepsilon} \left[(A + \varepsilon H) - A \right] A^{-1}$$

implies (6.8.4) immediately. Expanding $(A + \varepsilon H)^d$ yields (6.8.5). □

Lemma 6.9. *If A and H are $m \times m$ matrices, we define the following adjoint action of A as*

$$\mathrm{ad}_A H = AH - HA; \quad \mathrm{ad}_A^n H = \mathrm{ad}_A(\mathrm{ad}_A^{n-1} H) \, ; \tag{6.8.6}$$

and

$$e^{\mathrm{ad}_A} H = \sum_{n=0}^{\infty} \frac{1}{n!} \mathrm{ad}_A^n H \, , \quad \forall \ H \in \mathbb{R}^{m^2} \, . \tag{6.8.7}$$

With these notations we have

$$e^{-A} H e^{A} = e^{-ad_A} H \, . \tag{6.8.8}$$

Proof. Differentiating $e^{-sA}He^{sA}$ with respect to s, we have

$$\frac{d}{ds}e^{-sA}He^{sA} = -e^{-sA}AHe^{sA} + e^{-sA}HAe^{sA} = -e^{-sA}\left(\mathrm{ad}_A H\right)e^{sA}.$$

Continuing to use this formula yields

$$\frac{d^n}{ds^n}e^{-sA}He^{sA} = (-1)^n e^{-sA}\left(\mathrm{ad}_A^n H\right)e^{sA}.$$

Or

$$\left.\frac{d^n}{ds^n}\right|_{s=0} e^{-sA}He^{sA} = (-1)^n\left(\mathrm{ad}_A^n H\right).$$

The Taylor expansion for $e^{-sA}He^{sA}$ gives then

$$e^{-sA}He^{sA} = \sum_{n=0}^{\infty}\frac{s^n}{n!}\left.\frac{d^n}{ds^n}\right|_{s=0} e^{-sA}He^{sA} = e^{-s\,\mathrm{ad}_A}H.$$

The lemma is proved if we let $s = 1$. □

Remark 6.11. The linear operator $\mathrm{ad}_A : \mathbb{R}^{n\otimes n} \to \mathbb{R}^{n\otimes n}$ can be considered as an element in $\mathbb{R}^{n\otimes n}\otimes\mathbb{R}^{n\otimes n}$. It is very useful in expressing the derivatives of some matrix valued functions.

Proposition 6.36. *If $f(x) = e^x$, we have*

$$\partial_H e^A = e^A\frac{1-e^{\mathrm{ad}_A}}{\mathrm{ad}_A}H, \tag{6.8.9}$$

where ad_A *denotes the adjoint representation,* $\mathrm{ad}_A H = AH - HA$.

Proof. There is a simple proof of this formula in [Tuynman (1995)] (see also [Hall (2003)]). We follow [Hu (1990)], which is also very simple. Let $U_t = e^{tA}$ and $V_t = e^{t(A+H)}$. Then

$$\frac{d}{dt}V_t = (A+H)V_t = AV_t + HV_t = AV_t + F_t\,, \quad \text{where } F_t = HV_t\,.$$

Thus, by Duhamel's formula $V_t = I + \int_0^t U_{t-s}F_s ds = I + \int_0^t U_{t-s}HV_s ds$. Iterating this procedure we obtain a series expansion for V_t and then letting $t = 1$ we have

$$e^{A+H} = I + \sum_{n=1}^{\infty}\int_{0<s_1<\cdots<s_n<1} e^{(1-s_n)A}He^{(s_n-s_{n-1})A}H\cdots He^{s_1A}ds_1\cdots ds_n\,. \tag{6.8.10}$$

Replace H by εH in Equation (6.8.10) we have

$$e^{A+\varepsilon H} = I + \varepsilon^n\sum_{k=1}^{\infty}\int_{0<s_1<\cdots<s_n<1} e^{(1-s_n)A}He^{(s_n-s_{n-1})A}H\cdots He^{s_1A}ds_1\cdots ds_n\,.$$

$$\partial_H e^A = \int_0^1 e^{(1-s)A} H e^{sA} ds = e^A \int_0^1 e^{-sA} H e^{sA} ds$$
$$= e^A \int_0^1 e^{-s\,\mathrm{ad}_A} H ds = e^A \frac{1 - e^{-\mathrm{ad}_A}}{\mathrm{ad}_A} H\,,$$

which is the equation (6.8.9). □

Fixed H_1, as a function of A, if $\partial_{H_1} f(A)$ is differentiable along the direction H_2, then we denote the derivative by $\partial_{H_2}\partial_{H_1} f(A)$. We can define $\partial_{H_k} \cdots \partial_{H_1} f(A)$.

Example 6.16. We see that by (6.8.4) that $\left(\partial_{H_1} A^{-1}\right) = -A^{-1}H_1A^{-1}$. Use this formula again to obtain

$$\partial_{H_2}\left(\partial_{H_1} A^{-1}\right) = A^{-1}H_2A^{-1}H_1A^{-1} + A^{-1}H_1A^{-1}H_2A^{-1}\,.$$

From this example, we see that $\partial_{H_2}\left(\partial_{H_1} A^{-1}\right)$ is symmetric in H_1 and H_2. In general, $\partial_{H_k} \cdots \partial_{H_1} A^{-1}$ is a symmetric function of $H_1, \cdots, H_k$ and it is also given by

$$\partial_{H_k} \cdots \partial_{H_1} f(A) = \frac{\partial^k}{\partial\varepsilon_1 \cdots \partial\varepsilon_k}\Big|_{\varepsilon_1=\cdots=\varepsilon_k=0} f(A + \varepsilon_1 H_1 + \cdots + \varepsilon_k H_k)\,. \tag{6.8.11}$$

Lemma 6.10. *If $f(x) = e^x$, then*

$$\partial_{H_k} \cdots \partial_{H_1} e^A = \frac{1}{k!} \sum_{(i_1,\cdots,i_k)\in\Sigma(k)} e^A \frac{1 - e^{\mathrm{ad}_A}}{\mathrm{ad}_A} H_{i_1} \frac{1 - e^{\mathrm{ad}_A}}{\mathrm{ad}_A} H_{i_2} \cdots \frac{1 - e^{\mathrm{ad}_A}}{\mathrm{ad}_A} H_{i_k}\,, \tag{6.8.12}$$

where $\Sigma(k)$ denotes the set of all permutations of $\{1, 2, \cdots, k\}$.

Proof. Presumably, one can iterate (6.8.9). But the following proof seems simpler. From Equation (6.8.10) it follows

$$e^{A+\varepsilon_1 H_1+\cdots+\varepsilon_k H_k} = I + \sum_{k=1}^{\infty} \int_{0<s_1<\cdots<s_n<1} e^{(1-s_n)A}(\varepsilon_1 H_1 + \cdots + \varepsilon_k H_k)$$
$$e^{(s_n-s_{n-1})A}(\varepsilon_1 H_1 + \cdots + \varepsilon_k H_k)$$
$$\cdots (\varepsilon_1 H_1 + \cdots + \varepsilon_k H_k) e^{s_1 A} ds_1 \cdots ds_n\,.$$

It is then easy to deduce

$$\partial_{H_k} \cdots \partial_{H_1} e^A = \sum_{(i_1,\cdots,i_k)\in\Sigma(k)} e^A \int_{0<s_1<\cdots<s_k<1} e^{-s_k A} H_{i_1} e^{(s_k-s_{k-1})A} H_{i_2}$$
$$\cdots H_{i_k} e^{s_1 A} ds_1 \cdots ds_k\,,$$

where $\Sigma(k)$ denotes the set of all permutations of $\{1, 2, \cdots, k\}$. Denote $g_{i_1,\cdots,i_k}(s_1, \cdots, s_k) = e^{-s_n A} H_{i_1} e^{(s_k - s_{k-1})A} H_{i_2} \cdots H_{i_k} e^{s_1 A}$. Then, due to the summation, the expression $\sum_{(i_1,\cdots,i_k)\in\Sigma(k)} g_{i_1,\cdots,i_k}(s_1, \cdots, s_k)$ is symmetric in $s_1, \cdots, s_k \in [0, 1]$. Hence,

$$\begin{aligned}
\partial_{H_k} \cdots \partial_{H_1} e^A &= \frac{1}{k!} \sum_{(i_1,\cdots,i_k)\in\Sigma(k)} e^A \int_{[0,1]^n} \left[e^{-s_k A} H_{i_1} e^{s_k A}\right] \left[e^{-s_{k-1} A} H_{i_2} e^{s_{k-1} A}\right] \\
&\quad \cdots \left[e^{-s_1 A} H_{i_k} e^{s_1 A}\right] ds_1 \cdots ds_k \\
&= \frac{1}{k!} \sum_{(i_1,\cdots,i_k)\in\Sigma(k)} e^A \int_{[0,1]^n} e^{-s_k \mathrm{ad}_A} H_{i_1} e^{-s_{k-1} \mathrm{ad}_A} H_{i_2} \\
&\quad \cdots e^{-s_1 \mathrm{ad}_A} H_{i_k} ds_1 \cdots ds_k \,,
\end{aligned}$$

which is the right hand side of (6.8.12) after integration with respect to $s_1, \cdots, s_k$. □

Next, we want to find a formula for $\partial_H \log A$. From one form of the Baker-Campbell-Hausdorff formula (see for example [Hall (2003)], equation (3.6))

$$\log(e^X e^Y) = X + \int_0^1 g(e^{\mathrm{ad}_X} e^{t\mathrm{ad}_Y})(Y) dt \,, \quad \text{where } g(z) = \frac{\log z}{1 - \frac{1}{z}} \,,$$

we see that

$$\log(e^X e^Y) = X + g(e^{\mathrm{ad}_X})(Y) + O(Y^2) = X + \frac{\mathrm{ad}_X e^{\mathrm{ad}_X}}{e^{\mathrm{ad}_X} - 1} Y + O(Y^2) \,.$$

We shall use this expansion to compute the derivative $\partial_H \log A$. We assume that the eigenvalues of A are strictly positive and then the same holds true for $A + \varepsilon H$ for ε in a neighborhood of 0. Denote $X = \log A$ and denote

$$Y_\varepsilon = \log(I + \varepsilon A^{-1} H) = \varepsilon A^{-1} H - \varepsilon^2 \frac{(A^{-1} H)^2}{2} + \varepsilon^3 \frac{(A^{-1} H)^3}{3} + \cdots \,.$$

Then

$$\begin{aligned}
\log(A + \varepsilon H) &= \log A(I + \varepsilon A^{-1} H) = \log(e^X e^{Y_\varepsilon}) \\
&= X + \frac{\mathrm{ad}_X e^{\mathrm{ad}_X}}{e^{\mathrm{ad}_X} - 1} \left[\varepsilon A^{-1} H - \varepsilon^2 \frac{(A^{-1} H)^2}{2} + \cdots\right] \\
&\quad + O\left(\left[\varepsilon A^{-1} H - \varepsilon^2 \frac{(A^{-1} H)^2}{2} + \cdots\right]^2\right) .
\end{aligned}$$

This means

$$\frac{1}{\varepsilon} \left[\log(A + \varepsilon H) - \log A\right] = \frac{\mathrm{ad}_X e^{\mathrm{ad}_X}}{e^{\mathrm{ad}_X} - 1} A^{-1} H + o(\varepsilon) \,.$$

Taking the limit yields the following proposition.

Proposition 6.37. *Assume that the eigenvalues of A are strictly positive. Denote $X = \log A$. Then*

$$\partial_H \log A = \frac{\mathrm{ad}_X e^{\mathrm{ad}_X}}{e^{\mathrm{ad}_X} - 1} A^{-1} H \,. \tag{6.8.13}$$

From the above, we see that for any of the functions $x^n, x^{-1}, e^x, \log x$, its lifting to a function from $\mathbb{R}^{m\times m}$ to $\mathbb{R}^{m\times m}$ is differentiable and their derivatives can be explicitly expressed by using the adjoint action operator (6.8.6). It is natural to ask the following general question: if $f : \mathbb{R} \to \mathbb{R}$ is continuously differentiable, then does $\partial_H f(A)$ exist and if it exists, how to express this derivative?

Here we want to answer this question in a slightly more general form. If f is k-times continuously differentiable, then by Theorem 7.1 $f(A)$ is well-defined, does its lifting $f(A)$ is also k-times differentiable? We will give a definitive answer following [Hu (1995)]. But before we proceed with this, we need to recall some basic notions and results on the divided differences.

Given some real numbers $\lambda_1, \cdots, \lambda_n$, we define the first order and k-th order divided differences (see [Donoghue (1974)]) by the following recursive formula:

$$\{\lambda_2, \lambda_1\} f = \begin{cases} \frac{f(\lambda_2)-f(\lambda_1)}{\lambda_2-\lambda_1} & \text{if } \lambda_2 \neq \lambda_1 \\ f'(\lambda_2) & \text{if } \lambda_2 = \lambda_1 \,; \end{cases} \tag{6.8.14}$$

$$\{\lambda_{k+1}, \lambda_k, \cdots, \lambda_1\} f = \{\lambda_{k+1}, \lambda_k\} \{\cdot, \lambda_{k-1}, \cdots, \lambda_1\} f \,. \tag{6.8.15}$$

We need to use the following proposition.

Proposition 6.38. *Suppose that $f : [a, b] \to \mathbb{R}$ is continuously differentiable up to the order $n-1$ and suppose that $\lambda_1, \cdots, \lambda_n \in [a, b]$. Then, there is a $\xi \in [a, b]$ such that*

$$\{\lambda_n, \cdots, \lambda_1\} f = \frac{f^{(n-1)}(\xi)}{(n-1)!} \,. \tag{6.8.16}$$

Proof. First, we assume that all $\lambda_1, \cdots, \lambda_n$ are distinct. If $f(x) = \frac{1}{c-x}$ for a certain complex constant c and if λ_1, λ_2 are different than c, then $\{\lambda_2, \lambda_1\} f = \frac{1}{(c-\lambda_1)(c-\lambda_2)}$. In general, we have

$$\{\lambda_n, \cdots, \lambda_1\} f = \frac{1}{(c-\lambda_1)\cdots(c-\lambda_n)}\,, \quad \text{if } f(x) = \frac{1}{c-x} \,.$$

If f is an analytic function, then by Cauchy's integral formula

$$f(x) = \frac{1}{2\pi\iota} \int_C \frac{f(z)}{z-x} dz \,,$$

where C is a curve containing all points $\lambda_1, \cdots, \lambda_n$ in the interior of the domain surrounded by C. Thus, we have

$$\{\lambda_n, \cdots, \lambda_1\} f = \int_C \frac{f(z)}{(z-\lambda_1)\cdots(z-\lambda_n)} dz\,. \tag{6.8.17}$$

From the calculus of residues we have

$$\{\lambda_n, \cdots, \lambda_1\} f = \sum_{k=1}^{n} \frac{f(\lambda_k)}{\prod_{j\neq k}(\lambda_k - \lambda_j)}\,. \tag{6.8.18}$$

Denote

$$\begin{aligned} P_0(x) &= 1\,, \\ P_k(x) &= (x-\lambda_1)\cdots(x-\lambda_k)\,, k = 1, 2, \cdots, n\,. \end{aligned}$$

Set

$$p(x) = \sum_{k=0}^{n-1} (\{\lambda_{k+1}, \cdots, \lambda_1\} f)\ P_k(x)\,,$$

which coincides with $f(x)$ at all points $\lambda_j\,, j = 1, \cdots, n$.

Denote $a = \min \lambda_j$ and $b = \max \lambda_j$. Then, the function

$$F(x) = f(x) - p(x)$$

has n distinct zeros in $[a, b]$. By Rolle's theorem, its derivative has $n-1$ such zeros in that interval. Continuing this way, we see that the $(n-1)$-th derivative of $F(x)$ has one zero in the interval $[a, b]$. Namely, there is a $\xi \in [a, b]$ such that

$$0 = F^{(n-1)}(\xi) = f^{(n-1)}(\xi) - p^{(n-1)}(\xi) = f^{(n-1)}(\xi) - n!\,(\{\lambda_n, \cdots, \lambda_1\} f)\,. \tag{6.8.19}$$

Since

$$P_k^{(n-1)}(x) = \begin{cases} n!\,\{\lambda_n, \cdots, \lambda_1\} f & \text{if } k = n-1 \\ 0 & \text{if } k < n-1\,, \end{cases}$$

equation (6.8.19) becomes (6.8.16) when all $\lambda_1, \cdots, \lambda_n$ are distinct.

Now if not all λ_j are distinct, there is a sequence of distinct $\lambda_{j,m}$ such that $\lambda_{j,m} \to \lambda_j$ as $m \to \infty$. From (6.8.16), there is a sequence of ξ_m such that

$$\{\lambda_{n,m}, \cdots, \lambda_{1,m}\} f = \frac{f^{(n-1)}(\xi_m)}{(n-1)!}\,. \tag{6.8.20}$$

Since $\xi_m \in [a, b]$, there is a subsequence ξ_{k_m} converging to a $\xi \in [a, b]$. Taking the limit on both sides of (6.8.20) yields the proposition since $f^{(n-1)}$ is continuous and $\{\lambda_n, \cdots, \lambda_1\} f$ is a continuous function of $\lambda_1, \cdots, \lambda_n$. □

Proposition 6.39. *If $f : \mathbb{R} \to \mathbb{R}$ is continuously differentiable up to order n, then its lifting to a function $f : \mathbb{R}^{m\hat{\otimes}m} \to \mathbb{R}^{m\hat{\otimes}m}$ is also differentiable up to order n, where $\mathbb{R}^{m\hat{\otimes}m}$ is the set of $m \times m$ symmetric matrices.*

Proof. We follow [Hu (1995)]. From (6.8.16) we see that if f is continuously differentiable up to order n on the interval $[a, b]$, then

$$\sup_{\lambda_1,\cdots,\lambda_n\in[a,b]} |\{\lambda_n, \cdots, \lambda_1\} f| \le c_n \sum_{k=0}^{n-1} \sup_{x\in[a,b]} \left|f^{(k)}(x)\right| . \tag{6.8.21}$$

If $f(x) = x^d$ for some positive integer d, then

$$\{\lambda_2, \lambda_1\} f = \frac{\lambda_2^d - \lambda_2^d}{\lambda_2 - \lambda_1} = \lambda_1^{d-1} + \lambda_1^{d-2}\lambda_2 + \cdots \lambda_2^{d-1} .$$

In general, we have if $f(x) = x^d$ for some positive integer d, then

$$\{\lambda_n, \cdots, \lambda_1\} f = \begin{cases} \sum_{m_1+\cdots+m_n=d-n+1} \lambda_1^{m_1} \cdots \lambda_n^{m_n} & \text{if } n \le d \\ 1 & \text{if } n = d+1 \\ 0 & \text{if } n \ge d+2 , \end{cases} \tag{6.8.22}$$

where $m_1, \cdots, m_n = 0, 1, \cdots$.

Now assume $f(x) = x^d$ for some positive integer d. Then

$$\begin{aligned} \partial_H^n f(A) &= \frac{d^n}{d^n\varepsilon}\Big|_{\varepsilon=0}(A + \varepsilon H)^d \\ &= \sum_{m_1+\cdots+m_{n+1}=d-n} A^{m_1} H A^{m_2} H \cdots H A^{m_{n+1}} . \end{aligned} \tag{6.8.23}$$

Suppose that A is symmetric and $A = \Pi^{-1} D \Pi$, where $D = \mathrm{diag}(\lambda_1, \cdots, \lambda_m)$ is diagonal and $\Pi^T = (\pi_{ij})$ is orthogonal, namely, $\Pi^{-1} = \Pi^T$. Denote $\eta = \Pi H \Pi^T = (\eta_{ij})$. Then, we can write (6.8.23) as

$$\partial_H^n f(A) = \sum_{m_1+\cdots+m_{n+1}=d-n} \Pi^T D^{m_1} \Pi H \Pi^T D^{m_2} \Pi H \Pi^T \cdots \Pi H \Pi^T D^{m_{n+1}} \Pi .$$

Its (i,j)-th entry can be written as

$$\begin{aligned}
&(\partial_H^n f(A))_{ij} \\
&= \sum_{i_1,\cdots,i_{n+1}=1}^{m} \sum_{m_1+\cdots+m_{n+1}=d-n} \pi_{i_1 i}\lambda_{i_1}^{m_1}\eta_{i_1i_2}\lambda_{i_2}^{m_2}\eta_{i_2i_3}\cdots\eta_{i_ni_{n+1}}\lambda_{i_{n+1}}^{m_{n+1}}\pi_{i_{n+1}j} \\
&= \sum_{i_1,\cdots,i_{n+1}=1}^{m} \pi_{i_1 i}\eta_{i_1i_2}\eta_{i_2i_3}\cdots\eta_{i_ni_{n+1}}\pi_{i_{n+1}j} \\
&\qquad \sum_{m_1+\cdots+m_{n+1}=d-n} \lambda_{i_1}^{m_1}\lambda_{i_2}^{m_2}\cdots\lambda_{i_{n+1}}^{m_{n+1}} \\
&= \sum_{i_1,\cdots,i_{n+1}=1}^{m} \pi_{i_1 i}\eta_{i_1i_2}\eta_{i_2i_3}\cdots\eta_{i_ni_{n+1}}\pi_{i_{n+1}j}\left\{\lambda_{i_{n+1}},\cdots,\lambda_{i_1}\right\}f\,. \qquad (6.8.24)
\end{aligned}$$

Since (6.8.24) holds true for any monomial $f(x)=x^d$ it holds true for any polynomial. Moreover, from (6.8.24) and from (6.8.16) it follows that

$$\begin{aligned}
\sup_{1\le i,j\le m} |(\partial_H^n f(A))_{ij}| &\le C_n\|\eta\|^n \sup_{\lambda_1,\cdots,\lambda_n} \left|\left\{\lambda_{i_{n+1}},\cdots,\lambda_{i_1}\right\}f\right| \\
&\le C_n\|H\|^n \sup_{a\le x\le b,1\le k\le n} |f^{(k)}(x)|\,.
\end{aligned}$$

Now if f is any continuously differentiable function on $[a,b]$, we can approximate f by polynomials $f_\ell(x)$ such that $\sup_{a\le x\le b,1\le k\le n}|f_\ell^{(k)}(x)-f^{(k)}(x)|\to 0$ as $\ell\to\infty$. Applying (6.8.24) to f_ℓ and then taking the limit yields the proposition. □

If $A=\text{diag}(\lambda_1,\cdots,\lambda_m)$ is a diagonal matrix, then (6.8.24) yields

$$(\partial_H f(A))_{ij} = h_{ij}\left\{\lambda_j,\lambda_i\right\}f = h_{ij}\frac{f(\lambda_j)-f(\lambda_i)}{\lambda_j-\lambda_i}\,. \qquad (6.8.25)$$

Definition 6.15. If $A=(a_{ij})$ and $B=(b_{ij})$ are two $n\times n$ matrices, then the Hadamard product of $A\circ B$, denoted by $C=(c_{ij})$, is defined by

$$c_{ij}=a_{ij}b_{ij}\,.$$

With this definition, we can write the identity (6.8.25) as

Proposition 6.40. *Let f be continuously differentiable and let $A=\text{diag}(\lambda_1,\cdots,\lambda_m)$ be a diagonal matrix. Denote*

$$f^{[1]}(A)=\left(\frac{f(\lambda_j)-f(\lambda_i)}{\lambda_j-\lambda_i}\right)_{1\le i,j\le m}\,. \qquad (6.8.26)$$

[When $\lambda_j=\lambda_i$, we replace $\frac{f(\lambda_j)-f(\lambda_i)}{\lambda_j-\lambda_i}$ by $f'(\lambda_i)$]. Then,

$$\partial_H f(A)=f^{[1]}(A)\circ H\,, \tag{6.8.27}$$

where $\circ$ denotes the Hadamard product of two matrices.

In general, to express the derivative, we use the tensor notation. If $A_1=\left(a^1_{ij}\right),\cdots,A_n=\left(a^n_{ij}\right)\in\mathbb{R}^{m\times m}$ are n matrices, then we define the tensor product of these matrices as

$$A_1\otimes\cdots\otimes A_n=\left(a^1_{i_1i_2}\cdots a^n_{i_{2n-1}i_{2n}}\right)_{1\le i_1,\cdots,i_{2n}\le m}$$

which is an element in $(\mathbb{R}^{m\times m})^{\otimes n}$. The cyclic contraction of the tensor product $A_1\otimes\cdots\otimes A_n$ is defined as

$$A_1\oslash\cdots\oslash A_n=\left(a^1_{i_1i_2}a^2_{i_2i_3}\cdots a^n_{i_ni_{n+1}}\right)_{1\le i_1,\cdots,i_{n+1}\le m}\,.$$

This is an element in $\mathbb{R}^{m(n+1)}$. If $A_1=\cdots=A_n=A$, then we denote

$$A^{\oslash n}=\overbrace{A\oslash\cdots\oslash A}^{n}\,.$$

If $B=\left(b_{i_1i_2\cdots i_{n+1}}\right)$ is also an element in $\mathbb{R}^{m(n+1)}$, then we define the matrix-valued (or $\mathbb{R}^{m^2}$-valued) Hadamard product of

$$(A_1\oslash\cdots\oslash A_n)\circ B=(c_{ij})_{1\le i,j\le m}\,, \tag{6.8.28}$$

where

$$c_{ij}=\sum_{i_2,\cdots i_n=1}^{m}a^1_{ii_2}a^2_{i_2i_3}\cdots a^n_{i_{n-1}i_n}a^n_{i_nj}b_{ii_2i_3\cdots i_nj}\,. \tag{6.8.29}$$

With these notations we can write (6.8.24) as

Proposition 6.41. *Suppose that A is symmetric and has the decomposition $A=\Pi^{-1}D\Pi$, where $D=\mathrm{diag}(\lambda_1,\cdots,\lambda_m)$ is diagonal and $\Pi=(\pi_{ij})$ is orthogonal, namely, $\Pi^{-1}=\Pi^T$. Denote*

$$\Lambda_{n+1}=(a_{i,i_1,\cdots,i_{n+1},j})\,,\quad \textit{where}\ a_{i,i_1,\cdots,i_{n+1},j}=\left\{\lambda_{i_{n+1}},\cdots,\lambda_{i_1}\right\}f\,.$$

Then,

$$\partial^n_H f(A)=\left(\Pi^T\oslash(\Pi H\Pi^T)^{\oslash n}\oslash\Pi\right)\circ\Lambda_{n+1}\,. \tag{6.8.30}$$

By the polarization technique (5.2.3) we have

Proposition 6.42. *Suppose that A is symmetric and has the decomposition $A = \Pi^{-1}D\Pi$, where $D = \mathrm{diag}(\lambda_1, \cdots, \lambda_m)$ is diagonal and $\Pi = (\pi_{ij})$ is orthogonal, namely, $\Pi^{-1} = \Pi^T$. Denote*

$$\Lambda_{n+1} = (a_{i,i_1,\cdots,i_{n+1},j}), \quad \textit{where} \quad a_{i,i_1,\cdots,i_{n+1},j} = \{\lambda_{i_{n+1}}, \cdots, \lambda_{i_1}\} f .$$

Then,

$$\begin{aligned} &\partial^n_{H_1,\cdots,H_n} f(A) \qquad (6.8.31) \\ &= \frac{1}{n!}\sum_{\sigma\in\Sigma} \left(\Pi^T \oslash \left(\Pi H_{\sigma(1)}\Pi^T\right) \oslash \cdots \oslash \left(\Pi H_{\sigma(n)}\Pi^T\right) \oslash \Pi\right) \circ \Lambda_{n+1} , \end{aligned}$$

where Σ is the set of all permutations of $\{1, \cdots, n\}$.

A consequence of the above theorem is the following corollary which is quite useful.

Corollary 6.3. *Suppose that $\gamma \in \mathbb{D}_{1,\mathrm{loc}}$ is an $m \times m$ random matrix such that γ is invertible a.s. Then, $\gamma^{-1} \in \mathbb{D}_{1,\mathrm{loc}}$ and*

$$D\gamma^{-1} = -\gamma^{-1}(D\gamma)\gamma^{-1} , \qquad (6.8.32)$$

6.9 Stop Brownian Motion at Anticipative Exit Time

Let $(W_t\,, t \ge 0)$ be a one dimensional Brownian motion on the canonical Wiener $(\Omega\,, \mathcal{F}, P)$ with the natural filtration $\mathcal{F}_t$, $t \ge 0$. If $\tau : \Omega \to \mathbb{R}_+$ is a (bounded) stopping time, then it is well-known that $\mathbb{E}(W_\tau) = 0$. But if τ is an arbitrary positive random variable (may not be a stopping time), then $\mathbb{E}(W_\tau)$ may not be 0. We are interested in computing $\mathbb{E}(W_\tau)$ for some anticipative exit time τ. The general computation will be difficult. We shall show how to use analysis on Wiener space to do some computations for some specific anticipative exit times.

Fix a finite positive number T. Given a real number $a > 0$, let $\tau_a = \inf\{t > 0\,;\ W_t = a\}$. Then τ_a is a stopping time. But for any given $\delta > 0$, the random variable $\tau - \delta$ is not a stopping time. It is an $\mathcal{F}_{\cdot+\delta}$ stopping time (an anticipative time). Let us consider

$$\beta_{\delta,a,T} := (\tau_a - \delta)^+ \wedge T . \qquad (6.9.1)$$

We would like first to compute $\mathbb{E}\left[W_{\beta_{\delta,a,T}}\right]$ and $\mathbb{E}\left[e^{-\rho\delta,a,T}\left(W_{\beta_{\delta,a,T}} - c\right)\right]$.

Denote

$$M_t = \max_{0\le s\le t} W_s\,, \qquad \text{and} \qquad \theta_t = \arg\max_{0\le s\le t} W_s$$

(the unique time such that $W_{\theta_t} = \max_{0\le s\le t} W_s$). We have

$$D_s M_t = I_{\{s\le\theta_t\}}\,. \tag{6.9.2}$$

Since $\beta_{\delta,a,T} > s$ is equivalent to $M_{s+\delta} < a$, we have

$$I_{[0,\beta_{\delta,a,T})}(s) = \chi_{(-\infty,a)}(M_{s+\delta})\,.$$

Thus

$$\begin{aligned} D_{s+}I_{[0,\beta_{\delta,a,T})}(s) &= D_{s+}I_{(-\infty,a)}(M_{s+\delta}) \\ &= -\delta_a(M_{s+\delta})D_{s+}M_{s+\delta} \\ &= -\delta_a(M_{s+\delta})I_{[0,\theta_{s+\delta}]}(s)\,. \end{aligned}$$

Thus, we have by the above identity

$$\begin{aligned} \mathbb{E}(W_{\beta_{\delta,a,T}}) &= \mathbb{E}\left[\int_0^{\beta_{\delta,a,T}} dW_s\right] = \mathbb{E}\left[\int_0^T I_{[0,\beta_{\delta,a,T})}(s)dW_s\right] \\ &= \mathbb{E}\left[\int_0^T D_s I_{[0,\beta_{\delta,a,T})}(s)ds\right] \\ &= -\int_0^T \mathbb{E}\left\{\delta_a(M_{s+\delta})I_{[0,\theta_{s+\delta}]}(s)\right\}ds\,. \end{aligned} \tag{6.9.3}$$

To compute the above expectation, we use the following explicit form of the joint pdf of M_t and θ_t (see [Borodin and Salminen (2002), p.209, Formula 1.13.4])

$$P(M_t\in dy, \theta_t\in dv) = \frac{ye^{-\frac{y^2}{2v}}}{\pi\sqrt{(t-v)v^3}}dvdy\,.$$

Thus, the joint pdf of $M_{s+\delta}$ and $\theta_{s+\delta}$ is

$$q_{s,\delta}(v,y) = \frac{ye^{-\frac{y^2}{2v}}}{\pi\sqrt{(s+\delta-v)v^3}}\,.$$

Therefore, we have

$$\begin{aligned} \mathbb{E}(W_{\beta_{\delta,a,T}}) &= \int_0^T\int_{\mathbb{R}}\int_0^{s+\delta} \delta(y-a)I_{[s,s+\delta]}(v)q_{s,\delta}(v,y)dydvds \\ &= \int_0^T\int_0^{s+\delta} I_{[s,s+\delta]}(v)q_{s,\delta}(v,a)dvds \\ &= \int_0^T\int_{v-\delta}^{v} q_{s,\delta}(v,a)dsdv = \frac{\sqrt{\delta}}{\pi}\int_{\frac{a^2}{2T}}^\infty \frac{e^{-u}}{\sqrt{u}}du\,. \end{aligned}$$

The above computation can be stated as

Theorem 6.14. *Let $T > 0$, $a > 0$, and $\delta > 0$ be given and define*

$$\tau_a = \inf\{s > 0\,, W_s = a\}\,, \quad \text{and} \quad \beta_{\delta,a,T} = (\tau_a - \delta)^+ \wedge T\,.$$

Then, for any $\delta > 0$ and $a > 0$, we have

$$\mathbb{E}(W_{\beta_{\delta,a,T}}) = \frac{\sqrt{\delta}}{\pi}\int_{\frac{a^2}{2T}}^{\infty} \frac{e^{-u}}{\sqrt{u}}du\,.$$

It is clear from this computation that when $\delta = 0$, $\mathbb{E}(W_{\beta_{\delta,a,T}}) = 0$ and for fixed $\delta > 0$, the maximum is attained when $a = 0$. One also see that $\left|\mathbb{E}(W_{\beta_{\delta,a,T}})\right| \le C\sqrt{\delta}$ (as $\delta \to 0$).

Next, we turn to the computation of $\mathbb{E}\left[e^{-\rho\beta_{\delta,a,T}}\left(W_{\beta_{\delta,a,T}} - c\right)\right]$, which is slightly more complicated.

First, we compute $\mathbb{E}\left[e^{-\rho\beta_{\delta,a,T}}\right]$. We need to know the distribution of $\beta_{\delta,a,T}$. From [Borodin and Salminen (2002), p.197, Formula (1.14)], we see

$$\begin{aligned} P(\beta_{\delta,a,T} > \xi) &= P(M_{\xi+\delta} < a) \\ &= 1 - P(\sup_{0\le s\le \xi+\delta} W_s \ge a) \\ &= 1 - \tilde{q}_{\delta,a}(\xi)\,, \end{aligned}$$

where

$$\tilde{q}_{\delta,a}(\xi) = \text{Erfc}\left(\frac{a}{\sqrt{2(\xi+\delta)}}\right),$$

with $\text{Erfc}\,(x) = \frac{1}{\sqrt{2\pi}}\int_{-\infty}^{x} e^{-t^2/2}dt$ being the error function. Therefore

$$\begin{aligned} 1 - \mathbb{E}\left[e^{-\rho\beta_{\delta,a,T}}\right] &= \rho\int_0^{\beta_{\delta,a,T}} e^{-\rho\xi}d\xi = \rho\mathbb{E}\int_0^T e^{-\rho\xi}\mathbb{E}I_{\{\beta_{\delta,a,T}>\xi\}}d\xi \\ &= \rho\int_0^T e^{-\rho\xi}\int_{\frac{a}{\sqrt{2(\xi+\delta)}}}^{\infty} e^{-\frac{x^2}{2}}dxd\xi \\ &= \rho\int_{\frac{a}{\sqrt{2(T+\delta)}}}^{\infty}\int_{\frac{a^2}{2x^2}-\delta}^{T} e^{-\rho\xi}d\xi\; e^{-\frac{x^2}{2}}dx \\ &= \int_{\frac{a}{\sqrt{2(T+\delta)}}}^{\infty}\left[e^{\rho\delta - \frac{\rho a^2}{2x^2}} - e^{-\rho T}\right]e^{-\frac{x^2}{2}}dx\,. \end{aligned}$$

This is summarized to

Proposition 6.43. *Let $T > 0$, $a > 0$, and $\delta > 0$ be given and define*

$$\tau_a = \inf\{s > 0\,, W_s = a\}\,, \quad \text{and} \quad \beta_{\delta,a,T} = (\tau_a - \delta)^+ \wedge T\,.$$

Then

$$\mathbb{E}\left[e^{-\rho\beta_{\delta,a,T}}\right] = 1 - \int_{\frac{a}{\sqrt{2(T+\delta)}}}^{\infty} \left[e^{\rho\delta - \frac{\rho a^2}{2x^2}} - e^{-\rho T}\right] e^{-\frac{x^2}{2}} dx\,. \qquad (6.9.4)$$

Letting $T \to \infty$, we can obtain the computation for $\beta_{\delta,a} = (\tau_a - \delta)^+$ which we shall not repeat here.

Now, we shall compute $\mathbb{E}\left[e^{-\rho\beta_{\delta,a}} W_{\beta_{\delta,a}}\right]$, which is more sophisticated. A cleverer technique is needed. We have

$$\begin{aligned}
1 - \mathbb{E}\left[e^{-\rho\beta_{\delta,a}} W_{\beta_{\delta,a}}\right] &= \rho\mathbb{E}\left[\int_0^\infty e^{-\rho u} \chi_{\{u \le \beta_{\delta,a}\}} du W_{\beta_{\delta,a}}\right] \\
&= \rho\mathbb{E}\left[\int_{-\delta}^\infty \int_0^\infty e^{-\rho u} \chi_{\{u \le \beta_{\delta,a}\}} I_{\{s \le \beta_{\delta,a}\}} du dW_s\right] \\
&= \rho\mathbb{E}\left[\int_{-\delta}^\infty \int_0^\infty e^{-\rho u} \chi_{\{\beta_{\delta,a} \ge u \vee s\}} du dW_s\right] \\
&= \rho\mathbb{E}\left[\int_0^\infty \int_0^\infty e^{-\rho u} \chi_{(-\infty,a)}(M_{u\vee s+\delta}) du dW_s\right] \\
&= \rho \int_0^\infty \int_0^\infty e^{-\rho u} \mathbb{E}\left[D_{s+} I_{(-\infty,a)}(M_{u\vee s+\delta})\right] du ds \\
&= \rho \int_0^\infty \int_0^\infty e^{-\rho u} \mathbb{E}\left[\delta_a\left(M_{u\vee s+\delta}\right) \chi_{[0,\theta_{u\vee s+\delta}]}\right] du ds\,.
\end{aligned}$$

The joint pdf of $M_{u\vee s+\delta}$ and $\theta_{u\vee s+\delta}$ is

$$\tilde{q}_{s,u,\delta}(v,y) = \frac{y e^{-\frac{y^2}{2v}}}{\pi\sqrt{(u\vee s+\delta-v)v^3}}\,, \qquad v < u \vee s + \delta\,.$$

Therefore

$$\begin{aligned}
&1-\mathbb{E}\left[e^{-\rho\beta_{\delta,a}}W_{\beta_{\delta,a}}\right]\\
&=\rho\int_0^\infty\int_0^\infty\int_0^\infty\int_0^\infty e^{-\rho u}\delta_a(y)\,\chi_{[0,v]}(s)\tilde{q}_{s,u,\delta}(v,y)dydvduds\\
&=\rho\int_0^\infty\int_0^\infty\int_0^\infty e^{-\rho u}\chi_{[0,v]}(s)\tilde{q}_{s,u,\delta}(v,a)dvduds\\
&=\rho\int\int\int_{\{s<u\,,\ s<u\,,\ v<u+\delta\}} e^{-\rho u}\frac{ae^{-\frac{a^2}{2v}}}{\pi\sqrt{(u+\delta-v)v^3}}dsdvdu\\
&\quad+\rho\int\int\int_{\{s<u\,,\ s>u\,,\ v<u+\delta\}} e^{-\rho u}\frac{ae^{-\frac{a^2}{2v}}}{\pi\sqrt{(s+\delta-v)v^3}}dsdvdu\\
&=\frac{a\rho}{\pi}\int\int_{\{\ v<u+\delta\}}\frac{e^{-\rho u-\frac{a^2}{2v}}}{\sqrt{(u+\delta-v)v^3}}dvdu\\
&\quad+\frac{2a\rho}{\pi}\int\int_{\{u<v\}}\left[\sqrt{\delta}-\sqrt{u\vee(v-\delta)+\delta-v}\right]e^{-\rho u-\frac{a^2}{2v}}dvdu\,.
\end{aligned}$$

Thus, we have

Theorem 6.15. *Let $T>0$, $a>0$, and $\delta>0$ be given and define*

$$\tau_a=\inf\{s>0\,,W_s=a\}\,,\quad \text{and}\quad \beta_{\delta,a,T}=(\tau_a-\delta)^+\wedge T\,.$$

Then, for any real numbers ρ, a, and c. we have

$$\begin{aligned}
&\mathbb{E}\left[e^{-\rho\beta_{\delta,a}}\left(W_{\beta_{\delta,a}}-c\right)\right]\\
&=1+c-\frac{a\rho}{\pi}\int\int_{\{\ v<u+\delta\}}\frac{e^{-\rho u-\frac{a^2}{2v}}}{\sqrt{(u+\delta-v)v^3}}dvdu\\
&\quad-\frac{2a\rho}{\pi}\int\int_{\{u<v\}}\left[\sqrt{\delta}-\sqrt{u\vee(v-\delta)+\delta-v}\right]e^{-\rho u-\frac{a^2}{2v}}dvdu\\
&\quad-c\int_{\frac{a}{\sqrt{2(T+\delta)}}}^\infty\left[e^{\rho\delta-\frac{\rho a^2}{2x^2}}-e^{-\rho T}\right]e^{-\frac{x^2}{2}}dx\,.
\end{aligned}\tag{6.9.5}$$

It is clear from the above theorem that $\mathbb{E}\left[e^{-\rho\beta_{\delta,a}}\left(W_{\beta_{\delta,a}}-c\right)\right]$ is a continuous function of a. Thus, it has maximum on any compact set of a. This solves the problem of finding an optimal a so that the optimal anticipative stopping time to maximize $\mathbb{E}\left[e^{-\rho\beta_{\delta,a}}\left(W_{\beta_{\delta,a}}-c\right)\right]$ among the class of exiting times of level a. Numerical computation is needed to find the maximum.

Finally, we shall compute $\mathbb{E}\left[\int_0^{\beta_{\delta,a}} e^{-\rho t} W_t dt\right]$. We use

$$\mathbb{E}\left[\int_0^{\beta_{\delta,a}} e^{-\rho t} W_t dt\right] = \int_0^\infty e^{-\rho t}\mathbb{E}\left[I_{\{\beta_{\delta,a}>t\}} W_t\right] dt\,. \tag{6.9.6}$$

Denote again $M_t = \sup_{0\le r\le t} W_r$. Now

$$\begin{aligned}\mathbb{E}\left[I_{\{\beta_{\delta,a}>t\}} W_t\right] &= \int_0^t \mathbb{E}\left[D_s\chi_{\{\beta_{\delta,a}>t\}}\right] ds \\ &= \int_0^t \mathbb{E}\left[D_s\chi_{\{M_t<a\}}\right] ds = \int_0^t \mathbb{E}\left[\delta_a(M_t) I_{\{s<\theta_t\}}\right] ds\,.\end{aligned}$$

Since the joint pdf of M_t and θ_t is

$$\tilde{q}_t(v,y) = \frac{y e^{-\frac{y^2}{2v}}}{\pi\sqrt{(t+\delta-v)v^3}}\,,\quad v<t+\delta\,, \tag{6.9.7}$$

we have

$$\begin{aligned}\mathbb{E}\left[I_{\{\beta_{\delta,a}>t\}} W_t\right] &= -\int_0^t\int\int \delta_a(y) I_{\{s<v\}}\frac{y e^{-\frac{y^2}{2v}}}{\pi\sqrt{(t+\delta-v)v^3}} ds dv dy \\ &= -\int_{0<s<v<t+\delta}\frac{a e^{-\frac{a^2}{2v}}}{\pi\sqrt{(t+\delta-v)v^3}} ds dv \\ &= -\int_{0<s<v<t+\delta}\frac{a e^{-\frac{a^2}{2v}}}{\pi\sqrt{v}\sqrt{t+\delta-v}} dv\,.\end{aligned} \tag{6.9.8}$$

Thus

$$\begin{aligned}\mathbb{E}\left[\int_0^{\beta_{\delta,a}} e^{-\rho t} W_t dt\right] &= -\int_{0<v<t+\delta<\infty} e^{-\rho t}\frac{a e^{-\frac{a^2}{2v}}}{\pi\sqrt{v}\sqrt{t+\delta-v}} ds dv \\ &= -\frac{a g(\rho,\delta)}{\pi}\int_0^\infty e^{-\rho v-\frac{a^2}{2v}} v^{-1/2} dv\,,\end{aligned} \tag{6.9.9}$$

where

$$g(\rho,\delta) = \int_0^\infty \frac{e^{-\rho s+\rho\delta}}{\sqrt{s}} ds = \Gamma(\frac{1}{2}) e^{\rho\delta}\rho^{-\frac{1}{2}} = \sqrt{\pi} e^{\rho\delta}\rho^{-\frac{1}{2}}\,. \tag{6.9.10}$$

By [Gradshteyn and Ryzhik (2000), Formula 3.471 (9)], we have

$$\int_0^\infty e^{-\rho v-\frac{a^2}{2v}} v^{-1/2} dv = 2\left(\frac{a^2}{2\rho}\right)^{1/4} K_{1/2}\left(2\sqrt{\frac{a^2\rho}{2}}\right)$$

and again by [Gradshteyn and Ryzhik (2000), Formula 8.469 (3)], we have

$$K_{1/2}\left(2\sqrt{\frac{a^2\rho}{2}}\right) = \sqrt{\frac{\pi}{2\cdot 2\sqrt{\frac{a^2\rho}{2}}}} e^{-2\sqrt{\frac{a^2\rho}{2}}}\,.$$

Therefore

$$\int_0^\infty e^{-\rho v - \frac{a^2}{2v}} v^{-1/2} dv = \frac{\sqrt{\pi}}{\sqrt{\rho}} e^{-\sqrt{2a^2\rho}} \,.$$

Now, we can summarize the above computation as

Theorem 6.16. *Let $T > 0$, $a > 0$, $\rho > 0$, and $\delta > 0$ be given and define*

$$\tau_a = \inf\{s > 0\,, W_s = a\}\,, \quad \text{and} \quad \beta_{\delta,a,T} = (\tau_a - \delta)^+ \wedge T\,.$$

Then

$$\begin{aligned} \mathbb{E}\left[\int_0^{\beta_{\delta,a}} e^{-\rho t} W_t dt\right] &= -\frac{a g(\rho,\delta)}{\pi} \frac{\sqrt{\pi}}{\sqrt{\rho}} e^{-\sqrt{2a^2\rho}} \\ &= -\frac{a e^{\rho\delta} e^{-a\sqrt{2\rho}}}{\rho}\,. \end{aligned} \tag{6.9.11}$$

From this, the minimum is easily seen to be attained at $a = \frac{1}{\sqrt{2\rho}}$.

Chapter 7

Some Inequalities

7.1 Complex Hypercontractivity

Let $(\mathbb{B}_1, \mathbb{H}_1, \mathbb{P}_1)$ and $(\mathbb{B}_2, \mathbb{H}_2, \mathbb{P}_2)$ be two abstract Wiener spaces. Let $\mathbb{H}_1^c$ and $\mathbb{H}_2^c$ be the complexifications of $\mathbb{H}_1$ and $\mathbb{H}_2$. Let $A : \mathbb{H}_1^c \to \mathbb{H}_2^c$ be a bounded complex linear operator with norm ≤ 1. A can be extended to a bounded complex linear operator (denoted by $A^{\otimes n}$) from $(\mathbb{H}_1^c)^{\otimes n}$ to $(\mathbb{H}_2^c)^{\otimes n}$ in the following manner:

$$A^{\otimes n}(e_1 \otimes \cdots \otimes e_n) = (Ae_1) \otimes \cdots \otimes (Ae_n) , \quad \forall\, e_1, \cdots, e_n \in \mathbb{H}_1^c .$$

This way the operator A can be extended to a bounded complex linear operator between the Fock spaces $\Gamma(\mathbb{H}_1^c)$ and $\Gamma(\mathbb{H}_2^c)$, which is called the second quantization operator of A and is denoted by $\Gamma(A)$.

Given a complex random variable $F \in L^2(\mathbb{B}_1, \mathbb{H}_1, \mathbb{P})$,

$$F = \sum_{n=0}^{\infty} I_n(f_n) , \quad f_n \in (\mathbb{H}_1^c)^{\hat{\otimes} n} ,$$

we define

$$\Gamma(A)F = \sum_{n=0}^{\infty} I_n(A^{\otimes n} f_n) , \tag{7.1.1}$$

where we use the same notation I_n for multiple integrals in two abstract Wiener spaces since there is no confusion. It is easy to check that $\Gamma(A)$ is a bounded linear operators with norm ≤ 1 from $L^2(\mathbb{B}_1, \mathbb{H}_1^c, \mathbb{P}_1)$ and $L^2(\mathbb{B}_2, \mathbb{H}_2^c, \mathbb{P}_2)$.

Example 7.1. If $F = \mathcal{E}(f) \in \mathcal{E}$, $f \in \mathbb{H}_1^c$, then

$$\Gamma(A)F = \mathcal{E}(Af) = \exp\left\{ \widetilde{Af} - \frac{\langle Af, Af\rangle_{\mathbb{H}_2}}{2} \right\} .$$

Remark 7.1. With an appropriate choice of the topology in the Fock space or chaos expansion, one can extend the definition of the second quantization to an unbounded operator A.

Lemma 7.1. *If $f : D \to \mathbb{C}$ is a continuous complex valued function, where D is a compact subset of $\mathbb{C}$, then there is a sequence of polynomials $f_n(x, \bar{x})$ such that $\lim_{n\to\infty} \sup_{x\in D} |f_n(x, \bar{x}) - f(x)| = 0$.*

Proof. Write $f(x) = f'(x_1, x_2) + \imath f''(x_1, x_2)$, where $x = x_1 + \imath x_2 \in D$. Since D is compact, by the real version Stone-Weierstrass theorem (see for example [Lang (1993), p.54, Corollary 1.3]) there exist sequences of polynomials $f_n'(x_1, x_2)$ and $f_n''(x_1, x_2)$ such that $(f_n'(x_1, x_2), f_n''(x_1, x_2))$ converges to $(f'(x_1, x_2), f''(x_1, x_2))$ uniformly on D. We can express $x_1 = (x + \bar{x})/2$ and $x_2 = (x - \bar{x})/2$. Thus, the polynomials $f_n(x, \bar{x}) = f_n'(x_1, x_2) + \imath f_n''(x_1, x_2)$ converges to $f(x)$ uniformly on D. □

Since $\sqrt{x}$ is a continuous function on a bounded domain, we may approximate it by polynomial functions $f_n(x, \bar{x})$.

Definition 7.1. Let $A : \mathbb{H}_1^c \to \mathbb{H}_2^c$ be a bounded linear operator. The Hilbertian adjoint $\bar{A} : \mathbb{H}_2^c \to \mathbb{H}_1^c$ is defined as

$$\langle Ax, y \rangle = \langle x, \bar{A}y \rangle , \quad \forall\, x \in \mathbb{H}_1^c,\ y \in \mathbb{H}_2^c .$$

A is called normal if $A : \mathbb{H}^c \to \mathbb{H}^c$ is a linear bounded operator and $A\bar{A} = \bar{A}A$.

Proposition 7.1. *If $A : \mathbb{H}^c \to \mathbb{H}^c$ is normal, then $\sqrt{A}$ is well-defined.*

Proof. Take polynomial approximations $\{f_n(x, \bar{x})\}$ of $\sqrt{x}$. Then, $A_n = f_n(A, \bar{A})$ is well-defined since $A\bar{A} = \bar{A}A$. It is easy to see that A_n is a Cauchy sequence with respect to the operator norm. The limit is defined as $\sqrt{A}$. It is also easy to see that $\sqrt{A}$ is independent of the choice of the sequence $\{f_n\}$. □

Now let $A : \mathbb{H}_1^c \to \mathbb{H}_2^c$ be a bounded linear operator with norm ≤ 1. We shall give an explicit expression of its second quantization $\Gamma(A)$.

Let $(\tilde{\mathbb{B}}_2, \tilde{\mathbb{H}}_2, \tilde{\mathbb{P}}_2)$ be an identical and independent copy of $(\mathbb{B}_2, \mathbb{H}_2, \mathbb{P}_2)$. We assume that $\mathbb{H}_1^c$ can be embedded into $\mathbb{H}_2^c$ with the mapping $\tau : \mathbb{H}_1^c \to \mathbb{H}_2^c$: $\langle \tau(\varphi), \tau(\psi) \rangle_{\mathbb{H}_2} = \langle \varphi, \psi \rangle_{\mathbb{H}_1}$, $\forall \varphi, \psi \in \mathbb{H}_1$. Given an element $e \in \mathbb{H}_1^c$, $\langle \cdot, e \rangle$ is a centered Gaussian random variable in $L^2(\mathbb{B}_1)$. We define the lift of this Gaussian variable $\langle \cdot, e \rangle$ to $(\mathbb{B}_2 \times \tilde{\mathbb{B}}_2, \mathbb{H}_2 \times \tilde{\mathbb{H}}_2, \mathbb{P}_2 \times \tilde{\mathbb{P}}_2)$ by

$$A^{\Uparrow} \langle x, e \rangle = \langle x, Ae \rangle + \langle \tilde{x}, \tau(\sqrt{I - \bar{A}A}e) \rangle , \quad x \in \mathbb{B}_2,\ \tilde{x} \in \tilde{\mathbb{B}}_2 .$$

For any two vectors e_1 and $\tilde{e}_2$ in $\mathbb{H}_1$,

$$\begin{aligned}
&\mathbb{E}\left[\left(A^{\Uparrow}\langle x, e_1\rangle\right)\left(A^{\Uparrow}\langle x, e_2\rangle\right)\right] \\
&= \mathbb{E}\left[\langle x, Ae_1\rangle\langle x, Ae_2\rangle\right] + \mathbb{E}\left[\langle\tilde{x}, \tau(\sqrt{I-\bar{A}A}e_1)\rangle\langle\tilde{x}, \tau(\sqrt{I-\bar{A}A}e_2)\rangle\right] \\
&= \langle Ae_1\,, Ae_2\rangle + \langle\tau(\sqrt{I-\bar{A}A}e_1)\,, \tau(\sqrt{I-\bar{A}A}e_2)\rangle \\
&= \langle Ae_1\,, Ae_2\rangle + \langle\sqrt{I-\bar{A}A}e_1\,, \sqrt{I-\bar{A}A}e_2\rangle \\
&= \langle e_1\,, \bar{A}Ae_2\rangle + \langle e_1\,, 1-\bar{A}Ae_2\rangle = \langle e_1, e_2\rangle. \qquad (7.1.2)
\end{aligned}$$

This means that $\left\{A^{\Uparrow}\langle x, e\rangle\,, e \in \mathbb{H}_1\right\}$ is a system of Gaussian random variables whose covariance structure is the same as that of $\{\langle\cdot, e\rangle, e \in \mathbb{H}_1\}$.

If F is a nonlinear functional on $\mathbb{B}_1$ having the following form:

$$F(x) = \sum_{finite} f(\langle x, e_1\rangle\,, \cdots, \langle x, e_n\rangle)\,,\ \ x \in \mathbb{B}_1\,,$$

then we define its lifting on $(\mathbb{B}_2 \times \tilde{\mathbb{B}}_2, \mathbb{H}_2 \times \tilde{\mathbb{H}}_2, \mathbb{P}_2 \times \tilde{\mathbb{P}}_2)$ by A as follows

$$(A^{\Uparrow}F)(x, \tilde{x}) = \sum_{finite} f(A^{\Uparrow}\langle x, e_1\rangle\,, \cdots, A^{\Uparrow}\langle x, e_n\rangle)\,, \quad x \in \mathbb{B}_2,\ \tilde{x} \in \tilde{\mathbb{B}}_2\,.$$

From the isometry of (7.1.2), we see that

$$\mathbb{E}^{\mathbb{P}_2\times\tilde{\mathbb{P}}_2}\left[|\Gamma(A)F|^2\right] = \mathbb{E}^{\mathbb{P}_1}(|F|^2)\,.$$

Thus, we have

Proposition 7.2. $A^{\Uparrow} : L^2(\mathbb{B}_1) \to L^2(\mathbb{B}_2 \times \tilde{\mathbb{B}}_2)$ *is an isometry.*

Example 7.2. If $h \in \mathbb{H}_1^c$ and $\mathcal{E}(h) = \exp\left\{\langle\cdot, h\rangle_{\mathbb{H}_1} - \frac{\|h\|^2_{\mathbb{H}_1}}{2}\right\}$ is the exponential vector, then

$$A^{\Uparrow}\mathcal{E}(h) = \exp\left\{\langle x, Ah\rangle_{\mathbb{H}_2} + \langle\tilde{x}\,, \tau(\sqrt{1-\bar{A}A}h)\rangle_{\mathbb{H}_2} - \frac{\|h\|^2_{\mathbb{H}_1}}{2}\right\}.$$

The following is a Mehler formula for the second quantization of a normal operator.

Theorem 7.1. *Let* $A : \mathbb{H}_1^c \to \mathbb{H}_2^c$ *be a bounded linear operator with norm* ≤ 1. *Then, for any* $F \in L^2(\mathbb{B}_1)$ *we have*

$$\Gamma(A)F(x) = \mathbb{E}^{\tilde{\mathbb{P}}_2}\left[A^{\Uparrow}F(x, \tilde{x})\right]\,, \qquad (7.1.3)$$

where the expectation is with respect to the "variable" $\tilde{x}$ *when* x *is considered as fixed.*

Remark 7.2.

(i) The above expectation can also be considered as the conditional expectation with respect to x, namely, the σ-algebra generated by $\{\langle x, e\rangle, e \in \mathbb{H}_2\}$.

(ii) If we define formally $A : \mathbb{B}_2 \to \mathbb{B}_1$ and $\bar{\tau} : \tilde{\mathbb{B}}_2 \to \mathbb{B}_1$ by

$$\langle \bar{A}x, h\rangle_{\mathbb{H}_1} = \langle x, Ah\rangle_{\mathbb{H}_2}, \quad \langle \bar{\tau}\tilde{x}, h\rangle_{\mathbb{H}_1} = \langle \tilde{x}, \tau(h)\rangle_{\tilde{\mathbb{H}}_2}, \quad \forall x \in \mathbb{B}_2,\ \tilde{x} \in \tilde{\mathbb{B}}_2,\ h \in \mathbb{H}_1, \tag{7.1.4}$$

then (7.1.3) may be formally written as

$$\Gamma(A)F(x) = \mathbb{E}^{\tilde{\mathbb{P}}_2}\left[F(\bar{A}x + \sqrt{1 - A\bar{A}}\,\bar{\tau}\tilde{x})\right]. \tag{7.1.5}$$

Proof. Since the set $\mathcal{E}$ is dense in $L^2(\mathbb{B}_1)$ it suffices to show (7.1.3) for $F \in \mathcal{E}$. If $F = \mathcal{E}(h) = \exp\left\{\langle \cdot, h\rangle_{\mathbb{H}_1} - \frac{\|h\|^2_{\mathbb{H}_1}}{2}\right\}$ is the exponential vector, then

$$\begin{aligned}
\mathbb{E}^{\tilde{\mathbb{P}}_2}\left[A^{\Uparrow}F(x,\cdot)\right] &= \mathbb{E}^{\tilde{\mathbb{P}}_2}\exp\left\{A^{\Uparrow}\langle \cdot, h\rangle_{\mathbb{H}_1} - \frac{\|h\|^2_{\mathbb{H}_1}}{2}\right\} \\
&= \mathbb{E}^{\tilde{\mathbb{P}}_2}\exp\left\{\langle x, Ah\rangle_{\mathbb{H}_2} + \langle \tilde{x}, \tau(\sqrt{1 - \bar{A}A}h)\rangle_{\mathbb{H}_2} - \frac{\|h\|^2_{\mathbb{H}_1}}{2}\right\} \\
&= \mathbb{E}^{\tilde{\mathbb{P}}_2}\exp\left\{\langle x, Ah\rangle_{\mathbb{H}_2} + \frac{\|\sqrt{1 - \bar{A}A}h\|^2_{\mathbb{H}_1}}{2} - \frac{\|h\|^2_{\mathbb{H}_1}}{2}\right\} \\
&= \mathbb{E}^{\tilde{\mathbb{P}}_2}\exp\left\{\langle x, Ah\rangle_{\mathbb{H}_2} - \frac{\|Ah\|^2_{\mathbb{H}_2}}{2}\right\} \\
&= \Gamma(A)\mathcal{E}(h).
\end{aligned}$$

By linearity $\mathbb{E}^{\tilde{\mathbb{P}}_2}\left[A^{\Uparrow}F(x,\cdot)\right] = \Gamma(A)F$ for any $F \in \mathcal{E}$. The theorem is proved by a limiting argument. □

The following proposition is the Mehler formula when F is a functional of finite many variables. We shall write

$f(X) = f(X_1, \cdots, X_n)$ if $X = (X_1, \cdots, X_n)^T$ is an n-dimensional vector.

Proposition 7.3. *Let $A : \mathbb{H}_1^c \to \mathbb{H}_2^c$ be a bounded linear operator with norm ≤ 1. Let $e_1, \cdots, e_n$ be a subset of orthonormal vectors of $\mathbb{H}_1^c$. Then, there are orthonormal vectors $\{\varepsilon_1, \cdots, \varepsilon_n\}$ in $\mathbb{H}_2^c$ such that*

$$Ae_i = \sum_{j=1}^n a_{ij}\varepsilon_j, \quad i = 1, \cdots, n. \tag{7.1.6}$$

Let F have the form $F = f(\langle\cdot, e_1\rangle_{\mathbb{H}_1}, \cdots, \langle\cdot, e_n\rangle_{\mathbb{H}_1})$. Then

$$\begin{aligned}\Gamma(A)F(x) &= \mathbb{E}^Y \left[f(\langle x\,, Ae\rangle_{\mathbb{H}_2} + \sqrt{I - a_A\bar{a}_A}Y\right] \\ &= \mathbb{E}^Y \left[f(a_A\varepsilon_x + \sqrt{I - a_A\bar{a}_A}Y\right], \quad x \in \mathbb{B}_2 \qquad (7.1.7)\end{aligned}$$

where $Y = (Y_1, \cdots, Y_n)$ is an n-dimensional standard Gaussian; $\varepsilon_x = (\langle x, \varepsilon_1\rangle_{\mathbb{H}_2}, \cdots, \langle x, \varepsilon_n\rangle_{\mathbb{H}_2})^T$; $a_A = (a_{ij})_{1\le i,j\le n}$ is the matrix defined by (7.1.6); and the expectation $\mathbb{E}^Y$ is with respect to the Gaussian Y.

Proof. Since the dimension of the vector space spanned by $\{Ae_1, \cdots, Ae_n\}$ is at most n, it is obvious that there are orthonormal vectors $\{\varepsilon_1, \cdots, \varepsilon_n\}$ (we can take and fix any of these vectors) such that (7.1.6) holds.

Since $\bar{a}_A a_A$ is positive definite with norm ≤ 1, the matrix $\sqrt{I - a_A\bar{a}_A}$ is well-defined.

To show (7.1.7) it suffices to show it for functional of the exponential vector form

$$F = \exp\left\{\sum_{j=1}^n \lambda_j\langle\cdot, e_j\rangle_{\mathbb{H}_1} - \frac{1}{2}\sum_{j=1}^n \lambda_j^2\right\} = \exp\left\{\langle\lambda, \tilde{e}\rangle - \frac{1}{2}\lambda^2\right\},$$

where we use the notations

$$\tilde{e}_j = \langle\cdot, e_j\rangle, \quad \tilde{e} = (\tilde{e}_1, \cdots, \tilde{e}_n)^T, \quad \lambda\cdot\tilde{e} = \sum_{j=1}^n \lambda_j\tilde{e}_j, \quad \lambda^2 = \sum_{j=1}^n \lambda_j^2.$$

We also use this type of abbreviations for other vectors. We have

$$\begin{aligned}&\mathbb{E}^Y \left[f(\langle x\,, Ae\rangle_{\mathbb{H}_2} + \sqrt{I - a_A\bar{a}_A}Y\right] \\ &= \mathbb{E}^Y \left[\exp\left\{\lambda\cdot\widetilde{Ae} + \lambda\cdot\sqrt{1 - a_A\bar{a}_A}Y - \lambda^2/2\right\}\right] \\ &= \exp\left\{\lambda\cdot\widetilde{Ae} + \frac{1}{2}\lambda^T(1 - a_A\bar{a}_A)\lambda - \lambda^2/2\right\} \\ &= \exp\left\{\lambda\cdot\widetilde{Ae} - \frac{1}{2}\lambda^T a_A\bar{a}_A\lambda\right\} \\ &= \exp\left\{\sum_{j=1}^n \lambda_j\langle\cdot, Ae_j\rangle_{\mathbb{H}_1} - \frac{1}{2}\sum_{i,j=1}^n \lambda_i\lambda_j\langle Ae_i\,, Ae_j\rangle_{\mathbb{H}_2}\right\}.\end{aligned}$$

Thus, (7.1.7) holds true for exponential vectors. Then, it is true for linear combinations of exponential vectors. A density argument proves the proposition. □

Lemma 7.2. *Let $\{e_1, \cdots, e_n\}$ be orthonormal vectors in $\mathbb{H}_1$ and let $F = f(\langle\cdot, e_1\rangle_{\mathbb{H}_1}, \cdots, \langle\cdot, e_2\rangle_{\mathbb{H}_1})$ for some polynomial function f. Then*

$$\Gamma(A) : F := : f(\langle\cdot, Ae_1\rangle_{\mathbb{H}_2}, \cdots, \langle\cdot, Ae_2\rangle_{\mathbb{H}_2}) : \tag{7.1.8}$$

Proof. We shall use the notation $e = (e_1, \cdots, e_n)^T$ and we write $f(\langle\cdot, e\rangle_{\mathbb{H}_1}) = f(\langle\cdot, e_1\rangle_{\mathbb{H}_1}, \cdots, \langle\cdot, e_2\rangle_{\mathbb{H}_1})$ and we also use this abbreviation for other functionals. By the representation formula (6.7.6) we see that

$$: F := \int_{\mathbb{R}^n} f(\langle\cdot, e\rangle_{\mathbb{H}_1} + \imath z)p(dz) = \mathbb{E}^Z f(\langle\cdot, e\rangle_{\mathbb{H}_1} + \imath Z), \tag{7.1.9}$$

where $p(z) = (\sqrt{2\pi})^{-n/2} \exp\left(-|z|^2/2\right)$ and where Z is a n-dimensional standard normal. Let $\varepsilon_1, \cdots, \varepsilon_n$ and a_A be given as in (7.1.6). By the Mehler formula (7.1.7), we have

$$\Gamma(A) : F := \mathbb{E}^Z \mathbb{E}^Y \left[f(\langle x, Ae\rangle_{\mathbb{H}_2} + \sqrt{I - a_A \bar{a}_A} Y + \imath Z) \right], \tag{7.1.10}$$

where Y is an n-dimensional Gaussian, independent of Z, and $\varepsilon_x = \langle\cdot, \tilde{e}\rangle_{\mathbb{H}_2}$. Now $\langle x, Ae\rangle_{\mathbb{H}_2} = \alpha_A \varepsilon_x$ is n dimensional Gaussian which has the same law as $\alpha_A Z$, where we recall that we use Z to denote a standard Gaussian. Again by the representation formula (6.7.6) we have

$$: f(\langle\cdot, Ae_1\rangle_{\mathbb{H}_2}, \cdots, \langle\cdot, Ae_2\rangle_{\mathbb{H}_2}) := \mathbb{E}^Z f(\langle x, Ae\rangle_{\mathbb{H}_2} + \imath\alpha_A Z). \tag{7.1.11}$$

It is straightforward to verify that $\imath\alpha_A Z$ and $\sqrt{I - a_A \bar{a}_A} Y + \imath Z$ have the same law. From (7.1.10) and (7.1.11) the lemma follows. □

Remark 7.3. By (7.1.9) and (7.1.11) this lemma (formula 7.1.8) can be written as

$$\Gamma(A) \int_{\mathbb{R}^n} f(\langle\cdot, e\rangle_{\mathbb{H}_1} + \imath z)p(dz) = \int_{\mathbb{R}^n} f(\langle x, Ae\rangle_{\mathbb{H}_2} + \imath\alpha_A z)p(dz). \tag{7.1.12}$$

Theorem 7.2. *Let $1 \le p \le q < \infty$ and let $A : \mathbb{H}_1^c \to \mathbb{H}_2^c$ be a bounded linear mapping with norm ≤ 1. Then, $\Gamma(A)$ is hypercontractive, namely,*

$$\|\Gamma(A)F\|_q \le \|F\|_p, \quad \forall\, F \in L^p(\mathbb{B}_1), \tag{7.1.13}$$

if and only if

$$(q-1)\|\mathrm{Re}(A\xi)\|^2 + \|\mathrm{Im}(A\xi)\|^2 \le (p-1)\|\mathrm{Re}(\xi)\|^2 + \|\mathrm{Im}(\xi)\|^2, \quad \forall \xi \in \mathbb{H}_1^c. \tag{7.1.14}$$

Proof. The "only if" part is easy. It suffices to take

$$F = \exp\left[\langle\cdot, \xi\rangle - \frac{1}{2}\langle\xi, \xi\rangle_{\mathbb{H}_1}^2\right].$$

Now we show the "if" part. Let $\{e_1, \cdots, e_n\}$ be a set of orthonormal vectors in $\mathbb{H}_1$. We only need to show (7.1.13) for

$$F =: f(\tilde{e}_1, \cdots, \tilde{e}_n) := \int_{\mathbb{R}^n} f(\langle \cdot, e\rangle_{\mathbb{H}_1} + \imath z)p(dz)\,,$$

where f is a polynomial of n variables since the set of Hermite polynomials of $\tilde{e}_1, \cdots, \tilde{e}_n, \cdots$ spans $L^p(\mathbb{B}_1)$. Let $\varepsilon_1, \cdots, \varepsilon_n$ be a set of orthonormal vectors in $\mathbb{H}_2$ and let a_A be defined by (7.1.6). The condition (7.1.14) can be expressed in terms of a_A in the following manner. For any vector $\zeta = (\zeta_1, \cdots, \zeta_n)^T$, we consider $\xi = \zeta_1 e_1 + \cdots + \zeta_n e_n$ and we see

$$A\xi = \zeta_1 A e_1 + \cdots + \zeta_n A e_n = \sum_{i,j=1}^n \zeta_i a_{ij} \varepsilon_j\,.$$

Since $\varepsilon_j, j = 1, \cdots, n$ are orthonormal in $\mathbb{H}_2$, we see

$$\|A\xi\|_{\mathbb{H}_2}^2 = \sum_{i=j}^n |\sum_{i=1}^n \zeta_i a_{ij}|^2 = |a_A^T \zeta|^2\,.$$

Then, condition (7.1.14) becomes

$$(q-1)|\mathrm{Re}(a_A^T\zeta)|^2 + |\mathrm{Im}(a_A^T\zeta)|^2 \le (p-1)|\mathrm{Re}(\zeta)|^2 + |\mathrm{Im}(\zeta)|^2\,. \quad (7.1.15)$$

By (7.1.12), we have (recall we use the notation $p(dx) = (2\pi)^{-n/2}\exp\left(-|x|^2/2\right) dx$)

$$\Gamma(A)F = \int_{\mathbb{R}^n} f(\langle x, Ae\rangle_{\mathbb{H}_2} + \imath a_A z)p(dz) = \int_{\mathbb{R}^n} f(a_A\varepsilon_x + \imath a_A z)p(dz)\,.$$

Let X and Y be two independent n-dimensional standard Gaussian. Since $\tilde{e}$ and ε_x are n-dimensional standard Gaussian, to show (7.1.13) for our F it suffices to show

$$\left(\mathbb{E}^Y \left|\int_{\mathbb{R}^n} f(a_A Y + \imath a_A z)p(dz)\right|^q\right)^{p/q} \le \left\{\mathbb{E}^X \left|\int_{\mathbb{R}^n} f(X + \imath z)p(dz)\right|^p\right\}^{q/p}. \quad (7.1.16)$$

Denote $p_t(x) = (2\pi t)^{-n/2}\exp\left(-\frac{|x|^2}{2t}\right)$ and denote

$$\begin{cases} F = F(s,x,t,y) = \displaystyle\int_{\mathbb{R}^{2n}} f(x + a_A y + \imath z + \imath a_A w)p_s(dz)p_t(dw)\,; \\ Q = Q(s,x,t,y) = |F(s,x,t,y)|^2 + \varepsilon^2\,; \\ \Psi = \Psi(s,x,t) = \displaystyle\int_{\mathbb{R}^n} Q(s,x,t,y)^{q/2} p_t(dy)\,; \\ \Phi = \Phi(s,t) = \displaystyle\int_{\mathbb{R}^n} \Psi(s,x,t)^{p/q} p_s(dx)\,. \end{cases} \quad (7.1.17)$$

It is easy to see that

$$\Phi(0,1) = \left\{ \int_{\mathbb{R}^n} \left(\left| \int_{\mathbb{R}^n} f(a_A y + \imath a_A w) p(dw) \right|^2 + \varepsilon^2 \right)^{q/2} P_1(dy) \right\}^{p/q}$$

and

$$\Phi(1,0) = \int_{\mathbb{R}^n} \left\{ \left| \int_{\mathbb{R}^n} f(x + \imath z) p(dz) \right|^2 + \varepsilon^2 \right\}^{p/2} p(dx) .$$

If we can show that

$$\Phi(0,1) \le \Phi(1,0) , \tag{7.1.18}$$

then letting $\varepsilon \to 0$ will yield (7.1.16) and hence the theorem is proved. Our remaining task is then to show (7.1.18). This will be completed by showing that $\Phi(1-t,t)$ is a decreasing function of $t \in [0,1]$, which is achieved by differentiating in t.

To compute the derivatives, we shall use the following notations

$$\begin{cases} P_t^x f = \int_{\mathbb{R}^n} f(x+z) p_t(dz) , \\ \nabla_x f(x,y) = \left(\partial_{x_1} f(x,y) , \cdots , \partial_{x_n} f(x,y)\right)^T , \\ \nabla_y f(x,y) = \left(\partial_{y_1} f(x,y) , \cdots , \partial_{y_n} f(x,y)\right)^T , \\ \mathrm{Hess}_x(f) = \left(\partial^2_{x_i, x_j} f \right)_{1 \le i,j \le n} . \end{cases}$$

First, we have

$$\begin{aligned} \frac{\partial}{\partial s} F &= \int_{\mathbb{R}^{2n}} f(x + a_A y + \imath z + \imath a_A w) \partial_s p_s(dz) p_t(dw) \\ &= \int_{\mathbb{R}^{2n}} f(x + a_A y + \imath z + \imath a_A w) \frac{1}{2} \Delta_z p_s(dz) \partial_t p_t(dw) \\ &= \frac{1}{2} \int_{\mathbb{R}^{2n}} \Delta_z f(x + a_A y + \imath z + \imath a_A w) p_s(dz) \partial_t p_t(dw) \\ &= -\frac{1}{2} \int_{\mathbb{R}^{2n}} \Delta_x f(x + a_A y + \imath z + \imath a_A w) p_s(dz) \partial_t p_t(dw) \\ &= -\frac{1}{2} \Delta_x F . \end{aligned} \tag{7.1.19}$$

Similarly, we have

$$\frac{\partial}{\partial t} F = -\frac{1}{2} \Delta_y F . \tag{7.1.20}$$

We shall use the following simple facts. If $G = g(x + Ay), x, y \in \mathbb{R}^n$ for a $n \times n$ matrix A, then it is easy to verify

$$\nabla_y G = A^T \nabla_x G, \qquad \mathrm{Hess}_y(G) = A^T \mathrm{Hess}_x(G) A. \tag{7.1.21}$$

If $f : \mathbb{R}^n \to \mathbb{R}$ and $g : \mathbb{R} \to \mathbb{R}$, then

$$\Delta g(f(x)) = g'(f(x))\Delta f(x) + g''(f(x))(\nabla f)^2,$$

where

$$(\nabla f)^2 := \sum_{i=1}^{n} (\partial_{x_i} f)^2.$$

From the expression of F and from the above chain rule, we can express the derivatives with respect to y in terms the derivatives of x.

$$\begin{cases} \nabla_y F = a_A^T \nabla_x F, \\ \mathrm{Hess}_y(F) = a_A^T \mathrm{Hess}_x(F) a_A, \Delta_y(F) = \mathrm{Tr}\left(a_A a_A^T \mathrm{Hess}_x(F)\right). \end{cases}$$

A direct computation yields

$$\begin{cases} \partial_s Q = \bar{F}\partial_s F + F\partial_s \bar{F} = -\frac{1}{2}(\bar{F}\Delta_x F + F\Delta_x \bar{F}) & (7.1.22) \\ \partial_t Q = \bar{F}\partial_t F + F\partial_t \bar{F} = -\frac{1}{2}(\bar{F}\Delta_y F + F\Delta_y \bar{F}) & (7.1.23) \\ \nabla_x Q = \bar{F}\nabla_x F + F\nabla_x \bar{F} =: H & (7.1.24) \\ \Delta_x Q = \bar{F}\Delta_x F + F\Delta_x \bar{F} + 2|\nabla F|^2 & (7.1.25) \\ \nabla_y Q = \bar{F} a_A^T \nabla_x F + F\overline{a_A^T \nabla_x F} = 2\mathrm{Re}(\bar{F} a_A^T \nabla_x F) =: K & (7.1.26) \\ \Delta_y Q = \bar{F}\Delta_y F + F\overline{\Delta_y F} + 2|a_A^T \nabla_x F|^2. & (7.1.27) \end{cases}$$

Now we deal with Ψ. We use the fact that $\partial_t p_t(y) = \frac{1}{2}\Delta_y p_t(y)$ and integration by parts in the following computation.

$$\begin{aligned} \partial_t \Psi &= \int_{\mathbb{R}^n} p_t(dy) \left[\partial_t Q^{q/2}\right] + \int_{\mathbb{R}^n} \partial_t p_t(dy) \left[Q^{q/2}\right] \\ &= P_t \left[\partial_t Q^{q/2} + \frac{1}{2}\Delta_y Q^{q/2}\right] \\ &= P_t \left[\frac{q}{2} Q^{q/2-1}\partial_t Q + \frac{q}{4} Q^{q/2-1}\Delta_y Q + \frac{q(q-2)}{8} Q^{q/2-2}(\nabla Q)^2\right] \\ &= P_t \left[\frac{q}{2} Q^{q/2-1}\partial_t Q + \frac{q}{4} Q^{q/2-1}\Delta_y Q + \frac{q(q-2)}{8} Q^{q/2-2}(\nabla Q)^2\right] \\ &= \frac{q}{2} P_t \left[Q^{q/2-1}|a_A^T \nabla_x F|^2 + \frac{(q-2)}{4} Q^{q/2-2} K^2\right], \end{aligned} \tag{7.1.28}$$

where we recall $K = 2\mathrm{Re}(\bar{F}a_A^T\nabla_x F)$ and where in the last identity, we use (7.1.23) and (7.1.27). Thus, we can compute $\partial_t\Phi$

$$\begin{aligned}\partial_t\Phi &= P_s\left[\partial_t\Psi^{p/q}\right] = P_s\left[\frac{p}{q}\Psi^{p/q-1}\partial_t\Psi\right]\\ &= \frac{p}{2}P_s\left\{\Psi^{p/q-1}P_t\left[Q^{q/2-1}|a_A^T\nabla_x F|^2 + \frac{(q-2)}{4}Q^{q/2-2}K^2\right]\right\}.\end{aligned} \tag{7.1.29}$$

We turn to compute $\partial_s\Phi$. First, we have by (7.1.22), (7.1.24),

$$\partial_s\Psi = P_t\left[\partial_s Q^{q/2}\right] = -\frac{q}{4}P_t\left[Q^{q/2-1}\left(F\overline{\Delta_x F} + \bar{F}\Delta_x F\right)\right]. \tag{7.1.30}$$

$$\nabla_x\Psi = P_t\left[\nabla_x Q^{q/2}\right] = \frac{q}{2}P_t\left[Q^{q/2-1}\nabla_x Q\right] = \frac{q}{2}P_t\left[Q^{q/2-1}H\right]. \tag{7.1.31}$$

$$\begin{aligned}\Delta_x\Psi &= P_t\left[\Delta_x Q^{q/2}\right] = \frac{q}{2}P_t\left[Q^{q/2-1}\Delta Q + \frac{q-2}{2}Q^{q/2-2}(\nabla Q)^2\right]\\ &= \frac{q}{2}P_t\left[Q^{q/2-1}\left[\bar{F}\Delta_x F + F\Delta\bar{F} + 2|\nabla_x F|^2\right] + \frac{q-2}{2}Q^{q/2-2}H^2\right].\end{aligned} \tag{7.1.32}$$

Combining (7.1.30) and (7.1.32) we have

$$\partial_s\Psi + \frac{1}{2}\Delta_x\Psi = \frac{q}{2}P_t\left[Q^{q/2-1}|\nabla_x F|^2 + \frac{q-2}{4}Q^{q/2-2}H^2\right]. \tag{7.1.33}$$

As for (7.1.28), we can use (7.1.31) and (7.1.33) to obtain

$$\begin{aligned}\partial_s\Phi &= P_s\left[\partial_s\Psi^{p/q} + \frac{1}{2}\Delta\Psi^{p/q}\right]\\ &= \frac{p}{q}P_s\left[\Psi^{p/q-1}\partial_s\Psi + \frac{1}{2}\Psi^{p/q-1}\Delta_x\Psi + \frac{1}{2}(\frac{p}{q}-1)\Psi^{p/q-2}(\nabla_x\Psi)^2\right]\\ &= \frac{p}{2}P_s\Bigg[\Psi^{p/q-1}P_t\left[Q^{q/2-1}|\nabla_x F|^2 + \frac{q-2}{4}Q^{q/2-2}H^2\right]\\ &\qquad + \frac{(p-q)}{4}\Psi^{p/q-2}\left(P_t\left[Q^{q/2-1}H\right]\right)^2\Bigg].\end{aligned} \tag{7.1.34}$$

A Hölder inequality shows that

$$\left(P_t\left[Q^{q/2-1}H\right]\right)^2 \le P_t\left[Q^{q/2}\right]P_t\left[Q^{q/2-2}H^2\right] = \Psi P_t\left[Q^{q/2-2}H^2\right].$$

Substituting the above inequality into (7.1.34) and using the fact that $p \leq q$, we have

$$\begin{aligned}\partial_s \Phi \geq \frac{p}{2} P_s \Bigg[\Psi^{p/q-1} P_t \left[Q^{q/2-1} |\nabla_x F|^2 + \frac{q-2}{4} Q^{q/2-2} H^2 \right] \\ + \frac{(p-q)}{4} \Psi^{p/q-2} \Psi P_t \left[Q^{q/2-2} H^2 \right] \Bigg] \\ = \frac{p}{2} P_s \left[\Psi^{p/q-1} P_t \left[Q^{q/2-2} \left(Q|\nabla_x F|^2 + \frac{p-2}{4} H^2 \right) \right] \right]. \quad (7.1.35)\end{aligned}$$

Combining the above inequality with (7.1.29) we have

$$\begin{aligned}\frac{d}{dt}\Phi(1-t,t) \leq \frac{p}{2} P_{1-t} \Bigg\{ \Psi^{p/q-1} P_t \Bigg[Q^{q/2-2} \Bigg(Q|a_A^T \nabla_x F|^2 + \frac{(q-2)}{4} K^2 \\ - Q|\nabla_x F|^2 - \frac{p-2}{4} H^2 \Bigg) \Bigg] \Bigg\} \\ = \frac{p}{2} P_{1-t} \Bigg\{ \Psi^{p/q-1} P_t \Bigg[Q^{q/2-2} \Bigg((|F|^2 + \varepsilon^2) |a_A^T \nabla_x F|^2 + \frac{(q-2)}{4} K^2 \\ - (|F|^2 + \varepsilon^2) |\nabla_x F|^2 - \frac{p-2}{4} H^2 \Bigg) \Bigg] \Bigg\} \\ \leq \frac{p}{2} P_{1-t} \Bigg\{ \Psi^{p/q-1} P_t \Bigg[Q^{q/2-2} \Bigg(|F|^2 |a_A^T \nabla_x F|^2 + \frac{(q-2)}{4} K^2 \\ - |F|^2 |\nabla_x F|^2 - \frac{p-2}{4} H^2 \Bigg) \Bigg] \Bigg\}, \quad (7.1.36)\end{aligned}$$

where we use the definition $Q = |F|^2 + \varepsilon^2$ and A is a bounded operator of norm ≤ 1, which implies that $|a_A \xi| \leq |\xi|$ for any n dimensional vector ξ. Now we denote $\zeta = A\nabla_x F$. Then, $H = 2\mathrm{Re}(\zeta)$ and $K = 2\mathrm{Re}(a_A^T \zeta)$. The inequality (7.1.36) can be written as

$$\begin{aligned}\frac{d}{dt}\Phi(1-t,t) \leq \frac{p}{2} P_{1-t} \Bigg\{ \Psi^{p/q-1} P_t \Bigg(Q^{q/2-2} \Bigg[|a_A^T \zeta|^2 + (q-2)(\mathrm{Re}(a_A^T \zeta))^2 \\ - |\zeta|^2 - (p-2)(\mathrm{Re}(\zeta))^2 \Bigg] \Bigg) \Bigg\}. \quad (7.1.37)\end{aligned}$$

The condition (7.1.15) implies that the expression inside the above square bracket [ ] is negative. This means $\frac{d}{dt}\Phi(1-t,t) \leq 0$. Thus, $\Phi(1-t,t)$ is non-increasing, which implies (7.1.18). The theorem is then proved. □

Remark 7.4. We can consider a few special cases of the theorem.

(i) If $\mathbb{H}_1 = \mathbb{H}_2$ and $A = \alpha I$ (multiplication by α) for some real number α with $|\alpha| \le 1$, then it is easy to see that condition (7.1.14) becomes

$$\alpha^2 \le (p-1)/(q-1)\,. \tag{7.1.38}$$

This is the usual hypercontractivity.

(ii) If $\mathbb{H}_1 = \mathbb{H}_2$ and $A = \imath\alpha$, then condition (7.1.14) becomes

$$\alpha^2 \le (p-1) \quad \text{and} \quad \alpha^2 \le \frac{1}{q-1}\,. \tag{7.1.39}$$

In the case both equalities holds $\alpha^2 \le (p-1)$ and $\alpha^2 \le \frac{1}{q-1}$, then $\alpha = \sqrt{p-1}$ and $q = p'$ (p' denotes the conjugate number of p, namely, $1/p + 1/p' = 1$). This is a result of Beckner [Beckner (1975)]. See related work [Epperson (1989); Weissler (1979)].

(iii) The original proof is in [Janson (1997)] who use the stochastic analysis, in particular the Itô formula. The current proof without the use of Itô calculus is inspired by the work in [Driver and Hu (1996); Hu (1997, 2000)].

A direct consequence of the hypercontractivity is

Proposition 7.4. *If $F = \sum_{k=0}^n I_k(f_k)$ is linear combination of finitely many multiple Wiener-Itô integrals, then for any $1 < p, q < \infty$, there is a constant $c_{p,q,n}$ such that*

$$\|F\|_q \le c_{p,q,n} \|F\|_p\,. \tag{7.1.40}$$

Proof. First, we prove (7.1.40) for $p = 2$. This is true when $q \le 2$. Suppose $q > 2$. Denote $\alpha = \sqrt{\frac{1}{q-1}} \in (0,1)$ and $F_\alpha = \sum_{k=0}^n (\alpha)^{-k} I_k(f_k)$. Then

$$\begin{aligned} \|F\|_q &= \|\Gamma(\alpha)F_\alpha\|_q \le \|F_\alpha\|_p \le \|F_\alpha\|_2 \\ &= \left(\sum_{k=0}^n (\alpha)^{-2k} \|I_k(f_k)\|_2^2\right)^{1/2} \le \alpha^{-n}\|F\|_2\,. \end{aligned} \tag{7.1.41}$$

On the other hand, we can prove $\|F\|_2 \le c_{p,n}\|F\|_p$ for any $p > 1$. Clearly, this is true for $p \ge 2$. When $p \in (1,2)$, we can take $\alpha = \sqrt{p-1} \in (0,1)$ and $F_\alpha = \sum_{k=0}^n (\alpha)^{-k} I_k(f_k)$ and follow the same argument as for (7.1.41). Now for any $1 < p, q < \infty$, we have

$$\|F\|_q \le c_{2,q,n}\|F\|_2 \le c_{2,q,n} c_{p,n} \|F\|_p\,,$$

proving the proposition. □

7.2 Meyer's Inequality

Let (X, μ) be a measurable space. We define $L^p(X) = L^p(X, \mu)$ as the set of all measurable functions from X to $\mathbb{R}$ such that $\int_X |f(x)|^p \mu(dx) < \infty$ and we denote $\|f\|_p = \left(\int_X |f(x)|^p \mu(dx)\right)^{1/p}$.

The following is the famous Riesz-Thorin interpolation theorem. Its proof can be found in many standard text book. See for example [Grafakos (2008)].

Theorem 7.3. *Let (X, μ) and (Y, ν) be two measure spaces. Let T be a linear operator defined on the set of all simple functions on X and taking values in the set of measurable functions on Y. Let $1 \le p_0, p_1, q_0, q_1 \le \infty$ be given positive numbers and let p and q satisfying*

$$\frac{1}{p} = \frac{\theta}{p_0} + \frac{1-\theta}{p_1}, \qquad \frac{1}{q} = \frac{\theta}{q_0} + \frac{1-\theta}{q_1}$$

for some $\theta \in [0, 1]$. If

$$\|TF\|_{q_0} \le M_0 \|F\|_{p_0}, \qquad \|TF\|_{q_1} \le M_1 \|F\|_{p_1}$$

for all simple function F on X, then T can be extended to a bounded linear operator from $L^p(X)$ to $L^q(Y)$ and

$$\|TF\|_q \le M_0^{\theta} M_1^{1-\theta} \|F\|_p, \quad \forall\, F \in L^p(X). \tag{7.2.1}$$

The Hilbert transform of a function f is a very important concept in analysis. If $f : \mathbb{R} \to \mathbb{R}$, then the Hilbert transform of f is defined as

$$Hf(t) = \int_{\mathbb{R}} \frac{f(s)}{t-s} ds = \lim_{\varepsilon \to 0} \int_{|t-s| > \varepsilon} \frac{f(s)}{t-s} ds, \quad t \in \mathbb{R} \tag{7.2.2}$$

The Fourier symbol of the transform is

$$\widehat{Hf}(\xi) = -\imath \,\mathrm{sgn}(\xi) \hat{f}(\xi).$$

From here we see that

$$\|Hf\|_{L^2(\mathbb{R}, dx)} = \|f\|_{L^2(\mathbb{R}, dx)}.$$

We have the following theorem.

Theorem 7.4. *For any $1 < p < \infty$, there is a constant C_p such that*

$$\|Hf\|_{L^p(\mathbb{R}, dx)} \le C_p \|f\|_{L^p(\mathbb{R}, dx)}, \quad \forall f \in L^p(\mathbb{R}, dx). \tag{7.2.3}$$

Proof. We follow [Grafakos (2008)]. First, we prove the following identity

$$H(f)^2 = f^2 + 2H(fH(f))\,, \quad \forall \text{ real valued Schwartz function } f\,. \tag{7.2.4}$$

We shall use Fourier transformation to prove the above identity. Denote $m(\xi) = -\imath\,\mathrm{sgn}(\xi)$. Then, it is easy to check

$$1 + m(\xi)\,[m(\eta) + m(\xi - \eta)] = m(\eta)m(\xi - \eta)\,. \tag{7.2.5}$$

Now we have

$$\begin{aligned}\widehat{f^2}(\xi) + 2[H(fH(f))]^\wedge(\xi) &= \hat{f} * \hat{f}(\xi) + 2m(\xi)\left\{\hat{f} * \widehat{Hf}\right\}(\xi)\\ &= \int \hat{f}(\eta)\hat{f}(\xi-\eta)d\eta + 2m(\xi)\int \hat{f}(\eta)m(\xi-\eta)\hat{f}(\xi-\eta)d\eta\\ &= \int \hat{f}(\eta)\hat{f}(\xi-\eta)d\eta + 2m(\xi)\int \hat{f}(\eta)m(\eta)\hat{f}(\xi-\eta)d\eta\,.\end{aligned}$$

Averaging the above last two expressions, we have

$$\begin{aligned}&\widehat{f^2}(\xi) + 2[H(fH(f))]^\wedge(\xi)\\ &\qquad= \int \hat{f}(\eta)\hat{f}(\xi-\eta)\left\{1 + m(\xi)\,[m(\eta) + m(\xi-\eta)]\right\}d\eta\,.\end{aligned}$$

On the other hand,

$$\left[H(f)^2\right]^\wedge(\xi) = \int \hat{f}(\eta)f(\xi-\eta)m(\eta)m(\xi-\eta)d\eta\,.$$

Noticing (7.2.5), we see that the Fourier transform of both sides of (7.2.4) are the same and hence the identity is proved.

Now we can prove the inequality (7.2.3) for $p = 2^k$ by induction on k. It is obvious that the inequality is true (in fact we have identity now) when $k = 1$ (from the Parseval theorem). Assume the equality holds for $p = 2^k$. We want to show the inequality for $2p = 2^{k+1}$. Using (7.2.4), we have

$$\begin{aligned}\|Hf\|_{2p} = \|(Hf)^2\|_p^{1/2} &= \|f^2 + H(fHf)\|_p^{1/2}\\ &\le C_p\left(\|f^2\|_p^{1/2} + \|H(fHf)\|_p^{1/2}\right)\\ &\le C_p\|f\|_{2p} + C_p\|fHf\|_p^{1/2}\\ &\le C_p\|f\|_{2p} + C_p\|f\|_{2p}\|Hf\|_{2p}\,,\end{aligned}$$

where the second inequality follows from the induction assumption and the last inequality follows from Cauchy-Schwarz inequality. Now we use $ab \le \varepsilon a^2 + \frac{1}{\varepsilon}b^2$ for any $\varepsilon > 0$ and we have

$$\|Hf\|_{2p} \le C_p\|f\|_{2p} + 2C_p^2\|f\|_{2p} + \frac{1}{2}\|Hf\|_{2p}\,.$$

This completes the induction. The theorem is then proved for all $p = 2^k$. The general case $p \ge 2$ is deduced from the Riesz-Thorin theorem 7.3. The case $p \le 2$ can be proved by duality argument. □

Let us recall the use of derivative operator D, divergence operator δ and Ornstein-Uhlenbeck operator $L = -\delta D$.

Theorem 7.5 (Meyer's inequalities). *Let $p > 1$. There exist two positive constants c_p and C_p such that*

$$c_p \|DF\|_p \le \|(-L)^{1/2} F\|_p \le C_p \|DF\|_p \,, \quad \forall\ F \in \mathbb{D}_{1,p}\,. \tag{7.2.6}$$

Proof. It suffices to show that for $F \in \mathcal{P}$ with $\mathbb{E}(F) = 0$, the above inequalities hold. The following proof is based on a proof originally given in [Pisier (1988)]. We shall divide the proof into several steps.

Step 1 We introduce another probability space $(\mathbb{B}', \mathbb{H}', \mathbb{P}')$ which is an identical copy of $(\mathbb{B}, \mathbb{H}, \mathbb{P})$. For any function of the form

$$F(x) = f(\langle x, e_1\rangle, \cdots, \langle x, e_n\rangle) \in \mathcal{P}\,,$$

we define

$$\begin{aligned} R_\theta F(x, x') &= F(x\cos\theta + x'\sin\theta) \\ &= f(\langle x, e_1\rangle \cos\theta + \langle x', e_1\rangle \sin\theta\,, \cdots, \langle x, e_n\rangle \cos\theta + \langle x', e_n\rangle \sin\theta)\,. \end{aligned}$$

Denote

$$\phi(\theta) = \frac{1}{\sqrt{2\pi |\log\cos^2\theta|}} \operatorname{sign}(\theta)\,, \quad \theta \in (-\pi/2, 0) \cup (0, \pi/2)\,.$$

We claim

$$D(-L)^{-1/2} F = \int_{-\frac{\pi}{2}}^{\frac{\pi}{2}} \mathbb{E}'(D' R_\theta F) \phi(\theta) d\theta\,, \tag{7.2.7}$$

where D' denotes the derivative on the random element x' and $\mathbb{E}'$ denotes the expectation with respect to x'.

In fact, we have from the chain rule of D,

$$\mathbb{E}' D' R_\theta F(x, x') = \sin\theta\, \mathbb{E}' R_\theta (DF)(x, x') = \sin\theta\, \Gamma(\cos\theta)(DF)\,.$$

Thus

$$\begin{aligned} \int_{-\frac{\pi}{2}}^{\frac{\pi}{2}} \mathbb{E}'(D' R_\theta F)\phi(\theta) d\theta &= \int_{-\frac{\pi}{2}}^{\frac{\pi}{2}} \sin\theta\, \Gamma(\cos\theta)(DF)\phi(\theta) d\theta \\ &= \sum_{n=0}^{\infty} \int_{-\frac{\pi}{2}}^{\frac{\pi}{2}} \sin\theta\, \Gamma(\cos\theta)(J_n(DF))\phi(\theta) d\theta \\ &= \sum_{n=0}^{\infty} a_n\, J_n(DF)\,, \end{aligned}$$

where

$$\begin{aligned} a_n &= \int_{-\frac{\pi}{2}}^{\frac{\pi}{2}} \sin\theta \, \cos^n\theta \phi(\theta) d\theta \\ &= 2\int_0^{\frac{\pi}{2}} \sin\theta \, \cos^n\theta \frac{1}{\sqrt{4\pi|\log\cos\theta|}} d\theta \,. \end{aligned}$$

Making the substitution $t = |\log\cos\theta| = -\log\cos\theta$, we have

$$\begin{aligned} a_n &= \frac{1}{\sqrt{\pi}} \int_0^\infty e^{-(n+1)t} t^{-1/2} dt \\ &= \frac{1}{\sqrt{n+1}} \frac{1}{\sqrt{\pi}} \int_0^\infty e^{-s} s^{-1/2} ds = \frac{1}{\sqrt{n+1}} \,. \end{aligned}$$

Thus

$$\int_{-\frac{\pi}{2}}^{\frac{\pi}{2}} \mathbb{E}'(D'R_\theta F)\phi(\theta)d\theta = \sum_{n=0}^\infty \frac{1}{\sqrt{n+1}} J_n(DF)$$

$$= \sum_{n=0}^\infty \frac{1}{\sqrt{n+1}} DJ_{n+1}F = \sum_{n=0}^\infty D(-L)^{-1/2} J_{n+1}F = D(-L)^{-1/2}F \,.$$

This proves (7.2.7).

Step 2 We claim that

$$\|D(-L)^{-1/2}F\|_p \le C_p \|F\|_p \,, \quad \forall F \in \mathcal{P} \,. \tag{7.2.8}$$

In fact, from the identity (7.2.7) and using

$$\int_{-\frac{\pi}{2}}^{0} \mathbb{E}'(D'R_{-\theta}F)\phi(\theta)d\theta = -\int_0^{\frac{\pi}{2}} \mathbb{E}'(D'R_{-\theta}F)\phi(\theta)d\theta \,,$$

we have

$$\begin{aligned} &\| |D(-L)^{-1/2}F|_{\mathbb{H}} \|_p^p \\ &= \| | \int_0^{\frac{\pi}{2}} \mathbb{E}'(D'R_\theta F)\phi(\theta)d\theta - \int_0^{\frac{\pi}{2}} \mathbb{E}'(D'R_{-\theta}F)\phi(\theta)d\theta |_{\mathbb{H}} \|_p^p \\ &= \mathbb{E} \left| \mathbb{E}'D' \left(\int_0^{\frac{\pi}{2}} [R_\theta - R_{-\theta}] F\phi(\theta)d\theta \right) \right|_{\mathbb{H}}^p \\ &\le C_p \mathbb{E} \left| \int_0^{\frac{\pi}{2}} [R_\theta F - R_{-\theta}F] \phi(\theta)d\theta \right|^p \,. \end{aligned}$$

Let us make rotation transformation on $\mathbb{B} \times \mathbb{B}'$:

$$(x, x') \to R_\xi(x, x') = (x\cos\xi + x'\sin\xi, -x\sin\xi + x'\cos\xi) \,,$$

which is obviously a measure preserving transformation with respect to the product measure $(\mathbb{B}\times\mathbb{B}', \mathbb{P}\times\mathbb{P}')$. We also define $R_\xi F(x,x') = F(R_\xi(x,x'))$. Thus, we have for any $\xi \in (-\pi/2, \pi/2)$,

$$\| |D(-L)^{-1/2}F|_H \|_p^p \le C_p \mathbb{E} \left| \int_0^{\frac{\pi}{2}} \left[R_{\xi+\theta}F - R_{\xi-\theta}F\right] \phi(\theta) d\theta \right|^p .$$

Integration on ξ yields (noticing that we use C_p to denote a generic constant which may be different in different appearances)

$$\| |D(-L)^{-1/2}F|_H \|_p^p \le C_p \mathbb{E} \int_{-\frac{\pi}{2}}^{\frac{\pi}{2}} \left| \int_0^{\frac{\pi}{2}} \left[R_{\xi+\theta}F - R_{\xi-\theta}F\right] \phi(\theta) d\theta \right|^p d\xi .$$

It is easy to see that there is a continuous and bounded function $\psi(\theta)$ on $(-\pi/2, \pi/2)$ such that $\phi(\theta) = \frac{1}{\sqrt{2\pi}\theta} + \psi(\theta)$. Hence, we have

$$\| |D(-L)^{-1/2}F|_H \|_p^p \le C_p \left[I_1 + I_2\right] , \tag{7.2.9}$$

where

$$I_1 = \mathbb{E} \int_{-\frac{\pi}{2}}^{\frac{\pi}{2}} \left| \int_0^{\frac{\pi}{2}} \frac{\left[R_{\xi+\theta}F - R_{\xi-\theta}F\right]}{\theta} d\theta \right|^p d\xi$$

and

$$I_2 = \mathbb{E} \int_{-\frac{\pi}{2}}^{\frac{\pi}{2}} \left| \int_0^{\frac{\pi}{2}} \left[R_{\xi+\theta}F - R_{\xi-\theta}F\right] \psi(\theta) d\theta \right|^p d\xi .$$

Obviously, we have $I_2 \le C_p \|F\|_P^p$.

Now

$$\begin{aligned} I_1 &\le \mathbb{E} \int_{-\frac{\pi}{2}}^{\frac{\pi}{2}} \left| \int_0^{\frac{\pi}{2}} \frac{\left[R_{\xi+\theta}F - R_{\xi-\theta}F\right]}{\theta} I_{\{|\xi+\theta|\le\pi\}} d\theta \right|^p d\xi \\ &\le \mathbb{E} \int_{\mathbb{R}} \left| \int_{\mathbb{R}} \frac{R_{\xi+\theta}F}{\theta} I_{\{|\xi+\theta|\le\pi\}} d\theta \right|^p d\xi \\ &\qquad + \mathbb{E} \int_{-\frac{\pi}{2}}^{\frac{\pi}{2}} \left| \int_{[-\pi/2,\pi/2]^c} \frac{R_{\xi+\theta}F}{\theta} I_{\{|\xi+\theta|\le\pi\}} d\theta \right|^p d\xi \\ &=: I_{11} + I_{12} . \end{aligned}$$

Since $|\theta| > \pi/2$ and $\|R_{\xi+\theta}F\|_p = \|F\|_p$, it is easy to see that $I_{12} \le C_p \|F\|_p^p$. Using the inequality (7.2.3) for Hilbert transform, we have

$$I_{11} \le \mathbb{E} \int_{\mathbb{R}} \left| R_\xi F I_{\{|\xi|\le\pi\}} \right|^p d\xi = \int_{-\pi}^{\pi} \mathbb{E} |R_\xi F|^p d\xi \le C_p \|F\|_p^p .$$

Combining the estimates for I_{11}, I_{12} and I_2 and noticing (7.2.9), we have (7.2.8). □

From Meyer's inequality, we see that the Sobolev space $\mathbb{D}_{k,p}$ can be described in the following way through chaos expansion.

Proposition 7.5. *Denote by $\mathcal{C}_n$ the n-th chaos of $L^2(\mathbb{B})$. For any $k \in \mathbb{R}$ and $p \in (1, \infty)$ we have*

$$\mathbb{D}_{k,p} = \left\{ F\,; F = \sum_{k=0}^{\infty} F_n\,, F_n \in \mathcal{C}_n \ \ and \ \sum_{n=0}^{\infty} (n+1)^{k/2} F_n \in L^p(\mathbb{B}) \right\}. \tag{7.2.10}$$

In paricular, when $p = 2$, we have

$$\mathbb{D}_{k,2} = \left\{ F\,; F = \sum_{k=0}^{\infty} F_n\,, F_n \in \mathcal{C}_n \ and \ \sum_{n=0}^{\infty} (n+1)^{k/2} \|F_n\|^2_{L^2(\mathbb{B})} < \infty \right\}. \tag{7.2.11}$$

An immediate consequence of Meyer's inequality is the following.

Theorem 7.6. *Let $\{F_n, n = 1, 2, \cdots\} \subseteq \mathbb{D}_{s,p}$ for some $s > 0$ and $p > 1$. If F_n converges to F in $L^p(\mathbb{B})$ and $\sup_{n\ge 1} \|F_n\|_{s,p} < \infty$, then $F \in \mathbb{D}_{s,p}$.*

Proof. Let q be the conjugate number of p. By Meyer's inequality, the assumption means that $\sup_{n\ge1} \|(I - L)^{-s/2} F_n\|_p$ is a bounded sequence. Thus, there is subsequence $(I - L)^{-s/2} F_{i_n}$ convergent weakly in $\sigma(L^p, L^q)$ to an element $G \in L^p$. For any $Y \in \mathcal{E}$, we have

$$\begin{aligned} \mathbb{E}(F(I-L)^{s/2}Y) &= \lim_{n\to\infty} \mathbb{E}\left[F_{i_n}(I-L)^{s/2}Y)\right] \\ &= \lim_{n\to\infty} \mathbb{E}\left[(I-L)^{s/2}F_{i_n}Y)\right] = \mathbb{E}(GY)\,. \end{aligned}$$

This means $G = (I - L)^{s/2} F$. Namely, $F \in \mathbb{D}_{s,p}$. □

7.3 Multiplier Theorem

Let $(\mathbb{B}, \mathbb{H}, \mathbb{P})$ be an abstract Wiener space. From the chaos expansion theorem, we see that

$$F = \sum_{k=0}^{\infty} F_k\,.$$

F_k is called the k-th chaos of F. We define $J_k F = F_k$ to be the projection of F to its k-th chaos. Then, we have

Theorem 7.7. *For any $k \ge 0$, $p > 1$ and $q > 1$, J_k can be extended to a bounded mapping from L^p to L^q. Namely, any $k \ge 0$, $p > 1$ and $q > 1$ there is a constant $C_{p,q,k}$ such that*

$$\|J_k F\|_q \le C_{p,q,k} \|F\|_p\,, \quad \forall F \in L_p\,. \tag{7.3.1}$$

Proof. Since $L_p \subseteq L_q$ for $p \le q$, it is clear that we can assume that $1 < p < 2 < q < \infty$. Let $\alpha = \frac{1}{\sqrt{q-1}}$ and $\beta = \sqrt{p-1}$. Using $J_kF = (q-1)^k\Gamma(\alpha)J_kF = \frac{1}{(p-1)^{k/2}}\Gamma(\beta)J_kF$ and the hypercontractivity, we obtain

$$\begin{aligned}\|J_kF\|_q &= (q-1)^{\frac{k}{2}}\|\Gamma(\alpha)J_kF\|_q \le (q-1)^{\frac{k}{2}}\|J_kF\|_2\\ &= \left(\frac{q-1}{p-1}\right)^{\frac{k}{2}}\|\Gamma(\beta)J_kF\|_2 \le \left(\frac{q-1}{p-1}\right)^{\frac{k}{2}}\|\Gamma(\beta)F\|_2\\ &\le \left(\frac{q-1}{p-1}\right)^{\frac{k}{2}}\|F\|_p\,.\end{aligned}$$

This proves the theorem. □

We also define

$$\mathbb{J}_n = J_0 + J_1 + \cdots + J_n\,.$$

Theorem 7.8. *Fix any $n \ge 0$. Then, all the L^p norms of $\mathbb{J}_nF$ are equivalent. That means that for any $p > 1$ and $q > 1$ there are two positive constants $c_{p,q,n}$ and $C_{p,q,n}$ such that*

$$c_{p,q,n}\|\mathbb{J}_nF\|_p \le \|\mathbb{J}_nF\|_q \le C_{p,q,n}\|\mathbb{J}_nF\|_p\,. \tag{7.3.2}$$

Proof. We only need to show that $\|\mathbb{J}_nF\|_q \le C_{p,q,n}\|\mathbb{J}_nF\|_p$ for $1 < p < 2 < q < \infty$. From Theorem 7.7, we see that there is a constant $C_{q,n}$ such that

$$\|\mathbb{J}_nF\|_q \le \sum_{k=0}^n \|J_kF\|_q \le \sum_{k=0}^n C_{q,k}\|F\|_p \le C_{q,n}\|\mathbb{J}_nF\|_2\,.$$

Let $\alpha = 1/\sqrt{q-1}$ and $\beta = \sqrt{p-1}$ and let $G = \sum_{k=0}^n \frac{1}{\alpha^k}J_kF$. Then

$$\begin{aligned}\|\mathbb{J}_nF\|_q &= \|\Gamma(\alpha)G\|_q \le \|G\|_2 = \sum_{k=0}^n \frac{1}{\alpha^k}\|J_kF\|_2\\ &\le \frac{1}{\alpha^n}\sum_{k=0}^n \|J_kF\|_2 = (q-1)^{n/2}\|\mathbb{J}_nF\|_2\\ &= \sum_{k=0}^n \frac{(q-1)^{n/2}}{(p-1)^{k/2}}\|\Gamma(\beta)J_kF\|_2 \le \frac{(q-1)^{n/2}}{(p-1)^{n/2}}\sum_{k=0}^n \|\Gamma(\beta)J_kF\|_2\\ &\le \frac{(q-1)^{n/2}}{(p-1)^{n/2}}\|\Gamma(\beta)F\|_2 \le \frac{(q-1)^{n/2}}{(p-1)^{n/2}}\|F\|_p\,.\end{aligned}$$

This proves the theorem. □

Definition 7.2. A smooth function from $(0,\infty) \to [0,\infty)$ is called a completely monotone function if it is infinitely differentiable and if $(-1)^k \frac{d^k}{d\lambda^k} f(\lambda) \ge 0$ for all $\lambda > 0$ and $k = 1, 2, \cdots$.

Definition 7.3. A smooth function $f : (0,\infty) \to [0,\infty)$ is called a Bernstein function if $(-1)^k \frac{d^k}{d\lambda^k} f(\lambda) \le 0$ for all $\lambda > 0$ and $k = 1, 2, \cdots$.

It is well-known that if $\psi(x)$ is a Bernstein function, then for any $t > 0$, $f(x) = e^{-t\psi(x)}$ is a completely monotone function.

Theorem 7.9 (Bernstein). *Let $f : (0,\infty) \to \mathbb{R}$ be a measurable function. Then $f(\lambda)$ is completely monotone iff there is a positive measure μ with $\mu(0,\infty) < \infty$ such that*

$$f(\lambda) = \int_0^\infty e^{-\lambda s} \mu(ds) .$$

The proof of this theorem is referred to [Schilling, Song and Vondracek (2012); Widder (1941)].

Example 7.3. It is easy to see that for any $t > 0$ and $0 \le \alpha \le 1$, $\psi_t(\lambda) = t\lambda^\alpha$, $\lambda > 0$, is a Bernstein function. Usually it is hard to find the explicit measure μ even in the case $\psi_t(\lambda) = t\lambda^\alpha$. But when $\psi(\lambda) = t\sqrt{\lambda}$ we have

$$\mu_t(ds) = \frac{t}{2\sqrt{\pi}} e^{-t^2/4s} s^{-3/2} ds . \tag{7.3.3}$$

Namely,

$$\int_0^\infty e^{-\lambda s} \mu_t(ds) = e^{-t\sqrt{\lambda}} . \tag{7.3.4}$$

Theorem 7.10 (Multiplier theorem). *Let $h(x) = \sum_{k=0}^\infty a_k x^k$ be an analytic function in a neighborhood of 0. Namely, there is a $r_0 \in (0,\infty)$ such that $\sum_{k=0}^\infty |a_k| r_0^k < \infty$. Suppose $\psi : (0,\infty) \to [0,\infty)$ is a Bernstein function and $\lim_{\lambda\to\infty} \psi(\lambda) = \infty$. Assume that there is n_0 such that $\phi(n) = h(\frac{1}{\psi(n)})$ for all $n \ge n_0$. Define*

$$T_\phi F = \sum_{n=0}^\infty \phi(n) J_n F , \quad \forall\ F \in \mathcal{P} . \tag{7.3.5}$$

Then, for any $p > 1$, there is a constant c_p such that

$$\|T_\phi F\|_p \le c_p \|F\|_p , \quad \forall\ F \in \mathcal{P} . \tag{7.3.6}$$

Consequently, T_ϕ can be extended to L_p satisfying the above inequality.

Proof. Since h is analytic in a neighborhood of 0 and $\lim_{\lambda\to\infty}\psi(\lambda)=\infty$, there is an n_0' such that

$$\sum_{k=0}^{\infty}|a_k|\left(\frac{1}{\psi(n_0')}\right)^k<\infty\,. \tag{7.3.7}$$

We can choose n_0 such that (7.3.7) holds and $\phi(n)=h(\frac{1}{\psi(n)})$ for all $n\ge n_0$. Write

$$T_\phi F=\sum_{k=0}^{n_0-1}\phi(n)J_kF+\sum_{n_0}^{\infty}\phi(n)J_kF=:T_\phi^{(1)}F+T_\phi^{(2)}F\,.$$

From Theorem 7.8 we see that

$$\|T_\phi^{(1)}F\|_p\le c_p\|F\|_p\,.$$

Step 1 We claim that for any $\beta\in(0,1)$

$$\|\Gamma(\beta)(I-\mathbb{J}_{n_0-1})F\|_p\le C_{p,n_0}\beta^{n_0}\|F\|_p\,,$$

where C_{p,n_0} does not depend on β. The case $p=2$ is obvious. We prove this claim by considering two cases: $p>2$ and $p<2$. If $p>2$, then from the hypercontractivity theorem with $\beta_0=1/\sqrt{p-1}$, we have

$$\begin{aligned}\|\Gamma(\beta)(I-\mathbb{J}_{n_0-1})F\|_p&=\|\Gamma(\beta_0)\Gamma(\frac{\beta}{\beta_0})\ (I-\mathbb{J}_{n_0-1})F\|_p\\ \le\|\Gamma(\frac{\beta}{\beta_0})\ (I-\mathbb{J}_{n_0-1})F\|_2&\le\frac{\beta^{n_0}}{\beta_0^{n_0}}\|(I-\mathbb{J}_{n_0-1})F\|_2\le C_{p,n_0}\beta^{n_0}\|F\|_p\,.\end{aligned}$$

If $1<p<2$, then the claim follows by a duality argument since $I-\mathbb{J}_{n_0-1}$ is adjoint.

Step 2 Let $\mu_t^\psi(ds)$ be the positive measure on $[0,\infty)$ whose Laplace transform is $e^{-\psi(\lambda)t}$:

$$\int_0^\infty e^{-\lambda s}\mu_t^\psi(ds)=e^{-\psi(\lambda)t}\,.$$

Define $T_s=\Gamma(e^{-s})$ and

$$Q_t^\psi:=\sum_{n=0}^{\infty}e^{-\psi(n)t}J_n=\int_0^\infty T_s\mu_t^\psi(ds)\,.$$

Then, $\{Q_t^\psi,t\ge0)$ is a semigroup. From the first step, we have

$$\begin{aligned}\|Q_t^\psi(I-\mathbb{J}_{n_0-1})F\|_p&\le\int_0^\infty\|T_s(I-\mathbb{J}_{n_0-1})F\|_p\mu_t^\psi(ds)\\&\le C_{p,n_0}\int_0^\infty\|F\|_pe^{-n_0s}\mu_t^\psi(ds)=C_{p,n_0}e^{-\psi(n_0)t}\|F\|_p\,.\end{aligned} \tag{7.3.8}$$

Step 3 Define

$$R_{n_0} = \int_0^\infty Q_t^\psi (I - \mathbb{J}_{n_0-1}) dt\,.$$

Then, from equation (7.3.8), we have

$$\|R_{n_0} F\|_p = \int_0^\infty \|Q_t^\psi (I - \mathbb{J}_{n_0-1}) F\|_p dt \le \frac{C}{\psi(n_0)} \|F\|_p\,.$$

It is easy to see that

$$\begin{aligned} R_{n_0}^2 &= \int_0^\infty \int_0^\infty Q_t^\psi (I - \mathbb{J}_{n_0-1}) Q_s (I - \mathbb{J}_{n_0}) dt ds \\ &= \int_0^\infty \int_0^\infty Q_{t+s}^\psi (I - \mathbb{J}_{n_0-1}) dt ds\,. \end{aligned}$$

This can be used to show easily

$$\|R_{n_0}^2 F\|_p \le \frac{C}{\psi(n_0)^2} \|F\|_p\,.$$

In general we have

$$\|R_{n_0}^k F\|_p \le \frac{C}{\psi(n_0)^k} \|F\|_p\,, \quad k = 1, 2, \cdots \tag{7.3.9}$$

On the other hand, we have if $n \ge n_0$,

$$R_{n_0} J_n F = \int_0^\infty Q_t^\psi J_n F dt = \int_0^\infty e^{-\psi(n)t} J_n F dt = \frac{1}{\psi(n)} J_n F\,.$$

We also have

$$R_{n_0}^k J_n F = \frac{1}{\psi(n)^k} J_n F\,.$$

Thus

$$T_\phi^{(2)} F = \sum_{n=n_0}^\infty \sum_{k=0}^\infty a_k R_{n_0}^k J_n F = \sum_{k=0}^\infty a_k R_{n_0}^k F\,.$$

Therefore by (7.3.9), we have

$$\|T_\phi^{(2)} F\|_p \le \sum_{k=0}^\infty |a_k| \|R_{n_0}^k F\|_p \le C \sum_{k=0}^\infty |a_k| \frac{1}{\psi(n_0)^k} \|F\|_p$$

which yields the result of (7.3.7). □

For any $0 \le \alpha \le 1$, $\psi(\lambda) = \lambda^\alpha$ is a Bernstein function. Then, we have the following multiplier theorem (due to [Meyer (1982a)]).

Corollary 7.1. *Let $h(x) = \sum_{k=0}^\infty a_k x^k$ be an analytic function in a neighborhood of 0. Namely, there is a $r_0 \in (0,\infty)$ such that $\sum_{k=0}^\infty |a_k| r_0^k < \infty$. Let $\phi(n) = h(\frac{1}{n^\alpha})$ for all $n \ge n_0$ for some n_0 and for a certain $\alpha \in (0,1)$. Define*

$$T_\phi F = \sum_{n=0}^\infty \phi(n) J_n F\,, \quad \forall\ F \in \mathcal{P}\,. \tag{7.3.10}$$

Then, for any $p > 1$, there is a constant c_p such that

$$\|T_\phi F\|_p \le c_p \|F\|_p\,, \quad \forall\ F \in \mathcal{P}\,. \tag{7.3.11}$$

As a consequence, T_ϕ can be extended to L_p satisfying the above inequality.

Example 7.4. For any $\alpha > 0$, choose $h(x) = \left(\frac{x}{1+x}\right)^\alpha$. Then, $\phi(n) = h(\frac{1}{n}) = (1+n)^{-\alpha}$.

For any real number $k > 0$ the function $h(x) = \sqrt{1-kx}$, $|x| < \frac{1}{k}$ is an analytic function in a neighborhood of 0. Thus, $\phi(n) = h(\frac{1}{n}) = \sqrt{1 - \frac{k}{n}}$. Thus, we have

Corollary 7.2. *Let $R_k = \sum_{n \ge k} \sqrt{1 - \frac{k}{n}} J_n$. Then, for any $1 < p < \infty$, there is a positive constant $C_{p,k}$ such that*

$$\|R_k F\|_p \le C_{p,k} \|F\|_p\,. \tag{7.3.12}$$

7.4 Littlewood-Paley-Stein-Meyer Theory

The following two lemmas are straightforward and are useful later.

Lemma 7.3. *For any $F \in \mathbb{D}_\infty$, we have*

$$T_t DF = e^t D T_t F$$

$$LDF = F(L+1)F\,.$$

Lemma 7.4. *For any convex function $\psi : \mathbb{R} \to \mathbb{R}$, we have*

$$\phi(\Gamma(\alpha)F) \le \Gamma(\alpha)\phi(F)\,, \quad \forall\ F \in L^{1+}\,. \tag{7.4.1}$$

If E is a Banach space with norm $|||\cdot|||$ and if $F : \mathbb{B} \to E$ is such that $\mathbb{E}|||F|||^p < \infty$ for some $p > 1$, then we have

$$|||\Gamma(\alpha)F||| \le \Gamma(\alpha)|||F|||\,. \tag{7.4.2}$$

Let $\{B_t, t \geq 0\}$ be a one dimensional standard Brownian motion starting at 0 and denote by τ_a the hitting time of the Brownian motion to a point $a \in \mathbb{R}$:

$$\tau_a = \inf\{t > 0\,,\ B_t = a\}\,, \quad \forall a \in \mathbb{R} \quad \text{and } a \neq 0\,; \tag{7.4.3}$$

$$\tilde{\tau}_a = \inf\{t > 0\,,\ |B_t| = a\}\,, \quad a > 0\,. \tag{7.4.4}$$

Proposition 7.6. *The Laplace transforms of the law of τ_a and $\tilde{\tau}_a$ are given by*

$$\mathbb{E}e^{-\lambda\tau_a} = e^{-|a|\sqrt{2\lambda}}\,, \quad \forall\ \lambda > 0\,, \quad \forall a \in \mathbb{R} \quad \text{and } a \neq 0\,; \tag{7.4.5}$$

$$\mathbb{E}e^{-\lambda\tilde{\tau}_a} = \frac{2}{e^{a\sqrt{2\lambda}} + e^{-a\sqrt{2\lambda}}}\,, \quad \forall\lambda > 0\,, \quad \forall a > 0\,. \tag{7.4.6}$$

Proof. It suffices to consider the case $a > 0$. For any $\gamma > 0$, define $M_t^\gamma := e^{\gamma B_t - \frac{\gamma^2 t}{2}}$. Then, $\{M_t^\gamma, t \geq 0\}$ is a martingale with $M_0 = 1$. Hence $\{M_{t\wedge\tau_a}^\gamma, t \geq 0\}$ is a martingale bounded $e^{\gamma a}$. Using the Doob's optional stopping theorem, we have

$$\mathbb{E}M_{\tau_a}^\gamma = \mathbb{E}(M_0^\gamma) = 1\,,$$

This means that

$$\mathbb{E}e^{-\frac{\gamma^2\tau_a}{2}} = e^{-\gamma a}\,.$$

The identity (7.4.5) is proved with $\gamma = \sqrt{2\lambda}$. The second identity (7.4.6) can be proved in the similar way with M_t^γ replaced by the martingale

$$\frac{M_t^\gamma + M_t^{-\gamma}}{2} = \frac{e^{\gamma B_t} + e^{-\gamma B_t}}{2} e^{-\frac{\gamma^2 t}{2}}\,, \quad t \geq 0\,.$$

□

Let $(\mathbb{B}, \mathbb{H}, \mathbb{P})$ be an abstract Wiener space and K is a separable Hilbert space. Let the second quantization operator $\Gamma(\alpha)$ and the Ornstein-Uhlenbeck operator $T_t = \Gamma(e^{-t})$ be defined as before. Define

$$Q_t^{(\alpha)} = \int_0^\infty e^{-\alpha s} T_s \mu_t^{1/2}(ds)\,, \quad t \geq 0\,, \tag{7.4.7}$$

where $\mu_t^{1/2}(ds) = \mu_t(ds)$ is defined by (7.3.3). It is easy to see that $\{Q_t^{(\alpha)}, t \geq 0\}$ is a semigroup on $L^p(\mathbb{B}, K)$. From (7.3.4) the generator $\{Q_t^{(\alpha)}, t \geq 0\}$ can be formally written as $-\sqrt{\alpha - L}$.

From the Mehler's formula we see that $Q_t^{(\alpha)}$ has also an integral kernel representation:

$$Q_t^{(\alpha)} F(x) = \int_B F(y) Q_t^{(\alpha)}(x, dy)\,, \tag{7.4.8}$$

where

$$Q_t^{(\alpha)}(x,dy) = \int_0^\infty e^{-\alpha s} P_s(x,dy)\mu_t^{1/2}(ds)\,, \tag{7.4.9}$$

with $P_s(x,dy)$ being the transition probability measure of the Ornstein-Uhlenbeck process. From the definition of $Q_t^{(\alpha)}$, the contractivity of T_t and (7.3.4), we see that

$$\|Q_t^{(\alpha)}F(x)\|_p \le \int_0^\infty e^{-\alpha s}\mu_t^{1/2}(ds)\|F\|_p = e^{-\sqrt{\alpha}t}\|F\|_P\,.$$

For any $F \in \mathcal{P}(\mathbb{B}, K)$, we define

$$g_F^{\rightarrow}(x,t) = \left|\frac{\partial}{\partial t}Q_t^{(\alpha)}f(x)\right|_K\,, \tag{7.4.10}$$

$$g_F^{\uparrow}(x,t) = \left|DQ_t^{(\alpha)}f(x)\right|_{HS}\,, \tag{7.4.11}$$

$$g_F(x,t) = \sqrt{g_F^{\rightarrow}(x,t)^2 + g_F^{\uparrow}(x,t)^2}\,, \tag{7.4.12}$$

where D is the Gross-Sobolev derivative and $|\cdot|_{HS}$ is the Hilbert-Schmidt norm of an operator in $\mathcal{L}_{(2)}(\mathbb{H},K)$. In the definitions of (7.4.10)-(7.4.12), we omit the dependence of g's on α to simplify notation. Now we define the following quantities which are the central objects of this section.

$$G_F^{\rightarrow}(x,t) = \left\{\int_0^\infty t g_F^{\rightarrow}(x,t)^2 dt\right\}^{1/2}\,, \tag{7.4.13}$$

$$G_F^{\uparrow}(x,t) = \left\{\int_0^\infty t g_F^{\uparrow}(x,t)^2 dt\right\}^{1/2}\,, \tag{7.4.14}$$

$$G_F(x,t) = \left\{\int_0^\infty t g_F(x,t)^2 dt\right\}^{1/2}\,. \tag{7.4.15}$$

The Littlewood-Paley-Stein-Meyer theory is about the relationship among the L^p norms of $G_F^{\rightarrow}(x,t)$, $G_F^{\uparrow}(x,t)$, $G_F(x,t)$ and f. In the finite dimensional Euclidean space with the Lebesgue measure, such inequalities are very powerful (see [Stein (1970)]). In [Stein (1970)], Stein also treats the general spaces, including the case of general measure space. However, the general case is hard to allow a rich class of inequalities. In a series of work [Meyer (1976)], Meyer proved a larger set of such inequalities for Wiener space case. As an application, Meyer also proved the equivalence of various norms on Wiener space ([Meyer (1982a)]), which are now called the Meyer's inequalities.

Let K be a separable Hilbert space with norm $|\cdot| = |\cdot|_K$. Denote

$$L^p(\mathbb{B},K) = \left\{F:\mathbb{B}\to K\,;\ \|F\|_p^p = \mathbb{E}\left[|F|^p\right] < \infty\right\}\,.$$

The main result of this section is the following theorem.

Theorem 7.11. *Let* $1 < p < \infty$.

(i) If $\alpha > 0$*, then there is a positive constant* C_p *such that*

$$\|G_F\|_p \le C_p \|F\|_p \,, \quad \forall\ F \in L^p(\mathbb{B}, K)\,, \tag{7.4.16}$$

$$\|F\|_p \le C_p \|G_F^{\rightarrow}\|_p \,, \quad \forall\ F \in L^p(\mathbb{B}, K)\,. \tag{7.4.17}$$

(ii) If $\alpha = 0$*, then there is a positive constant* C_p *such that*

$$\|G_F\|_p \le C_p \|F - J_0(F)\|_p \,, \quad \forall\ F \in L^p(\mathbb{B}, K)\,, \tag{7.4.18}$$

$$\|F - J_0(F)\|_p \le C_p \|G_F^{\rightarrow}\|_p \,, \quad \forall\ F \in L^p(\mathbb{B}, K)\,, \tag{7.4.19}$$

$$\|F - J_0(F)\|_p \le C_p \|G_F^{\uparrow}\|_p \,, \quad \forall\ F \in L^p(\mathbb{B}, K)\,. \tag{7.4.20}$$

The Meyer's original proof of the theorem is a probabilistic one. We need first to represent the above norms by probabilistic terminologies.

Since L is a self-adjoint operator in $L^2(\mathbb{B}, \mathbb{P}, K)$, we have a spectral decomposition for L:

$$-L = \int_0^\infty \lambda dE_\lambda \,,$$

where $\{E_\lambda, \lambda \ge 0\}$ is a right continuous resolution of the identity with $E_0 = 0$. Namely, $\{E_\lambda, \lambda \ge 0\}$ is a family of orthogonal projection operators acting on $L^2(\mathbb{B}, \mathbb{P}, K)$, such that

(i) $E_{\lambda_1} \subseteq E_{\lambda_2}$ if $0 \le \lambda_1 < \lambda_2 < \infty$;
(ii) $\{E_\lambda, \lambda \ge 0\}$ is strongly right continuous;
(iii) E_λ converges strongly to the identity operator on the space $L^2(\mathbb{B}, \mathbb{P}, K)$.

With this resolution of identity we have

$$Q_t^{(\alpha)} = \int_0^\infty \int_0^\infty e^{-\alpha s} e^{-\lambda s} \mu_t^{1/2}(ds) dE_\lambda$$
$$= \int_0^\infty e^{-\sqrt{\alpha+\lambda}t} dE_\lambda \,. \tag{7.4.21}$$

Using the general notation of function of self-adjoint operator, we can write

$$Q_t^{(\alpha)} = \exp\left\{-\sqrt{\alpha - L}t\right\} . \tag{7.4.22}$$

To prove (7.4.16)-(7.4.20) one needs only to prove these inequalities for $F \in \mathcal{P}$. Thus, F can be considered as a function from $\mathbb{R}^d$ to $\mathbb{R}$. If we can show that all the above inequalities hold true for the finite dimensional

Gaussian space $(\mathbb{R}^d, \mu(dx))$ and the constants appeared are independent of dimension d, then by a density argument, the inequalities are proved for any abstract Wiener space. So from now we can assume $\mathbb{B} = \mathbb{R}^d$.

The probabilistic idea to prove the above inequalities is first to represent the above norms by some stochastic processes.

Let X_t be the solution of the following Langevin stochastic differential equation

$$dX_t = -X_t dt + \sqrt{2} dW_t \,,$$

where $\{W_t = (W_t^1, \cdots, W_t^d), t \geq 0\}$ is a d dimensional Brownian motion. Denote $\partial_j^k = \frac{\partial^k}{\partial x_j^k}$ for $k = 1, 2, \cdots$. Let $\mathcal{F}_t = \mathcal{F}_t^X = \sigma(W_s, 0 \leq s \leq t) = \sigma(X_s, 0 \leq s \leq t)$ be the σ-algebra generated by X_t. It is easy to see that $\{X_t, t \geq 0\}$ is a Markov process called Orsntein-Uhlenbeck process and

$$p_t(x, y) = \frac{1}{\sqrt{2\pi(1 - e^{-2t})}} \exp\left\{ -\frac{|y - e^{-t}x|^2}{2(1 - e^{-2t})} \right\}$$

is the transition density function associated with the Ornstein-Uhlenbeck process X_t. For any $F \in \mathcal{P}$, it is easy to see that

$$T_t F(x) := \Gamma(e^{-t}) F(x) = \int_{\mathbb{R}^d} p_t(x, y) F(y) dy \,.$$

Now we define the decreasing σ-algebra $\mathcal{G}_t = \sigma(X_s, s \geq t)$. The $\{\mathcal{G}_t, t \geq 0\}$ is time reverse filtration.

Lemma 7.5. *Let X_0 be a standard normal distribution on $\mathbb{R}^d$. Then $T_t F(X_t)$ is a time reverse martingale with respect to $\mathcal{G}_t$ and*

$$T_t F(X_t) = \mathbb{E}\left[F(X_0) | \mathcal{G}_t\right] \,, \quad \forall\, t \geq 0 \,. \tag{7.4.23}$$

Proof. It suffices to show that for any $0 < t < t_1 < \cdots < t_n$ and $F_0, F_1, \cdots, F_n \in \mathcal{P}$,

$$\begin{aligned} &\mathbb{E}\left[T_t F(X_t) F_0(X_t) F_1(X_{t_1}) \cdots F_n(X_{t_n})\right] \\ &\quad = \mathbb{E}\left[F(X_0) F_0(X_t) F_1(X_{t_1}) \cdots F_n(X_{t_n})\right] \,. \end{aligned} \tag{7.4.24}$$

First, we notice that X_t has standard Gaussian measure μ as the invariant measure and T_t is symmetric. Thus, we have

$$\begin{aligned} \text{LHS} &= \mathbb{E}\left[T_t F(X_t)\ F_0(X_t) T_{t_1 - t}\left(F_1 \cdots \left(F_{n-1} T_{t_n - t_{n-1}} F_n\right) \cdots\right)(X_t)\right] \\ &= \int_{\mathbb{R}^d} \mathbb{E}\left[T_t F(x)\ F_0(x) T_{t_1 - t}\left(F_1 \cdots \left(F_{n-1} T_{t_n - t_{n-1}} F_n\right) \cdots\right)(x)\right] \mu(dx) \\ &= \int_{\mathbb{R}^d} \mathbb{E}\left[F(x)\ T_t\left[F_0 T_{t_1 - t}\left(F_1 \cdots \left(F_{n-1} T_{t_n - t_{n-1}} F_n\right) \cdots\right)\right](x)\right] \mu(dx) \\ &= \mathbb{E}\left[F(X_0)\ F_0(X_t)\left[T_{t_1 - t}\left(F_1 \cdots \left(F_{n-1} T_{t_n - t_{n-1}} F_n\right) \cdots\right)\right](X_t)\right] \,, \end{aligned}$$

which is the right hand side of (7.4.24). This proves the lemma. □

Theorem 7.12. *Let K be a Hilbert space. For any $F \in \mathcal{F}(K)$, set*

$$F^*(x) = \sup_{t\geq 0} |T_t F(x)| . \tag{7.4.25}$$

Then, for any $1 < p < \infty$, we have

$$\|F^*\|_p \leq \frac{p}{p-1} \|F\|_p , \forall F \in L^p(\mathbb{B}, K) . \tag{7.4.26}$$

Proof. It suffices to consider the case that $K = \mathbb{R}$. Let $M_t = T_t F(X_t)$. Then, M_t is a continuous martingale and $\mathbb{E}\left[M_t \middle| \mathcal{F}_0\right] = T_{2t}F(X_0)$. Moreover,

$$|T_{2t}F(X_0)| \leq \mathbb{E}\left[|M_t| \middle| \mathcal{F}_0\right] \leq \mathbb{E}\left[\sup_{t\geq 0} |M_t| \middle| \mathcal{F}_0\right] .$$

Taking the sup on t yields

$$|F^*(X_0)| \leq \mathbb{E}\left[\sup_{t\geq 0} |M_t| \middle| \mathcal{F}_0\right] .$$

By Doob's martingale inequality, we have

$$\begin{aligned}\|F^*\|_p^p &= \mathbb{E}|F^*(X_0)|^p \leq \mathbb{E}\left|\mathbb{E}\left[\sup_{t\geq 0} |M_t| \middle| \mathcal{F}_0\right]\right|^p \leq \mathbb{E}\sup_{t\geq 0}|M_t|^p \\ &\leq \left(\frac{p}{p-1}\right)^p \mathbb{E}|M_0|^p \leq \left(\frac{p}{p-1}\right)^p \mathbb{E}|F(X_0)|^p = \left(\frac{p}{p-1}\right)^p \|F\|_p^p ,\end{aligned}$$

proving the theorem. □

Let $\{B_t^{(0)}, t \geq 0\}$ be another one dimensional Brownian motion starting at 0, independent of W and denote $B_t^a = a + \sqrt{2}B_t^{(0)}$. Consider the space $\mathbb{R}^{d+1}$. A point in $\mathbb{R}^{d+1}$ is denoted by (x, x_{d+1}), where $x = (x_1, \cdots, x_d)$. For any smooth function $f : \mathbb{R}^{d+1} \to K$ with polynomial growth, the Itô formula yields

$$\begin{aligned}f(X_t, B_t^a) &= f(X_0, B_0^a) + \sqrt{2}\sum_{j=1}^d \int_0^t \partial_j f(X_s, B_s^a) dW_s^j \\ &\quad + \sqrt{2}\int_0^t \partial_{d+1} f(X_s, B_s^a) dB_s^{(0)} + \int_0^t \left[L_x f + \partial_{d+1}^2 f\right](X_s, B_s^a) ds ,\end{aligned} \tag{7.4.27}$$

where L_x is the generator associated with the process X_t, namely

$$L_x f(x, x_{d+1}) = \sum_{j=1}^d \partial_j^2 f(x, x_{d+1}) - \sum_{j=1}^d x_j \partial_j f(x, x_{d+1}) .$$

It is easy to see that

$$L_x H_n(x) = n H_n(x)\,, \tag{7.4.28}$$

where H_n is the Hermite polynomial of order n. Set

$$M_t = f(X_t, B_t^a) - f(X_0, B_0^a) - \int_0^t \left[L_x f + \partial_{d+1}^2 f\right](X_s, B_s^a)ds\,.$$

Then, we see that $\{M_t, t \ge 0\}$ is a martingale and

$$\langle M\rangle_t = 2\int_0^t \left[|\partial_j f(X_s, B_s^a)|^2 + |\partial_{d+1} f(X_s, B_s^a)|^2\right] ds\,.$$

For a function $F \in \mathcal{P}(\mathbb{R}^d)$ consider the function $u(x,t) = Q_t^{(\alpha)}F(x)$. If $F(x) = H_n(x)$, then $u(x,t) = e^{-\sqrt{n+\alpha}t}H_n(x)$. It is easy to check from (7.4.28) that in this case we have

$$\begin{cases} L_x u(x,t) + \partial_t^2 u(x,t) = \alpha u(x,t) \\ u(x,0) = F(x)\,. \end{cases} \tag{7.4.29}$$

By a linearity argument, we see that the above equation holds true for any function $F \in \mathcal{P}(\mathbb{R}^d)$.

From (7.4.29), we see that if we could apply the Itô formula to $f(x,r) = Q_r^{(\alpha)}F(x)$ (we shall denote $r = x_{d+1}$), then $\{Q_{B_s^a}^{(\alpha)}F(X_s), s \ge 0\}$ would be a martingale. However, $Q_{B_s^a}F$ is defined only for $B_s^a \ge 0$. To get around this difficulty, we assume $a > 0$ and we introduce the stopping time

$$\tau = \inf\{t > 0, B_t^a = 0\}\,.$$

Then, when $0 \le s \le \tau$, $B_s^a \ge 0$. The Itô formula (7.4.27) applied to $f(X_s, B_s^a) = Q_{B_s^a}^{(\alpha)}F(X_s)$ yields that for any $t \ge 0$,

$$\begin{aligned} M_t^F &:= Q_{B_{t\wedge\tau}^a}^{(\alpha)}F(X_{t\wedge\tau}) - \alpha\int_0^{t\wedge\tau} Q_{B_s^a}^{(\alpha)}F(X_s)ds \\ &= Q_a^{(\alpha)}F(x) + \sqrt{2}\sum_{j=1}^d \int_0^{t\wedge\tau} \partial_j Q_{B_s^a}^{(\alpha)}F(X_s)dW_s^j \\ &\quad +\sqrt{2}\int_0^{t\wedge\tau} \partial_t Q_{B_s^a}^{(\alpha)}F(X_s)dB_s^{(0)} \end{aligned} \tag{7.4.30}$$

is a local martingale and the quadratic variation of this local martingale is

$$\langle M\rangle_t = 2\int_0^{t\wedge\tau} g_F^2(X_s, B_s^a)ds\,. \tag{7.4.31}$$

In (7.4.30), letting $t \to \infty$ and noticing $B^a_\tau = 0$, we obtain

$$M^F_\infty := Q^{(\alpha)}_{B^a_\tau} F(X_\tau) - \alpha \int_0^\tau Q^{(\alpha)}_{B^a_s} F(X_s) ds = F(X_\tau) - \alpha \int_0^\tau Q^{(\alpha)}_{B^a_s} F(X_s) ds .$$

Thus

$$M^F_t = \mathbb{E}\left[F(X_\tau) - \alpha \int_0^\tau Q^{(\alpha)}_{B^a_s} F(X_s) ds \big| \mathcal{F}^{W,B}_t \right] . \tag{7.4.32}$$

As a consequence when $\alpha = 0$ we have

$$\mathbb{E}\left[F(X_\tau) \big| \mathcal{F}^{W,B}_t \right] = M^F_t = Q^{(0)}_{B^a_{t\wedge\tau}} F(X_{t\wedge\tau}) . \tag{7.4.33}$$

Lemma 7.6. *For any measurable function $f : \mathbb{R}^d \to \mathbb{R}$ with polynomial growth we have*

$$\mathbb{E}^\mu \left[f(X_\tau) \right] = \int_{\mathbb{R}^d} f(x) \mu(dx) . \tag{7.4.34}$$

Proof. Let $\nu^\tau(dt)$ be the probability density function of τ. Then

$$\begin{aligned} \mathbb{E}^\mu \left[f(X_\tau) \right] &= \int_0^\infty \mathbb{E}^\mu f(X_r) \nu^\tau(dr) \\ &= \int_0^\infty \int_{\mathbb{R}^d} f(x) \mu(dx) \nu^\tau(dr) = \int_{\mathbb{R}^d} f(x) \mu(dx) . \end{aligned}$$

This proves the lemma. □

Lemma 7.7. *Let h be a non-negative measurable function on $\mathbb{R}_+$. Then*

$$\mathbb{E} \int_0^\tau h(B^a_s) ds = \int_0^\infty (a \wedge u) h(u) du . \tag{7.4.35}$$

Proof. It suffices to show (7.4.35) for $h \in C_0^\infty(\mathbb{R}_+)$. Let $g(x) = \int_0^\infty (x \wedge u) h(u) du$. Then, it is easy to verify $g''(x) = -h(x)$ and $g(0) = 0$. By the Itô formula we have

$$g(B^a_{t\wedge\tau}) - g(a) - \int_0^{t\wedge\tau} g'(B^a_s) dB^a_s = \int_0^{t\wedge\tau} g''(B^a_s) ds = - \int_0^{t\wedge\tau} h(B^a_s) ds .$$

Thus

$$\mathbb{E}\left[g(B^a_{t\wedge\tau}) \right] - g(a) = - \int_0^{t\wedge\tau} h(B^a_s) ds .$$

Letting $t \to \infty$ yields

$$\mathbb{E} \int_0^\tau h(B^a_s) ds = g(a) ,$$

which proves the lemma. □

Lemma 7.8. *Let $f : \mathbb{R}^d \times \mathbb{R}_+ \to \mathbb{R}_+$ be a Borel measurable function. Then*

$$\mathbb{E}\int_0^{\mathcal{T}} f(X_t, B_t^a)dt = \int_{\mathbb{R}^d} \mu(dx) \int_0^\infty (r\wedge a) f(x,r)dr\,. \tag{7.4.36}$$

Proof. First, we consider the function of the form $f(x,r) = k(x)h(r)$, where $k : \mathbb{R}^d \to \mathbb{R}$ and $h : \mathbb{R}_+ \to \mathbb{R}$ are smooth with compact support. In this case we have

$$\begin{aligned}
&\mathbb{E}\int_0^{\mathcal{T}} f(X_t, B_t^a)dt = \int_0^\infty k(X_t)h(B_t^a)I_{[0,\mathcal{T}]}(t)dt \\
&= \int_0^\infty \mathbb{E}\left[k(X_t)\right]\mathbb{E}\left[h(B_t^a)I_{[0,\mathcal{T}]}(t)\right]dt \\
&= \int_{\mathbb{R}^d} k(x)\mu(dx)\int_0^\infty \mathbb{E}\left[h(B_t^a)I_{[0,\mathcal{T}]}(t)\right]dt \\
&= \int_{\mathbb{R}^d} k(x)\mu(dx)\mathbb{E}\left[\int_0^{\mathcal{T}} h(B_t^a)\right]dt = \int_{\mathbb{R}^d} k(x)\mu(dx)\int_0^\infty (a\wedge r)h(r)dr \\
&= \int_{\mathbb{R}^d} \mu(dx)\int_0^\infty (a\wedge r)f(x,r)dr
\end{aligned}$$

which is the lemma when $f(x,r) = k(x)h(r)$. By linearity, the identity (7.4.36) holds true for $f(x,r) = \sum_{\text{finite}} k_i(x)h_i(r)$. A density argument shows (7.4.36) for all functions. □

Lemma 7.9. *Let $f : \mathbb{R}^d \times \mathbb{R}_+ \to \mathbb{R}_+$ be a Borel measurable function. Then*

$$\mathbb{E}\left[\int_0^{\mathcal{T}} f(X_t, B_t^a)dt \Big| X_{\mathcal{T}}\right] = \int_0^\infty (r\wedge a)Q_r^{(0)} f(X_r, r)dr\,. \tag{7.4.37}$$

Proof. Since the right hand side of (7.4.37) is a measurable function of $X_{\mathcal{T}}$ it suffices to show that for any smooth function $g : \mathbb{R}^d \to \mathbb{R}$ with compact support we have

$$\mathbb{E}\left[g(X_{\mathcal{T}})\int_0^{\mathcal{T}} f(X_t, B_t^a)dt\right] = \int_0^\infty (r\wedge a)\mathbb{E}\left[g(X_{\mathcal{T}})Q_r^{(0)} f(X_r, r)\right]dr\,. \tag{7.4.38}$$

From (7.4.34), the right hand side of (7.4.38) is

$$\text{RHS} = \int_0^\infty (a\wedge r)\int_{\mathbb{R}^d}\left[g(x)Q_r^{(0)} f(x,r)\right]\mu(dx)dr\,. \tag{7.4.39}$$

The left hand side of (7.4.38) is

$$\begin{aligned}\text{LHS} &= \int_0^\infty \mathbb{E}\left[g(X_\tau)f(X_t, B_t^a)I_{[0,\tau]}(t)\right] dt \\ &= \int_0^\infty \mathbb{E}\left\{f(X_t, B_t^a)I_{[0,\tau]}(t)\mathbb{E}\left[g(X_\tau)|\mathcal{F}_t^{W,B}\right]\right\} dt \\ &= \int_0^\infty \mathbb{E}\left\{f(X_t, B_t^a)I_{[0,\tau]}(t)Q^{(0)}_{B^a_{t\wedge\tau}}g(X_{t\wedge\tau})\right\} dt \\ &= \mathbb{E}\left\{\int_0^\tau f(X_t, B_t^a)Q^{(0)}_{B_t^a}g(X_t)\right\} dt \\ &= \int_0^\infty (a\wedge t)\int_{\mathbb{R}^d} f(x,r)Q_r^{(0)}g(x)\mu(dx)dr \\ &= \int_0^\infty (a\wedge t)\int_{\mathbb{R}^d} (Q_r^{(0)}f)(x,r)g(x)\mu(dx)dr\,,\end{aligned}$$

where the third equality follows from (7.4.33), the second last equality follows from (7.4.36) and the last equation follows from the symmetry of $Q_r^{(0)}$. Combining this with (7.4.39) yields the lemma. □

From the equation (7.4.31) we have $\langle M\rangle_\infty = 2\int_0^\tau g_F^2(X_s, B_s^a)ds$. By the above lemma 7.9, we see

$$\mathbb{E}\left[\langle M^F\rangle_\infty | X_\tau\right] = \int_0^\infty (a\wedge r)(Q_r^{(0)}g_F^2)(X_\tau, r)dr\,. \tag{7.4.40}$$

Taking the expectation, we have

$$\begin{aligned}\mathbb{E}\left[\langle M^F\rangle_\infty\right] &= 2\int_{\mathbb{R}^d}\int_0^\infty (a\wedge r)(Q_r^{(0)}g_F^2)(x,r)\mu(dx)dr \\ &= 2\int_0^\infty (a\wedge r)\int_{\mathbb{R}^d} g_F^2(x,r)\mu(dx)dr \\ &\le 2\int_{\mathbb{R}^d}\int_0^\infty r g_F^2(x,r)dr\mu(dx) = 2\|G_F\|_2^2 < \infty\,.\end{aligned} \tag{7.4.41}$$

Proof of Theorem 7.11. We shall divide the proof into several steps.

Step 1. First, we prove the case $p=2$. From (7.4.21), we have

$$\frac{\partial}{\partial t}Q_t^{(\alpha)} = \int_0^\infty \sqrt{\alpha+\lambda}e^{-\sqrt{\alpha+\lambda}t}dE_\lambda\,.$$

We claim the following statements.

(i) If $\alpha>0$, then there is a positive constant C_p such that

$$2\|G_F\|_2 = \|F\|_2\,, \quad \forall\, F\in L^2(\mathbb{B},\mathbb{P},K)\,. \tag{7.4.42}$$

(ii) If $\alpha = 0$, then there is a positive constant C_p such that

$$2\|G_F^{\rightarrow}\|_2 = \|F - J_0(F)\|_2\,, \quad \forall\ F \in L^2(\mathbb{B}, \mathbb{P}, K) \tag{7.4.43}$$

$$2\|G_F^{\uparrow}\|_2 = \|F - J_0(F)\|_2\,, \quad \forall\ F \in L^2(\mathbb{B}, \mathbb{P}, K)\,. \tag{7.4.44}$$

In fact, we have

$$\begin{aligned}
\|G_F^{\rightarrow}\|_2^2 &= \int_B \mu(dx) \int_0^\infty t \left|\frac{\partial}{\partial t} Q_t^{(\alpha)} F\right|_K^2 dt = \int_0^\infty t \left\|\frac{\partial}{\partial t} Q_t^{(\alpha)} F\right\|_2^2 dt \\
&= \int_0^\infty t \int_0^\infty (\lambda + \alpha) e^{-2\sqrt{\lambda+\alpha}t} d\langle E_\lambda F, F\rangle dt \\
&= \int_0^\infty \left[\int_0^\infty (\lambda + \alpha) t e^{-2\sqrt{\lambda+\alpha}t} dt\right] d\langle E_\lambda F, F\rangle\,.
\end{aligned}$$

Noticing the following simple fact

$$\int_0^\infty (\lambda + \alpha) t e^{-2\sqrt{\lambda+\alpha}} dt = \begin{cases} \frac{1}{4} & \lambda + \alpha > 0 \\ 0 & \lambda + \alpha = 0\,, \end{cases}$$

we obtain

$$\|G_F^{\rightarrow}\|_2^2 = \begin{cases} \frac{1}{4}\|F\|_2^2 & \alpha > 0 \\ \frac{1}{4}\|F - J_0(F)\|_2^2 & \alpha = 0\,. \end{cases}$$

This proves (7.4.43). Now we are going to prove (7.4.44). We have

$$\begin{aligned}
\|G_F^{\uparrow}\|_2^2 &= \int_B \mu(dx) \int_0^\infty t \left|DQ_t^{(\alpha)} F\right|_{HS}^2 dt \\
&= \int_0^\infty t \int_B \mu(dx) \left|DQ_t^{(\alpha)} F\right|_{HS}^2 \mu(dx) dt \\
&= \int_0^\infty t \int_B \langle -LQ_t^{(\alpha)} F, Q_t^{(\alpha)}\rangle \mu(dx) dt \\
&= \int_0^\infty \left[\int_0^\infty \lambda t e^{-2\sqrt{\lambda+\alpha}} dt\right] d\langle E_\lambda F, F\rangle \\
&= \int_0^\infty \frac{\lambda}{4(\lambda + \alpha)} d\langle E_\lambda F, F\rangle
\end{aligned}$$

which is $\frac{1}{4}\|F - J_0(F)\|_2$ when $\alpha = 0$.

Step 2. Now we prove the upper bound in the case $1 \le p \le 2$. For any $F \in \mathcal{P}$ denote $u(x,t) = Q_t^{(\alpha)} F(x)$. Then, $u(x,t)$ satisfies (7.4.29). Using the martingale M_t^F defined by (7.4.30), we have

$$u(X_{t\wedge\tau}, B^a_{t\wedge\tau}) = M_t^F + \alpha \int_0^{t\wedge\tau} u(X_s, B^a_s) ds\,.$$

Since $f(x) = |x|^p$ is not twice continuously differentiable at 0, we cannot directly apply the Itô formula to $|u|^p$. We shall use its approximation $(|u|^2+\varepsilon)^{p/2}$. First, from (7.4.31) it follows

$$\begin{aligned} d|u|^2 &= 2\langle u, dM^F\rangle + 2\alpha|u|^2dt + \langle dM^F, dM^F\rangle \\ &= 2\langle u, dM^F\rangle + (2\alpha|u|^2 + 2g_F^2)dt\,, \end{aligned} \tag{7.4.45}$$

where we omit the dependence of u, g_F and so on on their variables X_t and B_t^a. Thus, the Itô formula yields

$$\begin{aligned} d(|u|^2+\varepsilon)^{p/2} &= p(|u|^2+\varepsilon)^{p/2-1}\langle u, dM^F\rangle + p(|u|^2+\varepsilon)^{p/2-1}(\alpha|u|^2+g_F^2)dt \\ &\quad + \frac{p(p-2)}{2}(|u|^2+\varepsilon)^{p/2-2}\langle udM^F, udM^F\rangle\,. \end{aligned}$$

Here if $u = (u_1, \cdots, u_d)$ and $M^F = (M_1^F, \cdots, M_d^F)$, then

$$\begin{aligned} \langle udM^F, udM^F\rangle &:= \sum_{i,j=1}^d \langle u_i dM_i^F, u_j dM_j^F\rangle = \sum_{i,j=1}^d u_i u_j \langle dM_i^F, dM_j^F\rangle \\ &\le \sum_{i,j=1}^d |u_i||u_j|\sqrt{\langle dM_j^F, dM_j^F\rangle}\sqrt{\langle dM_i^F, dM_i^F\rangle} \\ &\le \sqrt{\sum_{i,j=1}^d |u_i|^2\langle dM_j^F, dM_j^F\rangle}\sqrt{\sum_{i,j=1}^d |u_j|^2\langle dM_i^F, dM_i^F\rangle} \\ &= 2|u|^2 g_F^2 dt\,, \end{aligned}$$

where the first inequality follows from Kunita-Watanabe inequality and the second one follow the Hölder inequality. Noticing that $p \le 2$, we have

$$\begin{aligned} d(|u|^2+\varepsilon)^{p/2} &\ge p(|u|^2+\varepsilon)^{p/2-1}\langle u, dM^F\rangle + p\alpha(|u|^2+\varepsilon)^{p/2-1}|u|^2dt \\ &\quad + p(|u|^2+\varepsilon)^{p/2-1}g_F^2dt + p(p-2)(|u|^2+\varepsilon)^{p/2-2}|u|^2g_F^2dt \\ &= p(|u|^2+\varepsilon)^{p/2-1}\langle u, dM^F\rangle + p\alpha(|u|^2+\varepsilon)^{p/2-1}|u|^2dt \\ &\quad + p(|u|^2+\varepsilon)^{p/2-2}g_F^2\left[(p-1)|u|^2+\varepsilon\right]dt \\ &\ge p(|u|^2+\varepsilon)^{p/2-1}\langle u, dM^F\rangle + p\alpha(|u|^2+\varepsilon)^{p/2-1}|u|^2dt \\ &\quad + p(p-1)(|u|^2+\varepsilon)^{p/2-1}g_F^2dt \\ &\ge p(|u|^2+\varepsilon)^{p/2-1}\langle u, dM^F\rangle + p(p-1)(|u|^2+\varepsilon)^{p/2-1}g_F^2dt\,. \end{aligned}$$

Therefore, we have

$$\begin{aligned} p(p-1)\mathbb{E}\int_0^\tau (|u|^2+\varepsilon)^{p/2-1}g_F^2dt &\le \mathbb{E}\left[(|u(X_\tau, B_\tau^a)|^2+\varepsilon)^{p/2}\right] \\ &= \mathbb{E}\left[(|F(X_\tau)|^2+\varepsilon)^{p/2}\right] = \int_{\mathbb{R}^d}(|F(x)|^2+\varepsilon)^{p/2}\mu(dx)\,. \end{aligned}$$

On the other hand, we have

$$|u(x,a)| = |Q_a^{(\alpha)}F(x)| \le \int_0^\infty e^{-\alpha s}|T_sF(x)|\mu_t(ds)$$
$$\le \int_0^\infty e^{-\alpha s}F^*(x)\mu_t(ds) \le F^*(x),$$

where $F^*(x) = \sup_{s\ge 0}|T_sF(x)|$. By (7.4.36), we have

$$\|(|F(x)|^2+\varepsilon)^{1/2}\|_p^p \ge p(p-1)\mathbb{E}\int_0^\tau (|u|^2+\varepsilon)^{p/2-1}g_F^2 dt$$
$$= p(p-1)\int_{\mathbb{R}^d}\int_0^\infty (|u(x,r)|^2+\varepsilon)^{p/2-1}g_F^2(x,r)(a\wedge r)dr\mu(dx)$$
$$\ge p(p-1)\int_{\mathbb{R}^d}\int_0^\infty (|F^*(x)|^2+\varepsilon)^{p/2-1}g_F^2(x,r)(a\wedge r)dr\mu(dx)$$

since $p/2-1\le 0$. Letting $a\to\infty$, we see

$$\|(|F(x)|^2+\varepsilon)^{1/2}\|_p^p$$
$$\ge p(p-1)\int_{\mathbb{R}^d}(|F^*(x)|^2+\varepsilon)^{p/2-1}\int_0^\infty g_F^2(x,r)rdr\mu(dx)$$
$$= p(p-1)\int_{\mathbb{R}^d}(|F^*(x)|^2+\varepsilon)^{p/2-1}G_F^2(x)\mu(dx)$$
$$= p(p-1)\|(|F^*|^2+\varepsilon)^{(p-2)/4}G_F\|_2^2.$$

Now we are ready to show the inequality (7.4.16).

$$\|G_F\|_p = \|(|F^*|^2+\varepsilon)^{(2-p)/4}(F^*+\varepsilon)^{(p-2)/4}G_F\|_p$$
$$\le \|(|F^*|^2+\varepsilon)^{(2-p)/4}\|_{\frac{2p}{2-p}}\|(F^*+\varepsilon)^{(p-2)/4}G_F\|_2$$
$$\le C_p\|(F^*+\varepsilon)^{1/2}\|_p^{(2-p)/2}\|(|F(x)|^2+\varepsilon)^{1/2}\|_p^{p/2}.$$

Letting $\varepsilon\to 0$ yields

$$\|G_F\|_p \le C_p\|F^*\|_p^{(2-p)/2}\|F(x)\|_p^{p/2}$$
$$\le C_p\|F\|_p^{(2-p)/2}\|F(x)\|_p^{p/2} = C_p\|F(x)\|_p,$$

where the second inequality follows from Theorem 7.12. This proves (7.4.16) for $p\le 2$ and $\alpha>0$. The case $\alpha=0$ can be proved by taking $F-J_0F$ instead of F.

Step 3. In this step we prove the following bound

$$\|H_F\|_p \le C_p\|F\|_p, \tag{7.4.46}$$

where

$$H_F(x) = \left\{ \int_0^\infty t Q_t^{(0)} g_F(x,t)^2 dt \right\}^{1/2} .$$

We continue to use the notation in Step 2. The equation (7.4.45) gives

$$\begin{aligned} |u(X_{t\wedge\tau}, B_{t\wedge\tau})|^2 = |u(X_0, B_0^a)|^2 + 2\int_0^{t\wedge\tau} \langle u(X_s, B_s^a), dM_s^F \rangle \\ +2\int_0^{t\wedge\tau} (\alpha|u(X_s, B_s^a)|^2 + g_F(X_s, B_s^a)^2) ds . \end{aligned}$$

It is easy to see from the above argument that $(|u(X_{t\wedge\tau}, B_{t\wedge\tau})|^2, t \geq 0)$ is a submartingale. From Doob's inequality, we see that

$$\begin{aligned} &\mathbb{E}\left\{ 2\int_0^{t\wedge\tau} (\alpha|u(X_s, B_s^a)|^2 + g_F(X_s, B_s^a)^2) ds \right\}^{p/2} \\ &\leq C_p \mathbb{E}\left[|u(X_{t\wedge\tau}, B_{t\wedge\tau})|^2 - |u(X_0, B_0^a)|^2 \right]^{p/2} \\ &= C_p \mathbb{E}\left[|Q_0^{(\alpha)} F(X_{t\wedge\tau})|^2 - |Q_a^{(\alpha)} F(X_{t\wedge\tau})|^2 \right]^{p/2} \\ &\leq C_p \left[\|F\|_p^p + \|Q_a^{(\alpha)} F\|_p^p \right] \leq C_p \|F\|_p^p . \end{aligned} \tag{7.4.47}$$

From the definition of H_F, we have

$$\begin{aligned} \|H_F\|_p^p &= \int_{\mathbb{R}^d} \left| \int_0^\infty r Q_r^{(0)} g_F(x,r)^2 dr \right|^{p/2} \mu(dx) \\ &= \lim_{a\to\infty} \int_{\mathbb{R}^d} \left| \int_0^\infty (r\wedge a) Q_r^{(0)} g_F(x,r)^2 dr \right|^{p/2} \mu(dx) \\ &= \lim_{a\to\infty} \mathbb{E}\left\{ \mathbb{E}\left[\int_0^\tau g_F(X_r, B_r^a)^2 ds \middle| X_\tau \right] \right\}^{p/2} \\ &\leq \lim_{a\to\infty} \mathbb{E}\left\{ \int_0^\tau g_F(X_r, B_r^a)^2 ds \right\}^{p/2} \\ &\leq \lim_{a\to\infty} \mathbb{E}\left\{ \int_0^\tau \left[\alpha|u(X_r, B_r^a)|^2 + g_F(X_r, B_r^a)^2 \right] dr \right\}^{p/2} \\ &\leq C_p \|F\|_p , \end{aligned}$$

where the third identity follows from Lemma 7.9, the first inequality follows from Jessen's inequality, and the last one follows from (7.4.47). This is (7.4.46).

Step 4. Define

$$H_F^{\uparrow}(x) = \left\{ \int_0^\infty tQ_t^{(0)} \left([g_F^{\rightarrow}]^2 \right)(x,t)dt \right\}^{1/2} ; \tag{7.4.48}$$

$$H_F^{\rightarrow}(x) = \left\{ \int_0^\infty tQ_t^{(0)} \left([g_F^{\rightarrow}]^2 \right)(x,t)dt \right\}^{1/2} . \tag{7.4.49}$$

Now we compare $H_F^{\uparrow}$ and $H_F^{\rightarrow}$ with $G_F^{\uparrow}$ and $G_F^{\rightarrow}$. We claim for any $F \in \mathcal{P}$,

$$G_F^{\uparrow}(x) \le 2H_F^{\uparrow}(x), \qquad \forall x \in \mathbb{R}^d ; \tag{7.4.50}$$

$$G_F^{\rightarrow}(x) \le 2H_F^{\rightarrow}(x), \qquad \forall x \in \mathbb{R}^d . \tag{7.4.51}$$

In fact, from the definition of $Q_t^{(\alpha)}$, we have

$$\begin{aligned} |Q_t^{(\alpha)} F(x)|^2 &= \left| \int_0^\infty e^{-\alpha s} T_s F(x) \lambda_t(ds) \right|^2 \\ &\le \int_0^\infty e^{-\alpha s} \left| T_s F(x) \right|^2 \lambda_t(ds) \\ &\le \int_0^\infty e^{-\alpha s} T_s(|F|^2)(x) \lambda_t(ds) = Q_t^{(0)}(|F|^2)(x) . \end{aligned} \tag{7.4.52}$$

From the semigroup property, we see

$$\frac{\partial}{\partial r} Q_r^{(\alpha)} F \big|_{r=2t} = Q_t^{(\alpha)} \frac{\partial}{\partial t} Q_t^{(\alpha)} F .$$

Thus

$$\begin{aligned} g_F^{\rightarrow}(\cdot, 2t)^2 &= \left| \frac{\partial}{\partial r} Q_r^{(\alpha)} F \big|_{r=2t} \right|^2 = \left| Q_t^{(\alpha)} \frac{\partial}{\partial t} Q_t^{(\alpha)} F) \right|^2 \\ &\le Q_t^{(0)} \left(\left| \frac{\partial}{\partial t} Q_t^{(\alpha)} F \right|^2 \right) = Q_t^{(0)} \left([g_F^{\rightarrow}]^2 \right)(\cdot, t) , \end{aligned}$$

where the inequality follows from (7.4.52). Now we can use the above inequality to prove (7.4.51).

$$\begin{aligned} G_F^{\rightarrow}(x)^2 &= \int_0^\infty t g_F^{\rightarrow}(\cdot, t)^2 dt = 4 \int_0^\infty t g_F^{\rightarrow}(\cdot, 2t)^2 dt \\ &\le 4 \int_0^\infty tQ_t^{(0)} \left([g_F^{\rightarrow}]^2 \right)(x,t)dt = 4H_F^{\rightarrow}(x)^2 . \end{aligned} \tag{7.4.53}$$

This proves (7.4.51). Now we prove (7.4.50). First, for any vector $n = (n_1, \cdots, n_d)$ we have

$$\partial_i Q_t^{(\alpha)} H_n(x) = D_i e^{-\sqrt{\alpha + |n|} t} H_n(x) = e^{-\sqrt{\alpha + n t}} H_{n-e_i}(x) ,$$

where $e_i = (0, \cdots, 0, 1, 0, \cdots, 0)$ (the i coordinate is 1 and others are 0), $H_n(x) = H_{n_1}(x_1) \cdots H_{n_d}(x_d)$ and $n = n_1 + \cdots + n_d$. We also have

$$Q_t^{(\alpha+1)} \partial_i H_n(x) = Q_t^{(\alpha+1)} H_{n-e_i}(x) = e^{-\sqrt{\alpha+|n|}t} H_{n-e_i}(x) .$$

Thus, we have $\partial_i Q_t^{(\alpha)} H_n(x) = Q_t^{(\alpha+1)} \partial_i H_n(x)$. By linearity and density argument we have

$$DQ_t^{(\alpha)} = Q_t^{(\alpha+1)} D .$$

With this identity, we see

$$\begin{aligned} g_F^{\uparrow}(x, 2t)^2 &= |DQ_{2t}^{(\alpha)} F|^2 = |DQ_t^{(\alpha)} Q_t^{(\alpha)} F|^2 = |Q_t^{(\alpha+1)} DQ_t^{(\alpha)} F|^2 \\ &\le Q_t^{(0)} \left(\left| DQ_t^{\alpha)} F \right|^2 \right) = Q_t^{(0)}([g_F^{\uparrow}]^2) . \end{aligned}$$

Now we can prove (7.4.50) in exactly the same way as (7.4.53).

Step 5. We prove the theorem in the case $p > 2$. From (7.4.46), (7.4.50), (7.4.51), we have the upper bound (7.4.16).

As routine we shall use the duality argument to show the lower bound (7.4.17). First, from (7.4.42) and a polarization technique, we have for any F and $\tilde{F}$ in $\mathcal{P}$,

$$\langle F, \tilde{F} \rangle = 4 \int_{\mathbb{R}^d} \mu(dx) \int_0^\infty t \left\langle \frac{\partial}{\partial t} Q_t^{(\alpha)} F(x), \frac{\partial}{\partial t} Q_t^{(\alpha)} \tilde{F}(x) \right\rangle_K dt . \quad (7.4.54)$$

Let p and q be the conjugate numbers, namely, $1/p + 1/q = 1$. By the Cauchy-Schwarz inequality and then the Hölder inequality, we have

$$\begin{aligned} \left| \langle F, \tilde{F} \rangle \right| &\le 4 \int_{\mathbb{R}^d} \mu(dx) \int_0^\infty t \left| \frac{\partial}{\partial t} Q_t^{(\alpha)} F(x) \right|_K \left| \frac{\partial}{\partial t} Q_t^{(\alpha)} \tilde{F}(x) \right|_K dt \\ &\le 4 \int_{\mathbb{R}^d} \mu(dx) \left\{ \int_0^\infty t \left| \frac{\partial}{\partial t} Q_t^{(\alpha)} F(x) \right|_K^2 dt \right\}^{1/2} \left\{ \int_0^\infty t \left| \frac{\partial}{\partial t} Q_t^{(\alpha)} \tilde{F}(x) \right|_K^2 dt \right\}^{1/2} \\ &= 4 \int_{\mathbb{R}^d} G_F^{\rightarrow}(x) G_{\tilde{F}}^{\rightarrow}(x) \mu(dx) \le 4 \|G_F^{\rightarrow}\|_p \|G_{\tilde{F}}^{\rightarrow}\|_q \le C_p \|G_F^{\rightarrow}\|_p \|\tilde{F}\|_q , \end{aligned}$$

where the last inequality follows from (7.4.16) for $q < 2$. This immediately implies (7.4.17) for $\alpha > 0$. The case $\alpha = 0$ can be proved similarly with F replaced by $F - J_0 F$. □

7.5 Meyer's Inequalities Revisited

In this section we first give another proof of Meyer's inequality and then we show the high order Meyer's inequality. We restate the Meyer's inequality in the following.

Theorem 7.13. *For any real number $p \in (1, \infty)$, there are positive constants c_p and C_p such that*

$$c_p \| |DF| \|_p \le \|(-L)^{1/2} F\|_p \le C_p \| |DF| \|_p . \tag{7.5.1}$$

Proof. We shall use C to denote $(-L)^{1/2}$. From Lemma 7.4 and then from Lemma 7.3, we have

$$T_t \left[\sum_{k=1}^N (D_k F)^2 \right]^{1/2} \ge \left[\sum_{k=1}^N (T_t D_k F)^2 \right]^{1/2} \ge e^t \left[\sum_{k=1}^N (D_k T_t F)^2 \right]^{1/2} .$$

Or we can write

$$T_t \sqrt{\langle DF, DF \rangle} \ge e^t \sqrt{\langle DT_t F, DT_t F \rangle}$$

which also implies (by replacing F by $T_t F$)

$$T_t \sqrt{\langle DT_t F, DT_t F \rangle} \ge e^t \sqrt{\langle DT_{2t} F, DT_{2t} F \rangle} .$$

Thus

$$\begin{aligned} \psi_F^2 &= \int_0^\infty \left[T_t \sqrt{\langle DT_t F, DT_t F \rangle} \right]^2 dt \ge \int_0^\infty e^{2t} \langle DT_{2t} F, DT_{2t} F \rangle dt \\ &= \frac{1}{2} \int_0^\infty e^t \langle DT_t F, DT_t F \rangle dt . \end{aligned}$$

From Theorem 7.11, we see

$$\|F\|_p \ge C_p \left\| \left[\int_0^\infty e^t \langle DT_t F, DT_t F \rangle dt \right]^{1/2} \right\|_p . \tag{7.5.2}$$

For any real number $u \ge 0$, substituting F by $T_u F$ in the above inequality and then making substitution $t + u \to t$ in the integration we have

$$e^{u/2} \|T_u F\|_p \ge C_p \left\| \left[\int_u^\infty e^t \langle DT_t F, DT_t F \rangle dt \right]^{1/2} \right\|_p .$$

Therefore, we have

$$\int_0^\infty e^u\|T_uF\|_p du \geq C_p \int_0^\infty e^{u/2}\left\|\left[\int_u^\infty e^t\langle DT_tF, DT_tF\rangle dt\right]^{1/2}\right\|_p du$$

$$= C_p \int_0^\infty e^{u/2}\left\|\left[\int_0^\infty \chi_{\{u<t\}}e^t\langle DT_tF, DT_tF\rangle dt\right]^{1/2}\right\|_p du$$

$$\geq C_p \left\|\int_0^\infty e^{u/2}\left[\int_0^\infty \chi_{\{u<t\}}e^t\langle DT_tF, DT_tF\rangle dt\right]^{1/2} du\right\|_p$$

$$= C_p \left\|\int_0^\infty e^{u/2}\left[\int_0^\infty \left(\chi_{\{u<t\}}e^{t/2}\sqrt{\langle DT_tF, DT_tF\rangle}\right)^2 dt\right]^{1/2} du\right\|_p$$

$$\geq C_p \left\|\left\{\int_0^\infty \left[\int_0^\infty e^{u/2}\chi_{\{u<t\}}e^{t/2}\sqrt{\langle DT_tF, DT_tF\rangle}du\right]^2 dt\right\}^{1/2}\right\|_p$$

$$= 4C_p \left\|\left\{\int_0^\infty \left[(e^t - e^{t/2})\sqrt{\langle DT_tF, DT_tF\rangle}\right]^2 dt\right\}^{1/2}\right\|_p$$

$$\geq C_p \left\|\left\{\int_0^\infty e^{2t}\langle DT_tF, DT_tF\rangle dt\right\}^{1/2}\right\|_p$$

$$-C_p \left\|\left\{\int_0^\infty e^{t}\langle DT_tF, DT_tF\rangle dt\right\}^{1/2}\right\|_p .$$

Combining this with (7.5.2), we have

$$\left\|\left\{\int_0^\infty e^{2t}\langle DT_tF, DT_tF\rangle dt\right\}^{1/2}\right\|_p \leq C_p\|F\|_p + C_p\int_0^\infty e^u\|T_uF\|_p du .$$

If $(J_0 + J_1)F = 0$, then $\|T_uF\|_p \leq e^{-2u}\|F\|_p$. Thus, if $(J_0 + J_1)F = 0$, we have

$$\left\|\left\{\int_0^\infty e^{2t}\langle DT_tF, DT_tF\rangle dt\right\}^{1/2}\right\|_p \leq C_p\|F\|_p . \tag{7.5.3}$$

Therefore, we have

$$|||DF|||_p^p = \left\|\left(\sum_{k=1}^\infty (D_kF)^2\right)^{1/2}\right\|_p \leq C_p\left\|\left\{\sum_{k=1}^\infty G_{D_kF}^2\right\}^{1/2}\right\|_p$$

$$= C_p\left\|\left\{\sum_{k=1}^\infty \int_0^\infty t\left[\frac{d}{dt}Q_tD_kF\right]^2 dt\right\}^{1/2}\right\|_p . \tag{7.5.4}$$

By Lemma 7.3 we have

$$Q_t D_i = D_i \tilde{Q}_t\,, \quad \text{where} \quad \tilde{Q}_t = \sum_{n=1}^{\infty} e^{-\sqrt{n-1}t} J_n\,.$$

Thus

$$\frac{d}{dt} Q_t D_i F = D_i \frac{d}{dt} \tilde{Q}_t F = D_i \tilde{Q}_t C R F\,,$$

where

$$RF = \sum_{n=0}^{\infty} \sqrt{1 - \frac{1}{n}} J_n F\,.$$

Substituting this into (7.5.4)

$$|||DF|||_p^p \le C_p \left\| \left\{ \int_0^\infty t \left| D\tilde{Q}_t CRF \right|^2 dt \right\}^{1/2} \right\|_p\,.$$

Now we use $\tilde{Q}_t = \int_0^\infty e^s T_s \mu_t(ds)$ to obtain

$$\begin{aligned} \left| D\tilde{Q}_t CRF \right| &= \left| D \int_0^\infty e^s T_s CRF \mu_t(ds) \right| \le \int_0^\infty e^s |DT_s CRF| \mu_t(ds) \\ &\le \left[\int_0^\infty e^{2s} |DT_s CRF|^2 \mu_t(ds) \right]^{1/2}\,. \end{aligned}$$

From the expression (7.3.3) we have

$$\int_0^\infty t \frac{\mu_t(ds)}{ds} dt = \int_0^\infty \frac{t^2}{2\sqrt{\pi}} e^{-t^2/4s} s^{-3/2} dt = 1\,, \quad \forall\ s > 0\,.$$

Hence, we have

$$\begin{aligned} |||DF|||_p^p &\le C_p \left\| \left\{ \int_0^\infty t \int_0^\infty e^{2s} |DT_s CRF|^2 \mu_t(ds) dt \right\}^{1/2} \right\|_p \\ &= C_p \left\| \left\{ \int_0^\infty e^{2s} |DT_s CRF|^2 ds \right\}^{1/2} \right\|_p \le C_p \|CRF\|_p\,, \end{aligned}$$

where the last inequality follows from (7.5.3), $CR = RC$ and the fact that $\|RF\|_p \le \|F\|_p$. This implies that if $(J_0 + J_1)F = 0$, then $|||DF|||_p \le C_p \|CF\|_p$. It is straightforward that

$$|||D(J_0 + J_1)F|||_p \le C_p \|F\|_p \le C_p \|CF\|_p\,.$$

Thus, we have proved

$$|||DF|||_p \le C_p \|CF\|_p\,, \quad \forall\ \ F \in \mathcal{P}\,. \tag{7.5.5}$$

Now we show the converse of above inequality (7.5.5). For any $F, G \in \mathcal{P}$, we have (denoting $\tilde{G} = C^{-1}(I - J_0)G$)

$$\begin{aligned}|\mathbb{E}[(CF)G]| &= |\mathbb{E}[(CF)(I - J_0)G]| = \left|\mathbb{E}[(CF)(C\tilde{G})]\right| \\ &= \left|\mathbb{E}[(C^2F)\tilde{G}]\right| = \left|\mathbb{E}[(LF)\tilde{G}]\right| = \mathbb{E}\langle DF, D\tilde{G}\rangle \\ &\le \||DF|\|_p \||D\tilde{G}|\|_q \le C_q \||DF|\|_p \||C\tilde{G}|\|_q \\ &= C_q \||DF|\|_p \||(I - J_0)G|\|_q \le C_q \||DF|\|_p \||G|\|_q \,,\end{aligned}$$

where the inequality in the second last line follows from (7.5.5). This implies

$$\|CF\|_p \le C_p \||DF|\|_p \,, \quad \forall \; F \in \mathcal{P} \,. \tag{7.5.6}$$

□

Theorem 7.14 (Meyer's inequality for higher derivatives). *For any real number $p \in (1, \infty)$ and positive integer k, there are positive constants $c_{p,k}$ and $C_{p,k}$ such that*

$$c_{p,k} \||D^k F|\|_p \le \|(-L)^{k/2} F\|_p \le C_{p,k} \left[\|F\|_p + \||D^k F|\|_p\right] \,. \tag{7.5.7}$$

Proof. We shall prove these two inequalities by induction on k. Let $\gamma_n(\omega)$ be the Rademacher random sequence indexed by the multi-indices $m = (i_1, \cdots, i_k) \in \mathbb{N}^k$ on some probability space $(\Omega, \mathcal{F}, P)$. Denote $D_m = D_{i_1} D_{i_1} \cdots D_{i_k}$. Then

$$|D^k F|^2 = \sum_{m \in \mathbb{N}^k} |D_m F|^2 \,.$$

Set

$$\gamma_F(\omega) = \sum_{m \in \mathbb{N}^k} \gamma_m(\omega) D_m F \,.$$

Then

$$D_i \gamma_F(\omega) = \sum_{m \in \mathbb{N}^k} \gamma_m(\omega) D_m F \quad \text{and} \quad C\gamma_F(\omega) = \sum_{m \in \mathbb{N}^k} \gamma_m(\omega) C D_m F \,.$$

From Theorem 7.13, we have

$$\mathbb{E}^\omega |D\gamma_F(\omega)|^p \le C_p \mathbb{E}^\omega |C\gamma_F(\omega)|^p \,, \quad \forall \, \omega \in \Omega' \,, \quad \forall \; F \in \mathcal{P} \,.$$

Thus

$$\mathbb{E}|D\gamma_F|^p \le C_p \mathbb{E}|C\gamma_F|^p \,, \quad \forall \; F \in \mathcal{P} \,. \tag{7.5.8}$$

But from the Kahane-Khintchine inequality (4.1.9), we have

$$\mathbb{E}|D\gamma_F|^p \geq c_p \left\| \left(\sum_{1\leq i<\infty, m\in\mathbb{N}^k} |D_i D_m F|^2 \right)^{1/2} \right\|_p^p = c_p \| |D^{k+1}F| \|_p^p . \quad (7.5.9)$$

Again from the Kahane-Khintchine inequality (4.1.9), we have

$$\begin{aligned}
\mathbb{E}|C\gamma_F|^p &= \mathbb{E}\left| \sum_{m\in\mathbb{N}^k} \gamma_m(\omega) C D_m F \right|^p \geq C_p \mathbb{E} \left| \left(\sum_{m\in\mathbb{N}^k} [C D_m F]^2 \right)^{1/2} \right|^p \\
&= C_p \mathbb{E} \left| \left(\sum_{m\in\mathbb{N}^k} [D_m C R_k F]^2 \right)^{1/2} \right|^p = C_p \mathbb{E} \left| D^k C R_k \right|^p ,
\end{aligned}$$

where $R_k = \sum_{n=k}^{\infty} \sqrt{1-\frac{k}{n}} J_n$ which is bounded from L^p to L^p by (7.3.12). By induction, we have

$$\mathbb{E}|C\gamma_F|^{\leq} C_p \mathbb{E} \left| C^{k+1} R_k F \right|^p = C_p \mathbb{E} \left| R_k C^{k+1} F \right|^p \leq C_p \|C^{k+1}F\|^p . \quad (7.5.10)$$

Substituting (7.5.9) and (7.5.10) into (7.5.8), we have

$$\| |D^{k+1}F| \|_p \leq C_p \|C^{k+1}F\|^p . \quad (7.5.11)$$

This proves the first inequality in (7.5.7). The other inequality in (7.5.7) can be proved by a duality argument as in the proof of (7.13). □

From the Meyer's inequalities we have

Theorem 7.15. *Let $F = \sum_{n=0}^{\infty} F_n$ be the chaos expansion of F, where F_n is the n-th chaos of F. Then, F is in $\mathbb{D}_{\theta,2}$, $\theta \in \mathbb{R}$ if and only if*

$$\sum_{n=0}^{\infty} (n+1)^\theta \|F_n\|^2 < \infty . \quad (7.5.12)$$

Proposition 7.7. *Let $F \in \mathbb{D}_{k,p}$ and let $G \in \mathbb{D}_{k,q}$. If $1/p + 1/g \leq 1$, then $FG \in \mathbb{D}_{k,r}$, where $1/p + 1/g = 1/r$.*

Proof. Let $F, G \in \mathcal{S}_{\exp}$ and let $j \in [1,k]$ be an integer. It is easy to establish the following Leibniz rule for the j derivative:

$$D^j(FG) = \sum_{\ell=0}^{j} \binom{j}{\ell} D^\ell F \otimes D^{j-\ell} G .$$

Thus

$$\begin{aligned} |D^j(FG)|_{\mathbb{H}^{\otimes j}} &\le \sum_{\ell=0}^{j} \binom{j}{\ell} |D^\ell F \otimes D^{j-\ell} G|_{\mathbb{H}^{\otimes j}} \\ &\le \sum_{\ell=0}^{j} \binom{j}{\ell} |D^\ell F|_{\mathbb{H}^{\otimes \ell}} |D^{j-\ell} G|_{\mathbb{H}^{\otimes j-\ell}} . \end{aligned}$$

By Hölder inequality, this implies that

$$\| |D^j(FG)|_{\mathbb{H}^{\otimes j}} \|_r \le C \| |F|_{\mathbb{H}^{\otimes j}} \|_p \| |G|_{\mathbb{H}^{\otimes j}} \|_q .$$

□

Denote the following Meyer-Watanabe test and distribution spaces:

$$\mathbb{D}_\infty = \bigcap_{k\ge 0, p\ge 1} \mathbb{D}_{k,p} \quad and \quad \mathbb{D}_{-\infty} = \bigcup_{k\le 0, p\ge 1} \mathbb{D}_{k,p} \tag{7.5.13}$$

They are topological spaces with respect to the induced topologies. It is clear that the Meyer-Watanabe test functional space $\mathbb{D}_\infty$ is an algebra (closed with respect to linear combination and multiplication).

For any $F, G \in \mathcal{S}_{\exp}$, we have

$$\mathbb{E}(FG) = \mathbb{E}(F(I-L)^{-k/2}(I-L)^{k/2}G) = \mathbb{E}((I-L)^{-k/2}F(I-L)^{k/2}G) .$$

The Hölder inequality yields

$$\begin{aligned} |\mathbb{E}(FG)| &\le \left\{ \mathbb{E}(|(I-L)^{-k/2}F|^q) \right\}^{1/q} \left\{ \mathbb{E}(|(I-L)^{k/2}F|^p) \right\}^{1/p} \\ &\le C \|F\|_{\mathbb{D}_{-k,q}} \|F\|_{\mathbb{D}_{k,p}} . \end{aligned}$$

Thus, we can consider $\mathbb{D}_{-k,q}$ as the dual (the space of bounded linear functionals) of $\mathbb{D}_{k,p}$ with the pairing $\mathbb{E}(FG)$.

Proposition 7.8. *We can identify $\mathbb{D}_{-k,q}$ as the dual of $\mathbb{D}_{k,p}$ with the pairing $\mathbb{E}(FG)$. Namely, we can write $\mathbb{D}_{-k,q} = (\mathbb{D}_{k,p})'$. Thus, the expectation $\mathbb{E}(FG)$ is well-defined for $F \in \mathbb{D}_{-k,q}$ and $G \in \mathbb{D}_{k,p}$.*

An application of Meyer's inequality is the Meyer's inequality for divergence (or stochastic integral).

Corollary 7.3. *Let V be a separable Hilbert space.*

(i) For every $p > 0$ and every $s \in \mathbb{R}$, the derivative operator D is continuous from $\mathbb{D}_{s,p}(V)$ to $\mathbb{D}_{s-1,p}(V \otimes \mathbb{H})$.

(ii) *For every $p > 0$ and every $s \in \mathbb{R}$, the divergence operator δ is continuous from $\mathbb{D}_{s,p}(V \otimes \mathbb{H})$ to $\mathbb{D}_{s-1,p}(V)$. This means that there is a constant $c_{s,p}$ such that*

$$\|\delta(u)\|_{s,p} \le c_{s,p}\|u\|_{s-1,p}\,, \quad \forall u \in \mathbb{D}_{s,p}(V \otimes \mathbb{H})\,, \tag{7.5.14}$$

where the norm $\|\cdot\|$ is with respect to the appropriate spaces.

Proof. For simplicity we consider the case $V = \mathbb{R}$. For any $F \in \mathcal{P}$, we have

$$(I-L)^{(s-1)/2}DF = DR(I-L)^{(s-1)/2}F\,, \tag{7.5.15}$$

where

$$R = \sum_{n=1}^{\infty}\left(\frac{n}{n+1}\right)^{(s-1)/2} J_n\,.$$

By Theorem 7.10, we know that R is a bounded operator from L^p to itself. Since R commute with L, we see that R is also bounded from $\mathbb{D}_{s,p}$ to itself for any $s \in \mathbb{R}$ and $p > 1$. Thus, we have

$$\begin{aligned}
\|DF\|_{s-1,p} &= \|(I-L)^{(s-1)/2}DF\|_p = \|DR(I-L)^{(s-1)/2}F\|_p \\
&\le \|R(I-L)^{(s-1)/2}F\|_{1,p} \le c_p\|(I-L)^{(s-1)/2}F\|_{1,p} \\
&= c_p\|(I-L)^{1/2}(I-L)^{(s-1)/2}F\|_p = c_p\|(I-L)^{s/2}F\|_p\,.
\end{aligned}$$

This proves (i). Let q be the conjugate of p. From the self-adjoint property of L; the commutativity between L and R; identity (7.5.15), and the boundedness of $D(I-L)^{-1/2}$, we have

$$\begin{aligned}
\|\delta(u)\|_{s-1,p} &= \sup_{\|F\|_q\le 1} \mathbb{E}\left(\left[(I-L)^{(s-1)/2}\delta(u)\right]F\right) \\
&= \sup_{\|F\|_q\le 1} \mathbb{E}\left(\left[D(I-L)^{(s-1)/2}F\right]u\right) \\
&= \sup_{\|F\|_q\le 1} \mathbb{E}\left(\left[DR(I-L)^{s/2}(I-L)^{-1/2}R^{-1}F\right]u\right) \\
&= \sup_{\|F\|_q\le 1} \mathbb{E}\left(\left[DR(I-L)^{s/2}(I-L)^{-1/2}R^{-1}F\right]u\right) \\
&= \sup_{\|F\|_q\le 1} \mathbb{E}\left(\left[(I-L)^{s/2}D(I-L)^{-1/2}R^{-1}F\right]u\right) \\
&= \sup_{\|F\|_q\le 1} \mathbb{E}\left(\left[D(I-L)^{-1/2}R^{-1}F\right](I-L)^{s/2}u\right) \\
&\le \sup_{\|F\|_q\le 1} \|D(I-L)^{-1/2}R^{-1}F\|_q\|(I-L)^{s/2}u\|_p \\
&\le \sup_{\|F\|_q\le 1} \|R^{-1}F\|_q\|u\|_{s,p} \le c_p\|u\|_{s,p}\,,
\end{aligned}$$

proving (ii). □

As an application of the Meyer's inequality, we have

Theorem 7.16. *Let $F \in \mathbb{D}^{2,s}$ such that $E[|F|^{2p}] < \infty$ and $E[\|DF\|_{\mathbb{H}}^{-2r}] < \infty$ for $p, r, s > 1$ satisfying $\frac{1}{p} + \frac{1}{r} + \frac{1}{s} = 1$. Denote*

$$w = \|DF\|_{\mathbb{H}}^2, \quad u = w^{-1} DF.$$

Then, $u \in \mathbb{D}^{1,p'}$ with $p' = \frac{p}{p-1}$ and F has a density given by

$$f_F(x) = \mathbb{E}\left[\mathbf{1}_{\{F>x\}} \delta(u)\right]. \tag{7.5.16}$$

Furthermore, $f_F(x)$ is bounded and Hölder continuous of order $\frac{1}{p}$, that is

$$f_F(x) \le C_p \left\|w^{-1}\right\|_r \|F\|_{2,s} \left(1 \wedge (|x|^{-2} \|F\|_{2p}^2)\right), \tag{7.5.17}$$

$$|f_F(x) - f_F(y)| \le C_p \left\|w^{-1}\right\|_r^{1+\frac{1}{p}} \|F\|_{2,s}^{1+\frac{1}{p}} |x-y|^{\frac{1}{p}} \tag{7.5.18}$$

for any $x, y \in \mathbb{R}$, where C_p is a constant depending only on p.

Proof. Note that

$$Du = w^{-1} D^2 F - 2w^{-2} \left(D^2 F \otimes_1 DF\right) \otimes DF.$$

Applying (7.5.14) and Hölder's inequality we have

$$\begin{aligned} \|\delta(u)\|_{p'} &\le C_p \|u\|_{1,p'} \le C_p(\|u\|_{p'} + \|Du\|_{p'}) \\ &\le C_p(\left\|w^{-1}\|DF\|_{\mathbb{H}}\right\|_{p'} + 3\left\|w^{-1}\left\|D^2F\right\|_{\mathbb{H}\otimes\mathbb{H}}\right\|_{p'}) \\ &\le 3C_p \left\|w^{-1}\right\|_r (\|DF\|_s + \left\|D^2 F\right\|_s). \end{aligned} \tag{7.5.19}$$

Then, $u \in \mathbb{D}^{1,p'}$ and the density formula (7.5.16) holds. From $E[\delta(u)] = 0$ and Hölder's inequality it follows that

$$\begin{aligned} \left|\mathbb{E}\left[\mathbf{1}_{\{F>x\}} \delta(u)\right]\right| &\le P(|F| > |x|)^{\frac{1}{p}} \|\delta(u)\|_{p'} \\ &\le \left(1 \wedge (|x|^{-2p} \|F\|_{2p}^{2p})\right)^{\frac{1}{p}} \|\delta(u)\|_{p'}. \end{aligned} \tag{7.5.20}$$

Then, (7.5.17) follows from (7.5.20) and (7.5.19).

Finally, for $x < y \in \mathbb{R}$, noticing that $\mathbf{1}_{\{F>x\}} - \mathbf{1}_{\{F>y\}} = \mathbf{1}_{\{x<F\le y\}}$, we have

$$|f_F(x) - f_F(y)| \le \left(E[\mathbf{1}_{\{x<F\le y\}}]\right)^{\frac{1}{p}} \|\delta(u)\|_{p'}.$$

Applying (7.5.17) and (7.5.19) with the fact that $E[\mathbf{1}_{\{x<F\le y\}}] = \int_x^y f_F(z)\, dz$ one gets (7.5.18). □

7.6 Interpolation Inequality

From Meyer's inequality (7.2.6), we see that $\mathbb{D}_{k,p}$ is given by

$$\mathbb{D}_{k,p} = \left\{ F : \|F\|_{k,p} = \|(1-L)^{k/2} F\|_p < \infty \right\} . \tag{7.6.1}$$

Since $(1-L)^{k/2}$ is well-defined for any real number k, we can use (7.6.1) to define the Sobolev space for all $k \in \mathbb{R}$ and $p \geq 1$.

Lemma 7.10. *For any $\alpha \geq 0$, $(1-L)^{-\alpha}$ is a bounded linear operator from $L^p(\mathbb{B})$ to itself with norm ≤ 1.*

Proof. For any $F \in \mathcal{S}_{\text{exp}}$, we have easily that $(1-L)^{-\alpha} F = \frac{1}{\Gamma(\alpha)} \int_0^\infty e^{-t} T_t t^{\alpha-1} F dt$, where T_t is the Ornstein-Uhlenbeck semigroup. Thus, by its contraction property, we have

$$\begin{aligned}
\|(1-L)^{-\alpha} F\|_p &= \frac{1}{\Gamma(\alpha)} \int_0^\infty e^{-t} \|T_t F\|_p t^{\alpha-1} dt \\
&\leq \frac{1}{\Gamma(\alpha)} \int_0^\infty e^{-t} t^{\alpha-1} dt \|F\|_p \leq \|F\|_p .
\end{aligned}$$

□

Proposition 7.9. *The Sobolev spaces satisfy the following relation.*

$$\mathbb{D}_{l,q} \subseteq \mathbb{D}_{k,p} , \quad \forall \ -\infty < k < l < \infty \text{ and } 1 \leq p \leq q < \infty . \tag{7.6.2}$$

Proof. For any $F \in \mathcal{S}_{\text{exp}}$, we have by the above lemma

$$\begin{aligned}
\|F\|_{\mathbb{D}_{l,q}} &= \|(1-L)^{-l/2} F\|_q = \|(1-L)^{-(l-k)/2} (1-L)^{-k/2} F\|_q \\
&\leq \|(1-L)^{-k/2} F\|_q \leq \|(1-L)^{-k/2} F\|_p = \|F\|_{\mathbb{D}_{k,p}} .
\end{aligned}$$

□

Theorem 7.17. *Denote*

$$K_\rho = \frac{(2\rho)^{1-\rho}}{\Gamma(2-\rho)} + \frac{2}{(2\rho)^\rho \Gamma(1-\rho)} , \quad \forall \rho \in (0,1) .$$

Then, for any $\rho \in (0,1)$ and $1 \leq p < \infty$ we have

$$\|(I-L)^\rho F\|_p \leq K_\rho \|F\|_p^{1-\rho} \|(I-L) F\|_p^\rho , \quad \forall \ F \in \mathbb{D}_{2,p} . \tag{7.6.3}$$

Proof. Denote $G = (I - L)F$. To show (7.6.3) it suffices to prove

$$\|(I-L)^{\rho-1}G\|_p \le K_\rho \|G\|_p^\rho \|(I-L)^{-1}G\|_p^{1-\rho} \tag{7.6.4}$$

for all $G \in L^p$. Set $Q_t = e^{-t}P_t$. Then

$$\begin{aligned}(I-L)^{\rho-1}G &= \frac{1}{\Gamma(1-\rho)}\int_0^\infty t^{-\rho}e^{-t}P_t G dt \\ &= \frac{1}{\Gamma(1-\rho)}\int_0^\infty t^{-\rho}Q_t G dt.\end{aligned}$$

For any $\varepsilon > 0$ we write

$$(I-L)^{\rho-1}G = \frac{1}{\Gamma(1-\rho)}\left[\int_0^\varepsilon t^{-\rho}Q_t G dt + \int_\varepsilon^\infty t^{-\rho}Q_t G dt\right].$$

Thus

$$\|(I-L)^{\rho-1}G\|_p \le \frac{1}{\Gamma(1-\rho)}\left[\left\|\int_0^\varepsilon t^{-\rho}Q_t G dt\right\|_p + \left\|\int_\varepsilon^\infty t^{-\rho}Q_t G dt\right\|_p\right]. \tag{7.6.5}$$

The above first term inside [ ] is bounded by

$$\begin{aligned}\left\|\int_0^\varepsilon t^{-\rho}Q_t G dt\right\|_p &\le \int_0^\varepsilon t^{-\rho}\|Q_t G\|_p dt \\ &\le \int_0^\varepsilon t^{-\rho}\|G\|_p dt = \frac{\varepsilon^{1-\rho}}{1-\rho}\|G\|_p.\end{aligned} \tag{7.6.6}$$

To bound the second term in (7.6.5), we use

$$\frac{dQ_t}{dt} = -Q_t(I-L) \quad \text{or} \quad Q_t = -\frac{dQ_t}{dt}(I-L)^{-1}.$$

Hence, we have

$$\begin{aligned}\int_\varepsilon^\infty t^{-\rho}Q_t G dt &= -\int_\varepsilon^\infty t^{-\rho}\frac{dQ_t}{dt}(I-L)^{-1}G dt \\ &= \varepsilon^{-\rho}Q_\varepsilon(I-L)^{-1}G - \rho\int_\varepsilon^\infty t^{-\rho-1}Q_t(I-L)^{-1}G dt.\end{aligned}$$

Now we use $\|Q_\varepsilon\|_{p,p} \le 1$, $\|Q_t\|_{p,p} \le 1$ to obtain

$$\begin{aligned}\left\|\int_\varepsilon^\infty t^{-\rho}Q_t G dt\right\|_p &\le \varepsilon^{-\rho}\|Q_\varepsilon(I-L)^{-1}G\|_p + \rho\int_\varepsilon^\infty t^{-\rho-1}\|Q_t(I-L)^{-1}G\|_p dt \\ &\le \varepsilon^{-\rho}\|(I-L)^{-1}G\|_p + \rho\int_\varepsilon^\infty t^{-\rho-1}\|(I-L)^{-1}G\|_p dt \\ &= 2\varepsilon^{-\rho}\|(I-L)^{-1}G\|_p.\end{aligned} \tag{7.6.7}$$

Substituting (7.6.6) and (7.6.7) into (7.6.5), we have

$$\|(I-L)^{\rho-1}G\|_p \le \frac{1}{\Gamma(1-\rho)}\left[\frac{\varepsilon^{1-\rho}}{1-\rho}\|G\|_p + 2\varepsilon^{-\rho}\|(I-L)^{-1}G\|_p\right].$$

Since the above inequality holds for all $\varepsilon > 0$, we take

$$\varepsilon = \frac{2\rho\|(I-L)^{-1}G\|_p}{\|G\|_p}$$

(which minimizes the above quantity as a function of ε) to obtain

$$\|(I-L)^{\rho-1}G\|_p \le \frac{1}{\Gamma(1-\rho)}\left[\frac{(2\rho)^{1-\rho}}{1-\rho} + \frac{2}{(2\rho)^\rho}\right]\|G\|_p^\rho\|(I-L)^{-1}G\|_p^{1-\rho}.$$

This is (7.6.4), proving the theorem. □

When $\rho = 1/2$, it is easy to see that $K_\rho = \frac{4}{\Gamma(1/2)}$. So, we have the following inequality from [Decreusefond, Hu and Üstünel (1993)].

Theorem 7.18. *Let $1 \le p < \infty$. For all $F \in L^p \cap \mathbb{D}^2_{1,p}$, we have*

$$\|(I-L)^{1/2}F\|_p \le \frac{4}{\Gamma(1/2)}\|F\|_p^{1/2}\|(I-L)F\|_p^{1/2}. \tag{7.6.8}$$

Remark 7.5.

(i) The constant K_ρ does not depend on p.
(ii) The theorem holds true for general operator which is the generator of a contractive semigroup, bounded from L^p to L^p.

The following inequality is also from [Decreusefond, Hu and Üstünel (1993)].

Theorem 7.19. *For all $p > 1$, there is a constant $C_p > 0$ such that*

$$\|DF\|_p \le C_p(\|F\|_p + \|F\|_p^{1/2}\|D^2F\|_p^{1/2}). \tag{7.6.9}$$

Proof. From the Meyer inequality and from

$$ab \le \varepsilon a^2 + \frac{1}{\varepsilon}b^2, \quad \forall\ a, b, \varepsilon \in \mathbb{R}_+,$$

we have

$$\begin{aligned}
\|DF\|_p &\le C_p\|(I-L)^{1/2}F\|_p \le C_p\|F\|_p^{1/2}\|(I-L)F\|_p^{1/2} \\
&\le C_p\|F\|_p^{1/2}\left(\|F\|_p + \|DF\|_p + \|D^2F\|_p\right)^{1/2} \\
&\le C_p\left[\|F\|_p + \|F\|_p^{1/2}\|DF\|_p^{1/2} + \|F\|_p^{1/2}\|D^2F\|_p^{1/2}\right] \\
&\le C_p\left[\|F\|_p + \frac{1}{\varepsilon}\|F\|_p + \varepsilon\|DF\|_p + \|F\|_p^{1/2}\|D^2F\|_p^{1/2}\right]
\end{aligned}$$

for any $\varepsilon > 0$. Choosing for example $\varepsilon C_p = 1/2$ proves the theorem. □

An immediate consequence of this theorem is

Corollary 7.4. *If* $(F_n, n \in \mathbb{N})$ *converges to* 0 *in* L^p *and if* $\sup_n \|D^2 F\|_p < \infty$, *then* $(DF_n, n \in \mathbb{N})$ *converges to* 0 *in* L^p.

7.7 Grothendieck Inequality

In this section we will discuss another important inequality called Grothendieck inequality.

Since its appearance in [Grothendieck (1956)], Grothendieck inequality has been restated in various forms and reproved in various ways by many mathematicians. Here, we present this inequality following Krivine [Krivine (1979)].

The Grothendieck inequality has a general form concerning a general Banach space and a Hilbert space. But we can reduce the problem to a concrete one. To shorten the presentation we begin with such particular Banach space and Hilbert space. Let T be a compact set on a topological space, $\mathbb{R}$ the set of real numbers and $C(T;\mathbb{R})$ the set of all the continuous functions from T to $\mathbb{R}$. $C(T;\mathbb{R})$ becomes a Banach space under the sup norm $\|\cdot\|_\infty$. The norm $\|U\|$ of a bounded operator U from $C(T;\mathbb{R})$ to $C(T;\mathbb{R})^*$ is defined as the smallest constant C satisfying

$$\langle Uf, g\rangle \le C\|f\|_\infty \|g\|_\infty, \quad \text{for any } f, g, \in C(T;\mathbb{R}). \tag{7.7.1}$$

The Hilbert space we consider is a real Hilbert space of Gaussian random variables (Gaussians), i.e. a closed subspace of $L^2(\mathbb{B}, \mathbb{H}, \mathbb{P})$. A function (defined on T taking values in $\mathbb{H}$) $f \in C(T;\mathbb{H}) : T \to \mathbb{H}$ can be written as $f(t,x)$ $(i \in T, x \in \mathbb{B})$. Now if U is a bounded operator from $C(T;\mathbb{R})$ to $C(T;\mathbb{R})^*$, we can extend it to be a bounded operator U_H from $C(T;\mathbb{H})$ to $C(T;\mathbb{H})^*$ (when x fixed $f(\cdot, x)$ is in $C(T;\mathbb{R})$ and then just apply U to this function). (Within tensor product we can write $U_H = U \otimes I$). Now $C(T;\mathbb{H})$ become a Banach space under the norm $|||f||| = \|(\mathbb{E}f^2)^{1/2}\|_\infty$, where $\mathbb{E}$ will note the expectation with respect to x (μ). The Grothendieck inequality says that the norm U_H is less than $C\|U\|$ with a suitable constant. The smallest constant is called Grothendieck constant. Its value is not yet known. We can state

Theorem 7.20. *[Grothendieck (1956)] There exists a universal constant* C *such that for any* $f, g \in C(T;\mathbb{H}), U \in L(C(T;\mathbb{R}); C(T;\mathbb{R})^*))$

$$|\mathbb{E}\langle U_H f, g\rangle| \le C\|U\| \|(\mathbb{E}f^2)^{1/2}\|_\infty \|(\mathbb{E}g^2)^{1/2}\|_\infty. \tag{7.7.2}$$

One estimate for this constant C is

$$K_G^R \le \pi/[2\log(1+\sqrt{2})] = 1.782\cdots \,. \qquad (7.7.3)$$

(K=Krivine, G=Grothendieck, R=real).

We will identify U_H and U (use the same notation U). Before proving this theorem we need some preparation. Denote by G the set of all functions $g(s,t)$ on $T \times T$ which can be written as

$$g(s,t) = \mathbb{E}[X_s Y_t] \quad \text{for some} \quad X, Y \in C(T;\mathbb{H}). \qquad (7.7.4)$$

G is a subset of $C(T;\mathbb{R}) \otimes C(T;\mathbb{R})$. Introduce a norm on G by

$$\|g\|_\otimes = \inf\{\mathbb{E}[\sup_{s\in T}|A_s| \sup_{t\in T}|B_t|]; A, B \in C(T;\mathbb{H}) \\ \text{such that} \quad \mathbb{E}(A_s B_t) = E(X_s Y_t)\}\,.$$

This does become a norm. The completion of G under this norm will be still denoted by G.

Lemma 7.11. *The dual of G can be identified with the space of the operators from $C(T;\mathbb{R})$ to $C(T;\mathbb{R})^*$ with the following correspondence:*

$$\langle T_U, f\otimes g\rangle = \langle Uf, g\rangle\,, \quad \forall \quad f, g \in C(T;\mathbb{R}), \qquad (7.7.5)$$

and

$$\|T_U\|_* = \|U\|. \qquad (7.7.6)$$

Proof. (7.7.5) is obvious. we need to show (7.7.6). Given U, T_U defined by (7.7.5) is bilinear and

$$\|T_U\|_* = \sup_{\|f\otimes g\|_\otimes \le 1} |\langle T_U, f\otimes g\rangle|.$$

Now $\|f\otimes g\| \le 1$ implies that for any $\varepsilon > 0$ there is $f^\varepsilon, g^\varepsilon \in C(T;\mathbb{H})$ such that $\mathbb{E}f_s g_t = \mathbb{E}f_s^\varepsilon g_t^\varepsilon$ and $\mathbb{E}[\sup_{s,t\in T}|f_s^\varepsilon g_t^\varepsilon|] \le 1+\varepsilon$. So

$$\|T_U\|_* = \sup_{\|f\otimes g\|_\otimes \le 1} |\mathbb{E}\langle Uf, g\rangle| \le |\mathbb{E}\langle Uf', g'\rangle| \\ = \|U\|\mathbb{E}[\sup_{s,t\in T}|f_s' g_t'|] \le \|U\|(1+\varepsilon)\,.$$

This gives $\|T_U\| \le \|U\|$. On the other hand we know easily $\|f\otimes g\|_\otimes \le \|f\|_\infty \|g\|_\infty$. Then

$$|\langle T_U, f\otimes g\rangle| \le \|T_U\|_* \|f\otimes g\|_\otimes \le \|T_U\|_* \|f\|_\infty \|g\|_\infty. \qquad (7.7.7)$$

So for fixed f, $\langle T_U, f\otimes g\rangle$ is continuous linear functional of g. By Riesz representation theorem, there exists a $Uf \in C(T;\mathbb{R})$ such that

$$\langle T_U, f\otimes g\rangle = \langle Uf, g\rangle.$$

By (7.7.7) we know U is a bounded operator and $\|U\| \le \|T_U\|_*$. □

Lemma 7.12. *The Grothendieck constant K_G^R is given in the following explicit form:*

$$K_G^R = \sup\{\|\mathbb{E}(X_s Y_t)\|_\otimes ; X_s, Y_t \textit{ are Gaussians with variance} \le 1\}\,. \tag{7.7.8}$$

Proof. Note by C_0 the right hand side of (7.7.8) . We want to prove $C_0 = K_G^R$. First, we have easily that for any Gaussians f_s, g_t

$$\|\mathbb{E}[f_s g_t]\|_\otimes \le C_0 \|(\mathbb{E}f^2)^{1/2}\|_\infty \|(\mathbb{E}g^2)1/2\|_\infty\,. \tag{7.7.9}$$

Let f_s, g_t be Gaussians. Then by (7.7.5) and (7.7.6)

$$\begin{aligned}|\mathbb{E}\langle Uf, g\rangle| &= |\mathbb{E}\langle T_U, f\otimes g\rangle| = |\langle T_U, \mathbb{E}[f\otimes g]\rangle| \\ &\le \|U\|\|\mathbb{E}[f\otimes g]\|_\otimes \quad (\text{by}(7.7.6)) \\ &\le C_0\|U\|\|(\mathbb{E}f^2)^{1/2}\|_\infty \|(\mathbb{E}g^2)1/2\|_\infty .\end{aligned}$$

So we know $K_G^R \le C_0$.

On the other hand, there exist X_s^0, Y_t^0 with variances ≤ 1 such that $\|\mathbb{E}[X^0 \otimes Y^0]\|_\otimes \ge C_0 - \varepsilon$. So there exists a T_U such that $\|T_U\| = 1$ and

$$\begin{aligned}C_0 - \varepsilon &\le \langle T_U, \mathbb{E}(X^0\otimes Y^0)\rangle = |\mathbb{E}\langle T_U, X^0 \otimes Y^0\rangle = |\mathbb{E}\langle UX^0, Y^0\rangle| \\ &\le (K_G^R + \varepsilon)\|U\| \quad (\text{by } (7.7.2))\,.\end{aligned}$$

Because ε is arbitrary, we have proved the lemma. □

Lemma 7.13. *Let X, Y be two Gaussians with variance 1. Then,*

$$\mathbb{E}[XY] = \sin[\frac{\pi}{2}\mathbb{E}(\text{sign}(X)(\text{sign}(Y))] \tag{7.7.10}$$

Proof. The following proof follows from [Pisier (1986)]. Let $\sin\theta_0 = \mathbb{E}[XY], \theta_0 \in [-\pi/2, \pi/2]$ (because $|\mathbb{E}[XY]| \le 1$). Then, the Gaussians (X, Y) and $(X, X\sin\theta_0 + Z\cos\theta_0)$ have the same law, where X and Z are two normal independent Gaussians. Consequently,

$$\mathbb{E}[\text{sign}(X)\text{sign}(Y)] = \int\int_{R^2} \text{sign}(x)\text{sign}(x\sin\theta_0 + z\cos\theta_0)e^{-[x^2+z^2]/2}\frac{dxdz}{2\pi}.$$

Using polar coordinates we obtain

$$\begin{aligned}\mathbb{E}[\text{sign}(X)\text{sign}(Y)] &= \int_0^\infty \int_=^{2\pi} \text{sign}(\cos\theta)\text{sign}(\sin(\theta+\theta_0))e^{-r^2/2}rdr\frac{d\theta}{2\pi} \\ &= \frac{1}{2\pi}\int_0^{2\pi} \text{sign}(\cos\theta)\text{sign}(\sin(\theta+\theta_0))d\theta = 2\theta_0/\pi,\end{aligned}$$

completing the proof. □

Proof of Theorem 7.20. Note $a = \pi/[2\log(1+\sqrt{2})]$. First, we want to show that for any Gaussian processes X_s, Y_t with variances smaller than 1, and $b < a$, there are two other Gaussians Z_s and Z'_t with variance 1 such that

$$\sin b\mathbb{E}[X_s Y_t] = \mathbb{E}\left(Z_s Z'_t\right) . \tag{7.7.11}$$

In fact, put

$$Z_s = \sum_{k\geq 0} \frac{(-1)^{k+1} b^{(2k+1)/2}}{\sqrt{(2k+1)!}} H_k(X_s)$$

and

$$Z'_t = \sum_{k\geq 0} \frac{b^{(2k+1)/2}}{\sqrt{(2k+1)!}} H_k(Y_t) ,$$

where $H_k(x)$ is the normalized Hermite polynomial of order k. Now X_s and Y_t are two normalized Gaussians of covariance $c = \mathbb{E}\left(X_s Y_t\right)$. For any σ and ε, $\sigma X_s + \varepsilon Y_t$ is a Gaussian of variance $\sigma^2 + \varepsilon^2 + 2c\sigma\varepsilon$. So $\mathbb{E}\exp\{\sigma X_s + \varepsilon Y_t - \sigma^2/2 - \varepsilon^2/2\} = \exp\{c\sigma\varepsilon\}$. Develop the above formula according to the Taylor series of σ and ε, namely, the Hermite polynomials of X_s and Y_t and compare the coefficients we get a formula for $\mathbb{E}H_m(X_s)H_k(Y_t)$. By this formula we can get the convergence in L^2 of Z_s and Z'_t and $\mathbb{E}|Z_s|^2 \leq 1, \mathbb{E}|Z'_t|^2 \leq 1$ (by the fact $b \leq a$). Moreover

$$\sin[b\mathbb{E}(X_s Y_t)] = \mathbb{E}(Z_s Z'_s). \tag{7.7.12}$$

Evidently Z_s and Z'_t are not Gaussian. But we can always identify a Hilbert space to a Hilbert space of Gaussians. That is to say we can find two Gaussians (note by the same notation) Z_s and Z'_t such that (7.7.11) is true. By an apparently easy technique, we can suppose that they have the variance 1. Now by (7.7.10)

$$\sin[b\mathbb{E}(X_s Y_t)] = \mathbb{E}[Z_s Z'_t] = \sin\frac{\pi}{2}[\mathbb{E}\text{sign}(Z_s)\text{sign}(Z'_t)]. \tag{7.7.13}$$

So

$$\mathbb{E}[X_s Y_t] = \pi\mathbb{E}[\text{sign}(Z_s)\text{sign}(Z'_t)]/(2b). \tag{7.7.14}$$

Consequently

$$\begin{aligned} K_G^R &\leq \pi\mathbb{E}\{\sup_s[\text{sign}(Z_s)]\sup_t[\text{sign}(Z'_t)]/(2b)\} \\ &= \pi/(2b). \end{aligned}$$

Because $b < a$ is arbitrary, we have then proved the theorem. □

Chapter 8

Convergence in Density

8.1 General Nonlinear Wiener Functional

For a sequence of random variables $\{F_n, n \geq 1\}$, there are many types of convergence of F_n to a random variable F. For example, almost sure convergence, convergence in L^p norm, convergence in probability, and convergence in law. In this section, we consider the convergence in density. We are concerned under what conditions, the densities $\rho_{F_n}(x)$ and $\rho_F(x)$ of F_n and F exist and ρ_{F_n} converges to ρ_F.

When we consider convergence in distribution or convergence in density, it is desirable to have a certain expression, or formula to express the difference of the distribution or density under consideration. However, when we consider the convergence in density or convergence in law, the random variables F_n and F may be defined in different probability spaces. At first glance, it seems impossible to find a formula to express the difference of associated distributions of two random variables. In this section, we shall establish the Stein's formula to express the difference between the probability densities of a Wiener functional and a normal distribution.

Lemma 8.1. *Let N be the standard Gaussian random variable. Namely, the probability density of N is $\phi(x) = \frac{1}{\sqrt{2\pi}} e^{-\frac{x^2}{2}}$. Then, for all integer $k \geq 0$, we have*

$$\phi^{(k)}(x) = (-1)^k \mathbb{E}\left[I_{\{N>x\}} H_{k+1}(N)\right] . \tag{8.1.1}$$

In particular, when $k = 0$, we have

$$\phi(x) = \mathbb{E}\left[I_{\{N>x\}} N\right] . \tag{8.1.2}$$

Proof. From the definition of Hermite polynomial (3.4.1) we have

$$\begin{aligned}\mathbb{E}\left[I_{\{N>x\}}H_{k+1}(N)\right] &= \int_x^\infty H_{k+1}(y)\phi(y)dy \\ &= \int_x^\infty \left[(-1)^{k+1}e^{\frac{y^2}{2}}\frac{d^{k+1}}{dy^{k+1}}e^{-\frac{y^2}{2}}\right]\phi(y)dy \\ &= (-1)^{k+1}\int_x^\infty \frac{d^{k+1}}{dy^{k+1}}\phi(y)dy = (-1)^k\phi^{(k)}(x)\,,\end{aligned}$$

yielding the lemma. □

Let F be an arbitrary random variable and let N be Gaussian random variable with variance >0. Assume that $h:\mathbb{R}\to\mathbb{R}$ is a given measurable function. Consider the following *Stein's equation* ([Stein (1972)])

$$f'(x) - xf(x) = h(x) - \mathbb{E}[h(N)]\,, \quad x\in\mathbb{R}\,. \tag{8.1.3}$$

Any solution f_h to Equation (8.1.3) verifies

$$\mathbb{E}(h(N)) = \mathbb{E}(h(F)) - \mathbb{E}[f_h'(F) - Ff_h(F)]\,. \tag{8.1.4}$$

The significance of the above identity is that if we can choose $h(y) = I_{\{y>x\}}H_{k+1}(y)$, then we can express the expectation of k-derivative $\phi(x)$ by some linear combination of the expectations of functions of F (which do not involve N). The function h can also be taken as the Dirac delta function $\delta_x(\cdot)$ at x. But then it is too singular.

First, we study the solution to equation (8.1.3) when $h(x) = I_{\{x>z\}}H_k(x)$ for any integer $k\ge 1$ and $z\in\mathbb{R}$, where $H_k(x)$ is the k-th Hermite polynomial.

Lemma 8.2. *Suppose $|h(x)|\le a|x|^k + b$ for some integer $k\ge 0$ and non-negative numbers a,b. Then, the solution f_h to the Stein's equation (8.1.3) is given by*

$$f_h(x) = e^{x^2/2}\int_{-\infty}^x \{h(y) - \mathbb{E}[h(N)]\}e^{-y^2/2}dy \tag{8.1.5}$$

and it satisfies

$$|f_h'(x)| \le aC_k\sum_{i=0}^k |x|^i + 4b$$

for all $x\in\mathbb{R}$, where C_k is a constant depending only on k.

Proof. It is clear that under the condition on h the unique solution to (8.1.3) is given by (8.1.5).

Since $\int_{-\infty}^{\infty}\{h(y)-\mathbb{E}[h(N)]\}e^{-y^2/(2)}dy = 0$, we have

$$\int_{-\infty}^{x}\{h(y)-\mathbb{E}[h(N)]\}e^{-y^2/2}dy = -\int_{x}^{\infty}\{h(y)-\mathbb{E}[h(N)]\}e^{-y^2/2}dy.$$

Hence

$$\left|\int_{-\infty}^{x}\{h(y)-\mathbb{E}[h(N)]\}e^{-y^2/2}dy\right| \le \int_{|x|}^{\infty}[ay^k+b+\mathbb{E}\,|h(N)|]e^{-y^2/2}dy.$$

By using the representation (8.1.5) of f_h and Stein's equation (8.1.3) we have

$$\begin{aligned}|f_h'(x)| &\le |h(x)-\mathbb{E}[h(N)]| + \ |x|\,e^{x^2/2}\left|\int_{-\infty}^{x}\{h(y)-\mathbb{E}[h(N)]\}e^{-y^2/2}dy\right| \\ &\le a\,|x|^k+b+\mathbb{E}\,|h(N)|+e^{x^2/2}\int_{|x|}^{\infty}y[ay^k+b+\mathbb{E}\,|h(N)|]e^{-y^2/2}dy \\ &= a\,|x|^k+(b+E\,|h(N)|)+as_{k+1}(x), \end{aligned} \tag{8.1.6}$$

where we let $s_k(x) = e^{x^2/2}\int_{|x|}^{\infty}y^k e^{-y^2/2}dy$ for any integer $k \ge 0$ and note that $s_1(x) = 1$.

Noticing that $\mathbb{E}\,|h(N)| \le a\mathbb{E}\,|N|^k + b \le C_k a + b$ and using integration by parts, we see by induction that for any integer $k \ge 1$,

$$\begin{aligned} s_{k+1}(x) &= e^{x^2/2}\int_{|x|}^{\infty}y^{k+1}e^{-y^2/2}dy \\ &= e^{x^2/2}\int_{|x|}^{\infty}y^k d(-e^{-y^2/2}) = [|x|^k + k\ s_{k-1}(x)]. \end{aligned}$$

Then, if $k \ge 1$ is even, we have

$$s_{k+1}(x) \le C_k[|x|^k + |x|^{k-2} + \cdots + s_1(x)] \le C_k\sum_{i=0}^{k}|x|^i.$$

If $k \ge 1$ is odd, we have

$$s_{k+1}(x) \le C_k[|x|^k + |x|^{k-2} + \cdots + (|x| + s_0(x))] \le C_k\sum_{i=0}^{k}|x|^i,$$

where we used the fact that $s_0(x) \le s_0(0) = \sqrt{\frac{\pi}{2}}$ for all $x \in \mathbb{R}$. Indeed, when $x \ge 0$ we have

$$s_0'(x) = xe^{x^2/2}\int_x^{\infty}e^{-y^2/2}dy - 1 \le e^{x^2/2}\int_x^{\infty}ye^{-\frac{y^2}{2}}dy - 1 = 0\,.$$

Similarly, when $x < 0$, we have $s_0'(x) \ge 0$. Substituting the above estimates into (8.1.6) we complete the proof. □

To deal with the third term on the right hand of (8.1.4) we use the following

$$\begin{aligned}\mathbb{E}[Ff(F)] &= \mathbb{E}[LL^{-1}Ff(F)] = \mathbb{E}[-\delta DL^{-1}Ff(F)] \\ &= \mathbb{E}[\langle -DL^{-1}F, Df(F)\rangle] = \mathbb{E}[f'(F)\langle -DL^{-1}F, DF\rangle_{\mathbb{H}}].\end{aligned}$$

Then, it follows that

$$\mathbb{E}[f'(F) - Ff(F)] = \mathbb{E}[f'(F)(1 - \langle DF, -DL^{-1}F\rangle_{\mathbb{H}})]. \tag{8.1.7}$$

Combining Equation (8.1.7) with (8.1.4) and Lemma 8.2 we obtain the following result.

Lemma 8.3. *Suppose $h : \mathbb{R} \to \mathbb{R}$ verifies $|h(x)| \le a\,|x|^k + b$ for some $a, b \ge 0$ and integer $k \ge 0$. Let $N \sim N(0,1)$ and let $F \in \mathbb{D}_{1,2k}$ with $\|F\|_{2k} \le c$ for some $c > 0$. Then, there exists a constant $C_{k,c}$ depending only on k and c such that*

$$|\mathbb{E}[h(F) - h(N)]| \le [aC_{k,c} + 4b]\,\left\|1 - \langle DF, DL^{-1}F\rangle_{\mathbb{H}}\right\|_2 . \tag{8.1.8}$$

Proof. It suffices to notice that $\left\|\sum_{i=0}^k |F|^i\right\|_2 \le \sum_{i=0}^k \|F\|_{2k}^i \le C_{k,c}$. □

8.2 Weak Convergence

Proposition 8.1. *Let $F_n = I_q(f_n)\,, n \ge 1$, where $f_n \in \mathbb{H}^{\hat{\otimes} q}$ and $q \ge 2$ is fixed. Assume that $\lim_{n\to\infty} \mathbb{E}\left[F_n^2\right] \to \sigma^2$. Denote by $N \sim N(0,\sigma^2)$ the Gaussian random variable with mean 0 and variance σ^2. Then, the following statements are equivalent.*

(i) F_n converges to $N = N(0,\sigma^2)$ in distribution.
(ii) For every $j \ge 3$, $\mathbb{E}\left[F_n^j\right] \to \mathbb{E}\left[N^j\right]$.

Proof. It is clear that $\mathbb{E}(F_n) = 0 = \mathbb{E}(N)$ and $\mathbb{E}(F_n^2) \to \sigma^2 = \mathbb{E}(N^2)$. Thus, (ii) means that all moments of F_n converge to the corresponding moments of N. Thus, F_n converges to N in distribution. □

Corollary 8.1. *Let $N \sim N(0,1)$ and let $F \in \mathbb{D}_{1,2k}$ with $\|F\|_{2k} \le c$ for some $c > 0$. If*

$$\lim_{n\to\infty} \mathbb{E}\left[\langle DF_n, DL^{-1}F_n\rangle_{\mathbb{H}} - 1\right]^2 = 0\,, \tag{8.2.1}$$

then F_n converges to N in distribution. In fact, F_n converges to N in Kolmogorov distance.

Proof. It suffices to take $h = I_{(-\infty,a]}$ in (8.1.8). $\square$

Theorem 8.1. *Let $F_n = I_q(f_n)$, $f \in \mathbb{H}^{\otimes q}$, $n \geq 1$, where $q \geq 2$ is fixed. Assume that $\mathbb{E}(F_n^2) \to 1$ as $n \to \infty$. Then, the following statements are equivalent.*

(i) F_n *converges to* $N = N(0,1)$ *in distribution.*
(ii) $\mathbb{E}\left[F_n^4\right] \to \mathbb{E}[N^4] = 3$, *where* $N \sim N(0,1)$ *(standard normal random variable).*
(iii) $\mathrm{Var}\left(|DF_n|^2\right) \to 0$.
(iv) $|f_n \hat{\otimes}_r f_n|_{\mathbb{H}^{\otimes(2q-2r)}} \to 0$ *for* $r = 1, \cdots, r-1$.
(v) $|f_n \otimes_r f_n|_{\mathbb{H}^{\otimes(2q-2r)}} \to 0$ *for* $r = 1, \cdots, r-1$.

Proof. "(i) $\Rightarrow$ (ii)" From $\mathbb{E}(F_n^2) \to \sigma^2$, we see that $\sup_n \mathbb{E}(F_n^2) < \infty$. From (7.1.40), we see that for any $p \geq 1$, $\sup_n \mathbb{E}(|F_n|^p) < \infty$. Thus, (i) implies that $\mathbb{E}(F_n^j) \to \mathbb{E}(N^j)$ for all $j \geq 1$, which implies (ii).

"(ii) $\Rightarrow$ (iii)" This follows from the first inequality of (6.2.20) (the n there is q).

"(iii) $\Rightarrow$ (iv)" This follows from the identity (6.2.17).

"(iv) $\Rightarrow$ (v)" This follows from (6.2.19).

"(v) $\Rightarrow$ (i)" By Corollary (8.1) we need to verify the condition (8.2.1). By the definition of $L = -\delta D$ we have

$$\langle DF_n, D(-L)^{-1}F_n\rangle_{\mathbb{H}} = \frac{1}{q}\langle DF_n, DF_n\rangle_{\mathbb{H}}\,.$$

Thus,

$$\mathbb{E}\left[\langle DF_n, D(-L)^{-1}F_n\rangle_{\mathbb{H}} - 1\right]^2 = \frac{1}{q^2}\left\{\mathrm{Var}\left(|DF_n|_{\mathbb{H}}^2\right) + \left(\mathbb{E}\left[|DF_n|_{\mathbb{H}}^2\right] - q\right)\right\}^2. \tag{8.2.2}$$

Equation (6.2.16) yields $\lim_{n\to\infty} \mathbb{E}\left[|DF_n|_{\mathbb{H}}^2\right] = q \lim_{n\to\infty} \mathbb{E}(F_n^2) = q$. This implies that the second term in (8.2.2) converges to 0. Using $|f_n \hat{\otimes}_r f_n|_{\mathbb{H}^{\otimes(2q-2r)}} \leq |f_n \otimes_r f_n|_{\mathbb{H}^{\otimes(2q-2r)}}$ and (6.2.17), we see that the first term in (8.2.2) converges to 0. Hence, the condition (8.2.1) holds. The theorem is proved. $\square$

The following proposition gives an estimate for the contraction $D^kF \otimes_1 D^lF$ with $k + l \geq 3$.

Proposition 8.2. *Let $F = I_q(f)$ be a random variable in the q-th Wiener chaos with $\mathbb{E}[F^2] = 1$. Then, for any integers $k \geq l \geq 1$ satisfying $k+l \geq 3$, there exists a constant $C_{k,l,q}$ depending only on k, l, q such that*

$$\left\|D^kF \otimes_1 D^lF\right\|_2 \leq C_{k,l,q}\left\|q - \|DF\|_{\mathbb{H}}^2\right\|_2. \tag{8.2.3}$$

Proof. Notice that $D^kF = q(q-1)\cdots(q-k+1)I_{q-k}(f)$. Applying (5.3.2), we get

$$D^kF \otimes_1 D^lF = q^2(q-1)^2\cdot\dots\cdot(q-l+1)^2(q-l)\cdots(q-k+1)$$
$$\times\sum_{r=0}^{q-k} r!\binom{q-k}{r}\binom{q-l}{r} I_{2q-k-l-2r}(f\widetilde{\otimes}_{r+1}f).$$

Taking into account the orthogonality of multiple integrals of different orders, we obtain

$$\mathbb{E}[\|D^kF \otimes_1 D^lF\|^2_{\mathbb{H}^{\otimes(k+l-2)}}] = \frac{(q!)^4}{(q-l)!^2\,(q-k)!^2}$$
$$\times\sum_{r=0}^{q-k} r!^2\binom{q-k}{r}^2\binom{q-l}{r}^2 (2q-k-l-2r)!\,\|f\widetilde{\otimes}_{r+1}f\|^2_{\mathbb{H}^{\otimes 2q-2-2r}}. \tag{8.2.4}$$

Notice that

$$\mathbb{E}[\|DF\|^4_{\mathbb{H}}] = \mathbb{E}[|DF\otimes_1 DF|^2] \tag{8.2.5}$$
$$= q^4\sum_{r=0}^{q-1} r!^2\binom{q-1}{r}^4 (2q-2-2r)!\,\|f\widetilde{\otimes}_{r+1}f\|^2_{\mathbb{H}^{\otimes 2q-2-2r}}$$
$$= q^4\sum_{r=0}^{q-2} r!^2\binom{q-1}{r}^4 (2q-2-2r)!\,\|f\widetilde{\otimes}_{r+1}f\|^2_{\mathbb{H}^{\otimes 2q-2-2r}} + q^2q!^2\,\|f\|^4_{\mathbb{H}^{\otimes q}}.$$

Taking into account that $\mathbb{E}[F^2] = q!\,\|f\|^2_{\mathbb{H}^{\otimes q}} = 1$, we obtain that for any $k+l\geq 3$, there exists a constant $C_{k,l,q}$ such that

$$\mathbb{E}[\|D^kF \otimes_1 D^lF\|^2_{\mathbb{H}^{\otimes(k+l-2)}}] \leq C^2_{k,l,q}\mathbb{E}[\|DF\|^4_{\mathbb{H}} - q^2].$$

Meanwhile, it follows from $\mathbb{E}[\|DF\|^2_{\mathbb{H}}] = q\,\|f\|^2_{\mathbb{H}^{\otimes q}} = q$ that

$$\mathbb{E}[\|DF\|^4_{\mathbb{H}} - q^2] = \mathbb{E}[\|DF\|^4_{\mathbb{H}} - 2q\,\|DF\|^2_{\mathbb{H}} + q^2] = \mathbb{E}[(\|DF\|^2_{\mathbb{H}} - q)^2]. \tag{8.2.6}$$

Combining the above computations, we have

$$\mathbb{E}[\|D^kF \otimes_1 D^lF\|^2_{\mathbb{H}^{\otimes(k+l-2)}}] \leq C^2_{k,l,q}\mathbb{E}[(\|DF\|^2_{\mathbb{H}} - q)^2],$$

which completes the proof. □

The following result is from [Hu and Nualart (2005)].

Theorem 8.2. *Consider a sequence of square integrable and centered random variables $\{F_k, k\geq 1\}$ with Wiener chaos expansions $F_k = \sum_{q=1}^{\infty} I_q(f_{q,k})$. Suppose that*

(i) $\lim_{N\to\infty}\limsup_{k\to\infty}\sum_{q=N+1}^{\infty} q!\,\|f_{q,k}\|^2_{\mathbb{H}^{\otimes q}} = 0.$
(ii) For every $q \geq 1$, $\lim_{k\to\infty} q!\,\|f_{q,k}\|^2_{\mathbb{H}^{\otimes q}} = \sigma_q^2.$
(iii) $\sum_{q=1}^{\infty} \sigma_q^2 = \sigma^2 < \infty.$
(iv) For all $q \geq 2$, $p = 1, ..., q-1$, $\lim_{k\to\infty} \|f_{q,k} \otimes_p f_{q,k}\|^2_{H^{\otimes 2(q-p)}} = 0.$

Then, F_k *converges in distribution to the normal law* $N(0, \sigma^2)$ *as* k *tends to infinity.*

Proof. By Theorem 8.1, conditions (ii) and (iv) imply that for each fixed $q \geq 1$ the sequence of multiple integrals $I_q(f_{q,k})$ converges in distribution to the normal law $N(0, \sigma_q^2)$ as k tends to infinity. Furthermore, by considering $\sum_{j=1}^{q} \lambda_j I_j(f_{j,k})$ for any real numbers $\lambda_j \in \mathbb{R}$ in the same way as in the proof of Theorem 8.1 we have that for each $q \geq 1$, the random vector

$$(I_1(f_{1,k}), \ldots, I_q(f_{q,k})) \tag{8.2.7}$$

converges in law to the vector $(N_1, \ldots, N_q)$, where $\{N_q, q \geq 1\}$ are independent centered Gaussian random variables with variances $\{\sigma_q^2, q \geq 1\}$ (see [Peccati and Tudor (2005), Theorem 1]). Now let f be a C^1 function such that $|f|$ and $|f'|$ are bounded by one. For every $N \geq 1$, set

$$F_k^{(N)} = \sum_{q=1}^{N} I_q(f_{q,k}).$$

Set $N^{(N)} = \sum_{q=1}^{N} N_q$, and $N = \sum_{q=1}^{\infty} N_q$. Then

$$\begin{aligned}
&|\mathbb{E}(f(F_k)) - \mathbb{E}(f(N))| \\
&\leq \left|\mathbb{E}(f(F_k)) - \mathbb{E}\left(f(F_k^{(N)})\right)\right| \\
&\quad + \left|\mathbb{E}\left(f(F_k^{(N)})\right) - \mathbb{E}\left(f(N^{(N)})\right)\right| + \left|\mathbb{E}\left(f(N^{(N)})\right) - \mathbb{E}(f(N))\right| \\
&\leq \left(\sum_{q=N+1}^{\infty} q!\,\|f_{q,k}\|^2_{H^{\otimes n}}\right)^{\frac{1}{2}} + \left|\mathbb{E}\left(f(F_k^{(N)})\right) - \mathbb{E}\left(f(N^{(N)})\right)\right| \\
&\quad + \left|\mathbb{E}\left(f(N^{(N)})\right) - \mathbb{E}(f(N))\right|.
\end{aligned}$$

Then, taking first the limit as k tends to infinity, and then the limit as N tends to infinity, and applying conditions (i), (iii) and the convergence in law of the vector (8.2.7) we get the desired result. □

Remark 8.1. A sufficient condition for (i) and (iii) is that for all $q \geq 1$, and $k \geq 1$,

$$q!\,\|f_{q,k}\|^2_{\mathbb{H}^{\otimes q}} \leq \delta_q,$$

where $\sum_{q=1}^{\infty} \delta_q < \infty$.

Remark 8.2. The assumptions of Theorem 8.2 imply the convergence in law of the projections on the Wiener chaos to a family of independent Gaussian random variables.

8.3 Representation of the Derivatives of the Density

Let us recall that if F is a random variable, then (if the following right hand side exists) its density exists and is given by (6.3.1), which can be rewritten as

$$p_F(x) = \mathbb{E}\left[I_{\{F>x\}}\,\delta\left(\frac{u}{\langle DF, u\rangle}\right)\right]$$

for an appropriately chosen u. Equation (6.3.3) may be used to obtain the higher derivatives of the density. However, they can only be used to show the existence or smoothness since the terms $H_\alpha(u,G)$ are not sufficiently precise so that we handle them for our purpose. In this section, we give a more precise expression for $H_\alpha(u,G)$ which shall be used ulteriorly.

Define

$$G_0 = 1\,, \quad G_{k+1} = \delta(G_k u)\,, k = 0, 1, \cdots\,,$$

where u is an $\mathbb{H}$-valued process and plays an essential role in the formula. The following technical lemma gives an explicit formula for the sequence G_k, relating it to Hermite polynomials. To simplify the notation, for an $\mathbb{H}$–valued random variable u, we denote

$$\delta_u = \delta(u),\ D_u G = \langle DG, u\rangle_{\mathbb{H}}\,,\ D_u^k G = \left\langle D\left(D_u^{k-1}G\right), u\right\rangle_{\mathbb{H}}. \tag{8.3.1}$$

Recall $H_k(x)$ denotes the k-th Hermite polynomial. For $\lambda > 0$ and $x \in \mathbb{R}$, we define the parametrized k-th Hermite polynomial as

$$H_k(\lambda, x) = \lambda^{\frac{k}{2}} H_k\left(\frac{x}{\sqrt{\lambda}}\right). \tag{8.3.2}$$

From the property $H_k'(x) = kH_{k-1}(x)$ it follows by induction that the k-th Hermite polynomials has the form $H_k(x) = \sum_{0\le i\le k/2} c_{k,i}x^{k-2i}$. Then, (8.3.2) implies

$$H_k(\lambda, x) = \sum_{0\le i\le k/2} c_{k,i}x^{k-2i}\lambda^i. \tag{8.3.3}$$

Lemma 8.4. *Fix an integer $m \ge 1$ and a number $p > m$. Suppose $u \in \mathbb{D}_{m,p}(\mathbb{H})$. We define recursively a sequence $\{G_k\}_{k=0}^m$ by $G_0 = 1$ and $G_{k+1} = \delta(G_k u)$. Then, these variables are well-defined and for $k = 1, 2, \ldots, m$, $G_k \in \mathbb{D}^{m-k,\frac{p}{k}}$ and*

$$G_k = H_k(D_u\delta_u, \delta_u) + T_k, \tag{8.3.4}$$

where we denote by T_k the higher order derivative terms which can be defined recursively as follows $T_1 = T_2 = 0$ and for $k \geq 2$,

$$T_{k+1} = \delta_u T_k - D_u T_k - \partial_\lambda H_k(D_u\delta_u, \delta_u) D_u^2 \delta_u. \tag{8.3.5}$$

Proof. By induction one can prove that the above sequence G_k is well-defined and $G_k \in \mathbb{D}_{m-k,\frac{p}{k}}$. Suppose first that $k = 1$. Then, Meyer's inequality implies that $G_1 = \delta_u \in \mathbb{D}_{m-1,p}$. Assume that for some $k \leq m-1$, $G_k \in \mathbb{D}_{m-k,\frac{p}{k}}$. Then, it follows from (7.5.14) that

$$\begin{aligned}\|G_{k+1}\|_{m-k-1,\frac{p}{k+1}} &\leq C_{m,p}\|G_k u\|_{m-k,\frac{p}{k+1}}\\ &\leq C'_{m,p}\|G_k\|_{m-k,\frac{p}{k}}\|u\|_{m-k,p} < \infty.\end{aligned}$$

Let us now show, by induction, the decomposition (8.3.4). When $k = 1$ (8.3.4) is true because $G_1 = \delta_u$ and $T_1 = 0$. Assume now (8.3.4) holds for $k \leq m-1$. Noticing that $\partial_x H_k(\lambda, x) = kH_{k-1}(\lambda, x)$ (since $H'_k(x) = kH_{k-1}(x)$), we get

$$D_u H_k(D_u\delta_u, \delta_u) = kH_{k-1}(D_u\delta_u, \delta_u)D_u\delta_u + \partial_\lambda H_k(D_u\delta_u, \delta_u)D_u^2\delta_u.$$

Hence, applying the operator D_u to both sides of (8.3.4),

$$D_u G_k = kH_{k-1}(D_u\delta_u, \delta_u)D_u\delta_u + \widetilde{T}_{k+1},$$

where

$$\widetilde{T}_{k+1} = D_u T_k + \partial_\lambda H_k(D_u\delta_u, \delta_u)D_u^2\delta_u. \tag{8.3.6}$$

From the definition of G_{k+1} and equation (6.2.15)

$$\begin{aligned}G_{k+1} &= \delta(uG_k) = G_k\delta_u - D_u G_k\\ &= \delta_u H_k(D_u\delta_u, \delta_u) + \delta_u T_k - kH_{k-1}(D_u\delta_u, \delta_u)D_u\delta_u - \widetilde{T}_{k+1}.\end{aligned}$$

Notice that $H_{k+1}(x) = xH_k(x) - kH_{k-1}(x)$ implies

$$xH_k(\lambda, x) - k\lambda H_{k-1}(\lambda, x) = H_{k+1}(\lambda, x).$$

Hence,

$$G_{k+1} = H_{k+1}(D_u\delta_u, \delta_u) + \delta_u T_k - \widetilde{T}_{k+1}.$$

The term $T_{k+1} = \delta_u T_k - \widetilde{T}_{k+1}$ has the form given in (8.3.5). This completes the proof. □

Lemma 8.5. *Follow the notation of Lemma 8.4. We have for $k \geq 3$*

$$T_k = \sum_{(i_0,\ldots,i_{k-1})\in J_k} a_{i_0,i_1,\ldots,i_{k-1}} \delta_u^{i_0} (D_u\delta_u)^{i_1} \left(D_u^2\delta_u\right)^{i_2} \cdots \left(D_u^{k-1}\delta_u\right)^{i_{k-1}}, \tag{8.3.7}$$

where the coefficients $a_{i_0,i_1,\ldots,i_{k-1}}$ are real numbers and J_k is the set of multi-indices $(i_0, i_1, \ldots, i_{k-1}) \in \mathbb{N}^k$ satisfying the following three conditions

$$(a)\ i_0 + \sum_{j=1}^{k-1} j i_j \leq k-1;\ (b)\ i_2 + \cdots + i_{k-1} \geq 1;\ (c)\ \sum_{j=1}^{k-1} i_j \leq \frac{k-1}{2}.$$

Proof. We shall prove these properties by induction. From $T_1 = T_2 = 0$, (8.3.3) and (8.3.5) we know that $T_3 = D_u^2\delta_u$, with $J_3 = \{(0,0,1)\}$; and $T_4 = \delta_u D_u^2\delta_u + D_u^3\delta_u$, with $J_4 = \{(1,0,1,0),(0,0,0,1)\}$. Now suppose the statement is true for all T_l with $l \le k-1$ for $k \ge 5$. We want to prove that the multi-indices of T_k satisfy (a)–(c). This will be done by studying the three operations, $\delta_u T_{k-1}$, $D_u T_{k-1}$ and $\partial_\lambda H_{k-1}(D_u\delta_u,\delta_u)D_u^2\delta_u$, in expression (8.3.5).

For the term $\partial_\lambda H_{k-1}(D_u\delta_u,\delta_u)D_u^2\delta_u$, we observe from (8.3.3) that

$$\partial_\lambda H_{k-1}(D_u\delta_u,\delta_u)D_u^2\delta_u = D_u^2\delta_u \sum_{1\le i\le (k-1)/2} i c_{k-1,i}\delta_u^{k-1-2i}(D_u\delta_u)^{i-1},$$

whose terms have multi-indices $(k-1-2i, i-1, 1, 0, \dots, 0) \in \mathbb{N}^k$ for $1 \le i \le \frac{k-1}{2}$. Then, it is straightforward to check that these multi-indices satisfy (a), (b) and (c).

The term $\delta_u T_{k-1}$ shifts the multi-index $(i_0, i_1, \dots, i_{k-2}) \in J_{k-1}$ to $(i_0+1, i_1, \dots, i_{k-2}, 0) \in \mathbb{N}^k$, which obviously satisfies (a), (b) and (c), due to the induction hypothesis.

The third term $D_u T_{k-1}$ shifts the multi-index $(i_0, i_1, \dots, i_{k-2}) \in J_{k-1}$ to either $\alpha = (i_0 - 1, i_1 + 1, \dots, i_{k-2}, 0) \in \mathbb{N}^k$ if $i_0 \ge 1$, or to

$$\beta = \begin{cases} (i_0, i_1, \dots, i_{j_0} - 1, i_{j_0+1} + 1, \dots, i_{k-2}, 0), & \text{for } 1 \le j_0 \le k-3; \\ (i_0, i_1, \dots, i_{j_0} - 1, 1), & \text{for } j_0 = k-2, \end{cases}$$

if $i_{j_0} \ge 1$. It is easy to check that β satisfies properties (a), (b) and (c) and α satisfies properties (a) and (b). We are left to verify that α satisfies property (c). That is, we want to show that

$$1 + \sum_{j=1}^{k-2} i_j \le \frac{k-1}{2}. \tag{8.3.8}$$

If k is odd, say $k = 2m+1$ for some $m \ge 2$, (8.3.8) is true because $(i_0, i_1, \dots, i_{k-2}) \in J_{k-1}$, which implies by induction hypothesis that $\sum_{j=1}^{k-2} i_j \le \frac{k-2}{2} = m - 1$. If k is even, say $k = 2m+2$, (8.3.8) is true because the following claim asserts that if $i_0 \ge 1$, then $\sum_{j=1}^{k-2} i_j < \frac{k-2}{2} = m$.

Now we claim that for $(i_0, i_1, \dots, i_{2m}) \in J_{2m+1}$ with $m \ge 1$, if $\sum_{j=1}^{2m} i_j = m$ then $i_0 = 0$.

Indeed, suppose $(i_0, i_1, \dots, i_{2m}) \in J_{2m+1}$, $\sum_{j=1}^{2m} i_j = m$ and $i_0 \ge 1$. We are going to show that leads to a contradiction. First notice that $i_1 \ge 1$, otherwise $i_1 = 0$ and $\sum_{j=2}^{2m} i_j = m$, which is not possible because

$$i_0 + 2m \le i_0 + \sum_{j=1}^{2m} j i_j \le 2m.$$

Also, we must have $i_{2m} = 0$, because otherwise property (a) implies $i_{2m} = 1$ and $i_0 = i_1 = \cdots = i_{2m-1} = 0$. Now we trace back to its parent multi-indices in J_{2m} by inverting the three operations. Of the three operations, we can exclude $\partial_\lambda H_{2m}(D_u\delta_u, \delta_u)D_u^2\delta_u$ and $\delta_u T_{2m}$, because $\partial_\lambda H_{2m}(D_u\delta_u, \delta_u)D_u^2\delta_u$ generates $(2m-2j, j-1, 1, 0, \ldots, 0)$ with $1 \le j \le m$, where j must be m; and $\delta_u T_{2m}$ traces it back to $(i_0 - 1, i_1, \ldots, i_{2m-1}) \in J_{2m}$, where $i_1 + \cdots + i_{2m-1} = m > \frac{2m-1}{2}$. Therefore, its parent multi-index in J_{2m} must come from the operation $D_u T_{2m}$ and hence must be $(i_0+1, i_1-1, \ldots, i_{2m-1}) \in J_{2m}$. Notice that for this multi-index, $i_1 - 1 + \cdots + i_{2m-1} = m-1$. Repeating the above process we will end up with $(i_0+i_1, 0, i_2 \ldots, i_{2m-i_1}) \in J_{2m+1-i_1}$, where $i_2 + \cdots + i_{2m-i_1} = m - i_1$, which contradicts the property (b) of J_{2m+1-i_1} because

$$i_0 + 2m - i_1 \le i_0 + i_1 + \sum_{j=2}^{2m-i_1} j i_j \le 2m - i_1.$$

This completes the proof of the lemma. □

Lemma 8.6. *Let $F = I_q(f)$ with $q \ge 2$ satisfying $\mathbb{E}[F^2] = 1$. For any $\beta \ge 1$ we denote $M_\beta = \left(\mathbb{E}\,\|DF\|_{\mathbb{H}}^{-\beta}\right)^{1/\beta}$ and $w = \|DF\|_{\mathbb{H}}^2$. Then, we have the following statements.*

(i) If $M_\beta < \infty$ for some $\beta \ge 6$, then for any $1 \le r \le \frac{2\beta}{\beta+6}$

$$\left\|D_{DF} w^{-1}\right\|_r \le C M_\beta^3 \left\|q - w\right\|_2. \tag{8.3.9}$$

(ii) If $k \ge 2$ and $M_\beta < \infty$ for some $\beta \ge 2k+4$, then for any $1 < r < \frac{2\beta}{\beta+2k+4}$

$$\left\|D_{DF}^k w^{-1}\right\|_r \le C\left(M_\beta^{k+2} \vee 1\right) \left\|q - w\right\|_2. \tag{8.3.10}$$

(iii) If $k \ge 1$and $M_\beta < \infty$ for any $\beta > k+2$, then for any $1 < r < \frac{\beta}{k+2}$

$$\left\|D_{DF}^k w^{-1}\right\|_r \le C\left(M_\beta^{k+2} \vee 1\right). \tag{8.3.11}$$

Proof. Notice that

$$D_{DF} w^{-1} = \left\langle Dw^{-1}, DF\right\rangle_{\mathbb{H}} = -2w^{-2}\left\langle D^2F \otimes_1 DF, DF\right\rangle.$$

Then

$$\left|D_{DF} w^{-1}\right| \le 2w^{-\frac{3}{2}} \left\|D^2F \otimes_1 DF\right\|_{\mathbb{H}}.$$

Applying Hölder's inequality with $\frac{1}{r} = \frac{1}{p} + \frac{1}{2}$ yields

$$\left\|D_{DF} w^{-1}\right\|_r \le 2\left(E(w^{-\frac{3p}{2}})\right)^{\frac{1}{p}} \left\|D^2F \otimes_1 DF\right\|_2,$$

which implies (8.3.9) by choosing $p \le \beta/3$ and taking into account (8.2.3). Notice that we need $1 \ge \frac{1}{r} \ge \frac{3}{\beta} + \frac{1}{2} = \frac{\beta+6}{2\beta}$.

Consider now the case $k \ge 2$. From the pattern indicated by the first three terms,

$$
\begin{aligned}
D_{DF}w^{-1} &= \langle Dw^{-1}, DF\rangle_{\mathbb{H}}, \\
D^2_{DF}w^{-1} &= \langle D^2w^{-1}, (DF)^{\otimes 2}\rangle_{\mathbb{H}^{\otimes 2}} + \langle Dw^{-1} \otimes DF, D^2F\rangle_{\mathbb{H}^{\otimes 2}}, \\
D^3_{DF}w^{-1} &= \langle D^3w^{-1}, (DF)^{\otimes 3}\rangle_{\mathbb{H}^{\otimes 3}} + 3\langle D^2w^{-1} \otimes DF, D^2F \otimes DF\rangle_{\mathbb{H}^{\otimes 3}} \\
&\quad + \langle Dw^{-1} \otimes D^2F, D^2F \otimes DF\rangle_{\mathbb{H}^{\otimes 3}} + \langle Dw^{-1} \otimes (DF)^{\otimes 2}, D^3F\rangle_{\mathbb{H}^{\otimes 3}},
\end{aligned}
$$

we can prove by induction that

$$
\left|D^k_{DF}w^{-1}\right| \le C\sum_{i=1}^{k} \left\|D^iw^{-1}\right\|_{\mathbb{H}^{\otimes i}} \|DF\|^i_{\mathbb{H}} \left(\sum_{\sum_{j=1}^k i_j = k-i} \prod_{j=1}^{k} \left\|D^jF\right\|^{i_j}_{\mathbb{H}^{\otimes j}} \right).
$$

By (7.1.40), for any $p > 1$, $\left\|D^jF\right\|_p \le C\|F\|_2 = C$. Applying Hölder's inequality and assuming that $s > r$, we have,

$$
\left\|D^k_{DF}w^{-1}\right\|_r \le C\sum_{i=1}^{k} \left\| \left\|D^iw^{-1}\right\|_{\mathbb{H}^{\otimes i}} \|DF\|^i_{\mathbb{H}} \right\|_s. \tag{8.3.12}
$$

We are going to see that $\|DF\|^i_{\mathbb{H}}$ will contribute to compensate the singularity of $\left\|D^iw^{-1}\right\|_{\mathbb{H}^{\otimes i}}$. First, by induction one can prove that for $1 \le i \le m$, D^iw^{-1} has the following expression

$$
D^iw^{-1} = \sum_{l=1}^{i}(-1)^l \sum_{(\alpha,\beta)\in I_{i,l}} w^{-(l+1)} \bigotimes_{j=1}^{l} \left(D^{\alpha_j}F \otimes_1 D^{\beta_j}F\right), \tag{8.3.13}
$$

where

$$
I_{i,l} = \{(\alpha,\beta) \in \mathbb{N}^{2l} \alpha_j + \beta_j \ge 3, \sum_{j=1}^{l}(\alpha_j + \beta_j) = i + 2l\}. \tag{8.3.14}
$$

In fact, for $i = 1$,

$$
Dw^{-1} = -2w^{-2}D^2F \otimes_1 DF,
$$

which is of the above form because $I_{1,1} = \{(1,2),(2,1)\}$. Suppose that

(8.3.13) holds for some $i \le m-1$. Then,

$$\begin{aligned}
& D^{i+1}w^{-1} \\
&= \sum_{l=1}^{i}(-1)^{l+1}2(l+1) \sum_{(\alpha,\beta)\in I_{i,l}} w^{-(l+2)}(D^2F\otimes_1 DF)\bigotimes_{j=1}^{l}\left(D^{\alpha_j}F\otimes_1 D^{\beta_j}F\right) \\
&+\sum_{l=1}^{i}(-1)^l \sum_{(\alpha,\beta)\in I_{i,l}} w^{-(l+1)}\sum_{h=1}^{l}(D^{\alpha_j+1}F\otimes_1 D^{\beta_j}F + D^{\alpha_j}F\otimes_1 D^{\beta_j+1}F) \\
&\times \bigotimes_{j=1, j\neq h}^{l} \left(D^{\alpha_j}F\otimes_1 D^{\beta_j}F\right),
\end{aligned}$$

which is equal to

$$\sum_{l=1}^{i+1}(-1)^l \sum_{(\alpha,\beta)\in I_{i+1,l}} w^{-(l+1)}\bigotimes_{j=1}^{l}\left(D^{\alpha_j}F\otimes_1 D^{\beta_j}F\right).$$

From (8.3.13) for any $i=1,\ldots,k$ we can write

$$\begin{aligned}
& \left\|D^i w^{-1}\right\|_{\mathbb{H}^{\otimes i}} \|DF\|_{\mathbb{H}}^{i} \qquad (8.3.15) \\
&\le \sum_{l=1}^{i} w^{-(l+1)+\frac{i}{2}} \sum_{(\alpha,\beta)\in I_{i,l}} \prod_{j=1}^{l} \left\|D^{\alpha_j}F\otimes_1 D^{\beta_j}F\right\|_{\mathbb{H}^{\otimes\alpha_j+\beta_j-2}},
\end{aligned}$$

where $I_{i,l}$ is defined by (8.3.14). Notice that by (7.1.40),

$$\left\|D^{\alpha_j}F\otimes_1 D^{\beta_j}F\right\|_p \le C\,\|F\|_2^2 = C$$

for all $p\ge 1$ and all α_j,β_j. This inequality will be applied to all but one of the contraction terms in the product $\prod_{j=1}^{l}\left\|D^{\alpha_j}F\otimes_1 D^{\beta_j}F\right\|_{\mathbb{H}^{\otimes\alpha_j+\beta_j-2}}$. We decompose the sum in (8.3.15) into two parts. If the index l satisfies $l\le \frac{i}{2}-1$, then the exponent of w is non-negative, and the p-norm of w can be bounded by a constant, while for $\frac{i}{2}-1<l$ this exponent is negative. Then, using Hölder's inequality and assuming that $\frac{1}{s}=\frac{1}{p}+\frac{1}{2}$, we obtain

$$\begin{aligned}
& \left\|\left\|D^i w^{-1}\right\|_{\mathbb{H}^{\otimes i}} \|DF\|_{\mathbb{H}}^{i}\right\|_s \\
&\le C\left(I_{\{i\ge 2\}} + \sum_{\frac{i}{2}-1<l\le i}\left\|w^{-(l+1)+\frac{i}{2}}\right\|_p\right)\left\|D^{\alpha_1}F\otimes_1 D^{\beta_1}F\right\|_2.
\end{aligned}$$

(8.3.16)

Notice that for $l\le i\le k$, $l+1-\frac{i}{2}\le\frac{k}{2}+1$. Therefore, for $\frac{i}{2}-1<l\le i$

$$\left\|w^{-(l+1)+\frac{i}{2}}\right\|_p = M_{2(l+1-\frac{i}{2})p}^{2l+2-i} \le M_{(k+2)p}^{2l+2-i} \le M_{(k+2)p}^{k+2}\vee 1.$$

Therefore, using (8.2.3) we obtain

$$\left\| \left\| D^i w^{-1} \right\|_{\mathbb{H}^{\otimes i}} \|DF\|_{\mathbb{H}}^i \right\|_s \leq C(M_{(k+2)p}^{k+2} \vee 1) \, \|q - w\|_2 \, . \tag{8.3.17}$$

Combining the above (8.3.17) with (8.3.12) and choosing p such that $(k+2)p \leq \beta$ we obtain (8.3.10). Notice that we need

$$1 > \frac{1}{r} > \frac{k+2}{\beta} + \frac{1}{2} = \frac{\beta + 2k + 4}{2\beta},$$

which holds if $1 < r < \frac{2\beta}{\beta+2k+4}$. The proof of part (iii) is similar and is omitted. □

Analogous to Lemma 8.6, we have the following lemma.

Lemma 8.7. *Let $F = I_q(f)$ with $q \geq 2$ satisfying $\mathbb{E}[F^2] = 1$. For any $\beta \geq 1$ we define $M_\beta = \left(\mathbb{E} \|DF\|_{\mathbb{H}}^{-\beta} \right)^{1/\beta}$ and denote $w = \|DF\|_{\mathbb{H}}^2$. Then, the following statements hold.*

(i) If $M_\beta < \infty$ for some $\beta > 3$, then for any $1 < s < \frac{\beta}{3}$,

$$\|\delta_u\|_s \leq C(M_\beta^3 \vee 1). \tag{8.3.18}$$

(ii) If $k \geq 1$ and $M_\beta < \infty$ for some $\beta > 3k+3$, then for any $1 < s < \frac{\beta}{3k+3}$,

$$\|D_u^k \delta_u\|_s \leq C(M_\beta^{3k+3} \vee 1). \tag{8.3.19}$$

(iii) If $k \geq 2$ and $M_\beta < \infty$ for some $\beta > 6k+6$, then for any $1 < s < \frac{2\beta}{\beta+6k+6}$

$$\|D_u^k \delta_u\|_s \leq C(M_\beta^{3k+3} \vee 1)\|q - w\|_2. \tag{8.3.20}$$

Proof. Recall that $\delta_u = qFw^{-1} - D_{DF}w^{-1}$. Then, for any $r > s$,

$$\|\delta_u\|_s \leq C \left(\|w^{-1}\|_r + \|D_{DF}w^{-1}\|_s \right).$$

Then, $\|w^{-1}\|_r = M_{2r}^2$ and the result follows applying Lemma 8.6 (iii) with $k = 1$ and choosing $r < \frac{\beta}{3}$.

To show (ii) and (iii) we need to find a useful expression for $D_u^k \delta_u$. Consider the operator $D_u = w^{-1} D_{DF}$. We claim that for any $k \geq 1$ the iterated operator D_u^k can be expressed as

$$D_u^k = \sum_{l=1}^{k} w^{-l} \sum_{\mathbf{i} \in I_{l,k}} b_{\mathbf{i}} \left[\prod_{j=1}^{k-l} D_{DF}^{i_j} w^{-1} \right] D_{DF}^{i_0}, \tag{8.3.21}$$

where $b_{\mathbf{i}} > 0$ is a real number and

$$I_{l,k} = \{ \mathbf{i} = (i_0, i_1, \ldots, i_l) i_0 \geq 1, i_j \geq 0 \; \forall \; j = 1, \ldots, l, \sum_{j=0}^{k-l} i_j = k \}.$$

In fact, this is clearly true for $k = 1$. Assuming (8.3.21) holds for a given k, then we have

$$\begin{aligned} D_u^{k+1} &= w^{-1} D_{DF} D^k u \\ &= \sum_{l=1}^{k} l w^{-l} D_{DF} w^{-1} \sum_{\mathbf{i} \in I_{l,k}} b_{\mathbf{i}} \left[\prod_{j=1}^{k-l} D_{DF}^{i_j} w^{-1} \right] D_{DF}^{i_0} \\ &\quad + \sum_{l=1}^{k} w^{-l-1} \sum_{\mathbf{i} \in I_{l,k}} b_{\mathbf{i}} \left[\sum_{h=1}^{k-l} D_{DF}^{i_h+1} w^{-1} \prod_{j=1, j \neq h}^{k-l} D_{DF}^{i_j} w^{-1} \right] D_{DF}^{i_0} \\ &\quad + \sum_{l=1}^{k} w^{-l-1} \sum_{\mathbf{i} \in I_{l,k}} b_{\mathbf{i}} \left[\prod_{j=1}^{k-l} D_{DF}^{i_j} w^{-1} \right] D_{DF}^{i_0+1}. \end{aligned}$$

Shifting the indices, this can be written as

$$\begin{aligned} D_u^{k+1} &= \sum_{l=1}^{k} l w^{-l} D_{DF} w^{-1} \sum_{\mathbf{i} \in I_{l,k}} b_{\mathbf{i}} \left[\prod_{j=1}^{k-l} D_{DF}^{i_j} w^{-1} \right] D_{DF}^{i_0} \\ &\quad + \sum_{l=2}^{k+1} w^{-l} \sum_{\mathbf{i} \in I_{l-1,k}} b_{\mathbf{i}} \left[\sum_{h=1}^{k+1-l} D_{DF}^{i_h+1} w^{-1} \prod_{j=1, j \neq h}^{k+1-l} D_{DF}^{i_j} w^{-1} \right] D_{DF}^{i_0} \\ &\quad + \sum_{l=2}^{k+1} w^{-l} \sum_{\mathbf{i} \in I_{l-1,k}} b_{\mathbf{i}} \left[\prod_{j=1}^{k+1-l} D_{DF}^{i_j} w^{-1} \right] D_{DF}^{i_0+1}, \end{aligned}$$

and it easy to check that this coincides with

$$\sum_{l=1}^{k+1} w^{-l} \sum_{\mathbf{i} \in I_{l,k+1}} b_{\mathbf{i}} \left[\prod_{j=1}^{k+1-l} D_{DF}^{i_j} w^{-1} \right] D_{DF}^{i_0}.$$

Also, notice that $\delta_u = qFw^{-1} + D_{DF} w^{-1}$ and

$$D_{DF}\delta_u = q + qF D_{DF} w^{-1} + D_{DF}^2 w^{-1}.$$

By induction we can show that for any $i_0 \geq 1$

$$D_{DF}^{i_0} \delta_u = q\delta_{1 i_0} + q \sum_{j=1}^{i_0-1} c_{i,j} D_{DF}^{i_0-1-j} w D_{DF}^{j} w^{-1} + qF D_{DF}^{i_0} w^{-1} + D_{DF}^{i_0+1} w^{-1}, \tag{8.3.22}$$

where δ_{1i_0} is the Kronecker symbol. Combining (8.3.21) and (8.3.22) we obtain

$$D_u^k \delta_u = \sum_{l=1}^{k} w^{-l} \sum_{\mathbf{i}\in I_{l,k}} b_{\mathbf{i}} \left[\prod_{j=1}^{k-l} D_{DF}^{i_j} w^{-1}\right] \times \left[q\delta_{1i_0} \right.$$
$$\left. +q \sum_{j=1}^{i_0-1} c_{i,0j} D_{DF}^{i_0-1-j} w D_{DF}^{j} w^{-1} + qF D_{DF}^{i_0} w^{-1} + D_{DF}^{i_0+1} w^{-1}\right].$$

Next, we shall apply Hölder's inequality to estimate $\left\|D_u^k \delta_u\right\|_s$. Notice that for $l = k$, $i_0 = k \geq 2$. Therefore,

$$\left\|D_u^k \delta_u\right\|_s \leq C \sum_{l=1}^{k-1} \sum_{\mathbf{i}\in I_{l,k}} \|w^{-l}\|_p \prod_{j=1}^{k-l} \|D_{DF}^{i_j} w^{-1}\|_{r_j}$$
$$|qquad \left(\delta_{1i_0} + \max_{1\leq h\leq i_0+1} \|D_{DF}^{h} w^{-1}\|_{r_0}\right)$$
$$+C\|w^{-k}\|_p \max_{1\leq h\leq k+1} \|D_{DF}^{h} w^{-1}\|_{\rho_0} = B_1 + B_2,$$

assuming that for $l = 1, \ldots, k-1$, $\frac{1}{s} > \frac{1}{p} + \sum_{j=0}^{k-l} \frac{1}{r_j}$ and $\frac{1}{s} > \frac{1}{p} + \frac{1}{\rho_0}$.

Let us consider first the term B_1. Notice that if $i_0 = 1$ there is at least one factor of the form $\|D_{DF}^{r_j} w^{-1}\|_{r_j}$ in the above product, because $\sum_{j=1}^{k-l} i_j = k-1 \geq 1$. Then, we can apply the inequality (8.3.10) to one of these factors and the inequality (8.3.11) to the remaining ones. The estimate (8.3.11) requires $\frac{1}{r_j} > \frac{i_j+2}{\beta}$ for $j = 1, \ldots, k-l$ and $\frac{1}{r_0} > \frac{i_0+3}{\beta}$. On the other hand, the estimate (8.3.10) requires $\frac{1}{r_j} > \frac{i_j+2}{\beta} + \frac{1}{2}$ for $j = 1, \ldots, k-l$ and $\frac{1}{r_0} > \frac{i_0+3}{\beta} + \frac{1}{2}$. Then, choosing p such that $2pl < \beta$, and taking into account that $\sum_{j=0}^{k-l} i_j = k$ we obtain the inequalities

$$\frac{1}{s} > \frac{1}{p} + \sum_{j=1}^{k-l} \frac{i_j+2}{\beta} + \frac{i_0+3}{\beta} + \frac{1}{2} > \frac{3k+3}{\beta} + \frac{1}{2}.$$

Hence, if $s < \frac{2\beta}{\beta+6k+6}$ we can write

$$B_1 \leq C \sum_{l=1}^{k-1} M_\beta^{2l} \prod_{j=1}^{k-l} (M_\beta^{i_j+2} \vee 1)(M_\beta^{i_0+3} \vee 1) \|q - w^{-1}\|_2$$
$$\leq C(M_\beta^{3k+3} \vee 1)\|q - w^{-1}\|_2.$$

As for the term B_2 we use the estimate (8.3.10) assuming $2pk < \beta$ and

$$\frac{1}{s} > \frac{1}{p} + \frac{k+3}{\beta} + \frac{1}{2} > \frac{3k+3}{\beta} + \frac{1}{2}.$$

This leads to the same estimate and the proof of (8.3.20) is complete. To show the estimate (8.3.19) we proceed as before but use the inequality (8.3.11) for all the factors. In this case the summand $\frac{1}{2}$ does not appear and we obtain (8.3.19). □

Now we are ready to present formula for the derivatives of the density function under some conditions on the random variable F. For a random variable F in $\mathbb{D}_{1,2}$ and for any $\beta \geq 1$ we are going to use the notation

$$M_\beta(F) = \left(\mathbb{E}[\|DF\|_{\mathbb{H}}^{-\beta}]\right)^{\frac{1}{\beta}}. \tag{8.3.23}$$

Proposition 8.3. *Fix an integer $m \geq 1$. Let F be in $\mathbb{D}_{m+2,\infty}$ such that $M_\beta(F) < \infty$ for some $\beta > 3m+3(\frac{m}{2}\vee 1)$. Denote $w = \|DF\|_{\mathbb{H}}^2$ and $u = \frac{DF}{w}$. Then, $u \in \mathbb{D}^{m+1,p}(\mathbb{H})$ for some $p > 1$, and the random variables $\{G_k\}_{k=0}^{m+1}$ introduced in Lemma 8.4 are well-defined. Under these assumptions, F has a density f of class C^m with derivatives*

$$f_F^{(k)}(x) = (-1)^k \mathbb{E}[I_{\{F>x\}} G_{k+1}] \tag{8.3.24}$$

for $k = 1, \ldots, m$.

Proof. It is sufficient to show that $\{G_k\}_{k=0}^{m+1}$ is well-defined, since it follows from Theorem 6.6 that the k-th derivative of the density of F is given by (8.3.24). To do this we will show that G_k defined in (8.3.4) are in $L^1(\Omega)$ for all $k = 1, \ldots, m+1$. From (8.3.4) we can write

$$\mathbb{E}[|G_k|] \leq \mathbb{E}[|H_k(D_u\delta_u, \delta_u)|] + \mathbb{E}[|T_k|].$$

Recall the explicit expression of $H_k(\lambda, x)$ in (8.3.3). Since $\beta > 3(m+1)$, we can choose $r_0 < \frac{\beta}{3}, r_1 < \frac{\beta}{6}$ such that

$$1 \geq \frac{k-2i}{r_0} + \frac{i}{r_1} > \frac{3\,(k-2i)}{\beta} + \frac{6i}{\beta} = \frac{3k}{\beta},$$

for any $0 \leq i \leq k/2$ and $1 \leq k \leq m+1$. Then, applying Hölder's inequality with (8.3.3), (8.3.18) and (8.3.19) we have

$$\mathbb{E}[|H_k(D_u\delta_u, \delta_u)|] \leq C_k \sum_{0\leq i\leq k/2} \|\delta_u\|_{r_0}^{k-2i} \|D_u\delta_u\|_{r_1}^{i} < \infty.$$

To prove that $E\left[|T_k|\right] < \infty$, applying Hölder's inequality to the expression (8.3.7) and choosing $r_j > 0$ for $0 \leq j \leq k-1$ such that

$$1 \geq \frac{i_0}{r_0} + \sum_{j=1}^{k-1} \frac{i_j}{r_j} > \frac{3i_0}{\beta} + \sum_{j=1}^{k-1} \frac{(3j+3)i_j}{\beta},$$

we obtain that (assuming $k \geq 3$, otherwise $T_k = 0$)

$$E\left[|T_k|\right] \leq C \sum_{(i_0,\dots,i_k)\in J_k} \|\delta_u\|_{r_0}^{i_0} \prod_{j=1}^{k-1} \left\|D_u^j \delta_u\right\|_{r_j}^{i_j}.$$

Due to (8.3.18) and (8.3.19), this expression is finite, provided $r_j < \frac{\beta}{3j+3}$ for $0 \leq j \leq k-1$. We can choose $(r_j, 0 \leq j \leq k-1)$ to satisfy the above conditions because $\beta > 3(k-1) + 3\frac{k-1}{2}$ for all $1 \leq k \leq m+1$. From properties (a) and (c) of J_k in Lemma 8.5 we have

$$\frac{3i_0}{\beta} + \sum_{j=1}^{k-1} \frac{(3j+3)i_j}{\beta} \leq \frac{3(k-1) + 3\frac{k-1}{2}}{\beta}.$$

This completes the proof. □

8.4 Random Variables in the q-th Wiener Chaos

Now we shall establish our main results on uniform estimates and uniform convergence of densities and their derivatives for random variables of a fixed chaos. We shall deal first with the convergence of densities and in next section we consider their derivatives.

Let $F = I_q(f)$ for some $f \in \mathbb{H}^{\hat{\otimes} q}$ and $q \geq 2$. To simplify the notation, along this section we denote

$$w = \|DF\|_{\mathbb{H}}^2, \quad u = w^{-1}DF.$$

Notice that $LF = -qF$ and using (6.2.15) and $L = -\delta D$, we can write

$$\delta_u = \delta(u) = qFw^{-1} - \left\langle Dw^{-1}, DF\right\rangle_{\mathbb{H}}. \tag{8.4.1}$$

Theorem 8.3. *Let $F = I_q(f)$, $q \geq 2$, $f \in \mathbb{H}^{\hat{\otimes} q}$ be a random variable in the q-th Wiener chaos with $\mathbb{E}[F^2] = 1$. Assume that $M_6(F) < \infty$, where $M_6(F)$ is defined in (8.3.23). Let $\phi(x)$ be the density of $N \sim N(0,1)$. Then, F has a density $f_F(x)$ given by*

$$f_F(x) = \mathbb{E}\left[I_{\{F>x\}}\delta(u)\right]. \tag{8.4.2}$$

Furthermore,

$$\sup_{x\in\mathbb{R}} |f_F(x) - \phi(x)| \leq C\sqrt{\mathbb{E}[F^4] - 3}, \tag{8.4.3}$$

where the constant C has the form $C = C_q\left(M_6(F)^2 + M_6(F)^3\right)$ and C_q depends only on q.

Proof. It follows from Theorem 7.16 that F admits a density $f_F(x) = E\left[I_{\{F>x\}}\delta(u)\right]$. By (8.1.1) with $k=1$ we can write $\phi(x) = \mathbb{E}[I_{\{N>x\}}N]$. Then, using (8.4.1), for all $x \in \mathbb{R}$ we obtain

$$\begin{aligned} & f_F(x) - \phi(x) = \mathbb{E}\left[I_{\{F>x\}}\delta(u)\right] - \mathbb{E}[I_{\{N>x\}}N] \\ &= \mathbb{E}[I_{\{F>x\}}(F(\frac{q}{w} - \langle Dw^{-1}, DF\rangle_{\mathbb{H}})] + \mathbb{E}\left[FI_{\{F>x\}} - NI_{\{N>x\}}\right] \\ &= A_1 + A_2. \end{aligned} \tag{8.4.4}$$

For the first term A_1, Hölder's inequality implies

$$\begin{aligned} |A_1| &= \left|\mathbb{E}[I_{\{F>x\}}(F(\frac{q}{w}-) - \langle Dw^{-1}, DF\rangle_{\mathbb{H}})]\right| \\ &\leq \mathbb{E}\left[\left|Fw^{-1}(w-q)\right|\right] + 2\mathbb{E}[w^{-\frac{3}{2}}\left\|D^2F \otimes_1 DF\right\|_{\mathbb{H}}] \\ &\leq \left\|w^{-1}\right\|_3 \|F\|_3 \|w-q\|_3 + 2\left\|w^{-\frac{3}{2}}\right\|_2 \left\|\left\|D^2F \otimes_1 DF\right\|_{\mathbb{H}}\right\|_2. \end{aligned}$$

Notice that (7.1.40) implies

$$\|w-q\|_3 \leq C\|w-q\|_2$$

and $\|F\|_3 \leq C\|F\|_2 = C$. Combining these estimates with (8.2.3) we obtain

$$|A_1| \leq C(\left\|w^{-1}\right\|_3 + \left\|w^{-1}\right\|_3^{\frac{3}{2}}) \|w-q\|_2. \tag{8.4.5}$$

For the second term A_2, applying Lemma 8.3 to the function $h(z) = zI_{\{z>x\}}$, which satisfies $|h(z)| \leq |z|$, we have

$$\begin{aligned} |A_2| &= \left|\mathbb{E}\left[FI_{\{F>x\}} - NI_{\{N>x\}}\right]\right| \\ &\leq C\left\|-\langle DF, -DL^{-1}F\rangle_{\mathbb{H}}\right\|_2 \leq C\|q-w\|_2. \end{aligned} \tag{8.4.6}$$

Combining (8.4.4) with (8.4.5)–(8.4.6) we obtain

$$\sup_{x\in\mathbb{R}} |f_F(x) - \phi(x)| \leq C(\left\|w^{-1}\right\|_3 + \left\|w^{-1}\right\|_3^{\frac{3}{2}} +) \|w-q\|_2.$$

Then, (8.4.3) follows from (6.2.20). This completes the proof. □

Using the estimates shown in Theorem 8.3 we can deduce the following uniform convergence and convergence in L^p of densities for a sequence of random variables in a fixed q-th Wiener chaos.

Corollary 8.2. *Let F be a random variables in the q-th Wiener chaos with $q \geq 2$. Assume $\mathbb{E}[F^2] = 1$ and define*

$$M = \left(\mathbb{E}[\|DF\|_{\mathbb{H}}^{-6}]\right)^{1/6} < \infty. \tag{8.4.7}$$

Let $\phi(x)$ be the density of the law $N(0,1)$. Then, F admits a density $f_F \in C(\mathbb{R})$ and there exists a constant C depending only on q, δ and M such that

$$\sup_{x\in\mathbb{R}} |f_F(x) - \phi(x)| \le C \left|\mathbb{E}[F^4] - 3\right|^{\frac{1}{2}}. \tag{8.4.8}$$

Furthermore, for any $p \ge 1$ and $\alpha \in (\frac{1}{2}, p)$,

$$\|f_F - \phi\|_{L^p(\mathbb{R})} \le C \left(\left|\mathbb{E}[F^4] - 3\right|^{\frac{1}{2}}\right)^{\frac{p-\alpha}{p}}, \tag{8.4.9}$$

where C is a constant depending on q, M, p, α.

Proof. Let $\phi(x)$ be the density of $N(0,1)$. Then, Theorem 8.3 implies that

$$\sup_{x\in\mathbb{R}} |f_F(x) - \phi(x)| \le C \left|\mathbb{E}[F^4] - 3\right|^{\frac{1}{2}}.$$

Then, (8.4.8) follows from triangle inequality. To show (8.4.9), first notice that (7.5.17) implies

$$f_F(x) \le C(1 \wedge |x|^{-2}).$$

Therefore, if $\alpha > \frac{1}{2}$ the function $(f_F(x)+\phi(x))^\alpha$ is integrable. Then, (8.4.9) follows from (8.4.8) and the inequality

$$|f_F(x) - \phi(x)|^p \le |f_F(x) - \phi(x)|^{p-\alpha} (f_F(x) + \phi(x))^\alpha.$$

This proves the corollary. □

8.5 Uniform Estimation of Difference of Derivatives of Densities

In this section, we establish the convergence for derivatives of densities of random variables to that of a normal distribution. We begin with the following theorem which estimates the uniform distance between the derivatives of the densities of a random variable F in the q-th Wiener chaos and the normal law $N(0,1)$.

Theorem 8.4. *Let $m \ge 1$ be an integer. Let F be a random variable in the q-th Wiener chaos, $q \ge 2$, with $\mathbb{E}[F^2] = 1$. Assume $M_\beta = M_\beta(F) < \infty$ for some $\beta > 6m + 6(\lfloor \frac{m}{2} \rfloor \vee 1)$ (Recall the definition of $M_\beta(F)$ in (8.3.23)). Let $\phi(x)$ be the density of $N \sim N(0,1)$. Then, F has a density $f_F(x) \in C^m(\mathbb{R})$ with derivatives given by (8.3.24). Moreover, for any $k = 1, \dots, m$*

$$\sup_{x\in\mathbb{R}} \left| f_F^{(k)}(x) - \phi^{(k)}(x) \right| \le C\sqrt{\mathbb{E}[F^4] - 3}, \tag{8.5.1}$$

where the constant C depends on q, β, m and M_β with polynomial growth in M_β.

To prove Theorem 8.4, we need some technical results. Recall the notation we introduced in (8.3.1), where we denote $\delta_u = \delta(u)$, $D_u\delta_u = \langle D\delta_u, u\rangle_{\mathbb{H}}$.

Lemma 8.8. *Let F be a random variable in the q-th Wiener chaos with $\mathbb{E}[F^2] = 1$. Let $w = \|DF\|_{\mathbb{H}}^2$ and $u = w^{-1}DF$.*

(i) If $M_\beta(F) < \infty$ for some $\beta > 6$, then for any $1 < r \le \frac{2\beta}{\beta+6}$

$$\|\delta_u - F\|_r \le C(M_\beta^3 \vee 1)\, \|q - w\|_2\,. \tag{8.5.2}$$

(ii) If $M_\beta(F) < \infty$ for some $\beta > 12$, then for any $1 < r < \frac{2\beta}{\beta+12}$

$$\|D_u\delta_u - 1\|_r \le C(M_\beta^6 \vee 1)\, \|q - w\|_2\,. \tag{8.5.3}$$

Proof. Recall that $\delta_u = qFw^{-1} - D_{DF}w^{-1}$. Using Hölder's inequality and (8.3.9) we can write

$$\begin{aligned}\|\delta_u - F\|_r &\le \left\|Fw^{-1}(q-w)\right\|_r + \left\|D_{DF}w^{-1}\right\|_r \\ &\le C\left(\|Fw^{-1}\|_s + (M_\beta^3 \vee 1)\right)\|q-w\|_2\,,\end{aligned}$$

provided $\frac{1}{r} = \frac{1}{s} + \frac{1}{2}$. By the hypercontractivity property (7.1.40) $\|F\|_\gamma \le C_{q,\gamma}\|F\|_2$ for any $\gamma \ge 2$. Thus, by Hölder's inequality, if $\frac{1}{s} = \frac{1}{\gamma} + \frac{1}{p}$, then

$$\|Fw^{-1}\|_s \le \|F\|_\gamma \|w^{-1}\|_p \le C_{q,\gamma}\sigma M_{2p}^2.$$

Choosing p such that $2p < \beta$ yields (8.5.2).

We can compute $D_u\delta_u$ as

$$D_u\delta_u = qw^{-1} + qFw^{-1}D_{DF}w^{-1} - w^{-1}D_{DF}^2 w^{-1} - w^{-1}\left\langle D^2F, DF \otimes Dw^{-1}\right\rangle_{\mathbb{H}}.$$

Applying Hölder's inequality we obtain

$$\begin{aligned}\|D_u\delta_u - 1\|_r &\le \left\|w^{-1}\left[(q-w) + qFD_{DF}w^{-1} - D_{DF}^2 w^{-1}\right]\right\|_r \\ &\le \|w^{-1}\|_{\frac{2r}{2-r}}\|q-w\|_2 + C\|w^{-1}\|_p\left(\left\|D_{DF}w^{-1}\right\|_s + \left\|D_{DF}^2 w^{-1}\right\|_s\right),\end{aligned}$$

if $\frac{1}{r} > \frac{1}{p} + \frac{1}{s}$. Then, using (8.3.9) and (8.3.10) with $k = 2$ and that $s < \frac{2\beta}{\beta+8}$ and $2p < \beta$ we obtain (8.5.3). □

Proof of Theorem 8.4. Proposition 8.3 implies that $f_F(x) \in C^{m-1}(\mathbb{R})$ and for $k = 0, 1, \dots, m-1$,

$$f_F^{(k)}(x) = (-1)^k \mathbb{E}[\mathbf{1}_{\{F>x\}}G_{k+1}],$$

where $G_0 = 1$ and $G_{k+1} = \delta(G_k u) = G_k\delta(u) - \langle DG_k, u\rangle_{\mathbb{H}}$. From (8.1.1),

$$\phi^{(k)}(x) = (-1)^k \mathbb{E}[\mathbf{1}_{\{N>x\}}H_{k+1}(N)].$$

Then, the identity $G_{k+1} = H_{k+1}(D_u\delta_u, \delta_u) + T_{k+1}$ (see Formula (8.3.4)) yields

$$\begin{aligned}\left|f_F^{(k)}(x) - \phi^{(k)}(x)\right| &= \left|\mathbb{E}[\mathbf{1}_{\{F>x\}}G_{k+1} - \mathbf{1}_{\{N>x\}}H_{k+1}(N)]\right| \\ &\le \left|\mathbb{E}[\mathbf{1}_{\{F>x\}}G_{k+1} - \mathbf{1}_{\{F>x\}}H_{k+1}(\sigma^{-2}, F)]\right| \\ &\quad + \left|\mathbb{E}[\mathbf{1}_{\{F>x\}}H_{k+1}(F) - \mathbf{1}_{\{N>x\}}H_{k+1}N)]\right| \\ &= A_1 + A_2 .\end{aligned}$$

We first estimate the term A_2. Notice that $\|F\|_{2k+2} \le C_{q,k}\|F\|_2 = C_{q,k}$ by the hypercontractivity property (7.1.40). Applying Lemma 8.3 with $h(z) = \mathbf{1}_{\{z>x\}}H_{k+1}(z)$, which satisfies $|h(z)| \le C_k(|z|^{k+1} + 1)$, we obtain

$$\begin{aligned}A_2 &= |\mathbb{E}[h(F) - h(N)]| \\ &\le C_{q,k}\left\|1 - \left\langle DF, -DL^{-1}F\right\rangle_{\mathbb{H}}\right\|_2 \\ &\le C_{q,k}\|q - w\|_2 ,\end{aligned} \tag{8.5.4}$$

where in the second inequality we use the fact that $\left\langle DF, -DL^{-1}F\right\rangle_{\mathbb{H}} = \frac{w}{q}$.

For the term A_1, Lemma 8.4 implies

$$A_1 \le \mathbb{E}[\left|H_{k+1}(D_u\delta_u, \delta_u) - H_{k+1}(\sigma^{-2}, F)\right|] + \mathbb{E}[|T_{k+1}|]. \tag{8.5.5}$$

To proceed with the first term above, applying (8.3.3) we have

$$\begin{aligned}&|H_{k+1}(D_u\delta_u, \delta_u) - H_{k+1}(F)| \\ &\le \sum_{0\le i\le\lfloor(k+1)/2\rfloor} |c_{k,i}|\left|\delta_u^{k+1-2i}(D_u\delta_u)^i - (F)^{k+1-2i}\right| \\ &\le \sum_{0\le i\le\lfloor(k+1)/2\rfloor} |c_{k,i}|\left[\left|\delta_u^{k+1-2i} - (F)^{k+1-2i}\right||D_u\delta_u|^i \right. \\ &\qquad \left. + |F|^{k+1-2i}\left|(D_u\delta_u)^i - 1\right|\right].\end{aligned} \tag{8.5.6}$$

Using the fact that $\left|x^k - y^k\right| \le C_k|x - y|\sum_{0\le j\le k-1}|x|^{k-1-j}|y|^j$ and applying Hölder's inequality and the hypercontractivity property (7.1.40) we obtain

$$\begin{aligned}&\mathbb{E}\left[\left|\delta_u^{k+1-2i} - (F)^{k+1-2i}\right||D_u\delta_u|^i\right] \\ &\le C_k\mathbb{E}\left[|\delta_u - F||D_u\delta_u|^i\sum_{0\le j\le k-2i}|\delta_u|^{k-2i-j}|F|^j\right] \\ &\le C_{q,k}\|\delta_u - F\|_r\|D_u\delta_u\|_s^i\sum_{0\le j\le k-2i}\|\delta_u\|_p^{k-2i-j},\end{aligned} \tag{8.5.7}$$

provided $1 \geq \frac{1}{r} + \frac{i}{s} + \frac{k-2i-j}{p}$, which is implied by $1 \geq \frac{1}{r} + \frac{i}{s} + \frac{k-2i}{p}$. In order to apply the estimates (8.5.2), (8.3.19) (with $k = 1$) and (8.3.18) we need $\frac{1}{r} > \frac{3}{\beta} + \frac{1}{2}$, $\frac{1}{s} > \frac{6}{\beta}$ and $\frac{1}{p} > \frac{3}{\beta}$, respectively. These are possible because $\beta > 6k+6$. Then, we obtain an estimate of the form

$$\mathbb{E}\left[\left|\delta_u^{k+1-2i} - (F)^{k+1-2i}\right| |D_u\delta_u|^i\right] \leq C_{q,k}(M_\beta^{3k+3} \vee 1)\|q-w\|_2. \quad (8.5.8)$$

Similarly, we have

$$\begin{aligned} &\mathbb{E}\left[|F|^{k+1-2i}\left|(D_u\delta_u)^i - 1\right|\right] \\ &\leq C_{q,k}\mathbb{E}\left[|F|^{k+1-2i}|D_u\delta_u - 1| \sum_{0\leq j\leq i-1} |D_u\delta_u|^j\right]. \\ &\leq C_{q,k}\|D_u\delta_u - 1\|_r \sum_{0\leq j\leq i-1} \|D_u\delta_u\|_s^j, \end{aligned} \quad (8.5.9)$$

provided $1 > \frac{1}{r} + \frac{j}{s}$. In order to apply the estimates (8.5.3) and (8.3.19) (with $k = 1$) we need $\frac{1}{r} > \frac{6}{\beta} + \frac{1}{2}$ and $\frac{1}{s} > \frac{6}{\beta}$, respectively. This implies

$$\frac{1}{r} + \frac{j}{s} > \frac{6+6j}{\beta} + \frac{1}{2}.$$

Notice that $6+6j \leq 6i \leq 3k+3$. So, we need $1 > \frac{1}{2} + \frac{3k+3}{\beta}$. The above r, s and p exist because $\beta > 6k+6$. Thus, we obtain the following estimate.

$$\mathbb{E}\left[|F|^{k+1-2i}\left|(D_u\delta_u)^i - 1\right|\right] \leq C_{q,k,\beta}(M_\beta^{3k+3} \vee 1)\|q-w\|_2. \quad (8.5.10)$$

Combining (8.5.8) and (8.5.10) we have

$$\mathbb{E}\left[|H_{k+1}(D_u\delta_u, \delta_u) - H_{k+1}(F)|\right] \leq C_{q,k,\beta}(M_\beta^{3k+3} \vee 1)\|q-w\|_2. \quad (8.5.11)$$

Applying Hölder's inequality to the expression (8.3.7) we obtain (assuming $k \geq 2$, otherwise $T_{k+1} = 0$)

$$\mathbb{E}\left[|T_{k+1}|\right] \leq C_{q,k,\beta} \sum_{(i_0,\dots,i_k)\in J_{k+1}} \|\delta_u\|_{r_0}^{i_0} \prod_{j=1}^k \left\|D_u^j\delta_u\right\|_{r_j}^{i_j},$$

where $1 = \frac{i_0}{r_0} + \sum_{j=1}^k \frac{i_j}{r_j}$. From property (b) in Lemma 8.5 there is at least one factor of the form $\left\|D_u^j\delta_u\right\|_{s_j}$ with $j \geq 2$. We apply the estimate (8.3.20) to one of these factors, and the estimate (8.3.19) to all the remaining factors. We also use the estimate (8.3.18) to control $\|\delta_u\|_{r_0}$. Notice that

$$1 = \frac{i_0}{r_0} + \sum_{j=1}^k \frac{i_j}{r_j} > \frac{3i_0}{\beta} + \sum_{j=1}^k \frac{i_j(3j+3)}{\beta} + \frac{1}{2},$$

and, on the other hand, using properties (a) and (c) in Lemma 8.5 we see

$$\frac{3i_0}{\beta} + \sum_{j=1}^{k} \frac{i_j(3j+3)}{\beta} + \frac{1}{2} \le \frac{3k+3\lfloor \frac{k}{2} \rfloor}{\beta} + \frac{1}{2}.$$

We can choose the r_j to satisfy the above properties because $\beta > 6k+6\lfloor \frac{k}{2} \rfloor$, and we obtain

$$\mathbb{E}\,|T_{k+1}| \le C_{q,k,\beta}(M_\beta^{3k+3\lfloor \frac{k}{2} \rfloor} \vee 1)\,\|q-w\|_2\,. \tag{8.5.12}$$

Combining (8.5.11) and (8.5.12) we complete the proof. □

Most of the time we assume that $\sigma_n = 1$ to simplify notation. Here we give a corollary for general σ_n to illustrate how the results look like in this general variance case.

Corollary 8.3. *Fix an integer $m \ge 1$. Let $\{F_n\}_{n\in\mathbb{N}}$ be a sequence of random variables in the q-th Wiener chaos with $q \ge 2$ and $\mathbb{E}[F_n^2] = \sigma_n^2$. Assume $\lim_{n\to\infty} \sigma_n = \sigma$, $0 < \delta \le \sigma_n^2 \le K$ for all n, $\lim_{n\to\infty} \mathbb{E}[F_n^4] = 3\sigma^4$ and*

$$M = \sup_n \left(\mathbb{E}[\|DF_n\|_{\mathfrak{H}}^{-\beta}]\right)^{\frac{1}{\beta}} < \infty \tag{8.5.13}$$

for some $\beta > 6(m)+6(\lfloor \frac{m}{2} \rfloor \vee 1)$. Let $\phi(x)$ be the density of $N(0,\sigma^2)$. Then, each F_n admits a probability density function $f_{F_n} \in C^m(\mathbb{R})$ with derivatives given by (8.3.24) and for any $k = 1,\dots,m$,

$$\sup_{x\in\mathbb{R}} \left| f_{F_n}^{(k)}(x) - \phi^{(k)}(x) \right| \le C\left(\sqrt{\mathbb{E}[F_n^4] - 3\sigma_n^4} + |\sigma_n - \sigma|\right),$$

where the constant C depends only on $q, m, \beta, M, \sigma, \delta$ and K.

Proof. Let $\phi_n(x)$ be the density of $N(0,\sigma_n^2)$. Then, Theorem 8.4 implies that

$$\sup_{x\in\mathbb{R}} \left| f_{F_n}^{(k)}(x) - \phi_n^{(k)}(x) \right| \le C_{q,m,\beta,M,\sigma} \sqrt{\mathbb{E}[F_n^4] - 3\sigma_n^4}.$$

On other hand, by the mean value theorem we can write

$$\begin{aligned} \left|\phi_n^{(k)}(x) - \phi^{(k)}(x)\right| &\le |\sigma_n - \sigma| \sup_{\gamma\in[\frac{\sigma}{2},2\sigma]} \left|\partial_\gamma \phi_\gamma^{(k)}(x)\right| \\ &= \frac{1}{2}|\sigma_n - \sigma| \sup_{\gamma\in[\frac{\sigma}{2},2\sigma]} \gamma \left|\phi_\gamma^{(k+2)}(x)\right|, \end{aligned}$$

where $\phi_\gamma(x)$ is the density of the law $N(0,\gamma^2)$. Then, using the expression

$$\phi_\gamma^{(k+2)}(x) = \mathbb{E}[\mathbf{1}_{N>x} H_{k+3}(\gamma^{-2}, \gamma^{-2}Z)],$$

where $Z \sim N(0,\gamma^2)$ and the explicit form of $H_{k+3}(\lambda, x)$, we obtain

$$\sup_{\gamma\in[\frac{\sigma}{2},2\sigma]} \gamma \left|\phi_\gamma^{(k+2)}(x)\right| \le C_{k,\sigma}.$$

Therefore,

$$\sup_{x\in\mathbb{R}} \left|\phi_n^{(k)}(x) - \phi^{(k)}(x)\right| \le C_{k,\sigma} \left|\sigma_n - \sigma\right|.$$

This completes the proof. □

Remark 8.3. We refer to [Hu, Lu and Nualart (2014)] for more discussions on bounding the difference between the normal density and the probability density of a more general Wiener functional (or a vector) by the moments of F or the derivatives of F.

8.6 Density Convergence of Higher Rank Hermite Polynomials

Let $X = \{X_k;\ k \ge 0\}$ be a centered Gaussian stationary sequence with unit variance. Denote the covariance function $\rho(v) = \mathbb{E}[X_0 X_{|v|}]$, $v \in \mathbb{Z}$. Define

$$f(x) = \sum_{j=d}^{\infty} a_j H_j(x), \qquad V_n = \frac{1}{\sqrt{n}} \sum_{k=0}^{n-1} f(X_k), \tag{8.6.1}$$

where $d \ge 1$ and $a_d \ne 0$.

Theorem 8.5 (Breuer-Major theorem). *Suppose that $\sum_{v\in\mathbb{Z}} |\rho(v)|^d < \infty$ and suppose $\sigma^2 = \sum_{j=d}^{\infty} j! a_j^2 \sum_{v\in\mathbb{Z}} \rho(v)^j \in (0,\infty)$. Then, the convergence*

$$V_n \xrightarrow{Law} N(0,\sigma^2) \tag{8.6.2}$$

holds true as n tends to infinity.

Proof. To prove the theorem we need to verify the conditions in Theorem 8.2. We can represent $X_k = \tilde{e}_k$, where e_k is an element in a Hilbert space $\mathbb{H}$ and $\langle e_k, e_j\rangle = \rho(k-j)$ and $|e_j| = 1$. Thus, we can write

$$V_n = \frac{1}{\sqrt{n}} \sum_{k=1}^{n} \sum_{q=d}^{\infty} H_q(\tilde{e}_k) = \sum_{q=d}^{\infty} I_q(f_{n,q}),$$

where

$$f_{n,q} = \sum_{k=1}^{n} e_k^{\otimes q}.$$

Step 1. Fix $q \geq d$. We have

$$\mathbb{E}\left[I_q(f_{n,q})\right]^2 = \frac{q!a_q^2}{n}\sum_{k,l=1}^{n}\rho(k-l) = q!a_q^2\sum_{v\in\mathbb{Z}}\rho(v)^q\left(1-\frac{|v|}{n}\right)I_{\{|v|<n\}}$$
$$= q!a_q^2\sum_{v\in\mathbb{Z}}\rho(v)^q\,.$$

This verifies (ii) of Theorem 8.2.
Step 2. By Cauchy-Schwarz inequality, we have $|\rho(v)| \leq 1$ for all $v \in \mathbb{Z}$. Thus,

$$\sum_{q=d}^{\infty} q!a_q^2\sum_{v\in\mathbb{Z}}\rho(v)^q \leq \sum_{v\in\mathbb{Z}}\rho(v)^d\sum_{q=d}^{\infty}q!a_q^2 = \sum_{v\in\mathbb{Z}}\rho(v)^d\mathbb{E}\left[f(X_1)^2\right] < \infty\,.$$

This verifies the condition (iii) in Theorem 8.2.
Step 3. Let $N \geq d$. Then

$$\sum_{q=N+1}^{\infty}\mathbb{E}\left[I_q(f_{n,q})\right]^2 = \frac{1}{n}\sum_{q=N+1}^{\infty}a_q q!\sum_{k,l=1}^{n}q!a_q^2\rho(k-l)^q$$
$$\leq \frac{1}{n}\sum_{q=N+1}^{\infty}a_q q!\sum_{v\in\mathbb{Z}}q!a_q^2\rho(k-l)^q$$
$$\leq \sum_{v\in\mathbb{Z}}q!a_q^2\rho(k-l)^d\sum_{q=N+1}^{\infty}a_q q!$$

This implies that $\lim_{N\to\infty}\sum_{n\geq 1}\sum_{q=N+1}^{\infty}\mathbb{E}\left[I_q(f_{n,q})\right]^2 \to 0$, verifying the condition (i).
Step 4. Fix $q \geq d$. For all $n \geq 1$ and $r = 1, \cdots, q-1$, we have

$$f_{n,q}\otimes_r f_{n,q} = \frac{a_q^2}{n}\sum_{k,l=1}^{n}\rho(k-l)e_k^{\otimes(q-r)}\otimes e_l^{\otimes(q-r)}\,.$$

Consequently, the contraction can be computed as follows.

$$|f_{n,q}\otimes_r f_{n,q}|^2_{\mathbb{H}^{\otimes(2q-2r)}} = \frac{a_q^4}{n^2}\sum_{i,j,k,l=1}^{n}\rho(k-l)^r\rho(i-j)^r\rho(k-i)^{q-r}\rho(l-j)^{q-r}\,.$$

We bound the factor $|\rho(k-l)^r\rho(k-i)^{q-r}|$ in the above summation by

$|\rho(k-l)^q| + |\rho(k-i)^q|$. Thus, we have

$$
\begin{aligned}
&|f_{n,q} \otimes_r f_{n,q}|^2_{\mathbb{H}^{\otimes(2q-2r)}} \\
&= \frac{a_q^4}{n^2} \sum_{i,j,k,l=1}^{n} |\rho(k-l)|^q \left[\rho(i-j)|^r |\rho(l-j)|^{q-r} + |\rho(i-j)|^{q-r} |\rho(l-j)|^r \right] \\
&= \frac{a_q^4}{n^2} \sum_{k\in\mathbb{Z}} |\rho(k)|^d \sum_{i,j,l=1}^{n} |\rho(k-l)|^q \\
&\qquad \left[\rho(i-j)|^r |\rho(l-j)|^{q-r} + |\rho(i-j)|^{q-r} |\rho(l-j)|^r \right] \\
&\le \frac{a_q^4}{n} \sum_{k\in\mathbb{Z}} |\rho(k)|^d \sum_{|i|<n} |\rho(i)|^r \sum_{|j|<n} |\rho(j)|^{q-r} \\
&= 2a_q^2 \sum_{k\in\mathbb{Z}} |\rho(k)|^d \left(n^{-1+\frac{r}{q}} \sum_{|i|<n} |\rho(i)|^r \right) \left(n^{-1+\frac{q-r}{q}} \sum_{|j|<n} |\rho(j)|^{q-r} \right).
\end{aligned}
$$

It is easy to see that both factors inside $(\cdot)$ are bounded. We need to show one of them (in fact, both) will go to 0. To explain, we consider the first factor $n^{-1+\frac{r}{q}} \sum_{|i|<n} |\rho(i)|^r$. Choose a fixed arbitrary $\delta > 0$. We use Hölder inequality $\sum_{i\in T} a_i \le (\sum a_i^q)^{r/q} (\#T)^{1-r/q}$ to obtain

$$
\begin{aligned}
n^{-1+\frac{r}{q}} \sum_{|i|<n} |\rho(i)|^r &= n^{-1+\frac{r}{q}} \sum_{|i|<[n\delta]} |\rho(i)|^r + n^{-1+\frac{r}{q}} \sum_{[n\delta]\le|i|<n} |\rho(i)|^r \\
&\le n^{-1+\frac{r}{q}} (2n\delta+1)^{1-\frac{r}{q}} \left(\sum_{i\in\mathbb{Z}} |\rho(i)|^q \right)^{r/q} \\
&\quad + n^{-1+\frac{r}{q}} (2n+1)^{1-\frac{r}{q}} \left(\sum_{[n\delta]\le|i|\le n} |\rho(i)|^q \right)^{r/q} \\
&\le \delta^{1-\frac{r}{q}} \left(\sum_{i\in\mathbb{Z}} |\rho(i)|^q \right)^{r/q} + n^{-1+\frac{r}{q}} \left(\sum_{i\in\mathbb{Z}} |\rho(i)|^q \right)^{r/q} \\
&\quad + n^{-1+\frac{r}{q}} (2n+1)^{1-\frac{r}{q}} \left(\sum_{[n\delta]\le|i|\le n} |\rho(i)|^q \right)^{r/q}.
\end{aligned}
$$

Choose δ to be sufficiently small we can make the first term uniformly small for all $n \ge 1$. For any arbitrarily fixed (though small) δ we can choose N sufficiently large such that the above second and third terms are uniformly small when $n \ge N$. This proves $|f_{n,q} \otimes_r f_{n,q}|^2_{\mathbb{H}^{\otimes(2q-2r)}} \to 0$, verifying the condition (iv) of Theorem 8.2. The theorem is proved. □

To obtain the density convergence we shall be interested in a particular case of Theorem 8.5 for finite linear combinations of Hermite polynomials, which is stated here for convenience.

Corollary 8.4. *Consider $2 \le d \le q < \infty$ and a family of real numbers $\{a_j;\, j = d, \dots, q\}$. Let H_j be the jth order Hermite polynomial, and assume that $\sigma^2 \in (0, \infty)$, where $\sigma^2 \equiv \sum_{j=d}^{q} j! a_j^2 \sum_{v\in\mathbb{Z}} \rho(v)^j$. Set*

$$V_n^{d,q} = \frac{1}{\sqrt{n}} \sum_{k=0}^{n-1} \sum_{j=d}^{q} a_j\, H_j(X_k). \tag{8.6.3}$$

Then, $V_n^{d,q} \xrightarrow{Law} \mathcal{N}(0, \sigma^2)$ as n tends to infinity. In particular, we have

$$\lim_{n\to 0} \mathbb{E}\left[\left(V_n^{d,q}\right)^4\right] = 3\sigma^4. \tag{8.6.4}$$

We shall mainly be concerned with the convergences of densities of $V_n^{d,q}$ to a normal density as $n \to \infty$. Before we state our main result, we need some preparation.

Definition 8.1. The spectral density of the Gaussian sequence is defined as

$$f_\rho(t) = \frac{1}{2\pi} \sum_{k=-\infty}^{\infty} \rho(k) e^{-\iota k t}\,, \quad -\pi \le t \le \pi\,.$$

We make the following hypothesis.

Hypothesis 8.1 The spectral density f_ρ satisfies $\log(f_\rho) \in L^1([-\pi, \pi])$.

Let us state a result from [Brockwell and Davis (1991), Theorem 7.5.2].

Proposition 8.4. *Let $X = \{X_n\,, n \ge 1\}$ be a Gaussian stationary sequence satisfying Hypothesis 8.1. Then, for each $k \in \mathbb{N} \cup \{0\}$ the random variable X_k can be decomposed as*

$$X_k = \sum_{j\ge 0} \psi_j\, w_{k-j}, \tag{8.6.5}$$

where $(w_k)_{k\in\mathbb{Z}}$ is a discrete Gaussian white noise and the coefficient ψ_j is deterministic. We can always choose the white noise and the coefficients in such a way that $\psi_0 > 0$.

Notice that from the expansion (8.6.5) we obtain

$$\rho(k_1 - k_2) = \sum_{l=-\infty}^{k_1\wedge k_2} \psi_{k_1-l}\, \psi_{k_2-l}, \quad \text{for all} \quad k_1, k_2 \in \mathbb{N}, \tag{8.6.6}$$

and this relation will be used in the proof of our main theorem.

Let us now turn to examples for which our standing assumptions of Hypothesis 8.1 are met.

Proposition 8.5. *Let ρ be the covariance function of X. We have the following statements.*

(i) *If $\rho \in \ell^1$, then the spectral density f_ρ exists and is a non-negative bounded function defined on $[-\pi, \pi]$.*

(ii) *Suppose that $\{\rho(k)k^\alpha, k > 0\}$ is positive and it is slowly varying at infinity for some $\alpha \in (0,1)$. That is, for every $\delta > 0$, for sufficiently large k, $\rho(k)k^{\alpha+\delta}$ is increasing and $\rho(k)k^{\alpha-\delta}$ is decreasing. Then, the spectral density exists and satisfies $\lim_{\lambda \to 0} |\lambda|^{1-\alpha} f_\rho(\lambda) = c_f$ for some constant $c_f > 0$ (see [Gubner (2005)]).*

We now give two specific and important examples which satisfy Hypothesis 8.1.

Example 8.1. The so-called autoregressive fractionally integrated moving-average (ARFIMA) processes are introduced in [Granger and Joyeux (1980)] and [Hosking (1981)]. Denote by B the one lag backward operator ($BX_k = X_{k-1}$). Let $\phi(z)$ and $\theta(z)$ be two polynomials which have no common zeros and such that the zeros of ϕ lie outside the closed unit disk $\{z\,,\ |z| \leq 1\}$. Suppose that X_k is given by

$$\phi(B)X_k = (\mathrm{Id} - B)^{-d}\theta(B)w_k\,, \tag{8.6.7}$$

where $-1 < d < 1/2$, and where the operator $(\mathrm{Id} - B)^{-d}$ is defined by

$$(\mathrm{Id} - B)^{-d} = \sum_{j=1}^{\infty} \eta_j B^j \quad \text{with} \quad \eta_j = \frac{\Gamma(d+j)}{\Gamma(j+1)\Gamma(d)}\,.$$

Also notice that in (8.6.7) the sequence $(w_k)_{k\in\mathbb{Z}}$ is a discrete Gaussian white noise. It is well-known (see [Palma (2007)], Theorem 3.4 and Equation (3.19)) that under the above conditions, $\{X_k, k \in \mathbb{N}\}$ admits a spectral density whose exact expression is

$$f(\lambda) = \frac{1}{2\pi}\left[2\sin\frac{\lambda}{2}\right]^{-2d} \frac{|\theta(e^{-i\lambda})|^2}{|\phi(e^{-i\lambda})|^2}.$$

It is thus readily checked that Hypothesis 8.1 is satisfied, and hence X_k has a causal representation.

Example 8.2. Our second example is the fractional Gaussian noise. Let $\{B_t, t \geq 0\}$ be a fractional Brownian motion of Hurst parameter $H \in (0,1)$. Then, $\{X_k = B_{k+1} - B_k\,, k \in \mathbb{N} \cup \{0\}\}$ is a stationary Gaussian process with correlation

$$\rho(k) = \frac{1}{2}\left[(k+1)^{2H} - 2k^{2H} + (k-1)^{2H}\right].$$

Its spectral density (see e.g. [Beran (1984)], Equation (2.17)) is

$$\begin{aligned} f(\lambda) &= \frac{1}{2\pi} \sum_{k=-\infty}^{\infty} \rho(|k|) e^{ik\lambda} \\ &= 2c_f(1-\cos(\lambda)) \sum_{j=-\infty}^{\infty} |2\pi j + \lambda|^{-2H-1}, \quad \forall\, \lambda \in [-\pi, \pi], \end{aligned}$$

where $c_f = (2\pi)^{-1} \sin(\pi H)\Gamma(2H+1)$. If $H \leq 1/2$, it is clear that $\sum_{k=-\infty}^{\infty} |\rho(|k|)| < \infty$. This implies

$$\sup_{\lambda \in [-\pi,\pi]} |f(\lambda)| < \infty.$$

If $1/2 < H < 1$, then

$$0 \leq f(\lambda) \leq 2c_f(1-\cos(\lambda))|\lambda|^{-2H-1} + 2c_f \sum_{j \neq 0} |2\pi j + \lambda|^{-2H-1},$$

where $\lambda \in [-\pi, \pi]$. The first term is in L^1 since $H < 1$. When $j \neq 0$, $\int_{-\pi}^{\pi} |2\pi j + \lambda|^{-2H-1} d\lambda \leq C j^{-2H}$ for some positive constant C. Thus, $\int_{-\pi}^{\pi} \sum_{j\neq 0} |2\pi j + \lambda|^{-2H-1}\, d\lambda < \infty$, owing to the fact that $H > 1/2$. Therefore, we have $f \in L^1$. Summarizing the above argument we have $f \in L^1$ for all $H \in (0,1)$. This also implies $\log^+ f(\lambda) \in L^1$. To see $\log^- f(\lambda) \in L^1$, we notice that

$$f(\lambda) \geq 2c_f(1-\cos(\lambda))|\lambda|^{-2H-1}.$$

So $\log^- f(\lambda) \leq C + \left|\log\left[(1-\cos(\lambda))|\lambda|^{-2H-1}\right]\right|$ which is in L^1. In conclusion, the sequence X satisfies Hypothesis 8.1.

The following inequality is due to Carbery and Wright [Carbery and Wright (2001), Theorem 8], which is recalled here for convenience.

Proposition 8.6. *Let $X = (X_1, \dots, X_n)$ be a Gaussian random vector in $\mathbb{R}^n$ and $Q : \mathbb{R}^n \to \mathbb{R}$ a polynomial of degree at most m. Then, there is a universal constant $c > 0$ such that*

$$\mathbb{E}[|Q(X_1,\dots,X_n)|]^{\frac{1}{m}}\, \mathbb{P}(|Q(X_1,\dots,X_n)| \leq x) \leq c\, m\, x^{\frac{1}{m}}, \quad \textit{for all} \quad x > 0. \tag{8.6.8}$$

Here is one of the main results of this section (when $q = d$).

Theorem 8.6. *Let X be a Gaussian stationary sequence whose spectral density f_ρ satisfies Hypothesis 8.1. Let $V_n^{d,d}$ be the random variable defined by (8.6.3), and assume the hypothesis of Corollary 8.4 is satisfied. Define $F_n = V_n^{d,d}/\sigma_n$, where $\sigma_n^2 = \mathbb{E}[(V_n^{d,d})^2]$. Then, for all $m \geq 0$ there exists an n_0 (depending on m) such that*

$$\sup_{n \geq n_0} \sup_{x \in \mathbb{R}} |p_{F_n}^{(m)}(x) - \phi^{(m)}(x)| \leq c_m \sqrt{\mathbb{E}[F_n^4] - 3}\,, \tag{8.6.9}$$

where $\phi(x) = (2\pi)^{-1/2} \exp(-|x|^2/2)$ is the density of the standard Gaussian distribution on $\mathbb{R}$.

From Theorem 8.4, we know that if

$$\sup_n \mathbb{E}\left[\|DF_n\|^{-p}\right] < \infty\,, \quad \forall\, p \geq 1\,,$$

then, for any $m \geq 0$, there exists a constant c_m such that for all $n \geq 1$,

$$\sup_{x \in \mathbb{R}} |p_{F_n}^{(m)}(x) - \phi^{(m)}(x)| \leq c_m \sqrt{\mathbb{E}[F_n^4] - 3}.$$

Thus, to show Theorem 8.6, we only need to show the following.

Theorem 8.7. *Let X be a Gaussian stationary sequence whose spectral density f_ρ satisfies $\log(f_\rho) \in L^1([-\pi, \pi])$. Let $V_n^{d,q}$ be the random variable defined by (8.6.3), and assume that the hypothesis of Corollary 8.4 is satisfied. Then, for any $p \geq 1$, there exists n_0 such that*

$$\sup_{n \geq n_0} \mathbb{E}\left[\|DV_n^{d,q}\|^{-p}\right] < \infty. \tag{8.6.10}$$

Proof. We divide the proof into several steps.

Step 1. Bound of the Malliavin covariance matrix. Recall that $V_n^{d,q}$ is defined by relation (8.6.3), and set for the moment $f = \sum_{j=d}^{q} a_j H_j$. Using the chain rule and the fact that $Dw_j = \varepsilon^j$, we have

$$DV_n^{d,q} = \frac{1}{\sqrt{n}} \sum_{k=0}^{n-1} f'(X_k) \left(\sum_{j \geq 0} \psi_j\, \varepsilon^{k-j} \right) = \frac{1}{\sqrt{n}} \sum_{l=-\infty}^{n-1} \left(\sum_{k=l^+}^{n-1} \psi_{k-l}\, f'(X_k) \right) \varepsilon^l\,, \tag{8.6.11}$$

where $l^+ = \max\{l, 0\}$. From the identity (8.6.6), we see

$$\langle DV_n^{d,q}, DV_n^{d,q} \rangle_{\mathbb{H}} = \frac{1}{n} \sum_{k_1, k_2 = 0}^{n-1} f'(X_{k_1}) \rho(k_1 - k_2) f'(X_{k_2}),$$

where ρ is the covariance function of the Gaussian stationary sequence $\{X_k;\, k \geq 0\}$. Since $\{\varepsilon^l;\, l \in \mathbb{Z}\}$ is an orthonormal basis of $\ell^2(\mathbb{Z})$ we have

$$\left\langle DV_n^{d,q},\, DV_n^{d,q}\right\rangle_{\mathbb{H}} = \frac{1}{n}\sum_{\ell=-\infty}^{n-1}\left(\sum_{k=\ell^+}^{n-1}\psi_{k-\ell}\, f'(X_k)\right)^2.$$

Rearranging terms (namely, change $k-\ell$ to k and then $n-\ell-1$ to m), we end up with

$$\begin{aligned}\left\langle DV_n^{d,q},\, DV_n^{d,q}\right\rangle_{\mathbb{H}} &\geq \frac{1}{n}\sum_{\ell=0}^{n-1}\left(\sum_{k=0}^{n-\ell-1} f'(X_{\ell+k})\,\psi_k\right)^2\\ &= \frac{1}{n}\sum_{m=0}^{n-1}\left(\sum_{k=0}^{m} f'(X_{n-1-(m-k)})\,\psi_k\right)^2 \equiv A_n.\end{aligned}$$

Since $X = \{X_k;\, k \in \mathbb{N}\cup\{0\}\}$ is a Gaussian stationary sequence, we see that A_n is identical in law to B_n with

$$B_n = \frac{1}{n}\sum_{m=0}^{n-1}\left(\sum_{k=0}^{m} f'(X_{m-k})\,\psi_k\right)^2 = \frac{1}{n}\sum_{m=0}^{n-1}\left(\sum_{k=0}^{m} f'(X_k)\,\psi_{m-k}\right)^2.$$

We will now bound the negative moments of B_n.

Step 2. Block decomposition. We plan to apply the Carbery-Wright inequality (8.6.8) in order to get bounds for negative moments of B_n. However, relation (8.6.8) only applies to moments of small order, and this is why we shall decompose B_n into smaller blocks.

Fix an integer $N \geq 1$ and let $M = [n/N]$ be the integer part of n/N. Then, $n \geq NM$ and as a consequence,

$$B_n = \frac{1}{n}\sum_{m=0}^{n-1}\left(\sum_{k=0}^{m} f'(X_k)\,\psi_{m-k}\right)^2 \geq \frac{1}{n}\sum_{i=0}^{N-1}\sum_{m=iM}^{(i+1)M-1}\left(\sum_{k=0}^{m} f'(X_k)\,\psi_{m-k}\right)^2.$$

For $i = 0, \ldots, N-1$, define

$$B_n^i = \frac{1}{n}\sum_{m=iM}^{(i+1)M-1}\left(\sum_{k=0}^{m} f'(X_k)\,\psi_{m-k}\right)^2$$

so that $B_n \geq \sum_{i=0}^{N-1} B_n^i$. Then, the following inequality holds true

$$(B_n)^{-\frac{p}{2}} \leq \prod_{i=0}^{N-1}(B_n^i)^{-\frac{p}{2N}}. \tag{8.6.12}$$

Recall once again the representation of the sequence X in (8.6.5), and denote by $\mathcal{F}_k$ the filtration generated by $\{w_\ell\,,\ \ell<k\}$. Then, starting from (8.6.12) we obtain

$$\mathbb{E}\left[(B_n)^{-\frac{p}{2}}\right]\le\mathbb{E}\left[\prod_{i=0}^{N-1}(B_n^i)^{-\frac{p}{2N}}\right]$$
$$=\mathbb{E}\left[\mathbb{E}\left[(B_n^{N-1})^{-\frac{p}{2N}}|\mathcal{F}_{(N-1)M}\right]\prod_{i=0}^{N-2}(B_n^i)^{-\frac{p}{2N}}\right]. \qquad (8.6.13)$$

Step 3. Application of Carbery-Wright. Let us go back to the particular situation of $f=\sum_{j=d}^q a_j\,H_j$, which means in particular that $f'=\sum_{j=d}^q j\,a_j\,H_{j-1}$. We are now in a position to apply a conditional version of inequality (8.6.8) to the block $(B_n^{N-1})^{-\frac{p}{2N}}$ in (8.6.13). First, notice

$$\mathbb{E}\left[(B_n^{N-1})^{-\frac{p}{2N}}|\mathcal{F}_{(N-1)M}\right]$$
$$\le 1+\frac{p}{2N}\int_0^1\mathbb{P}\left(B_n^{N-1}\le x|\,\mathcal{F}_{(N-1)M}\right)x^{-\frac{p}{2N}-1}dx\,. \qquad (8.6.14)$$

Since B_n^{N-1} is a polynomial of order $m=2(q-1)$, an application of Carbery-Wright's inequality (8.6.8) yields

$$\mathbb{P}\left(B_n^{N-1}\le x|\mathcal{F}_{(N-1)M}\right)\le\frac{c\,x^{\frac{1}{2(q-1)}}}{\left[\mathbb{E}\left(B_n^{N-1}|\mathcal{F}_{(N-1)M}\right)\right]^{\frac{1}{2(q-1)}}}. \qquad (8.6.15)$$

Step 4. Estimates for the conditional expectation. We now estimate the conditional expectation $\mathbb{E}[B_n^{N-1}|\mathcal{F}_{(N-1)M}]$. We have

$$\mathbb{E}\left[B_n^{N-1}|\mathcal{F}_{(N-1)M}\right]=\frac{1}{n}\sum_{m=(N-1)M}^{NM-1}\mathbb{E}\left[\left(\sum_{k=0}^{m}f'(X_k)\,\psi_{m-k}\right)^2\Big|\mathcal{F}_{(N-1)M}\right]$$
$$\ge\frac{1}{n}\sum_{m=(N-1)M}^{NM-1}A_m, \qquad (8.6.16)$$

where we have set

$$A_m=\mathrm{Var}\left(\sum_{k=(N-1)M}^{m}f'(X_k)\,\psi_{m-k}\Big|\mathcal{F}_{(N-1)M}\right).$$

Furthermore, notice that

$$f'(X_k)=f'\left(\sum_{\ell=-\infty}^{k}\psi_{k-i}\,w_i\right)=f'(Y_k+Z_k),$$

where $Y_k = \sum_{i=-\infty}^{(N-1)M-1} \psi_{k-i}\, w_i$ is $\mathcal{F}_{(N-1)M}$-measurable and $Z_k = \sum_{i=(N-1)M}^{k} \psi_{k-i}\, w_i$ is independent of $\mathcal{F}_{(N-1)M}$. Recall that $f' = \sum_{j=d}^{q} j\, a_j\, H_{j-1}$. The (3.4.20) and Proposition 3.15 yield

$$\begin{aligned}
&f'(X_k) - \mathbb{E}[f'(X_k)|\mathcal{F}_{(N-1)M}] \\
&= \sum_{j=d}^{q}\sum_{\ell=1}^{j-1} j a_j \binom{j-1}{\ell} \sigma_{Y_k}^{j-1-\ell} H_{j-1-\ell}(\widetilde{Y}_k)\sigma_{Z_k}^{\ell} H_\ell(\widetilde{Z}_k),
\end{aligned}$$

where $\sigma_{Y_k} = [\mathrm{Var}(Y_k)]^{1/2}$, $\sigma_{Z_k} = [\mathrm{Var}(Z_k)]^{1/2}$, $\widetilde{Y}_k = Y_k/\sigma_{Y_k}$ and $\widetilde{Z}_k = Z_k/\sigma_{Z_k}$. Therefore,

$$\begin{aligned}
A_m &= \mathbb{E}\left[\left(\sum_{\substack{(N-1)M\le k\le m\\ d\le j\le q\\ 1\le \ell\le j-1}} a_{j,\ell,k}\, H_{j-1-\ell}(\widetilde{Y}_k)\, H_\ell(\widetilde{Z}_k)\, \psi_{m-k}\right)^2 \Bigg|\mathcal{F}_{(N-1)M}\right] \\
&= \mathbb{E}\left[\left(\sum_{\substack{1\le \ell\le q-1\\ (N-1)M\le k\le m\\ (\ell+1)\vee d\le j\le q}} a_{j,\ell,k}\, H_{j-1-\ell}(\widetilde{Y}_k)\, H_\ell(\widetilde{Z}_k)\, \psi_{m-k}\right)^2 \Bigg|\mathcal{F}_{(N-1)M}\right],
\end{aligned}$$

where we have set $a_{j,\ell,k} = j a_j \binom{j-1}{\ell} \sigma_{Y_k}^{j-1-\ell}\sigma_{Z_k}^{\ell}$.

Recall that the random variable $\widetilde{Y}_k$ is $\mathcal{F}_{(N-1)M}$-measurable while the random variable $\widetilde{Z}_k$ is independent of $\mathcal{F}_{(N-1)M}$. By orthogonality properties of Hermite polynomials we have

$$A_m - \sum_{\ell=1}^{q-1} \mathbb{E}\left[\left(\sum_{\substack{(N-1)M\le k\le m\\ (\ell+1)\vee d\le j\le q}} a_{j,\ell,k}\, H_{j-1-\ell}(\widetilde{Y}_k)\, H_\ell(\widetilde{Z}_k)\, \psi_{m-k}\right)^2 \Bigg|\mathcal{F}_{(N-1)M}\right]$$

and we can bound this quantity from below by taking only the term corresponding to $\ell = q-1$. In this situation the sum $\sum_{j=(\ell+1)\vee d}^{q}$ is reduced to the term corresponding to $j = q$, and since $a_{q,q-1,k} = q a_q \sigma_{Z_k}^{q-1}$ we obtain

$$\begin{aligned}
A_m &\ge \mathbb{E}\left[\left(\sum_{k=(N-1)M}^{m} q\, a_q\, \sigma_{Z_k}^{q-1}\, H_{q-1}(\widetilde{Z}_k)\, \psi_{m-k}\right)^2 \Bigg|\mathcal{F}_{(N-1)M}\right] \\
&= q^2\, a_q^2\, \mathbb{E}\left[\left(\sum_{k=(N-1)M}^{m} \sigma_{Z_k}^{q-1}\, H_{q-1}(\widetilde{Z}_k)\, \psi_{m-k}\right)^2\right].
\end{aligned}$$

Using the identity $\mathbb{E}[H_p(\tilde{Z}_{k_1})H_p(\tilde{Z}_{k_2})] = p!(\mathbb{E}[\tilde{Z}_{k_1}\tilde{Z}_{k_2}])^p$ we obtain

$$A_m \geq qq!a_q^2 \sum_{k_1,k_2=(N-1)M}^{m} \sigma_{Z_{k_1}}^{q-1}\sigma_{Z_{k_2}}^{q-1}\,\mathbb{E}\left[\widetilde{Z}_{k_1}\,\widetilde{Z}_{k_2}\right]^{q-1} \psi_{m-k_1}\psi_{m-k_2}.$$

Furthermore, similarly to (8.6.6), it is easy to verify

$$\mathbb{E}\left[\widetilde{Z}_{k_1}\,\widetilde{Z}_{k_2}\right] = \frac{1}{\sigma_{Z_{k_1}}\sigma_{Z_{k_2}}} \sum_{i=(N-1)M}^{k_1\wedge k_2} \psi_{k_1-i}\,\psi_{k_2-i},$$

and thus

$$\begin{aligned}
A_m &\geq qq!a_q^2 \sum_{k_1,k_2=(N-1)M}^{m} \left(\sum_{i=(N-1)M}^{k_1\wedge k_2} \psi_{k_1-i}\,\psi_{k_2-i}\right)^{q-1} \psi_{m-k_1}\psi_{m-k_2} \\
&= qq!a_q^2 \sum_{i_1,\ldots,i_{q-1}=(N-1)M}^{m}\ \sum_{k_1,k_2=\max(i_1,\ldots,i_{q-1})}^{m} \psi_{m-k_1}\psi_{m-k_2}\prod_{j=1}^{q-1}\psi_{k_1-i_j}\,\psi_{k_2-i_j} \\
&= qq!a_q^2 \sum_{i_1,\ldots,i_{q-1}=(N-1)M}^{m} \left(\sum_{k=\max(i_1,\ldots,i_{q-1})}^{m} \psi_{m-k}\prod_{j=1}^{q-1}\psi_{k-i_j}\right)^2 .
\end{aligned}$$

Here again, this sum of squares is bounded below by taking only the term corresponding to $i_1 = \cdots = i_{q-1} = m$, which yields

$$A_m \geq c_{a,q,\psi} \quad \text{with} \quad c_{a,q,\psi} \equiv qq!a_q^2\,\psi_0^{2q} > 0. \tag{8.6.17}$$

Step 5. Conclusion. Plugging relation (8.6.17) into (8.6.16) and recalling that N is a given integer whose exact value will be fixed below, we get

$$\mathbb{E}\left[B_n^{N-1}|\mathcal{F}_{(N-1)M}\right] \geq \frac{M\,c_{a,q,\psi}}{n} \geq c_{a,q,\psi,N} > 0,$$

as long as N stays bounded. We then insert back this inequality into (8.6.14) and (8.6.15) in order to get

$$\begin{aligned}
\mathbb{P}\left(B_n^{N-1} \leq x|\mathcal{F}_{(N-1)M}\right) &\leq 1 + \frac{p\,c_{a,q,\psi,N}}{2N}\int_0^1 x^{\frac{1}{2(q-1)}-\frac{p}{2N}-1}dx \\
&= c_{a,q,\psi,N,p} < \infty,
\end{aligned}$$

where we have chosen N such that $\frac{p}{2N} < \frac{1}{2(q-1)}$. Substituting this bound into (8.6.13), and iterating this procedure we obtain

$$\mathbb{E}\left[(B_n)^{-\frac{p}{2}}\right] \leq c_{a,q,\psi,N,p}^{N} < \infty .$$

Finally, recall from Step 1 that $\mathbb{E}[(B_n)^{-\frac{p}{2}}] \geq \mathbb{E}[\|DV_n^{d,q}\|_{\mathbb{H}}^{-p}]$. This completes the proof. □

In the case $q \neq d$, Theorem 8.4 cannot be applied. Nevertheless, condition (8.6.10) allows us to derive the uniform convergence of the densities and their derivatives from a general result proved below (see Proposition 8.7) although in this case we have no information about the rate of convergence.

Lemma 8.9. *Let $p > 2$, and let $F_n \in \cup_{k=0}^{p} \mathbb{H}^{\hat{\otimes} k}$ be a sequence of finite sum of multiple Wiener-Itô integrals (of order up to a fixed integer p). Assume that F_n converges in distribution. Then*

$$\sup_n \mathbb{E}\left[|F_n|^r\right] < \infty \quad \forall\, r > 1. \tag{8.6.18}$$

Proof. From (7.1.40), we know that there are constants $c_{p,r,q}$ such that $\|F_n\|_r \le c_{p,r,q}\|F_n\|_q$ for any $r, q \in (1, \infty)$ and the constant $c_{p,r.q}$ does not depend on n. From the Paley-Zygmund inequality (A.30) (with $\theta = 1/2$ and $X = F_n^2$) and then from the equivalence of different norms for a finite sum of chaos, we have

$$\mathbb{P}\left(F_n^2 \ge \mathbb{E}(F_n^2)/2\right) \ge \frac{(\mathbb{E}(F_n^2))^2}{4\mathbb{E}(F_n^4)} \ge c_p\,.$$

Since F_n converges in distribution, it is tight. Namely, there is a $M > 0$ such that $\mathbb{P}(F_n > M) < c_p \le \mathbb{P}\left(F_n^2 \ge \mathbb{E}(F_n^2)/2\right)$. This immediately yields $\mathbb{E}(F_n^2) \le 2M$ since otherwise, $\mathbb{P}\left(F_n^2 \ge \mathbb{E}(F_n^2)/2\right) \le \mathbb{P}\left(F_n^2 \ge M\right) < c_p$, a contradiction. But M is independent of n, we have $\sup_n \mathbb{E}(F_n^2) < \infty$. Now the inequality (7.1.40) gives (8.6.18) for all $r > 1$. □

Proposition 8.7. *Let $\{F_n; n \in \mathbb{N}\}$ be a sequence of random variables belonging to a finite sum of Wiener chaos, which converges in law to a nonzero random variable F_∞. Suppose that, for all $p \ge 1$*

$$\sup_n \mathbb{E}[\|DF_n\|^{-p}] < \infty. \tag{8.6.19}$$

Then, for all $m \ge 0$ the derivative $p_n^{(m)}$ of the density of F_n, converges uniformly and in $L^p(\mathbb{R})$ for all $p \ge 1$ to the corresponding derivative of the density of F_∞.

Proof. First notice that by condition (8.6.19), the random variable F_n has an infinitely differentiable density p_n, whose derivatives can be expressed by

$$p_n^{(m)}(x) = \mathbb{E}\left[\mathbf{1}_{\{F>x\}}\, G_n^{(m)}\right], \tag{8.6.20}$$

where the random variables $G_n^{(m)}$ are defined recursively by $G_n^{(0)} = \delta\left(\frac{DF_n}{\|DF_n\|_H^2}\right)$ and

$$G_n^{(m)} = -\delta\left(\frac{G_n^{(m-1)}DF_n}{\|DF_n\|_H^2}\right),$$

for any $m \geq 1$. From this formula it follows that the derivatives $p_n^{(m)}$ are uniformly bounded and also uniformly bounded in $L^p(\mathbb{R})$ for all $p \geq 1$. Indeed, by (8.6.18) we have $\sup_n \mathbb{E}[|F_n|^r] < \infty$ for all $r \geq 1$. This uniform bound on the moments, together with the equivalence of the $\|\cdot\|_{m,p}$ norms in any finite sum of Wiener chaos and condition (8.6.19) imply that $\sup_n \|G_n^{(m)}\|_{L^p(\Omega)} = c_{m,p} < \infty$ for all $m \geq 0$. Then, from (8.6.20) we have

$$\sup_n \sup_x |p_n^{(m)}(x)| \leq \sup_n \mathbb{E}\left[|G_n^{(m)}|\right] = c_{m,1} < \infty, \tag{8.6.21}$$

and, using the fact that $\mathbb{E}[G_n^{(m)}] = 0$, we get

$$\begin{aligned}\sup_n |p_n^{(m)}(x)| &\leq \sup_n \left(\mathbb{P}(|F_n| > |x|)\,\beta\left[|G_n^{(m)}|^2\right]\right)^{\frac{1}{2}} \\ &\leq c_{m,2} \sup_n \sqrt{\mathbb{P}(|F_n| > |x|)} \\ &\leq c_{m,2} \sup_n \mathbb{E}[|F_n|^q]^{\frac{1}{2}} |x|^{-\frac{q}{2}} \leq c|x|^{-\frac{q}{2}},\end{aligned} \tag{8.6.22}$$

for all $q \geq 1$ and for some constant c depending on q and m.

By [Nourdin and Poly (2013), Theorem 3.1] the sequence F_n converges in total variation to F_∞, that is, the densities p_n converge in $L^1(\mathbb{R})$ to the density p_∞ of F_∞. The boundedness in $L^p(\mathbb{R})$ and the uniform bound of p_n imply that this convergence holds in $L^p(\mathbb{R})$ for any $p \geq 1$.

On the other hand, the estimates (8.6.21) and (8.6.22) imply that for any $m \geq 1$ and any $p \geq 1$, the sequence $p_n^{(m)}$ is uniformly bounded in $L^p(\mathbb{R})$. Therefore, for any $m \geq 1$ and any $p \geq 1$ the sequence $p_n^{(m)}$ is relatively compact in $L^p(\mathbb{R})$. Suppose that a subsequence $\{p_{n_k}^{(m)}, k \geq 1\}$ converges in $L^p(\mathbb{R})$ to some limit $\widetilde{p}_\infty^{(m)}$. This limit must coincide with the m-th derivative (in the distribution sense) of p_∞, and, therefore, it is unique. This implies that for any $m \geq 1$ and any $p \geq 1$ the sequence $p_n^{(m)}$ converges in $L^p(\mathbb{R})$ to the m-th derivative of p_∞.

Finally, the uniform convergence is also easy to establish from the convergence of the densities in $L^p(\mathbb{R})$ for all $p \geq 1$. □

Chapter 9

Local Time and (Self-) Intersection Local Time

9.1 Local Time of Brownian Motion

Lemma 9.1. *Let $-1 < \alpha_i < \infty$, $i = 1, \cdots, d$ and set $|\alpha| = \sum_{i=1}^d \alpha_i$ (which may be negative). Then, there is a constant κ such that*

$$\int_{0<r_1<\cdots<r_d<t} \prod_{i=1}^d (r_i - r_{i-1})^{\alpha_i} dr \le \frac{\kappa^d t^{|\alpha|+d}}{\Gamma(|\alpha| + d + 1)},$$

where by convention, $r_0 = 0$.

Proof. It is straightforward by induction. □

The Fourier transform of the Dirac delta function is

$$\int_{\mathbb{R}^d} e^{-\imath\xi x} \delta(x) dx = 1 .$$

By the inverse transform formula we have

$$\delta(x) = \frac{1}{(2\pi)^d} \int_{\mathbb{R}^d} e^{\imath\xi x} d\xi , \qquad \text{where } \xi x = \xi_1 x_1 + \cdots + \xi_d x_d .$$

This identity is formal. But we can justify it by using approximation, such as the hear kernel approximation

$$p_\varepsilon(x) = (2\pi)^{-d/2} e^{-\frac{|x|^2}{2\varepsilon}} = (2\pi)^{-d} \int_{\mathbb{R}^d} e^{\imath\xi x - \frac{\varepsilon|\xi|^2}{2}} d\xi , \quad \text{as } \varepsilon \to 0 .$$

Definition 9.1. The local time of the (one-dimensional) Brownian motion at x is defined as the following $L^2(\mathbb{B})$ limit (if exists)

$$\mathcal{L}_t(x) = \int_0^t \delta(B_r - x) dr = \lim_{\varepsilon \to} \int_0^t p_\varepsilon(B_r - x) dr . \tag{9.1.1}$$

To simplify the presentation, we shall operate on the Dirac delta function and its Fourier transform directly and leave the justification by heat kernel approximation to readers (see e.g. the proof of (9.3) or (9.10) for justification by approximation). We consider the case $x = 0$.

First, we shall obtain the chaos expansion of $\mathcal{L}_t = \mathcal{L}_t(0)$. We have

$$\begin{aligned}\int_0^t \delta(B_r)dr &= \frac{1}{2\pi}\int_0^t \int_{\mathbb{R}} e^{\imath B_r\xi} d\xi dr \\ &= \frac{1}{2\pi}\int_0^t \int_{\mathbb{R}} e^{-\frac{r|\xi|^2}{2}} e^{\imath B_r\xi + \frac{r|\xi|^2}{2}} d\xi dr \\ &= \sum_{n=0}^{\infty} \frac{1}{2\pi}\int_0^t \int_{\mathbb{R}} (\imath\xi)^n e^{-\frac{r|\xi|^2}{2}} d\xi \int_{0<r_1<\cdots<r_n<r} dB_{r_1}\cdots dB_{r_n} dr\,.\end{aligned}$$

An elementary computation gives us

Proposition 9.1. *The local time $\mathcal{L}_t$ of one dimensional Brownian motion have the following convergent chaos expansion (in L^2)*

$$\mathcal{L}_t = \int_0^t \delta(B_r)dr = \sum_{m=0}^{\infty} \mathcal{L}_t^{(2m)}\,, \tag{9.1.2}$$

where (if we denote $\mathcal{T}_{n,t} := \{(r_1,\cdots,r_n);\ 0<r_1<\cdots<r_n<t\}$)

$$\begin{cases} \mathcal{L}_t^{(2m)} := \displaystyle\int_{\mathcal{T}_{2m,t}} g_m(t;r_1,\cdots,r_{2m})dB_{r_1}\cdots dB_{r_{2m}} \\ g_m(t;r_1,\cdots,r_{2m}) := \dfrac{(-1)^m(2m)!}{\sqrt{2\pi}2^m m!(m-1/2)}\left[r_{2m}^{-m+1/2} - t^{-m+1/2}\right]\,. \end{cases}$$

Moreover, for any $k<1/2$, $\mathcal{L}_t \in \mathbb{D}_{k,2}$.

Proof. The explicit expression of g_m comes from the simple computation. Now

$$\begin{aligned}\mathbb{E}\left[\mathcal{L}_t^{(2m)}\right]^2 &= \int_{\mathcal{T}_{2m,t}} g_m^2(t;r_1,\cdots,r_{2m})dr_1\cdots dr_{2m} \\ &= \frac{[(2m)!]^2}{2\pi 2^{2m}(m!)^2(m-1/2)^2}\int_{\mathcal{T}_{2m,t}}\left[r_{2m}^{-m+1/2} - t^{-m+1/2}\right]^2 dr_1\cdots dr_{2m} \\ &\le \frac{[(2m)!]^2}{2\pi 2^{2m}(m!)^2(m-1/2)^2(2m-1)!}\int_{0<r_{2m}<t} r_{2m}^{2m-1} r_{2m}^{-2m+1} dr_{2m}\,.\end{aligned}$$

By the Stirling formula $m! \approx \sqrt{2\pi m}m^m e^{-m}$, we see that there is a constant C such hat

$$\mathbb{E}\left[\mathcal{L}_t^{(2m)}\right]^2 \le \frac{Ct}{m^{3/2}}\,.$$

By Meyer's theorem 7.15, this shows that $\mathcal{L}_t \in \mathbb{D}_{k,2}$ for any $k<1/2$. □

Now we turn to the exponential integrability of $\mathcal{L}_t$.

$$\begin{aligned}
\mathbb{E}\left[\mathcal{L}_t^n\right] &= (2\pi)^{-n}\mathbb{E}\left[\int_0^t \int_{\mathbb{R}} e^{\imath B_r \xi} d\xi dr\right]^n \\
&= (2\pi)^{-n} \int_{[0,t]^n} \int_{\mathbb{R}^n} \mathbb{E}\left[\exp\left\{\imath\left(\xi_1 B_{r_1} + \cdots + \xi_n B_{r_n}\right)\right\}\right] d\xi dr \\
&= (2\pi)^{-n} n! \int_{\mathcal{T}_{n,t}} \int_{\mathbb{R}^n} \mathbb{E}\left[\exp\left\{\imath\left(\eta_1 B_{r_1} + \cdots + \eta_n (B_{r_n} - B_{r_{n-1}})\right)\right\}\right] d\xi dr\,,
\end{aligned}$$

where we denote $dr = dr_1 \cdots dr_n$, $d\xi = d\xi_1 \cdots d\xi_n$ and $\eta_j = \xi_j + \xi_{j+1} + \cdots + \xi_n$, $j = 1, \cdots, n$. Hence,

$$\begin{aligned}
\mathbb{E}\left[\mathcal{L}_t^n\right] &= (2\pi)^{-n} n! \int_{\mathcal{T}_{n,t}} \int_{\mathbb{R}^n} \left[\exp\left\{-\frac{1}{2}\left(\eta_1^2 r_1 + \cdots + \eta_n^2 (r_n - r_{n-1})\right)\right\}\right] d\eta dr \\
&= (2\pi)^{-n/2} n! \int_{\mathcal{T}_{n,t}} r_1^{-1/2} (r_2 - r_1)^{-1/2} \cdots (r_n - r_{n-1})^{-1/2} dr \\
&\le \frac{C^n t^{n/2} n!}{\Gamma(\frac{n}{2}+1)} \le C^n t^{n/2} n^{n/2}\,. \qquad (9.1.3)
\end{aligned}$$

Using Hölder inequality, it is easy to see that the above inequality also holds true when n is any positive number greater than 1. If $\kappa \in [0,2]$ and $\alpha > 0$, then

$$\begin{aligned}
\mathbb{E}\exp\left\{\alpha \mathcal{L}_t^\kappa\right\} &= \sum_{n=0}^\infty \frac{\alpha^n}{n!} \mathbb{E}\left[\mathcal{L}_t^{\kappa n}\right] \le \sum_{n=0}^\infty \frac{\alpha^n C^{\kappa n} t^{\kappa n/2} \kappa^{\kappa n/2} n^{\kappa n/2}}{n!} \\
&\le \sum_{n=0}^\infty \frac{\alpha^n C^{\kappa n} t^{\kappa n/2} \kappa^{\kappa n/2}}{\Gamma\left((1-\kappa/2)n + 1\right)}\,.
\end{aligned}$$

If $\kappa = 2$, then there is an $\alpha_0 > 0$ such that for any $\alpha \in [0, \alpha_0]$, $\mathbb{E}\exp\left\{\alpha \mathcal{L}_t^\kappa\right\} < \infty$. If $\kappa \in (0,2)$, then for any $\alpha > 0$, by the asymptotic behavior of the Mittag-Leffler function, we have

$$\mathbb{E}\exp\left\{\alpha \mathcal{L}_t^\kappa\right\} \le c_1 \exp\left[c_2 \alpha^{\frac{2}{2-\kappa}} t^{\frac{\kappa}{2-\kappa}}\right]\,.$$

Summarizing the above we have

Theorem 9.1. *Let $\mathcal{L}_t$ be the local time at 0 of the one dimensional Brownian motion. Then, there is an $\alpha_0 > 0$ such that*

$$\mathbb{E}\exp\left\{\alpha \mathcal{L}_t^2\right\} < \infty\,, \quad \forall\, \alpha \in (0, \alpha_0)\,. \qquad (9.1.4)$$

For any $\kappa \in (0,2)$, there exist c_1 and c_2 (depending also on κ) such that

$$\mathbb{E}\exp\left\{\alpha \mathcal{L}_t^\kappa\right\} \le c_1 \exp\left[c_2 \alpha^{\frac{2}{2-\kappa}} t^{\frac{\kappa}{2-\kappa}}\right]\,, \quad \forall\ \alpha,\ t > 0\,. \qquad (9.1.5)$$

9.2 Chaos Expansion of Self-intersection Local Time

Let $B = \{B_t\,, 0 \le t \le T\}$ be the standard d-dimensional Brownian motion and let δ be the d-dimensional Dirac delta function,namely,

$$\int_{\mathbb{R}^d} f(x)\delta(x)dx = f(0)\,, \quad \forall \text{ smooth function } f \text{ of compact support}\,.$$

The random variable $\pounds(T,x) := \int_0^T\int_0^T \delta(B_t - B_s - x)dsdt$ is called the self-intersection local time of B at x. If B^1 and B^2 are two Brownian motions, then $\int_0^T\int_0^T \delta(B_t^1 - B_s^2 - x)dsdt$ is called the intersection local time of B^1 and B^2. We shall study the self-intersection local time at 0 in this section $\pounds(T) := \pounds(T,0) = \int_0^T\int_0^T \delta(B_t - B_s)dsdt$.

Since δ is not a classical function (it is a generalized function or distribution in the sense of Sobolev-Schwartz) we shall use the heat kernel to approximate the Dirac delta function δ on $\mathbb{R}^d$. Denote

$$\begin{aligned} p_t(x) &= (2\pi t)^{-d/2}e^{-\frac{|x|^2}{2t}}\,, \quad p_t(x,y) = p_t(x-y)\,, \\ \nabla_i f &:= \frac{\partial}{\partial x_i} f, \quad P_\varepsilon f(x) = \int_{\mathbb{R}^d} p_\varepsilon(x,y)f(y)dy\,. \end{aligned}$$

We shall define the self-intersection local time by the following approximate self-intersection local times:

$$\pounds_\varepsilon(T) = \int_0^T\int_0^T p_\varepsilon(B_t - B_s)dsdt\,. \tag{9.2.1}$$

Definition 9.2. If $\pounds_\varepsilon(T)$ converges to some random variable in probability as $\varepsilon \downarrow 0$, then we say that the self-intersection local time of B exists and the limit is denoted by $\pounds(T)$. If $\pounds_\varepsilon(T) - \mathbb{E}\left[\pounds_\varepsilon(T)\right]$ converges to some random variable in probability as $\varepsilon \downarrow 0$, then we say that the normalized self-intersection local time of B exists and the limit is denoted by $\tilde{\pounds} := \pounds(T) - \mathbb{E}\left[\pounds(T)\right]$

If B and $\tilde{B}$ are two Brownian motions, then we define the intersection local time as follows

Definition 9.3. If

$$\pounds_\varepsilon(T,B,\tilde{B}) = \int_0^T\int_0^T p_\varepsilon(B_t - \tilde{B}_s)dsdt$$

converges to some random variable in probability as $\varepsilon \downarrow 0$, then we say that the self-intersection local time of B exists and the limit is denoted by $\pounds(T,B,\tilde{B})$.

First, we are going to find the chaos expansion of $\mathcal{L}_\varepsilon(T)$. Given $u \ge s \ge 0$ and $f \in C^\infty$, consider the process

$$X_v := P_{u-v} f(B_v - B_s), \quad s \le v \le u\,.$$

Applying the Itô's formula to the above process and noticing $\partial_t p_t - \Delta p_t = 0$, we obtain

$$f(B_u - B_s) = P_{u-s} f(0) + \int_s^u \nabla_i P_{u-u_1} f(B_{u_1} - B_s) dB^i_{u_1}\,, \qquad (9.2.2)$$

where we used the Einstein's convention on summation for $i = 1, \cdots, d$, i.e. $a_i b^i := \sum_{i=1}^d a_i b^i$. Taking $f = p_\varepsilon$ in (9.2.2), we have

$$p_\varepsilon(B_t - B_s) = p_{t-s+\varepsilon}(0) + \int_s^t \nabla_i p_{t-u_1+\varepsilon}(B_{u_1} - B_s) dB^i_{u_1}\,. \qquad (9.2.3)$$

Applying (9.2.2) to the integrand $\nabla_i P_{t-u_1+\varepsilon}$ in (9.2.3), and repeating this procedure we obtain

Lemma 9.2. *The finite chaos expansion of $p_\varepsilon(B_t - B_s)$ with exact residual is given by the following.*

$$\begin{aligned}
p_\varepsilon(B_t - B_s) &= p_{t-s+\varepsilon}(0) + \int_s^t \nabla_i p_{t-u_1+\varepsilon}(0) dB^i_{u_1} \\
&\quad + \cdots + \int_{s \le u_1 < \cdots < u_n \le t} \nabla_{\alpha_1} \cdots \nabla_{\alpha_n} p_{t-s+\varepsilon}(0) dB^{\alpha_1}_{u_1} \cdots dB^{\alpha_n}_{u_n} \\
&\quad + \int_{s \le u_1 < \cdots < u_{n+1} \le t} \nabla_{\alpha_1} \cdots \nabla_{\alpha_{n+1}} p_{t-u_1+\varepsilon}(B_{u_1} - B_s) dB^{\alpha_1}_{u_1} \cdots dB^{\alpha_{n+1}}_{u_{n+1}} \\
&= p_{t-s+\varepsilon}(0) + I_1 + \cdots + I_n + \tilde{I}_{n+1}\,, \qquad (9.2.4)
\end{aligned}$$

where

$$I_j := \int_{s \le u_1 < \cdots < u_j \le t} \nabla_{\alpha_1} \cdots \nabla_{\alpha_j} p_{t-s+\varepsilon}(0) dB^{\alpha_1}_{u_1} \cdots dB^{\alpha_j}_{u_n}\,, \quad j = 1, 2, \cdots, n;$$

and

$$\tilde{I}_{n+1} := \int_{s \le u_1 < \cdots < u_{n+1} \le t} \nabla_{\alpha_1} \cdots \nabla_{\alpha_{n+1}} p_{t-u_1+\varepsilon}(B_{u_1} - B_s) dB^{\alpha_1}_{u_1} \cdots dB^{\alpha_{n+1}}_{u_{n+1}}\,.$$

For a fixed $0 \le j \le n$, I_j is orthogonal to I_k, $k = 0, 1, \cdots, j-1, j+1, \cdots, n+1$, i.e. $\mathbb{E}\big(I_j I_k\big) = 0$. Moreover, $\tilde{I}_{n+1}$ is orthogonal to all $I_j, j = 0, 1, \cdots, n$. So I_n is the n-th Itô-Wiener chaos of $p_\varepsilon(B_t - B_s)$. Now we compute $\nabla_{\alpha_1} \cdots \nabla_{\alpha_n} p_{t-s+\varepsilon}(0)$ in (9.2.4). We shall use the following explicit form.

$$\begin{aligned}
p_{t-s+\varepsilon}(x) &= [2\pi(t-s+\varepsilon)]^{-d/2} e^{-\frac{|x|^2}{2(t-s+\varepsilon)}} \\
&= [2\pi(t-s+\varepsilon)]^{-d/2} \sum_{n=0}^\infty \frac{(-1)^n |x|^{2n}}{2^n n! (t-s+\varepsilon)^n}\,. \qquad (9.2.5)
\end{aligned}$$

Notice that when n is odd, $\nabla_{\alpha_1}\cdots\nabla_{\alpha_n}p_{t-s+\varepsilon}(0)=0$. It suffices to consider the case n is even. We set for $j=1,\cdots,d$ and $\alpha=(\alpha_1,\cdots,\alpha_n)$

$$2n_j := 2n_j(\alpha) := \#\{\alpha_i;\alpha_i=j\}\,,\quad n=n_1+\cdots+n_d\,. \tag{9.2.6}$$

Readers may notice a little strange factor 2 in front of n_j. This is to simplify the notation below and noticing that if one of the quantities $\#\{\alpha_i;\alpha_i=j\}$ is odd, then $\nabla_{\alpha_1}\cdots\nabla_{\alpha_n}p_{t-s+\varepsilon}(0)=0$.

With these notations and (9.2.5), we have

$$\nabla_{\alpha_1}\cdots\nabla_{\alpha_n}p_{t-s+\varepsilon}(0)=\begin{cases}\frac{(-1)^n(2n_1)!\cdots(2n_d)!}{(2\pi)^{d/2}2^n n_1!\cdots n_d!(t-s+\varepsilon)^{n+d/2}} & \text{if } n \text{ is even}\\ 0 & \text{if } n \text{ is odd}\,.\end{cases} \tag{9.2.7}$$

Other derivatives vanish at 0. We set when $|\alpha|=0$,

$$C_\emptyset=(2\pi)^{-d/2}\,.$$

and

$$f_\emptyset^\varepsilon(T)=\begin{cases}2\left[(T+\varepsilon)\log(T+\varepsilon)-T-\varepsilon\log\varepsilon\right]-2T\log\varepsilon\\ \qquad\qquad \text{when } d=2;\\ 2\varepsilon^{-1}T-2\log\left(\frac{T+\varepsilon}{\varepsilon}\right)\\ \qquad\qquad \text{when } d=4;\\ 4(d-2)^{-1}\left[\varepsilon^{-d/2+1}T-2(d-4)^{-1}\left(\varepsilon^{-d/2+2}-(T+\varepsilon)^{-d/2+2}\right)\right]\\ \qquad\qquad \text{when } d\neq 2,4\,.\end{cases}$$

For $|\alpha|\geq 1$, we define

$$\begin{cases}C_\alpha=(-1)^n 2\dfrac{(2n_1)!\cdots(2n_d)!}{(2\pi)^{d/2}2^n n_1!\cdots n_d!}(n+d/2-1)^{-1}(n+d/2-2)^{-1}\,;\\ f_\alpha^\varepsilon(T)(u_1,\cdots,u_{2n})=\left[-(u_{2n}+\varepsilon)^{-n-d/2+2}+(T+\varepsilon)^{-n-d/2+2}\right.\\ \qquad\left.+(u_{2n}-u_1+\varepsilon)^{-n-d/2+2}-(T-u_1+\varepsilon)^{-n-d/2+2}\right].\end{cases}$$

Theorem 9.2. *The chaos expansion for the approximate self-intersection local time is given by*

$$\begin{cases}\mathcal{L}_\varepsilon(T)=\displaystyle\sum_{n=0}^\infty \mathcal{L}_{\varepsilon,2n}(T)\\ \mathcal{L}_{\varepsilon,2n}(T)=\displaystyle\sum_{|\alpha|=2n}C_\alpha J_\alpha(f_\alpha^\varepsilon(T))\,,\end{cases} \tag{9.2.8}$$

where $J_\alpha(f_\alpha^\varepsilon(T))$ *is defined as*

$$J_\alpha(f_\alpha^\varepsilon(T)) := \int_{0\le u_1<\cdots<u_{2n}\le T} f_\alpha^\varepsilon(T)(u_1,\cdots,u_{2n})dB_{u_1}^{\alpha_1}\cdots dB_{u_{2n}}^{\alpha_{2n}} \quad (9.2.9)$$

and $\sum_{|\alpha|=2n}$ *denotes the sum over all* $\alpha = (\alpha_1,\cdots,\alpha_{2n})$ *such that* $\#\{\alpha_i;\alpha_i=j\}=2n_j$ *is even.*

Proof. It is obvious that the chaos of odd orders is zero. Let $0\le u_1<\cdots<u_{2n}\le T$. Then, it is easy to show that

$$\int_0^T\int_0^t\int_{s\le u_1<\cdots<u_n\le t} \nabla_{\alpha_1}\cdots\nabla_{\alpha_n}p_{t-s+\varepsilon}(0)dB_{u_1}^{\alpha_1}\cdots dB_{u_n}^{\alpha_n} = C_\alpha J_\alpha(f_\alpha^\varepsilon(T)) . \quad (9.2.10)$$

Since $\int_0^T\int_0^t p_\varepsilon(B_t-B_s)dsdt$ is in L^2, it admits a chaos expansion according to the Wiener-Itô chaos expansion theorem. Since (9.2.10) is orthogonal to the other terms obtained from (9.2.4), we see that (9.2.10) is the $2n$-th chaos expansion of $\int_0^T\int_0^t p_\varepsilon(B_t-B_s)dsdt$. □

The coefficients c_α can be bounded as follows.

Lemma 9.3. *Let* C_α *be given as above. Then*

$$\sum_{|\alpha|=2n} C_\alpha^2 \approx (2n)!n^{\frac{d}{2}-5} . \quad (9.2.11)$$

Proof. By the Stirling formula $n! = (2\pi)^{1/2}n^{n+1/2}e^{-n}(1+O(1/n))$, i.e. $n!\approx n^{n+1/2}e^{-n}$ we see that

$$\frac{(2n_1)!\cdots(2n_d)!}{2^{2n}n_1!^2\cdots n_d!^2} \approx (n_1+1)^{-1/2}\cdots(n_d+1)^{-1/2} .$$

(We allow $n_1,\cdots,n_d$ to be 0). On the other hand it is easy to have

$$\sum_{n_1+\cdots+n_d=n} (n_1+1)^{-1/2}\cdots(n_d+1)^{-1/2}$$
$$\approx \int_{u_1,\cdots,u_{d-1}\ge 0} (u_1+1)^{-1/2}\cdots(n_{d-1}+1)^{-1/2}$$
$$(n-u_1-\cdots u_{d-1}+1)^{-1/2}du_1\cdots du_{d-1} \approx n^{d/2-1} .$$

Thus,

$$\sum_{|\alpha|=2n} C_\alpha^2 \approx \sum_{n_1+\cdots+n_d=n} \frac{(2n)!}{(2n_1)!\cdots(2n_d)!}\Big(\frac{(2n_1)!\cdots(2n_d)!}{2^n n_1!^2\cdots n_d!^2}\Big)^2 n^{-4}$$
$$\approx (2n)! \sum_{n_1+\cdots+n_d=n} (n_1+1)^{-1/2}\cdots(n_d+1)^{-1/2}$$
$$\approx (2n)!n^{d/2-5} ,$$

proving the lemma. □

Letting $\varepsilon \to 0$, we obtain

Theorem 9.3. *Let*

$$\begin{cases} C_\alpha = (-1)^n \dfrac{(2n_1)! \cdots (2n_d)!}{(2\pi)^{d/2} 2^n n_1! \cdots n_d!} (n+d/2-1)^{-1}(n+d/2-2)^{-1}\,; \\ f_\alpha(T)(u_1, \cdots, u_{2n}) = [-u_{2n}^{-n-d/2+2} + T^{-n-d/2+2} \\ \qquad + (u_{2n}-u_1)^{-n-d/2+2} - (T-u_1)^{-n-d/2+2}]\,. \end{cases}$$

When $d = 1, 2$ the chaos expansion of $\pounds(T)$ is given by

$$\begin{cases} \pounds(T) - \mathbb{E}\left[\pounds(T)\right] = \sum_{n=1}^{\infty} \pounds_{2n}(T) \\ \qquad \pounds_{2n}(T) = \sum_{|\alpha|=2n} C_\alpha J_\alpha(f_\alpha(T))\,, \end{cases} \tag{9.2.12}$$

where $J_\alpha(f_\alpha(T))$ is defined as

$$J_\alpha(f_\alpha(T)) := \int_{0 \le u_1 < \cdots < u_{2n} \le T} f_\alpha(T)(u_1, \cdots, u_{2n}) dB_{u_1}^{\alpha_1} \cdots dB_{u_{2n}}^{\alpha_{2n}}\,. \tag{9.2.13}$$

Remark 9.1. When $d = 1$, $\mathbb{E}\left[\pounds(T)\right] = \frac{8}{3}T^{3/2}$ exists and then we have the chaos expansion for $\pounds(T)$. When $d = 2$, $\mathbb{E}\left[\pounds(T)\right]$ does not exist and however, the above identity means that after subtracting the infinite quantity $\mathbb{E}\left[\pounds(T)\right]$, the self-intersection local time is an element in L^2.

Proof. Assume $d = 1, 2$. First, we estimate

$$\int_{0 \le u_1 < \cdots < u_{2n} \le T} |f_\alpha^\varepsilon(T)(u_1, \cdots, u_{2n})|^2 du_1 \cdots du_{2n}\,.$$

Denote for $n \ge 1$

$$g_n^\varepsilon(T)(u_1, \cdots, u_{2n}) := (u_{2n} - u_1 + \varepsilon)^{-n-d/2+2}$$

and

$$\begin{aligned} &G_n^\varepsilon(T)(u_1, \cdots, u_{2n}) \\ &\quad = -(u_{2n}+\varepsilon)^{-n-d/2+2} + (T+\varepsilon)^{-n-d/2+2} - (T-u_1+\varepsilon)^{-n-d/2+2} \end{aligned}$$

so that $f_\alpha^\varepsilon(T) = g_n^\varepsilon(T) + G_n^\varepsilon(T)$.

First, we have

$$
\begin{aligned}
&\int_{0\le u_1<\cdots<u_{2n}\le T} |g_\alpha^\varepsilon(T)(u_1,\cdots,u_{2n})|^2 du_1\cdots du_{2n} \\
&= \int_{0\le u_1<\cdots<u_{2n}\le T} (u_{2n}-u_1+\varepsilon)^{-2n-d+4} du_1\cdots du_{2n} \\
&= \frac{1}{(2n-2)!}\int_{0\le u_1<u_{2n}\le T} (u_{2n}-u_1+\varepsilon)^{-2n-d+4}(u_{2n}-u_1)^{2n-2} du_1 du_{2n} \\
&\le \frac{1}{(2n-2)!}\int_{0\le u_1<u_{2n}\le T} (u_{2n}-u_1)^{4-2d} du_1 du_{2n} \le \frac{C}{(2n-2)!}\,.
\end{aligned}
$$

This gives the estimate for the L^2 norm of $g_n^\varepsilon(T)$. The L^2 norms of the terms in $G_n^\varepsilon(T)$ can be estimated similarly.

$$
\int_{0\le u_1<\cdots<u_{2n}\le T} |G_n^\varepsilon(T)(u_1,\cdots,u_{2n})|^2 du_1\cdots du_{2n} \le \frac{C}{(2n-2)!}\,.
$$

By the orthogonality of J_α, we have

$$
\begin{aligned}
\mathbb{E}\left[\pounds^2_{\varepsilon,2n}(T)\right] &= \sum_{|\alpha|=2n} C_\alpha^2 \mathbb{E}\left[J_\alpha(f_\alpha^\varepsilon(T))\right]^2 \\
&\le \sum_{|\alpha|=2n} C_\alpha^2 \frac{C}{(2n-2)!} \le Cn^{\frac{d}{2}-5}(2n)(2n-1) \le Cn^{\frac{d}{2}-3}\,.
\end{aligned}
\tag{9.2.14}
$$

Thus,

$$
\sup_{\varepsilon>0}\sum_{n=0}^\infty \mathbb{E}\left[\pounds^2_{\varepsilon,2n}(T)\right] < \infty\,.
$$

On the other hand, it is easy to see that for any fixed $n\ge 1$, $\pounds^2_{\varepsilon,2n}(T)\to \pounds^2{}_{2n}(T)$ in L^2 as $\varepsilon\to 0$. This proves the theorem. □

From the estimate (9.2.14) and Meyer's theorem 7.15, we have immediately the following.

Corollary 9.1. *For any $\alpha < 2-d/2$, $d=1,2$, we have $\pounds(T)-\mathbb{E}\left[\pounds(T)\right]\in \mathbb{D}_{\alpha,2}$. More precisely, we have*

(i) When $d=1$, for any $\alpha<3/2$, $\pounds(T)\in\mathbb{D}_{\alpha,2}$.
(ii) When $d=2$, for any $\alpha<1$, $\pounds(T)-\mathbb{E}\left[\pounds(T)\right]\in\mathbb{D}_{\alpha,2}$.

9.3 Exponential Integrability

We need a lemma.

Lemma 9.4. *If Q is a symmetric positive definite matrix, then*

$$\int_{\mathbb{R}^d} \exp\left\{-\frac{1}{2}\xi^T Q\xi\right\} d\xi = (2\pi)^{d/2} \det(Q)^{-1/2} . \tag{9.3.1}$$

If Q_1 and Q_2 are two symmetric positive definite matrices, then

$$\det(Q_1 + Q_2) \geq \det(Q_1)^{1/2} \det(Q_2)^{1/2} . \tag{9.3.2}$$

Proof. The first inequality is easy. To show the second one. First, we notice that $\det(Q_1 + Q_2) \geq \det(Q_1)$ and $\det(Q_1 + Q_2) \geq \det(Q_2)$. Hence,

$$\det(Q_1 + Q_2) \geq \sqrt{\det(Q_1 + Q_2)}\sqrt{\det(Q_1 + Q_2)} \geq \sqrt{\det(Q_1)}\sqrt{\det(Q_2)} .$$

This completes the proof. In fact, we have also easily

$$\det(Q_1 + Q_2) \geq \det(Q_1)^{\alpha} \det(Q_2)^{\beta}$$

for any $\alpha, \beta \geq= 0$ and $\alpha + \beta = 1$. □

Let us first consider the intersection local time of two independent Brownian motions. Instead of using the chaos expansion, we use a formula given in [Albeverio; Hu and Zhou (1997)]. As in previous sections to simplify the presentation, we shall use the Fourier transform for the Dirac function directly, keeping the justification by the use of the above heat kernel approximation to the readers.

Theorem 9.4. *Let B and $\tilde{B}$ be two independent d-dimensional Brownian motions. Then, when $d \leq 3$, the intersection local time $\pounds(T, B, \tilde{B})$ (as a genius positive L^2 random variable) exists. Moreover, there is a constant $\alpha_{d,T} > 0$ such that*

$$\mathbb{E}\left[\exp\left\{\alpha_{d,T}\left[\pounds(T, B, \tilde{B})\right]^{2/d}\right\}\right] < \infty . \tag{9.3.3}$$

Proof. We only show (9.3.3). We use

$$\pounds(T, B, \tilde{B}) = \frac{1}{(2\pi)^d} \int_0^T \int_0^T \int_{\mathbb{R}^d} \exp\left\{\imath\xi(B_t - \tilde{B}_s)\right\} d\xi ds dt .$$

Then, we have

$$
\begin{aligned}
(2\pi)^{dn}\Theta_n &:= (2\pi)^{dn}\mathbb{E}\left[\mathcal{L}(T,B,\tilde{B})\right]^n \\
&= \int_{[0,T]^{2n}}\int_{\mathbb{R}^{dn}} \mathbb{E}\exp\left\{\imath\left[\sum_{k=1}^n \xi_k(B_{t_k}-\tilde{B}_{s_k})\right]\right\}d\xi ds dt \\
&= \int_{[0,T]^{2n}}\int_{\mathbb{R}^{dn}} \exp\left\{-\frac{1}{2}\mathbb{E}\left[\sum_{k=1}^n \xi_k(B_{t_k}-\tilde{B}_{s_k})\right]^2\right\}d\xi ds dt \\
&= \int_{[0,T]^{2n}}\int_{\mathbb{R}^{dn}} \exp\Bigg\{-\frac{1}{2}\mathbb{E}\left[\xi_1 B_{t_1}+\cdots+\xi_n B_{t_n}\right]^2 \\
&\qquad -\frac{1}{2}\mathbb{E}\left[\xi_1\tilde{B}_{s_1}+\cdots+\xi_n\tilde{B}_{s_n}\right]^2\Bigg\}d\xi ds dt \\
&= \int_{[0,T]^{2n}}\Bigg[\int_{\mathbb{R}^n}\exp\Bigg\{-\frac{1}{2}\mathbb{E}\left[\xi_1 B_{t_1}+\cdots+\xi_n B_{t_n}\right]^2 \\
&\qquad -\frac{1}{2}\mathbb{E}\left[\xi_1\tilde{B}_{s_1}+\cdots+\xi_n\tilde{B}_{s_n}\right]^2\Bigg\}d\xi\Bigg]^d ds dt\,,
\end{aligned}
$$

where $\xi_i=(\xi_i^1,\cdots,\xi_i^d)$ is a d-dimensional vector and $d\xi=d\xi_1\cdots d\xi_n$ etc. Where in the last identity inside [ ] are one dimensional and where we have used the fact that all the components of the Brownian motion are independent. In one dimensional case we have

$$
\mathbb{E}\left[\xi_1 B_{t_1}+\cdots+\xi_n B_{t_n}\right]^2=(\xi_1,\cdots,\cdots,\xi_n)^T Q_t(\xi_1,\cdots,\xi_n)\,,
$$

where $Q_t=(t_i\wedge t_j)_{1\le i,j\le n}$ is the covariance matrix of the one dimensional Brownian motion. Integration with respect to ξ and the use of (9.3.1) yield

$$
\begin{aligned}
\Theta_n &= C^n\int_{[0,T]^{2n}}\det(Q_t+Q_s)^{-d/2}ds dt \\
&\le \int_{[0,T]^{2n}}\det(Q_t)^{-d/4}\det(Q_s)^{-d/4}ds dt \\
&= C^n\left[\int_{[0,T]^n}\det(Q_t)^{-d/4}dt\right]^2 .
\end{aligned}
$$

It is well-known that if $0<t_1<\cdots<t_n<T$, then $\det(Q_t)=t_1(t_2-t_1)\cdots(t_n-t_{n-1})$. Thus, we have

$$
\begin{aligned}
\Theta_n &\le C^n\left[n!\int_{0<t_1<\cdots<t_n<T}\left[t_1(t_2-t_1)\cdots(t_n-t_{n-1})\right]^{-d/4}dt\right]^2 \\
&\le \frac{C^n T^{2n-dn/2}(n!)^2}{\Gamma((-d/4+1)n+1)^2}\le C^n T^{2n-dn/2}(n!)^{d/2}\,.
\end{aligned}
$$

From this estimate we know that there is an $\alpha_{d,T} > 0$ such that

$$\mathbb{E}\left[\exp\left\{\alpha_{d,T}\left[\pounds(T,B,\tilde{B})\right]^{2/d}\right\}\right] = \sum_{n=0}^{\infty}\frac{1}{n!}\alpha_{d,T}^{n}\mathbb{E}\left\{\left[\pounds(T,B,\tilde{B})\right]^{2n/d}\right\} < \infty\,.$$

□

Now we want to study the self-intersection local time.

Theorem 9.5. *Let $(B_t, 0 \le t \le T)$ be two dimensional Brownian motion. Then, there is a constant $\alpha_0 > 0$ such that for all $\alpha \in (0,\alpha_0)$,*

$$\mathbb{E}\exp\left\{\alpha\left(\pounds(T) - \mathbb{E}\pounds(T)\right)\right\} < \infty\,. \tag{9.3.4}$$

Proof. *Step 1.* For $k = 1,\dots,2^{n-1}$ we denote

$$A_{n,k} = \left[\frac{2k-2}{2^n}, \frac{2k-1}{2^n}\right] \times \left[\frac{2k-1}{2^n}, \frac{2k}{2^n}\right]$$

and define

$$\alpha_{n,k} = \int_{A_{n,k}} \delta(B_t - B_s)dsdt.$$

The random variables $\alpha_{n,k}$ have the following two properties:

(i) For every $n \ge 1$, the variables $\alpha_{n,1},\dots,\alpha_{n,2^{n-1}}$ are independent.

(ii) $\alpha_{n,k} \stackrel{d}{=} 2^{-n/2}\alpha_0$, where

$$\alpha_0 = \int_0^1\int_0^1 \delta(B_t - \tilde{B}_s)dsdt,$$

and $\tilde{B}$ is a (two dimensional) standard Brownian motion independent of B.

With these notations, we have

$$\pounds(T) - \mathbb{E}\left[\pounds(T)\right] = \sum_{n=1}^{\infty}\sum_{k=1}^{2^{n-1}}\left[\alpha_{n,k} - \mathbb{E}(\alpha_{n,k})\right]\,.$$

Step 2. To simplify notation we use notation $\bar{X} = X - \mathbb{E}(X)$. Define

$$\phi(\lambda) = \mathbb{E}\exp\left\{\lambda\left[\alpha_0 - \mathbb{E}(\alpha_0)\right]\right\}\,.$$

From Theorem 9.4 (when $d = 2$) we see that there is a $\lambda_0 > 0$ such that $\phi(\lambda) < \infty$ for all $\lambda < \lambda_0$. Since $\phi'(0) = 0$, we see that there is a $\lambda_1 > 0$ (we can assume $\lambda_1 < 2$) such that

$$\phi(\lambda) \le 1 + c\lambda^2\,,\quad \forall\,\lambda \in [0,\lambda_1]\,.$$

Fix a $a \in (0,1)$. For every $N \geq 1$ set

$$b_N = 2\lambda_1 \sum_{j=2}^{N} (1 - 2^{-a(j-1)})$$

and

$$c_N = 2^{a(N-1)} .$$

Thus, from the Hölder inequality, we have

$$\mathbb{E}\left\{\exp\left[b_N \sum_{n=}^{N} \sum_{k=1}^{2^{n-1}} \bar{\alpha}_{n,k}\right]\right\} \leq \mathbb{E}\left\{\exp\left[c_N b_N \sum_{k=1}^{2^{N-1}} \bar{\alpha}_{n,k}\right]\right\}^{\frac{1}{c_N}}$$

$$\mathbb{E}\left\{\exp\left[\frac{b_N}{1-\frac{1}{c_N}} \sum_{n=1}^{N-1} \sum_{k=1}^{2^{n-1}} \bar{\alpha}_{n,k}\right]\right\}^{1-\frac{1}{c_N}}$$

$$\leq \phi\left(b_N 2^{a(N-1)-N}\right)^{2^{(1-a)(N-1)}} \mathbb{E}\left\{\exp\left[\frac{b_N}{1-\frac{1}{c_N}} \sum_{n=1}^{N-1} \sum_{k=1}^{2^{n-1}} \bar{\alpha}_{n,k}\right]\right\}^{1-\frac{1}{c_N}} .$$

Notice that $b_N 2^{a(N-1)-N} \leq \lambda_1$. It follows the above second factor is bounded by

$$\begin{aligned}
\phi\left(b_N 2^{a(N-1)-N}\right)^{2^{(1-a)(N-1)}} &\leq \left[1 + c b_n^2 2^{2((a-1)N-a)}\right]^{2^{(1-a)(N-1)}} \\
&\leq \exp\left[c' 2^{(a-1)N}\right]
\end{aligned}$$

for some constant c', independent of N. By induction, we see

$$\begin{aligned}
\mathbb{E}\left\{\exp\left[b_N \sum_{n=}^{N} \sum_{k=1}^{2^{n-1}} \bar{\alpha}_{n,k}\right]\right\} &\leq \exp\left(c' \sum_{n=2}^{N} 2^{(a-1)n}\right) \mathbb{E}\left[\exp\left(b_1 \bar{\alpha}_{1,1}\right)\right] \\
&\leq \exp\left[c'(1 - 2^{a-1})^{-1}\right] \phi(\lambda_1) .
\end{aligned}$$

Letting $N \to \infty$ and by Fatou's lemma we see that $\mathbb{E}\exp\left[b_\infty \mathcal{L}(T)\right] < \infty$ for $b_\infty = 2\lambda_1 \prod_{j=1}^{\infty}(1 - 2^{-aj})$. Since $a \in (0,1)$ and $\lambda_1 \in (0,2)$ are arbitrary, we prove the theorem. □

Remark 9.2. One may be able to express the exact constant α_0 in the above theorem (see [Bass and Chen (2004); Chen (2010)]).

9.4 Renormalization When $d \geq 3$

In this section we need the following computation

Lemma 9.5. *Let $d \geq 3$ and $n \geq 1$. Then, when $d \geq 4$,*

$$\lim_{K\to\infty} K^{-1}\int_{0\leq u<v\leq K}(v-u+1)^{-2n-d+4}(v-u)^{2n-2}dudv$$
$$= \int_0^\infty (x+1)^{-2n-d+4}x^{2n-2}dx\,; \tag{9.4.1}$$

$$\lim_{K\to\infty} K^{-1}\int_{0\leq u<v\leq K}(v+1)^{-2n-d+4}(v-u)^{2n-2}dudv = 0\,; \tag{9.4.2}$$

$$\lim_{K\to\infty} K^{-1}\int_{0\leq u<v\leq K}(K-u+1)^{-2n-d+4}(v-u)^{2n-2}dudv = 0\,. \tag{9.4.3}$$

When $d = 3$, we have

$$\lim_{K\to\infty} (K\log K)^{-1}\int_{0\leq u<v\leq K}(v-u+1)^{-2n+1}(v-u)^{2n-2}dudv = 1\,; \tag{9.4.4}$$

$$\lim_{K\to\infty} (K\log K)^{-1}\int_{0\leq u<v\leq K}(v+1)^{-2n+1}(v-u)^{2n-2}dudv = 0\,; \tag{9.4.5}$$

$$\lim_{K\to\infty} (K\log K)^{-1}\int_{0\leq u<v\leq K}(K-u+1)^{-2n+1}(v-u)^{2n-2}dudv = 0\,. \tag{9.4.6}$$

Proof. Making the transformation $u = y$ and $v - u = x$, we have

$$\int_{0\leq u<v<K}(v-u+1)^{-2n-d+4}(v-u)^{2n-2}dudv$$
$$= \int_0^K dx\int_x^K dy(x+1)^{-2n-d+4}x^{2n-2}$$
$$= \int_0^K (K-x)(x+1)^{-2n-d+4}x^{2n-2}dx$$
$$= K\int_0^K (x+1)^{-2n-d+4}x^{2n-2}dx - \int_0^K (x+1)^{-2n-d+4}x^{2n-1}dx$$
$$=: A_K + B_K\,.$$

Let $K \to \infty$. When $d \geq 4$, we have

$$\lim_{K\to\infty} K^{-1}A_K = \int_0^\infty (x+1)^{-2n-d+4}x^{2n-2}dx$$

and $\lim_{K\to\infty} K^{-1}B_K = 0$. This gives (9.4.1). When $d = 3$, one can see that $\lim_{K\to\infty}(K\log K)^{-1}A_K = 1$ and $\lim_{K\to\infty}(K\log K)^{-1}B_K = 0$. This proves (9.4.4). Now

$$\begin{aligned}\int_{0\le u<v\le K} &(v+1)^{-2n-d+4}(v-u)^{2n-2}dudv \\ &\le \frac{2}{2n-2+1}\int_0^K (v+1)^{-d+3}dv \\ &\le \frac{3}{2n-2+1}(K+1)^{-d+4}\log(K+1)\,.\end{aligned}$$

This shows that when $d \ge 4$,

$$\lim_{K\to\infty} K^{-1}\int_{0\le u<v\le K}(v+1)^{-2n-d+4}(v-u)^{2n-2}dudv = 0\,,$$

proving (9.4.2). Similarly, we can prove that when $d = 3$, we have

$$\lim_{K\to\infty}(K\log K)^{-1}\int_{0\le u<v\le K}(v+1)^{-2n+1}(v-u)^{2n-2}dudv = 0\,,$$

proving (9.4.5). □

Now we estimate $\int_{0\le u_1<\cdots<u_{2n}\le T}|f_\alpha^\varepsilon(T)(u_1,\cdots,u_{2n})|^2du_1\cdots du_{2n}$. As in the previous section denote for $n\ge 1$

$$g_n^\varepsilon(T)(u_1,\cdots,u_{2n}) := (u_{2n}-u_1+\varepsilon)^{-n-d/2+2}$$

and

$$\begin{aligned}&G_n^\varepsilon(T)(u_1,\cdots,u_{2n}) \\ &\quad = -(u_{2n}+\varepsilon)^{-n-d/2+2} + (T+\varepsilon)^{-n-d/2+2} - (T-u_1+\varepsilon)^{-n-d/2+2}\end{aligned}$$

so that $f_\alpha^\varepsilon(T) = g_n^\varepsilon(T) + G_n^\varepsilon(T)$. First, we have

$$\begin{aligned}&\int_{0\le u_1<\cdots<u_{2n}\le T}|g_\alpha^\varepsilon(T)(u_1,\cdots,u_{2n})|^2du_1\cdots du_{2n} \\ &= \int_{0\le u_1<\cdots<u_{2n}\le T}(u_{2n}-u_1+\varepsilon)^{-2n-d+4}du_1\cdots du_{2n} \\ &= \frac{1}{(2n-2)!}\int_{0\le u_1<u_{2n}\le T}(u_{2n}-u_1+\varepsilon)^{-2n-d+4}(u_{2n}-u_1)^{2n-2}du_1du_{2n} \\ &= \frac{\varepsilon^{4-d}}{(2n-2)!}\int_{0\le u<v\le T/\varepsilon}(v-u+1)^{-2n-d+4}(v-u)^{-2n-2}dudv \\ &= \begin{cases}\frac{T}{(2n-2)!}\int_0^\infty (x+1)^{-2n-d+4}x^{2n-2}dx\cdot O(\varepsilon^{3-d}) & \text{when } d\ge 4 \\ \frac{T}{(2n-2)!}\cdot O(\log(\frac{1}{\varepsilon})), & \text{when } d=3\,,\end{cases}\end{aligned}$$

where the last identity follows from (9.4.1) and (9.4.4). This gives the estimate for the L^2 norm of $g_n^\varepsilon(T)$. Now we have to estimate the L^2 norm of $G_n^\varepsilon(T)$. It is easy to see that

$$|G_n(u_1,\cdots,u_{2n})| \le \mu[(u_{2n}+\varepsilon)^{-n-d/2+2}+(T-u_1+\varepsilon)^{-n-d/2+2}] \quad (9.4.7)$$

for some positive constant $0<\mu<\infty$. We should dominate the two terms in (9.4.7). When $d\ge 4$, we have for $n\ge 1$

$$\begin{aligned}&\int_{0\le u_1<\cdots<u_{2n}\le T}(u_{2n}+\varepsilon)^{-2n-d+4}du_1\cdots du_{2n}\\ &=\int_{0\le u_1<u_{2n}\le T}(u_{2n}+\varepsilon)^{-2n-d+4}(u_{2n}-u_1)^{2n-2}du_1du_{2n}\\ &=\frac{\varepsilon^{4-d}}{(2n-2)!}\int_{0\le u<v\le T/\varepsilon}(v+1)^{-2n-d+4}(v-u)^{-2n-2}dudv\,.\end{aligned}$$

By (9.4.2), we see that when $d\ge 4$

$$\lim_{\varepsilon\to 0}\varepsilon^{d-3}\int_{0\le u_1<\cdots<u_{2n}\le T}(u_{2n}+\varepsilon)^{-2n-d+4}du_1\cdots du_{2n}=0\,.$$

Similarly, we can prove by (9.4.5) that when $d=3$,

$$\frac{1}{\log(1/\varepsilon)}\int_{0\le u_1<\cdots<u_{2n}\le T}(u_{2n}+\varepsilon)^{-2n+1}du_1\cdots du_{2n}=0\,.$$

This gives the estimate arising from the first member of the RHS of (9.4.7). The same argument (using (9.4.3) and (9.4.6)) implies that the same conclusion holds for the second member of the RHS of (9.4.7). Thus, we have

$$\lim_{\varepsilon\to 0}\varepsilon^{d-3}\int_{0\le u_1<\cdots<u_{2n}\le T}|G_n^\varepsilon(T)|^2du_1\cdots du_{2n}=0 \text{ when } d\ge 4 \text{ and}$$

$$\frac{1}{\log(1/\varepsilon)}\int_{0\le u_1<\cdots<u_{2n}\le T}|G_n^\varepsilon(T)|^2du_1\cdots du_{2n}=0 \text{ when } d=3\,.$$

Summarizing the above computation yields

Theorem 9.6. *We have*

1) when $d=3$,

$$\lim_{\varepsilon\to 0}\frac{1}{\log(1/\varepsilon)}\int_{0\le u_1<\cdots<u_{2n}\le T}|f_\alpha^\varepsilon(T)(u_1,\cdots,u_{2n})|^2du_1\cdots du_{2n} =\frac{T}{(2n-2)!}\,. \quad (9.4.8)$$

2) When $d\ge 4$,

$$\lim_{\varepsilon\to 0}\varepsilon^{d-3}\int_{0\le u_1<\cdots<u_{2n}\le T}|f_\alpha^\varepsilon(T)(u_1,\cdots,u_{2n})|^2du_1\cdots du_{2n} =\frac{T}{(2n-2)!}\int_0^\infty(x+1)^{-2n-d+4}x^{2n-2}dx\,. \quad (9.4.9)$$

First we discuss the case $d = 3$. Let

$$\begin{aligned}\Phi_\varepsilon(T) &:= \frac{1}{\sqrt{\log(1/\varepsilon)}}\Big\{\mathcal{L}_\varepsilon(T) - \mathbb{E}\mathcal{L}_\varepsilon(T)\Big\} \\ &= \sum_{n=1}^{\infty}\sum_{|\alpha|=2n} C_\alpha J_\alpha\big(f_\alpha^\varepsilon(T)/\sqrt{\log(1/\varepsilon)}\big)\,. \end{aligned} \tag{9.4.10}$$

By Lemma 9.5, we have

$$\lim_{\varepsilon\to 0}\|J_\alpha\big(f_\alpha^\varepsilon(T)/\sqrt{\log(1/\varepsilon)}\big)\|^2 = \frac{T}{(2n-2)!}\,.$$

Thus,

$$\begin{aligned}&\|\sum_{|\alpha|=2n} C_\alpha J_\alpha\big(f_\alpha^\varepsilon(T)/\sqrt{\log(1/\varepsilon)}\big)\|^2 \\ \approx&\sum_{|\alpha|=2n} C_\alpha^2\|J_\alpha\big(f_\alpha^\varepsilon(T)/\sqrt{\log(1/\varepsilon)}\big)\|^2 \\ \approx&\frac{T}{(2n-2)!}\sum_{|\alpha|=2n} C_\alpha^2 \\ \approx&\frac{T}{(2n-2)!}(2n)!n^{3/2-5} \approx Tn^{-3/2}\,,\end{aligned}$$

where $A_\varepsilon \approx B_\varepsilon$ means that A_ε and B_ε have the same limit when $\varepsilon \to 0$. Thus, we see that for $\theta < 1/2$, $\Phi_\varepsilon(T)$ is bounded hence weakly compact in the Hilbert space $\mathbb{D}_{\theta,2}$ as $\varepsilon \to 0$. And when $\theta \geq 1/2$, $\Phi_\varepsilon(T)$ is unbounded in $\mathbb{D}_{\theta,2}$ as $\varepsilon \to 0$. Namely, we have

Theorem 9.7. *Let $d = 3$. When $\theta < 1/2$, $\Phi_\varepsilon(T)$ is weakly compact in $\mathbb{D}_{\theta,2}$ as $\varepsilon \to 0$. When $\theta \geq 1/2$, $\Phi_\varepsilon(T)$ is unbounded in $\mathbb{D}_{\theta,2}$ as $\varepsilon \to 0$.*

When $d \geq 4$, let

$$\begin{aligned}\Psi_\varepsilon(T) &:= \varepsilon^{\frac{d-3}{2}}\Big\{\mathcal{L}_\varepsilon(T) - \mathbb{E}\mathcal{L}_\varepsilon(T)\Big\} \\ &= \sum_{n=1}^{\infty}\sum_{|\alpha|=2n} C_\alpha J_\alpha\big(f_\alpha^\varepsilon(T)\varepsilon^{\frac{d-3}{2}}\big)\,.\end{aligned} \tag{9.4.11}$$

By Lemma 9.5, we obtain

$$\begin{aligned}&\|\sum_{|\alpha|=2n} C_\alpha J_\alpha\big(f_\alpha^\varepsilon(T)\varepsilon^{\frac{d-3}{2}}\big)\|^2 = \sum_{|\alpha|=2n} C_\alpha^2\|J_\alpha\big(f_\alpha^\varepsilon(T)\varepsilon^{\frac{d-3}{2}}\big)\|^2 \\ &\approx \frac{T}{(2n-2)!}\sum_{|\alpha|=2n} C_\alpha^2\int_0^\infty (x+1)^{-2n-d+4}x^{2n-2}dx \approx Tn^{d/2-3}\,.\end{aligned} \tag{9.4.12}$$

Therefore, we have

Theorem 9.8. *Let $d \geq 4$. When $\theta < \frac{4-d}{2}$, $\Psi_\varepsilon(T)$ is weakly compact in $\mathbb{D}_{\theta,2}$ as $\varepsilon \to 0$. And when $\theta \geq \frac{4-d}{2}$, $\Psi_\varepsilon(T)$ is unbounded in $\mathbb{D}_{\theta,2}$ as $\varepsilon \to 0$.*

The above results show that when $d \geq 3$, Therefore $\pounds(T)$ is not regular (i.e. in the sense that it is not in the Meyer-Watanabe test functional space). For any $\theta \in \mathbb{R}$, let us introduce

$$L^{-\theta}\Psi_\varepsilon(T) := \sum_{n=1}^{\infty} n^{-\theta} \sum_{|\alpha|=2n} C_\alpha J_\alpha(f_\alpha^\varepsilon(T))\,. \tag{9.4.13}$$

From the estimates in Section 3, we see easily that $\varepsilon^{\frac{d-3}{2}} L^{-\theta}\Psi_\varepsilon(T)$ is weakly compact in $\mathbb{D}_{\theta-d/2+2-\rho,2}$ for any $\rho > 0$. Thus, if $\theta > d/2-2$, then $\varepsilon^{\frac{d-3}{2}} L^{-\theta}\Psi_\varepsilon(T)$ is weakly compact in the Meyer-Watanabe test functional space $\mathbb{D}_{\alpha,2}$ for some $\alpha > 0$. Hence we think that $\varepsilon^{\frac{d-3}{2}} L^{-\theta}\Psi_\varepsilon(T)$ would be a better renormalization scheme for the self-intersection local time of higher dimensional Brownian motion ($d \geq 4$). Using the operator $\Gamma(e^{-t})$ of second quantization of e^{-t}, we have

$$L^{-\theta}\Psi_\varepsilon(T) = \int_0^\infty t^{\theta+1}\Gamma(e^{-t})\Big\{\pounds_\varepsilon(T) - \mathbb{E}\Big(\pounds_\varepsilon(T)\Big)\Big\}dt\,. \tag{9.4.14}$$

Now using the Mehler formula, we have

$$\begin{aligned}\Gamma(e^{-t})p_\varepsilon(B_t - B_s) &= \mathbb{E}'\Big\{p_\varepsilon(e^{-u}(B_t - B_s) + \sqrt{1-e^{-2u}}(B_t' - B_s'))\Big\}\\ &= P_{\varepsilon+(t-s)(1-e^{-2u})}(e^{-u}(B_t - B_s))\,,\end{aligned}$$

where B' is a d dimensional Brownian motion independent of B and $\mathbb{E}'$ means the expectation with respect to B'. Therefore

$$L^{-\theta}\Psi_\varepsilon(T) = \psi_{\varepsilon,\theta}(t-s, B_t - B_s)\,,$$

where

$$\psi_{\varepsilon,\theta}(\nu, x) = \int_0^\infty P_{\varepsilon+\nu(1-e^{-2u})}(e^{-u}x)u^{\theta+1}du\,. \tag{9.4.15}$$

Summarizing the above we have

Theorem 9.9. *Let $\psi_{\varepsilon,\theta}(\nu, x)$ be defined by(9.4.15). Then for any $\theta \in \mathbb{R}$,*

$$\tilde{\Psi}_{\varepsilon,\theta} := \varepsilon^{\frac{d-3}{2}}\Big\{\int_0^T\int_0^t \psi_{\varepsilon,\theta}(t-s, B_t - B_s)dsdt \tag{9.4.16}$$

$$-\mathbb{E}\Big(\int_0^T\int_0^t \psi_{\varepsilon,\theta}(t-s, B_t - B_s)dsdt\Big)\Big\} \tag{9.4.17}$$

is weakly compact in $\mathbb{D}_{\theta-d/2+4,2}$.

It is easy to see that $\|\tilde{\Psi}_{\varepsilon,\theta}\|^2 \to \mu T$ for some constant μ.

9.5 L^2-Modulus of Continuity of Local Time of Brownian Motion

The local time $\mathcal{L}_t(x) = \int_0^t \delta(B_s - x)ds$ of Brownian motion at x has been discussed earlier. As before we shall carry out operations directly on the Dirac delta function leaving the justification (by $\delta(x) \leftarrow p_\varepsilon(x) = (2\pi)^{-1/2}\exp\{-|x|^2/(2\varepsilon)\}$) to the readers. We have the following computation for any $h \in \mathbb{R}$,

$$\begin{aligned}
\mathbb{E}\int_{\mathbb{R}} \mathcal{L}_t(x+h)\mathcal{L}_t(x)dx &= \mathbb{E}\int_{\mathbb{R}}\int_0^t \delta(B_s - x - h)ds\int_0^t \delta(B_r - x)drdx \\
&= \int_0^t\int_0^t \mathbb{E}\int_{\mathbb{R}} \delta(B_s - x - h)\delta(B_r - x)dxdrds \\
&= \int_0^t\int_0^t \mathbb{E}\delta(B_s - B_r - h)drds \\
&= 2\int_0^t\int_0^s p_{s-r}(h)drds = 2\int_0^t\int_0^s p_r(h)drds\,.
\end{aligned}$$

From the above computation it follows

$$\begin{aligned}
\mathcal{E}_{\mathcal{L}} := \mathbb{E}\int_{\mathbb{R}} \left[\mathcal{L}_t(x+h) - \mathcal{L}_t(x)\right]^2 dx &= 4\int_0^t\int_0^s \left[p_r(0) - p_r(h)\right] drds \\
&= \sqrt{\frac{8}{\pi}}\int_0^t\int_0^s \left[r^{-1/2} - r^{-1/2}\exp\left\{-\frac{h^2}{2r}\right\}\right] dr\,.
\end{aligned}$$

Making substitution and denoting $\kappa = \int_0^\infty u^{-3/2}(1-e^{-u})du$, we have

$$\begin{aligned}
\mathcal{E}_{\mathcal{L}} &= \frac{2h}{\sqrt{\pi}}\int_0^t\int_{\frac{h^2}{2s}}^\infty u^{-3/2}(1-e^{-u})duds \\
&= \frac{2ht}{\sqrt{\pi}}\int_0^\infty u^{-3/2}(1-e^{-u})du - \frac{2h}{\sqrt{\pi}}\mathcal{E}_h\,, \qquad (9.5.1)
\end{aligned}$$

where $\mathcal{E}_h = \int_0^t\int_0^{\frac{h^2}{2s}} u^{-3/2}(1-e^{-u})duds$. The above integral can be computed by [Gradshteyn and Ryzhik (2000), Equation (3.434) with $\nu \downarrow 0$] and then the same reference with Equation (8.338)(2). We have

$$\kappa = \int_0^\infty u^{-3/2}(1-e^{-u})du = 2\Gamma(1/2) = 2\sqrt{\pi}\,.$$

On the other hand it is obvious that

$$\mathcal{E}_h \le C\int_0^t\int_0^{\frac{h^2}{2s}} u^{-1/2}duds = Ct^{1/2}h\,.$$

Combining the above computation with (9.5.1), we have

Proposition 9.2. *As $h \to 0$, we have*

$$\mathbb{E}\int_{\mathbb{R}} \left[\mathcal{L}_t(x+h) - \mathcal{L}_t(x)\right]^2 dx = 4th + O(h^2)\,. \tag{9.5.2}$$

Next, we establish a (non)central limit type theorem for the L^2 modulus of continuity of the local time: $\int_{\mathbb{R}} \left[\mathcal{L}_t(x+h) - \mathcal{L}_t(x)\right]^2 dx$. We need the following two lemmas later in our proof of the main theorem.

Lemma 9.6. *We have*

$$\int_0^t \int_0^r \frac{\left[(h - |B_r - B_s|)^+\right]^2}{h^3} ds dr \xrightarrow{L^2} \frac{4}{3}\pounds_t,$$

as h tends to zero, where we recall that $\pounds_t$ is the self-intersection local time at 0.

Proof. Notice that

$$\pounds_t = \int_{\mathbb{R}} \mathcal{L}_t(x)^2 dx = \int_0^t \int_{\mathbb{R}} L_r(x) L_{dr}(x) dx = \int_0^t L_r(B_r) dr,$$

and

$$\int_0^t \int_0^r \frac{\left[(h - |B_r - B_s|)^+\right]^2}{h^3} ds dr = \int_0^t \int_{\mathbb{R}} \frac{\left[(h - |B_r - x|)^+\right]^2}{h^3} L_r(x) dx.$$

As a consequence, taking into account that

$$\int_{\mathbb{R}} \frac{\left[(h - |B_r - x|)^+\right]^2}{h^3} dx = \int_{\mathbb{R}} \frac{[(h - |x|)^+]^2}{h^3} dx = \frac{4}{3},$$

we obtain

$$\begin{aligned}
&\left| \int_0^t \int_0^r \frac{\left[(h - |B_r - B_s|)^+\right]^2}{h^3} ds dr - \frac{4}{3}\pounds_t \right| \\
&\le \int_0^t \int_{\mathbb{R}} \frac{\left[(h - |B_r - x|)^+\right]^2}{h^3} \left| L_r(x) - L_r(B_r) \right| dx dr \\
&\le \frac{4}{3} \int_0^t \sup_{|x-y|<h} \left| L_r(x) - L_r(y) \right| dr,
\end{aligned}$$

which clearly converges to zero in L^2 by the properties of the Brownian local time (see, for instance, [Borodin (1989)]). □

Lemma 9.7. *We have that*

$$\frac{1}{h^6}\mathbb{E}\left[\left(\int_0^t\left(\int_0^r\left(\int_s^r(h-|B_r-B_u|)^+d\widehat{B}_u\right)(h-|B_r-B_s|)^+d\widehat{B}_s\right)dr\right)^2\right]$$

converges to 0 *as* h *tends to zero, where* $d\widehat{B}_s$ *denotes the backward stochastic integral (with respect to the future filtration).*

Proof. By the isometry property of the backward Itô integral we can write

$$B_h := \frac{1}{h^6}\mathbb{E}\Bigg[\Bigg(\int_0^t\Bigg(\int_0^r\left(\int_s^r(h-|B_r-B_u|)^+d\widehat{B}_u\right)$$
$$(h-|B_r-B_s|)^+d\widehat{B}_s\Bigg)dr\Bigg)^2\Bigg]$$
$$= \frac{1}{h^6}\mathbb{E}\left[\int_0^t\left(\int_s^t(h-|B_r-B_s|)^+\left(\int_s^r(h-|B_r-B_u|)^+d\widehat{B}_u\right)dr\right)^2ds\right]$$
$$\le \mathbb{E}\left[B_h^1B_h^2\right],$$

where

$$B_h^1 = \int_0^t\left(\int_s^t\frac{(h-|B_r-B_s|)^+}{h^2}\,dr\right)^2ds$$

and

$$B_h^2 = \sup_{0<s<r<t}\left|\int_s^r\frac{(h-|B_r-B_u|)^+}{h}d\widehat{B}_u\right|^2.$$

As in Lemma 9.6, we can show that B_h^1 converges to

$$\frac{9}{4}\int_0^t(\mathcal{L}_t(B_s)-L_s(B_s))^2\,ds,$$

and the convergence holds in L^p for any $p\ge 2$. Here we use

$$\int_{\mathbb{R}}\frac{(h-|x|)^+}{h^2}dx = \frac{3}{2}.$$

On the other hand, $\int_s^r\frac{(h-|B_r-B_u|)^+}{h}d\widehat{B}_u$ can be expressed by using again Tanaka's formula:

$$\frac{1}{h}\int_s^r(h-|B_r-B_u|)^+d\widehat{B}_u = \int_0^h\left(\int_s^r I_{y>|B_r-B_u|}d\widehat{B}_u\right)dy$$
$$= \frac{1}{h}\int_0^h(L_r(B_r-y)-L_r(B_r+y))\,dy$$
$$-\frac{1}{h}\int_0^h(L_s(B_r-y)-L_s(B_r+y))\,dy+\frac{h}{2}$$
$$+\frac{1}{h}\int_0^h\left[(B_r-B_s+y)^+-(B_r-B_s-y)^+\right]dy.$$

Therefore,

$$\left|\frac{1}{h}\int_s^r (h-|B_r-B_u|)^+ d\widehat{B}_u\right| \le \sup_{s<r<t}\ \sup_{|x-y|\le 2h} |L_r(x)-L_r(y)| + O(h),$$

which also converges to zero in L^p for any $p \ge 2$. □

Now we state our main theorem of this section.

Theorem 9.10. *Let $\{B_t, t\ge 0\}$ be a standard Brownian motion. For each fixed $t>0$,*

$$h^{-\frac{3}{2}}\left(\int_{\mathbb{R}}(\mathcal{L}_t(x+h)-\mathcal{L}_t(x))^2dx - 4th\right) \xrightarrow{\mathcal{L}} 8\sqrt{\frac{\pounds_t}{3}}\eta, \tag{9.5.3}$$

as h tends to zero, where

$$\pounds_t = \int_{\mathbb{R}}(\mathcal{L}_t(x))^2dx, \tag{9.5.4}$$

and η is a $N(0,1)$ random variable independent of B.

Proof. Denote

$$G_t(h) = \int_{\mathbb{R}}(\mathcal{L}_t(x+h)-\mathcal{L}_t(x))^2dx. \tag{9.5.5}$$

From Proposition 9.2 we known that $\mathbb{E}\,(G_t(h)) = 4th + O(h^2)$. Therefore, we can replace the term $4th$ in (9.5.3) by $\mathbb{E}\,(G_t(h))$.

The random variable $G_t(h) = \int_{\mathbb{R}}(\mathcal{L}_t(x+h)-\mathcal{L}_t(x))^2dx$ can be expressed in terms of the self-intersection local time of Brownian motion. In fact,

$$\begin{aligned} G_t(h) &= \int_{\mathbb{R}}\left[\int_0^t \delta(B_u+x+h)du - \int_0^t \delta(B_u+x)du\right]^2 dx \\ &= 2\left[\int_0^t\int_0^t \delta(B_v-B_u+h)dudv - \int_0^t\int_0^t \delta(B_v-B_u)dudv\right] \\ &= 2\int_0^t\int_0^v (\delta(B_v-B_u+h)+\delta(B_v-B_u-h)-2\delta(B_v-B_u))\,dudv\,. \end{aligned}$$

The rigorous justification of above argument can be made easily by approximating the Dirac delta function by the heat kernel $p_\varepsilon(x) = \frac{1}{\sqrt{2\pi\varepsilon}}e^{-x^2/2\varepsilon}$ as ε tends to zero. That is, $G_t(h)$ is the limit in L^2 as ε tends to zero of

$$\begin{aligned} G_t^\varepsilon(h) = 2\int_0^t\int_0^v \Big(&p_\varepsilon(B_v-B_u+h) \\ &+p_\varepsilon(B_v-B_u-h)-2p_\varepsilon(B_v-B_u)\Big)dudv. \end{aligned} \tag{9.5.6}$$

Applying Clark-Ocone formula we can derive the following stochastic integral representation for $G_t(h)$.

Step 1. We claim that the random variable $G_t(h)$ defined in (9.5.5) can be expressed as

$$G_t(h) = \mathbb{E}(G_t(h)) + \int_0^t u_{t,h}(r) dB_r \,,$$

where

$$u_{t,h}(r) = 4\int_0^r \int_0^h \left(p_{t-r}(B_r - B_u + \eta) - p_{t-r}(B_r - B_u - \eta)\right) d\eta du$$
$$+4\int_0^r \left(I_{[0,h]}(B_r - B_u) - I_{[0,h]}(B_u - B_r)\right) du. \tag{9.5.7}$$

We shall use the Clark-Ocone formula. For any $u < v$ and any $h \in \mathbb{R}$ we can write

$$D_r p_\varepsilon(B_v - B_u + h) = p'_\varepsilon(B_v - B_u + h) I_{[u,v]}(r),$$

and for any $u < r < v$

$$\begin{aligned}\mathbb{E}\left(D_r p'_\varepsilon(B_v - B_u + h)|\mathcal{F}_r\right) &= \mathbb{E}p'_\varepsilon(\sqrt{v-r}\eta + B_r - B_u + h)\\ &= p'_{v-r+\varepsilon}(B_r - B_u + h),\end{aligned}$$

where η denotes an $N(0,1)$ random variable independent of B. Therefore, from Clark-Ocone formula (6.1.22) and Equation (9.5.6) we obtain

$$G_t^\varepsilon(h) = \mathbb{E}(G_t^\varepsilon(h)) + \int_0^t u_{t,h}^\varepsilon(r) dB_r,$$

where

$$u_{t,h}^\varepsilon(r) = 2\int_r^t \int_0^r \Bigg(p'_{v-r+\varepsilon}(B_r - B_u + h) + p'_{v-r+\varepsilon}(B_r - B_u - h)$$
$$-2p'_{v-r+\varepsilon}(B_r - B_u)\Bigg) du dv.$$

This expression can be written as

$$u_{t,h}^\varepsilon(r) = 2\int_r^t \int_0^r \int_0^h \left((p''_{v-r+\varepsilon}(B_r - B_u + \eta) - p''_{v-r+\varepsilon}(B_r - B_u - \eta)\right) d\eta du dv.$$

Using the fact that $p''_t(x) = 2\frac{\partial p_t}{\partial t}(x)$ we obtain

$$u_{t,h}^\varepsilon(r) = 4\Bigg(\int_0^r \int_0^h \left((p_{t-r+\varepsilon}(B_r - B_u + \eta) - p_{t-r+\varepsilon}(B_r - B_u - \eta)\right) d\eta du$$
$$-\int_0^r \int_0^h \left((p_\varepsilon(B_r - B_u + \eta) - p_\varepsilon(B_r - B_u - \eta)\right) d\eta du\Bigg).$$

Letting ε tend to zero we get that $u^{\varepsilon}_{t,h}(r)$ converges in $L^2(\Omega \times [0,t])$ to $u_{t,h}(r)$ as h tends to zero, which implies the desired result.

In Equation (9.5.7) we can make the following decomposition

$$u_{t,h}(r) = \hat{u}_{t,h}(r) + \tilde{u}_{t,h}(r),$$

where

$$\begin{aligned}\hat{u}_{t,h}(r) &= 4\int_0^r \int_0^h \left(p_{t-r}(B_r - B_u + \eta) - p_{t-r}(B_r - B_u - \eta)\right) d\eta du \\ &= 4\int_0^r \int_0^h \int_{-\eta}^{\eta} p'_{t-r}(B_r - B_u + \xi) d\xi d\eta du \qquad (9.5.8)\end{aligned}$$

and

$$\tilde{u}_{t,h}(r) = 4\int_0^r \left(I_{[0,h]}(B_r - B_u) - I_{[0,h]}(B_u - B_r)\right) du.$$

As a consequence, we finally obtain

$$\begin{aligned}G_t(h) - \mathbb{E}(G_t(h)) &= 4\int_0^t \hat{u}_{t,h}(r) dB_r \\ &+4\int_0^t \left(\int_0^r \left(I_{[0,h]}(B_r - B_u) - I_{[0,h]}(B_u - B_r)\right) du\right) dB_r. \qquad (9.5.9)\end{aligned}$$

Step 2. We are going to show that there is a constant $C > 0$ such that

$$\mathbb{E}\left(\int_0^t |\hat{u}_{t,h}(r)|^2 dr\right) \le Ch^4, \quad \forall \ h > 0. \qquad (9.5.10)$$

From (9.5.8) we can write

$$\begin{aligned}\mathbb{E}\left(|\hat{u}_{t,h}(r)|^2\right) = \int_0^r \int_0^r \int_0^h \int_0^h \int_{-\eta_1}^{\eta_1} \int_{-\eta_2}^{\eta_2} \mathbb{E}(p'_{t-r}(B_r - B_{u_1} + \xi_1) \\ \times p'_{t-r}(B_r - B_{u_2} + \xi_2)) d\xi_1 d\xi_2 d\eta_1 d\eta_2 du_1 du_2.\end{aligned}$$

By a symmetry argument, it suffices to integrate in the region $0 < u_1 < u_2 < r$. Set

$$\Phi(u_1, u_2, \xi_1, \xi_2) = \mathbb{E}\left(p'_{t-r}(B_r - B_{u_1} + \xi_1) p'_{t-r}(B_r - B_{u_2} + \xi_2)\right).$$

Then,

$$\begin{aligned}\Phi(u_1, u_2, \xi_1, \xi_2) &= \mathbb{E}\left(p'_{t-r}(B_r - B_{u_2} + B_{u_2} - B_{u_1} + \xi_1) p'_{t-r}(B_r - B_{u_2} + \xi_2)\right) \\ &= \mathbb{E}\left(p'_{t-r+u_2-u_1}(B_r - B_{u_2} + \xi_1) p'_{t-r}(B_r - B_{u_2} + \xi_2)\right) \\ &= \int_{\mathbb{R}} p_{r-u_2}(z) p'_{t-r+u_2-u_1}(z + \xi_1) p'_{t-r}(z + \xi_2) dz \\ &\le \|p_{r-u_2}\|_{p_1} \|p'_{t-r+u_2-u_1}\|_{p_2} \|p'_{t-r}\|_{p_3},\end{aligned}$$

where $\frac{1}{p_1} + \frac{1}{p_2} + \frac{1}{p_3} = 1$. It is easy to see that

$$\begin{aligned} \|p_{r-u_2}\|_{p_1} &\le C(r-u_2)^{-\frac{1}{2}+\frac{1}{2p_1}}, \\ \|p'_{t-r+u_2-u_1}\|_{p_2} &\le C(t-r+u_2-u_1)^{-1+\frac{1}{2p_2}} \le C(u_2-u_1)^{-1+\frac{1}{2p_2}}, \\ \|p'_{t-r}\|_{p_3} &\le C(t-r)^{-1+\frac{1}{2p_3}}, \end{aligned}$$

for some constant $C > 0$. Thus

$$\begin{aligned} \mathbb{E}\left(|\hat{u}_{t,h}(r)|^2\right) &\le C \int_0^r \int_0^{u_2} \int_0^h \int_0^h \int_{-\eta_1}^{\eta_1} \int_{-\eta_2}^{\eta_2} (r-u_2)^{-\frac{1}{2}+\frac{1}{2p_1}} \\ &\quad \times (u_2-u_1)^{-1+\frac{1}{2p_2}} (t-r)^{-1+\frac{1}{2p_3}} d\xi_1 d\xi_2 d\eta_1 d\eta_2 du_1 du_2 \\ &\le C\, h^4 . \end{aligned}$$

This proves (9.5.10).

Step 3. Taking into account inequality (9.5.10) and Equation (9.5.9), in order to show Theorem 9.10 it suffices to show the following convergence in law:

$$h^{-\frac{3}{2}} \int_0^t \left(\int_0^r \left(I_{[0,h]}(B_r - B_u) - I_{[0,h]}(B_u - B_r)\right) du \right) dB_r \xrightarrow{\mathcal{L}} 2\eta \sqrt{\frac{\pounds_t}{3}} ,$$

where η is a standard normal random variable independent of B, and $\pounds_t$ has been defined in (9.5.4). Notice that

$$M_t^h = h^{-\frac{3}{2}} \int_0^t \left(\int_0^r \left(I_{[0,h]}(B_r - B_u) - I_{[0,h]}(B_u - B_r)\right) du \right) dB_r$$

is a martingale with quadratic variation

$$\left\langle M^h \right\rangle_t = h^{-3} \int_0^t \left(\int_0^r \left(I_{[0,h]}(B_r - B_u) - I_{[0,h]}(B_u - B_r)\right) du \right)^2 dr.$$

From the asymptotic version of Knight's theorem it suffices to show the following convergences in probability.

$$h^{-3} \int_0^t \left(\int_0^r \left(I_{[0,h]}(B_r - B_u) - I_{[0,h]}(B_u - B_r)\right) du \right)^2 dr \to \frac{4}{3} \pounds_t, \quad (9.5.11)$$

and

$$\left\langle M^h, B \right\rangle_t = h^{-3/2} \int_0^t \int_0^r \left(I_{[0,h]}(B_r - B_u) - I_{[0,h]}(B_u - B_r)\right) dudr \to 0, \quad (9.5.12)$$

as h tends to zero. In fact, let B^h be the Brownian motion such that $M_t^h = B^h_{\langle M^h \rangle_t}$. Then, from Theorem 2.3, p. 524 in [Revuz and Yor (1999)], and the convergences (9.5.11) and (9.5.12), we deduce that $(B, B^h, \langle M^h \rangle_t)$

converges in distribution to $(B, \beta, \frac{4}{3}\alpha_t)$, where β is a Brownian motion independent of B. This implies that $M_t^h = B^h_{\langle M^h \rangle_t}$ converges in distribution to $\beta_{\frac{4}{3}\alpha_t}$, which yields the desired result.

Step 4. Convergences (9.5.11) and (9.5.12) will be proved in the following two additional steps. The convergence (9.5.12) follows from the following statement.

For any $t \geq 0$, $\langle M^h, B \rangle_t$ converges to zero in L^2 as h tends to zero.

To prove the above statement, we fix $\varepsilon > 0$. The quadratic covariation $\langle M^h, B \rangle_t$ is the limit, almost surely and in L^2, of

$$\Phi_{\varepsilon,h} = h^{-3/2} \int_0^t \int_0^r \int_0^h \int_{-\eta}^{\eta} p_\varepsilon'(B_r - B_u + \xi) d\xi d\eta dr$$

as ε tends to zero. We have

$$\mathbb{E}\left(\Phi_{\varepsilon,h}^2\right) = h^{-3} \int_0^t \int_0^t \int_0^{r_1} \int_0^{r_2} \int_0^h \int_0^h \int_{-\eta_1}^{\eta_1} \int_{-\eta_2}^{\eta_2} \mathbb{E}\big(p_\varepsilon'(B_{r_1} - B_{u_1} + \xi_1)$$
$$p_\varepsilon'(B_{r_2} - B_{u_2} + \xi_2)\big) d\xi_1 d\xi_2 d\eta_1 d\eta_2 du_1 du_2 dr_1 dr_2.$$

We may assume that $0 < r_1 < r_2 < t$, and we divide the region of integration into three disjoint parts:

Case 1: $0 < u_1 < r_1 < u_2 < r_2 < t$. In this situation

$$\mathbb{E}\left[p_\varepsilon'(B_{r_1} - B_{u_1} + \xi_1) p_\varepsilon'(B_{r_2} - B_{u_2} + \xi_2)\right] = p'_{r_1 - u_1 + \varepsilon}(\xi_1) p'_{r_2 - u_2 + \varepsilon}(\xi_2),$$

which is an odd function of ξ_1 and ξ_2. Integration with respect to ξ_1 from $-\eta_1$ to η_1 (or with respect to ξ_2 from $-\eta_2$ to η_2) yields 0.

Case 2: $0 < u_1 < u_2 < r_1 < r_2 < t$. In this case,

$$\begin{aligned}
&\mathbb{E}\left[p_\varepsilon'(B_{r_1} - B_{u_1} + \xi_1) p_\varepsilon'(B_{r_2} - B_{u_2} + \xi_2)\right] \\
&= \mathbb{E}\left[p_\varepsilon'(B_{r_1} - B_{u_2} + B_{u_2} - B_{u_1} + \xi_1) p_\varepsilon'(B_{r_2} - B_{r_1} + B_{r_1} - B_{u_2} + \xi_2)\right] \\
&= \mathbb{E}\left[p'_{u_2 - u_1 + \varepsilon}(B_{r_1} - B_{u_2} + \xi_1) p'_{r_2 - r_1 + \varepsilon}(B_{r_1} - B_{u_2} + \xi_2)\right] \\
&= \int_{\mathbb{R}} p_{r_1 - u_2}(z) p'_{u_2 - u_1 + \varepsilon}(z + \xi_1) p'_{r_2 - r_1 + \varepsilon}(z + \xi_2) dz \,.
\end{aligned}$$

The technique used in the proof of the inequality (9.5.10) can be applied to show that in this situation, we have

$$\begin{aligned}
&h^{-3} \int_{0<u_1<u_2<r_1<r_2<t} \int_0^h \int_0^h \int_{-\eta_1}^{\eta_1} \int_{-\eta_2}^{\eta_2} \int_{\mathbb{R}} p_{r_1 - u_2}(z) \\
&\quad \times \left| p'_{u_2 - u_1 + \varepsilon}(z + \xi_1) p'_{r_2 - r_1 + \varepsilon}(z + \xi_2) \right| dz d\xi_1 d\xi_2 d\eta_1 d\eta_2 du_1 du_2 dr_1 dr_2 \\
&\leq C t^2 h \,.
\end{aligned}$$

Case 3: $0 < u_2 < u_1 < r_1 < r_2 < t$. In this case,

$$\begin{aligned}
&\mathbb{E}\left[p'_\varepsilon(B_{r_1} - B_{u_1} + \xi_1)p'_\varepsilon(B_{r_2} - B_{u_2} + \xi_2)\right] \\
&= \mathbb{E}\left[p'_\varepsilon(B_{r_1} - B_{u_1} + \xi_1)p'_\varepsilon(B_{r_2} - B_{r_1} + B_{r_1} - B_{u_1} + B_{u_1} - B_{u_2} + \xi_2)\right] \\
&= \mathbb{E}\left[p'_\varepsilon(B_{r_1} - B_{u_1} + \xi_1)p'_{r_2-r_1+u_1-u_2+\varepsilon}(B_{r_1} - B_{u_1} + \xi_2)\right] \\
&= \int_{\mathbb{R}} p_{r_1-u_2}(z)p'_\varepsilon(z+\xi_1)p'_{r_2-r_1+u_1-u_2+\varepsilon}(z+\xi_2)dz.
\end{aligned}$$

Hence, in this case we have

$$\begin{aligned}
&h^{-3}\int_{0<u_2<u_1<r_1<r_2<t}\int_0^h\int_0^h\int_{-\eta_1}^{\eta_1}\int_{-\eta_2}^{\eta_2}\int_{\mathbb{R}} p_{r_1-u_2}(z) \\
&\times p'_\varepsilon(z+\xi_1)p'_{r_2-r_1+u_1-u_2+\varepsilon}(z+\xi_2)dzd\xi_1d\xi_2d\eta_1d\eta_2du_1du_2dr_1dr_2 \\
&= h^{-3}\int_{0<u_2<u_1<r_1<r_2<t}\int_0^h\int_0^h\int_{\mathbb{R}} p_{r_1-u_2}(z)\left[p_\varepsilon(z+\eta_1) - p_\varepsilon(z-\eta_1)\right] \\
&\times\left[p_{r_2-r_1+u_1-u_2+\varepsilon}(z+\eta_2) - p_{r_2-r_1+u_1-u_2+\varepsilon}(z-\eta_2)\right]dzd\eta_1d\eta_2du_1du_2dr_1dr_2.
\end{aligned}$$

As ε tends to zero, this converges to

$$\begin{aligned}
A(h) &:= 2h^{-3}\int_{0<u_2<u_1<r_1<r_2<t}\int_0^h\int_0^h p_{r_1-u_2}(\eta_1) \\
&\quad\times\left[p_{r_2-r_1+u_1-u_2}(\eta_1-\eta_2) - p_{r_2-r_1+u_1-u_2}(\eta_1+\eta_2)\right]d\eta_1d\eta_2du_1du_2dr_1dr_2 \\
&= -2h^{-3}\int_{0<u_2<u_1<r_1<r_2<t}\int_0^h\int_0^h\int_{-\eta_2}^{\eta_2} p_{r_1-u_2}(\eta_1) \\
&\quad\times p'_{r_2-r_1+u_1-u_2}(\eta_1+\xi)d\xi d\eta_1d\eta_2du_1du_2dr_1dr_2.
\end{aligned}$$

Notice that

$$\begin{aligned}
&2h^{-3}\int_0^h\int_0^h\int_{-\eta_2}^{\eta_2} p_{r_1-u_2}(\eta_1)\left|p'_{r_2-r_1+u_1-u_2}(\eta_1+\xi)\right|d\xi d\eta_1d\eta_2 \\
&\le Ch^{-3}\int_0^h\int_0^h\int_{-\eta_2}^{\eta_2}(r_1-u_2)^{-\frac{1}{2}}(r_2-r_1+u_1-u_2)^{-1}d\xi d\eta_1d\eta_2 \\
&\le C(r_1-u_2)^{-\frac{1}{2}}(r_2-r_1+u_1-u_2)^{-1},
\end{aligned}$$

which is integrable in the region $\{0 < u_2 < u_1 < r_1 < r_2 < t\}$. On the other hand,

$$\begin{aligned}
&\lim_{h\downarrow 0} 2h^{-3}\int_0^h\int_0^h\int_{-\eta_2}^{\eta_2} p_{r_1-u_2}(\eta_1)p'_{r_2-r_1+u_1-u_2}(\eta_1+\xi)d\xi d\eta_1d\eta_2 \\
&= 2p_{r_1-u_2}(0)p'_{r_2-r_1+u_1-u_2}(0) = 0,
\end{aligned}$$

and, by the dominated convergence theorem, we deduce that $A(h)$ converges to zero as h tends to zero. This completes the proof of (9.5.12).

Step 5. The last task is to show the convergence (9.5.11) in probability. Define

$$\Psi_h(r) = \int_0^r \left(I_{[0,h]}(B_r - B_u) - I_{[0,h]}(B_u - B_r)\right) du.$$

We are going to show that

$$h^{-3} \int_0^t \Psi_h(r)^2 dr \xrightarrow{L^2} \frac{4}{3}\pounds_t, \tag{9.5.13}$$

as h tends to zero. Notice that

$$\alpha_t = 2\int_0^t \int_0^v \delta_0(B_v - B_u) du dv$$

is the self-intersection local time of B, and Equation (9.5.13) provides an approximation for this self-intersection local time which has its own interest.

By the occupation formula for the Brownian motion we can write

$$\begin{aligned}\Psi_h(r) &= \int_{\mathbb{R}} \left(I_{[0,h]}(B_r - x) - I_{[0,h]}(x - B_r)\right) L_r(x) dx \\ &= \int_0^h \left(L_r(B_r - y) - L_r(B_r + y)\right) dy.\end{aligned}$$

We can express the difference $L_r(B_r - y) - L_r(B_r + y)$ by means of Tanaka's formula for the Brownian motion $\{B_r - B_s, 0 \le s \le r\}$:

$$\begin{aligned}L_r(B_r - y) - L_r(B_r + y) &= -y - (B_r - y)^+ + (B_r + y)^+ \\ &\quad + \int_0^r \left(I_{B_r - B_s + y > 0} - I_{B_r - B_s - y > 0}\right) d\widehat{B}_s,\end{aligned}$$

where $d\widehat{B}_s$ denote the backward stochastic Itô integral and $y > 0$. Integrating in the variable y yields

$$\begin{aligned}\Psi_h(r) &= -\frac{h^2}{2} + \int_0^h \left[(B_r + y)^+ - (B_r - y)^+\right] dy \\ &\quad + \int_0^h \left(\int_0^r I_{y > |B_r - B_s|} d\widehat{B}_s\right) dy.\end{aligned} \tag{9.5.14}$$

By stochastic Fubini's theorem

$$\int_0^h \left(\int_0^r I_{y > |B_r - B_s|} d\widehat{B}_s\right) dy = \int_0^r (h - |B_r - B_s|)^+ d\widehat{B}_s. \tag{9.5.15}$$

Taking into account (9.5.14) and (9.5.15), the convergence (9.5.13) will follow from

$$h^{-3} \int_0^t \left(\int_0^r (h - |B_r - B_s|)^+ d\widehat{B}_s\right)^2 dr \xrightarrow{L^2} \frac{4}{3}\alpha_t, \tag{9.5.16}$$

as h tends to zero. By Itô's formula we can write

$$\left(\int_0^r (h-|B_r-B_s|)^+ d\widehat{B}_s\right)^2 = 2\int_0^r \left(\int_s^r (h-|B_r-B_u|)^+ d\widehat{B}_u\right)$$
$$\times (h-|B_r-B_s|)^+ d\widehat{B}_s + \int_0^r \left[(h-|B_r-B_s|)^+\right]^2 ds. \qquad (9.5.17)$$

Finally, (9.5.16) follows from (9.5.17) and Lemmas 9.6 and 9.7. □

Remark 9.3. For more work related to this theorem we refer to [Chen, Li, Marcus and Rosen (2010)], [Hu and Nualart (2009)], [Hu and Nualart (2010)], [Rosen (2006)], and [Rosen (2011)].

Chapter 10

Stochastic Differential Equation

There are many papers and books on stochastic differential equations (sde). In this and next sections we are also concerned with stochastic differential equations. We view the solution of an sde as a nonlinear functional on the canonical Wiener space. We shall be concerned with existence and uniqueness, non-explosion, exponential integrability, some inequality associated with the marginal distribution, chaos expansion, and existence of density of the solution. In next chapter, we shall concentrate on the approximation of the solution (by time discretization scheme or by Wong-Zakai approximation).

10.1 Existence, Uniqueness and Non-explosion

Lemma 10.1. *Let $\{g_n(t)\,,\ n=1,2,\cdots\}$ be a sequence of positive continuous functions and let a and b be two positive constants.*

(i) If $g_n(t)$ satisfies

$$g_{n+1}(t) \le b + a\int_0^t g_n(s)ds \quad \forall\, 0\le t\le T\,,\ \text{and}\ \ \forall\, n=1,2,\cdots\,, \tag{10.1.1}$$

then

$$g_n(t) \le Ke^{at} \quad \forall\, 0\le t\le T\,,\ n\ge 1\,, \tag{10.1.2}$$

where $K=\max\{b, \max_{0\le t\le T} g_1(t)\}$.

(ii) If $g_n(t)$ satisfies

$$g_{n+1}(t) \le a\int_0^t g_n(s)ds \quad \forall\, 0\le t\le T\,,\ \text{and}\ \ \forall\, n=1,2,\cdots\,, \tag{10.1.3}$$

then

$$g_n(t) \le \frac{(at)^{n-1}}{(n-1)!}\sup_{0\le t\le T}|g_1(t)| \quad \forall\, 0\le t\le T\,,\ n\ge 1\,. \tag{10.1.4}$$

Proof. We shall prove (10.1.2) by induction on n. The fact that the inequality is true for $n = 1$ follows from our definition of K. Assume that the inequality (10.1.2) holds true for n. Then by (10.1.1), we have

$$\begin{aligned} g_{n+1}(t) &\le b + a\int_0^t g_n(s)ds \le K + a\int_0^t Ke^{as}ds \\ &= Ke^{at}. \end{aligned}$$

This means that (10.1.2) holds true for $n+1$, proving part (i) of the lemma.

The part (ii) can be proved in the same way. □

Lemma 10.2. *Let a and b be two positive numbers and let $\beta \in (0,1)$. If a positive continuous function g_t, $t \in [0,T]$ satisfies*

$$g_t \le a + b\int_0^t g_s^\beta ds\,, \quad \forall\, t \in [0,T]\,, \tag{10.1.5}$$

then

$$g_t \le (a^{1-\beta} + (1-\beta)bt)^{\frac{1}{1-\beta}}\,. \tag{10.1.6}$$

Proof. Denote $h_t = a + b\int_0^t g_s^\beta ds$. Then, $\dot h_t = bg_t^\beta$. Condition (10.1.5) implies that $\dot h_t / h_t^\beta \le b$. Integrating this inequality and noticing $h_0 = a$ yield

$$\frac{1}{1-\beta}h_t^{1-\beta} \le bt + \frac{a^{1-\beta}}{1-\beta}\,,$$

which gives (10.1.6). □

Let us assume that $(\mathbb{B}, \mathbb{H}, \mathbb{P})$ is m-dimensional canonical Wiener space and let $W = (W^1, \cdots, W^m)$ be the m-dimensional standard Brownian motion. The filtration generated by W is denoted by $\mathcal{F}_t := \sigma(W_s, 0 \le s \le t)$.

Let $\sigma_j : [0,T] \times \mathbb{B} \times \mathbb{R}^d \to \mathbb{R}^d$, $j = 0, 1, \cdots, m$ satisfy the following conditions.

(1) For any fixed $(t,\omega) \in [0,T] \times \mathbb{B}$, $\sigma_j(t,\omega,\cdot)$ is uniformly (in t and ω) global Lipschitz. Namely, there is a positive constant K such that

$$|\sigma_j(t,\omega,x) - \sigma_j(t,\omega,y)| \le K|x-y| \quad \forall\, x, y \in \mathbb{R}^d\,, \quad j = 0, 1, \cdots, m\,. \tag{10.1.7}$$

(2) For any fixed $x \in \mathbb{R}^d$, $\sigma(\cdot,\cdot,x)$ is an $\mathcal{F}_t$ adapted stochastic process satisfying

$$\sup_{0\le t\le T} \mathbb{E}|\sigma_j(t,\omega,0)|^p \le K_p \quad \forall\, j = 0, 1, \cdots, m\,. \tag{10.1.8}$$

Consider the following stochastic differential equation

$$Y(t) = \alpha(t) + \sum_{j=0}^{m} \int_0^t \sigma_j(s, Y(s))dW_s^j\,, \tag{10.1.9}$$

where we denote $W_t^0 = t$.

Theorem 10.1. *Let $\alpha(t)$ be an m-dimensional adapted process such that $\alpha_p = \sup_{0\le t\le T} \mathbb{E}|\alpha(t)|^p \le \alpha_p$ for any $p \ge 1$. Then, equation (10.1.9) has a unique continuous and adapted solution $Y(t)$. Moreover,*

$$\mathbb{E}\left(\sup_{0\le t\le T} |Y(t)|^p\right) \le C < \infty \tag{10.1.10}$$

for all $p \ge 2$.

Remark 10.1. If we denote $\mathbf{L}^{p,T}$ the space of all continuous and $\mathcal{F}_t$-adapted processes $X(t)$ such that

$$\|X\|_{p,T} := \left(\mathbb{E} \sup_{0\le t\le T} |X(t)|^p\right)^{1/p} < \infty\,, \tag{10.1.11}$$

then $\mathbf{L}^{p,T}$ is a Banach space.

Proof. We use the standard technique of Picard iteration. Define $Y_0(t) = \alpha(t)$ and

$$Y_{n+1}(t) = \alpha(t) + \sum_{j=0}^{m} \int_0^t \sigma_j(s, Y_n(s))dW_s^j\,, \tag{10.1.12}$$

First, we show that

$$\sup_{n\ge 0} \mathbb{E}\left(\sup_{0\le t\le T} |Y_n(t)|^p\right) < \infty\,. \tag{10.1.13}$$

To show this we shall use

$$(a_1 + \cdots + a_k)^p \le k^p(a_1^p + \cdots + a_k^p) \quad \forall\; a_1, \cdots, a_k \ge 0\,,\;\; p \ge 1\,.$$

From the above inequality and from the Burkholder-Davis-Gundy inequality, it follows that for any $p \geq 2$ and for any $0 \leq r \leq T$,

$$\begin{aligned}
\mathbb{E}\left(\sup_{0\leq t\leq r}|Y_{n+1}(t)|^p\right) &\leq (m+1)^p\left\{\alpha_p + \left(\int_0^r \mathbb{E}|\sigma_0(s,Y_n(s)|ds\right)^p\right. \\
&\qquad \left. + \sum_{j=1}^m \mathbb{E}\left(\sup_{0\leq t\leq r}\left|\int_0^t \sigma_j(s,Y_n(s))dW_s^j\right|^p\right)\right\} \\
&\leq C_{p,T}\left(1 + \int_0^r \mathbb{E}|\sigma_0(s,Y_n(s)|^p ds\right) \\
&\qquad \left. + \sum_{j=1}^m \mathbb{E}\left[\int_0^r |\sigma_j(s,Y_n(s))|^2\, ds\right]^{p/2}\right\} \\
&\leq C_{p,T}\left\{1 + \sum_{j=0}^m \int_0^r \mathbb{E}\,|\sigma_j(s,Y_n(s))|^p\, ds\right\}.
\end{aligned}$$

Using the global Lipschitz condition (10.1.7) and the condition (10.1.8) we see that

$$\begin{aligned}
\mathbb{E}\left(\sup_{0\leq t\leq r}|Y_{n+1}(t)|^p\right) &\leq C_{p,T}\left\{1 + \int_0^r \mathbb{E}\,|Y_n(s)|^p\, ds\right\} \\
&\leq C_{p,T}\left\{1 + \int_0^r \mathbb{E}\sup_{0\leq u\leq s}|Y_n(u)|^p\, ds\right\} \quad \forall\, 0 \leq r \leq T.
\end{aligned}$$

Then, the part (i) of Lemma 10.1 shows (10.1.13).

Exactly following the same argument, we can obtain easily

$$\mathbb{E}\left(\sup_{0\leq t\leq r}|Y_{n+1}(t)-Y_n(t)|^p\right) \leq C_{p,T}\int_0^r \mathbb{E}\left(\sup_{0\leq t\leq s}|Y_n(t)-Y_{n-1}(t)|^p\right) ds. \tag{10.1.14}$$

The part (ii) of Lemma 10.1 yields

$$\mathbb{E}\left(\sup_{0\leq t\leq T}|Y_{n+1}(t)-Y_n(t)|^p\right) \leq \frac{C_{p,T}^n}{n!}\mathbb{E}\left(\sup_{0\leq t\leq T}|Y_1(t)-Y_0(t)|^p\right).$$

Consequently, we have

$$\sum_{n=0}^\infty \mathbb{E}\left(\sup_{0\leq t\leq T}|Y_{n+1}(t)-Y_n(t)|^p\right) < \infty.$$

This implies that $\{Y_n(t)\}$ is a Cauchy sequence in $\mathbf{L}^{p,T}$. It is easy to show that its limit $Y(t)$ in $\mathbf{L}^{p,T}$ satisfies (10.1.9).

If Y_1 and Y_2 are two solutions to (10.1.9), then in exactly the same way as for (10.1.14), we can obtain

$$\mathbb{E}\left(\sup_{0\leq t\leq r}|Y_2(t)-Y_2(t)|^p\right) \leq C_{p,T}\int_0^r \mathbb{E}\left(\sup_{0\leq t\leq s}|Y_2(t)-Y_1(t)|^p\right) ds. \tag{10.1.15}$$

The usual Gronwall lemma yields that $Y_1(t) = Y_2(t)$ for all $0 \le t \le T$. This is the uniqueness. $\square$

Let us assume that $A_j : [0,T] \times \mathbb{R}^d \to \mathbb{R}^d$, $j = 0, 1, \cdots, m$ are measurable functions satisfying the following uniformly (in t) global Lipschitz condition.

(H1) $\sup_{0\le j\le m} |A_j(t,x) - A_j(t,y)| \le K|x-y| \quad \forall\, t \in [0,T]\,,\ x, y \in \mathbb{R}^d$.
(H2) $\sup_{0\le j\le m} \sup_{0\le t\le r} |A_j(t,0)| < \infty$.

Consider the following stochastic differential equation

$$X(t) = x + \sum_{j=0}^m \int_0^t A_j(s, X(s))dW_s^j\,, \quad 0 \le t \le T\,. \tag{10.1.16}$$

A direct consequence of Theorem 10.1 is

Proposition 10.1. *Assume that* $A_j : [0,T] \times \mathbb{R}^d \to \mathbb{R}^d$ *satisfies the above conditions (H1) and (H2). Then, equation* (10.1.16) *has a unique solution* X. *Moreover, the solution* X *satisfies*

$$\mathbb{E}\left(\sup_{0\le t\le T} |X(s)|^p\right) \le C < \infty\,. \tag{10.1.17}$$

For the stochastic differential equation (10.1.16), we introduce for any function $f \in C^{1,2}(\mathbb{R}_+ \times \mathbb{R}^d, \mathbb{R})$ (the set of all functions $f(t,x)$ which is continuously differentiable with t and twice continuously differentiable with respect to x)

$$A(t,x) = (a_{ij}(t,x))_{1\le i,j\le d} = \sum_{\ell=1}^m \sigma_\ell(t,x)\sigma_\ell(t,x)^T\,;$$

$$\mathcal{L}f(t,x) = \frac{\partial}{\partial t} f(t,x) + \sum_{i=1}^d \langle \sigma_0(t,x)\,, \nabla f(t,x)\rangle + \frac{1}{2}\mathrm{Tr}\left[A(t,x)\nabla^2 f(t,x)\right]\,. \tag{10.1.18}$$

With these notations, we have the following Itô formula. For any function $f \in C^{1,2}(\mathbb{R}_+ \times \mathbb{R}^d, \mathbb{R})$ and $0 \le \rho < t < \infty$,

$$\begin{aligned} f(t, Y(t)) &= f(\rho, X_\rho) + \int_\rho^t \mathcal{L}f(s, Y(s))ds \\ &\quad + \sum_{\ell=1}^m \int_\rho^t \langle \sigma_\ell(s, Y(s)), \nabla f(s, Y(s))\rangle dW_s^\ell\,. \end{aligned} \tag{10.1.19}$$

Now we are going to consider stochastic differential equation which does not satisfy the global Lipschitz conditions. We shall replace the global Lipschitz condition (10.1.7) by the following local Lipschitz condition while the condition (10.1.8) remains the same.

Let $\sigma_j : [0,T] \times \mathbb{B} \times \mathbb{R}^d \to \mathbb{R}^d$, $j = 0, 1, \cdots, m$ satisfy that for any fixed $(t, \omega) \in [0,T] \times \mathbb{B}$, $\sigma_j(t, \omega, \cdot)$ is uniformly locally Lipschitz. Namely, there is a positive constant K_R such that

$$|\sigma_j(t,\omega,x) - \sigma_j(t,\omega,y)| \le K_R|x-y| \quad \forall\, j = 0, 1, \cdots, m\,, \text{and} \\ \forall\, x, y \in \mathbb{R}^d \text{ with } |x|, |y| \le R\,. \quad (10.1.20)$$

Theorem 10.2. *Let σ_j, $j = 0, 1, \cdots, m$ satisfy the local Lipschitz conditions* (10.1.20) *and* (10.1.8). *Assume there is a non-negatve (Lyapunov) function $V : \mathbb{R}_+ \times \mathbb{R}^d \to \mathbb{R}$ satisfying*

(i) $V \in C^{1,2}(\mathbb{R}_+ \times \mathbb{R}^d, \mathbb{R})$*;*
(ii) $\lim_{|x|\to\infty} \inf_{0\le t\le T} V(t,x) = \infty$ *for all* $T > 0$*;*
(iii) there is a constant c such that $\mathcal{L}V(t,x) \le cV(t,x)$ for all $t \in \mathbb{R}_+$ and $x \in \mathbb{R}^d$.

Then, the stochastic differential equation (10.1.9) *has a unique solution for all time $t \ge 0$.*

Proof. Let ϕ_R be a C^∞ function from $\mathbb{R}^d \to \mathbb{R}$ such that $\phi_R(x) = 1$ when $|x| \le R$ and $\phi_R(x) = 0$ when $|x| \ge R+1$. Then, $\sigma_j^R(t,x) = \sigma_j(t,x)\phi_R(x)$, $j = 0, 1, \cdots, m$, satisfy the global Lipschitz conditions (10.1.7) and (10.1.8). The corresponding equation (replace σ_j in (10.1.9) by σ_j^R) has a unique solution $Y_R(t)$. Define

$$\tau_R = \inf\{t > 0\,, Y_R(t) \in B(0,R)^c\}\,,$$

where $B(0,R)^c$ denotes the complement of the ball $B(0,R)$ of center 0 and radius R in $\mathbb{R}^d$. It is obvious that the stopping time τ_R is an (almost surely) increasing function of R and converges to a stopping time τ as $R \to \infty$. By the uniqueness we have that when $R \le R'$, then $Y_{R'}(t) = Y_R(t)$ for all $t \le \tau_R$. We can define the process $Y(t) = Y_R(t)$ if $t \le \tau_R$. Then, Y satisfies equation (10.1.16) when $t < \tau$. To show the solution exists for all $t \ge 0$ it suffices to show that $\tau_R \to \tau = \infty$ as $R \to \infty$. Denote $U(t,x) = V(t,x)e^{-ct}$ and denote $\mathcal{L}_R$ the operator $\mathcal{L}$ defined by (10.1.18) with σ_j replaced by σ_j^R.

By Itô's formula we have

$$\begin{aligned} e^{-ct}V(t, Y_R(t)) &= e^{-ct}V(0, Y_R(0)) \\ &+ \int_0^t e^{-cs}\left[\mathcal{L}_R V(s, Y_R(s)) - cV(s, Y_R(s))\right] ds \\ &+ \sum_{\ell=1}^m \int_0^t e^{-cs}\langle \sigma_\ell^R(s, Y_R(s)), \nabla V(s, Y_R(s))\rangle dW_s^j . \end{aligned} \tag{10.1.21}$$

Since when $|x| \le R$, $\mathcal{L}_R = \mathcal{L}$, we have by the condition (iii)

$$\begin{aligned} & e^{-c\tau_R}V(\tau_R, Y_R(\tau_R)) \\ &= V(0, Y_R(0)) + \int_0^{\tau_R} e^{-cs}\left[\mathcal{L}V(s, Y_R(s)) - cV(s, Y_R(s))\right] ds \\ &+ \sum_{\ell=1}^m \int_0^{\tau_R} e^{-cs}\langle \sigma_\ell^R(s, Y_R(s)), \nabla V(s, Y_R(s))\rangle dW_s^j \\ &\le V(0, Y_R(0)) + \sum_{\ell=1}^m \int_0^{\tau_R} e^{-cs}\langle \sigma_\ell^R(s, Y_R(s)), \nabla V(s, Y_R(s))\rangle dW_s^j . \end{aligned}$$

Taking the expectation yields

$$\mathbb{E}\left[e^{-c\tau_R}V(\tau_R, Y_R(\tau_R))\right] \le \mathbb{E}\left[V(0, Y_R(0))\right] . \tag{10.1.22}$$

Now we can show that for any positive integer m, $P(\tau \le m) = 0$. In fact, we have (denoting $A_m = \{\tau \le m\}$),

$$e^{-cm}\mathbb{E}\left[V(\tau_R, Y_R(\tau_R))I_{A_m}\right] \le \mathbb{E}\left[e^{-c\tau_R}V(\tau_R, Y_R(\tau_R))I_{A_m}\right] \le C < \infty .$$

Let $R \to \infty$. The condition (ii) implies that $P(A_m) = 0$. Thus, $P(\tau < \infty) = 0$. Or $P(\tau = \infty) = 1$.

If X and Y are two solutions to equation (10.1.16), then one can define $\tau_{X,R}$ and $\tau_{Y,R}$ as the exit times of the processes X and Y to the ball $B(0, R)$. Clearly, $\tau_R' = \tau_{X,R} \wedge \tau_{Y,R} \to \infty$ almost surely. But when $t \le \tau_R'$, $X(t) = Y(t)$ since on $\overline{B(0, R)}$ the coefficients are globally Lipschitz. The uniqueness then follows. □

10.2 Hörmander Theorem

Let us return to consider the following stochastic differential equation

$$X(t) = x + \sum_{j=0}^m \int_0^t A_j(s, X(s))dW_s^j , \quad 0 \le t \le T . \tag{10.2.1}$$

We always view $A_j(t, x) = \left(A_j^1(t, x), \cdots, A_j^d(t, x)\right)^T$ as a column vector.

Theorem 10.3. *Let $A_j : [0,T] \times \mathbb{R}^d \to \mathbb{R}^d$ be globally Lipschitz and satisfy the linear growth condition. Then, the solution $X(t)$ to (10.2.1) belongs to $\mathbb{D}_{1,\infty}$ for any $t \in [0,T]$. Moreover,*

$$\sup_{0\le r\le T} \mathbb{E}\left(\sup_{r\le s\le T} |D_r^{(\ell)} X(s)|^p\right) < \infty\,, \tag{10.2.2}$$

where $D_r^{(\ell)}$ denotes the derivative with respect to the ℓ-th Brownian motion, and it satisfies the following linear stochastic differential equation with random coefficients:

$$D_r^{(\ell)} X(t) = A_\ell(r, X(r)) + \sum_{j=0}^m \int_r^t \bar{A}_j(s) D_r^{(\ell)} X(s) dW_s^j\,, \quad t \ge r\,, \tag{10.2.3}$$

where $\bar{A}_j$, $j = 0, 1, \cdots, m$, are $d \times d$ uniformly bounded and $\mathcal{F}_t$-adapted matrix-valued processes. Moreover, if A_j are continuously differentiable, then

$$\begin{aligned} \bar{A}_j(s) &= \partial A_j(s, X(s)) = (\partial_1 A_j(s, X(s)), \cdots, \partial_d A_j(s, X(s))) \\ &= \left(A_{k,\ell}^i(s)\right)_{1\le i,k\le d}, \end{aligned} \tag{10.2.4}$$

with $A_{k,\ell}^i(s) = (\partial_k A_j^i)(s, X(s))^T$. If $t < r$, then $D_r^{(\ell)} X(t) = 0$.

Proof. Let us consider the Picard iteration given by

$$\begin{aligned} X_0(t) &= x \\ X_{n+1} &= x + \sum_{j=0}^m \int_0^t A_j(s, X_n(s)) dW_s^j\,. \end{aligned} \tag{10.2.5}$$

From the proof of Theorem 10.1, applied to our special deterministic case, we obtain

$$\mathbb{E}\left(\sup_{0\le t\le T} |X_n(t) - X(t)|^p\right) \to 0\,.$$

By Theorem 7.6, to show $X(t) \in \mathbb{D}_{1,\infty}$ and to prove (10.2.2), we only need to show the following

$$X_n(t) \in \mathbb{D}_{1,\infty} \text{ for all } n = 0, 1, \cdots \text{ and } t \in [0,T]\,; \tag{10.2.6}$$

$$\psi_n := \mathbb{E}\left(\sup_{0\le r\le t\le T} |D_r^{(\ell)} X_n(t)|^p\right) < \infty\,; \tag{10.2.7}$$

$$\psi_{n+1} \le c_1 + c_2 \int_0^t \psi_n(s) ds \tag{10.2.8}$$

for some constants c_1 and c_2. It is clear that (10.2.6)-(10.2.7) hold true for $n=0$. Now assume they hold true for n and we want to show that they also hold true for $n+1$. Since A_j are globally Lipschitz we know by Proposition 6.6 there are matrix valued adapted processes $\bar{A}_j(s)$ uniformly bounded by a constant such that

$$D_r^{(\ell)} A_j(s, X_n(s)) = \bar{A}_j(s) D_r^{(\ell)} X_n(s) I_{\{r \le s\}}\,.$$

This can be used to show (10.2.6) easily through the iteration formula (10.2.5) and through the property of taking the derivative of stochastic integrals. Applying $D_r^{(\ell)}$ to (10.2.5) yields

$$D_r^{(\ell)} X_{n+1}(t) = A_\ell(r, X_n(r)) + \sum_{j=0}^m \int_r^t \bar{A}_j(s) D_r^{(\ell)} X_n(s) dW_s^j\,, \quad t \ge r\,. \tag{10.2.9}$$

Since $X_n(r)$ is in $\mathbf{L}^{p,T}$ for any $p \ge 1$ and $|A_j(r, X_n(r))| \le C(1+|X_n(r)|)$. This shows that $\mathbb{E}\sup_{0\le r\le T} |A_\ell(r, X_n(r))|^p < \infty$ for all $p \ge 1$. This combined with the fact that $\bar{A}_j(s)$ is uniformly bounded by a constant and the identity (10.2.9) proves (10.2.7) and (10.2.8).

As long as we know that $X(s) \in \mathbb{D}_{1,\infty}$ and that (10.2.2) holds true, we just apply $D_r^{(\ell)}$ to the equation to obtain (10.2.3).

Using the usual chain rule we see that when A_j are continuously differentiable then $\bar{A}_j$ are given by the derivatives of A_j, which yields (10.2.4). □

Similar to Theorem 10.3 we have the following result for (10.1.9).

Theorem 10.4. *Let $\sigma_j : [0,T] \times \mathbb{B} \times \mathbb{R}^d \to \mathbb{R}^d$, $j = 0, 1, \cdots, m$ satisfy the global Lipschitz condition* (10.1.7) *and the linear growth condition* (10.1.8). *[We omit the explicit dependence of σ_j on $\omega \in \mathbb{B}$]. In addition, we assume the following conditions.*

(i) For any fixed $(t,x) \in [0,T] \times \mathbb{R}^d$, $\sigma_j(t,x) \in \mathbb{D}_{1,\infty}$ and for any $\ell = 1, \cdots, m$,

$$\sup_{r \le s \le T} |D_r^{(\ell)} \sigma_j(s,x)| \le \tilde{K}(1+|x|)^q\,, \tag{10.2.10}$$

where q is a positive number and $\tilde{K}$ is a random variable satisfying that for any $p \ge 1$,

$$\mathbb{E}|\tilde{K}|^p < \infty\,. \tag{10.2.11}$$

(ii) $(\alpha(t)\,, t\in[0,T]\}$ *is* $\mathcal{F}_t$*-adapted and* $\alpha(t)\in\mathbb{D}_{1,\infty}$*. Moreover,*

$$\mathbb{E}\left(\sup_{0\le t\le T}|\alpha(t)|^p\right)+\sup_{0\le\ell\le m,0\le r\le T}\mathbb{E}\left(\sup_{0\le t\le T}|D_r^{(\ell)}\alpha(t)|^p\right)<\infty\,. \tag{10.2.12}$$

Then, the solution $Y(t)$ *to (10.1.9) is in* $\mathbb{D}_{1,\infty}$ *and* $\left\{D_r^{(\ell)}Y(t)\,, r\le t\le T\right\}$ *satisfies the following linear stochastic differential equation (for any fixed* $r\in[0,T]$*) with random coefficients*

$$\begin{aligned} D_r^{(\ell)}Y(t) &= D_r^{(\ell)}\alpha(t)+\sigma_\ell(r,Y(r))+\sum_{j=0}^m\int_r^t(D_r^{(\ell)}\sigma_j)(s,Y(s))dW_s^j \\ &\quad+\sum_{j=0}^m\int_r^t\bar\sigma_j(s)D_r^{(\ell)}Y(s)dW_s^j\,,\quad r\le t\le T\,, \end{aligned} \tag{10.2.13}$$

where $\bar\sigma_j$, $\ell=0,1,\cdots,m$, *are* $d\times d$ *uniformly bounded and* $\mathcal{F}_t$*-adapted matrix-valued processes. If* $\sigma_j, j=0,1,\cdots,m$, *are continuously differentiable in* x, *then*

$$\bar\sigma_j(s)=\partial\sigma_j(s,X(s))=(\partial_1\sigma_j(s,X(s))\,,\cdots,\partial_d\sigma_j(s,X(s)))^T \tag{10.2.14}$$

with the meaning that $\partial_k\sigma_j(s,x)=\frac{\partial}{\partial x_k}\sigma_j(s,x)$*. If* $t<r$*, then* $D_r^{(\alpha)}X(t)=0$*.*

Proof. The proof is similar to that of Theorem 10.3. We only point out that conditions (10.2.10), (10.2.11), and estimate (10.1.10) show that

$$\mathbb{E}\left(|D_r^{(\ell)}\sigma_j)(s,Y(s))|^p\right)<\infty\,.$$

This implies that the stochastic integral $\int_0^t(D_r^{(\ell)}\sigma_j)(s,Y(s))dW_s^j$ is well-defined and has needed properties. □

If $A_j(t,x)$ is continuously differentiable in x, then $D_r^{(\ell)}X(t)$ satisfies

$$D_r^{(\ell)}X(t)=A_\ell(r,X(r))+\sum_{j=0}^m\int_r^t\partial A_j(s,X(s))D_r^{(\ell)}X(s)dW_s^j\,,\quad t\ge r\,. \tag{10.2.15}$$

This is a linear stochastic differential equation in $D_r^{(\ell)}X(t), t\ge r$ with random coefficients (depending on $X(t)$). If $A_j(t,x)$ is twice continuously differentiable in x with bounded first and second order derivatives, then the coefficients $\alpha(r):=A_\alpha(r,X(r))$ and $\sigma_j(t,x)=\partial A_j(s,X(s))x$ satisfy the conditions in Theorem 10.4. Thus, we know by Theorem 10.4 that

$D_r^{(\ell)}X(t)$ is in $\mathbb{D}_{1,\infty}$ for any $\ell = 1, \cdots, m$. This means that $X(t) \in \mathbb{D}_{2,\infty}$. Moreover, we can prove with routine approach that

$$\sup_{0\le r_1, r_2\le T} \mathbb{E}\left(\sup_{r_1\vee r_2\le t\le T} |D_{r_1,r_2}^{(\ell_1,\ell_2)}X(t)|^p\right) < \infty .$$

We may also write the (linear) stochastic differential equation with random coefficients for $D_{r_1,r_2}^{(\ell_1,\ell_2)}X(t)$ which we omit here.

Continue this way, we can prove easily the following theorem.

Theorem 10.5. *Let n be a positive integer. Let $A_j : [0,T]\times\mathbb{R}^d \to \mathbb{R}^d$ be n times continuously differentiable in x and let all the derivatives in x from first order to the n-th order be bounded. Then, the solution $X(t)$ to (10.2.1) belongs to $\mathbb{D}_{n,\infty}$ for any $t\in[0,T]$. Moreover,*

$$\sup_{0\le k\le n}\ \sup_{0\le r_1,\cdots,r_k\le T} \mathbb{E}\left(\sup_{r_1\vee\cdots\vee r_k\le s\le T} |D_{r_1,\cdots,r_k}^{(\ell_1,\cdots,\ell_k)}X(s)|^p\right) < \infty . \qquad (10.2.16)$$

If $t < r_1\vee\cdots\vee r_k$, then $D_{r_1,\cdots,r_k}^{(\ell_1,\cdots,\ell_k)}X(t) = 0$.

The following lemma will be useful in the future, whose proof is straightforward consequence of the usual Itô formula.

Lemma 10.3. *Let $A_j(t)$ and $B_j(t)$, $j = 0, 1, \cdots, m$, be $\mathcal{F}_t$ adapted matrix valued processes such that*

$$\sup_{1\le j\le m} \mathbb{E} \sup_{0\le t\le T} \left(|A_j(t)|^2 + |B_j(t)|^2\right) + \mathbb{E}\sup_{0\le t\le T}\left(|A_0(t)| + B_0(t)|\right) < \infty ,$$

where $|A| := (\sum_{i,j=1}^d |a_{ij}|^2)^{1/2}$ is the Hilbert-Schmidt norm of the matrix A. Denote

$$Y(t) = Y(0) + \sum_{j=0}^m \int_0^t A_j(s)dW_s^j \quad \text{and} \quad Z(t) = Z(0) + \sum_{j=0}^m \int_0^t B_j(s)dW_s^j \qquad (10.2.17)$$

(recall that $W_s^0 = s$). Then, the following matrix valued product formula holds

$$\begin{aligned} Y(t)Z(t) &= Y(0)Z(0) + \sum_{j=0}^m \int_0^t Y(s)B_j(s)dW_s^j \\ &\quad + \sum_{j=0}^m \int_0^t A_j(s)Z(s)dW_s^j + \sum_{j=1}^m \int_0^t A_j(s)B_j(s)ds . \end{aligned} \qquad (10.2.18)$$

Remark 10.2. The above lemma can be written in differential form

$$d\left[Y(t)Z(t)\right] = Y(t)dZ(t) + \left[dY(t)\right]Z(t) + \sum_{j=1}^{m} A_j(t)B_j(t)dt\,. \qquad (10.2.19)$$

Let us assume the condition of Theorem 10.3. Then, the solution of (10.2.1) $X(t) \in \mathbb{D}_{1,\infty}$ and $D_r^{(\ell)}X(t)$ satisfies the linear stochastic differential equation (10.2.3). To solve this equation let us consider the following $\mathbb{R}^{d^2}$ matrix-valued linear stochastic differential equation

$$\Phi(t) = I + \sum_{j=0}^{m} \int_0^t \bar{A}_j(s)\Phi(s)dW_s^j\,, \quad t \geq 0 \qquad (10.2.20)$$

and another matrix-valued linear stochastic differential equation

$$\begin{aligned}\Psi(t) = I - \sum_{j=1}^{m} \int_0^t \Psi(s)\bar{A}_j(s)dW_s^j& \qquad (10.2.21)\\ - \int_0^t \Psi(s)\left[\bar{A}_0(s) - \sum_{j=1}^{m}\bar{A}_j(s)^2\right] ds\,,& \quad t \geq 0\,.\end{aligned}$$

Using the product rule (10.2.19) we obtain

$$\begin{aligned}d\left[\Psi(t)\Phi(t)\right] &= \sum_{j=1}^{m} \Psi(t)\bar{A}_j(t)\Phi(t)dW_t^j + \Psi(t)\bar{A}_0(t)\Phi(t)dt\\ &- \sum_{j=1}^{m} \Psi(t)\bar{A}_j(t)\Phi(t)dW_t^j - \Psi(t)\left[\bar{A}_0(t) - \sum_{j=1}^{m}\bar{A}_j(t)^2\right]\Phi(t)dt\\ &- \sum_{j=1}^{m} \Psi(t)\bar{A}_j(t)^2\Phi(t)d = 0\,.\end{aligned}$$

This implies that $\Psi(t)\Phi(t) = \Psi(0)\Phi(0) = I$. As a consequence, we see that the matrix $\Phi(t)$ is invertible and $\Phi^{-1}(t) = \Psi(t)$.

$\Phi(t)$ is called the fundamental solution for the equation (10.2.3). Consider the process

$$\xi(t) = \Phi(t)\Phi^{-1}(r)A_\ell(r, X(r))\,.$$

From (10.2.20) it follows easily that

$$d\xi(t) = \sum_{j=0}^{m} \bar{A}_j(t)\xi(t)dW_t^j\,.$$

Moreover, $\xi(r) = A_\ell(r, X(r))$. By the uniqueness of the solution we see that the solution to (10.2.3) can be represented by

$$D_r^{(\ell)} X(t) = \Phi(t)\Phi^{-1}(r)A_\ell(r, X(r))\,. \tag{10.2.22}$$

Consider the Malliavin covariance matrix

$$\begin{aligned} Q(t) &= \left(\langle DX^i(t), DX^j(t)\rangle_{\mathbb{H}}\right)_{1\le i,j\le d} \\ &= \left(\sum_{\ell=1}^m \int_0^t D_r^{(\ell)} X^i(t) D_r^{(\ell)} X^j(t) dr\right)_{1\le i,j\le d}. \end{aligned}$$

For two $\mathbb{H}$-valued $d \times d$ matrices A and B, we denote the contraction

$$A \diamond_{\mathbb{H}} B = \left(\sum_{k=1}^d \langle a_{ik}, b_{kj}\rangle_{\mathbb{H}}\right)_{1\le i,j\le d}.$$

We may also write

$$Q(t) = DX(t) \diamond_{\mathbb{H}} (DX(t))^T = \sum_{\ell=1}^m \int_0^t D_r^{(\ell)} X(t) \left(D_r^{(\ell)} X(t)\right)^T dr\,.$$

From the expression (10.2.22) we see that

$$Q(t) = \Phi(t)C(t)\Phi(t)^T\,. \tag{10.2.23}$$

The matrix $C(t)$ in the above expression is given by

$$\begin{aligned} C(t) &= \sum_{\ell=1}^m \int_0^t \Phi^{-1}(s)A_\ell(s, X(s))A_\ell^T(s, X(s)) \left(\Phi^{-1}(s)\right)^T ds \\ &= \int_0^t \Phi^{-1}(s)\Sigma(s) \left(\Phi^{-1}(s)\right)^T ds\,, \end{aligned} \tag{10.2.24}$$

where

$$\Sigma(s) = \Sigma(s, X(s)) = \sum_{\ell=1}^m A_\ell(s, X(s))A_\ell^T(s, X(s))\,. \tag{10.2.25}$$

Theorem 10.6. *Let $A_j, j = 0, 1, \cdots, m$ satisfy the conditions of Theorem 10.3 and let $\{X(t), t \in [0, T]\}$ be the solution to (10.2.1). Define*

$$\tau = \inf\left\{t > 0\,,\ \int_0^t I_{\{\det(\Sigma(s,X(s)))\}} ds > 0\right\} \wedge T\,. \tag{10.2.26}$$

Then, conditioned to $\{t > \tau\}$ the law of $X(t)$ is absolutely continuous with respect to the Lebesgue measure.

Proof. We need to verify the two conditions in Theorem 6.5. Theorem 10.3 implies the condition (i).

Now we are going to check the condition (ii). If $\tau < t$, then there exists a measurable set $G \subseteq [0,t]$ of positive Lebesgue measure and a function $\lambda(s) > 0$ such that

$$v^T\Sigma(s, X(s))v \geq \lambda(s)|v|^2 \quad \forall\, s \in G\,,\ v \in \mathbb{R}^d\,.$$

Taking $v = (\Phi^{-1}(s))^T u$ and integrating over G, we see that

$$u^T C(t) u = \int_0^t u^T\Phi^{-1}(s)\Sigma(s, X(s))(\Phi^{-1}(s))^T u ds \geq |u|^2 \int_G \lambda(s) ds > 0\,.$$

This means that $C(t)$ is almost surely invertible on $\{t > \tau\}$ and consequently, $Q(t)$ is is almost surely invertible on $\{t > \tau\}$. The theorem is proved by Theorem 6.5. □

We assume that $A_j : \mathbb{R}^d \to \mathbb{R}^d, j = 0, 1, \cdots, m$ are smooth and their derivatives of order greater than or equal to 1 are bounded by a constant independent of t (the constant may depend on the order). For our purpose in this section we shall consider the Stratonovich type of stochastic differential equation:

$$X(t) = x_0 + \sum_{j=0}^m \int_0^t A_j(X(s)) d^\circ W_s^j\,, \quad 0 \leq t \leq T\,, \tag{10.2.27}$$

where d° denotes the Stratonovich integral. This equation can be written as Itô type as

$$X(t) = x_0 + \int_0^t \tilde{A}_0(X(s) ds + \sum_{j=1}^m \int_0^t A_j(X(s)) dW_s^j\,, \quad 0 \leq t \leq T \tag{10.2.28}$$

with $\tilde{A}_0(x) = A_0(x) + \frac{1}{2}\sum_{j=1}^m \lfloor \partial A_j(x)\rfloor A_j(x)$, where and in what follows $\partial A_j = \left(\partial_k A_j^i\right)_{1\leq i,k\leq d}$. We identify A_j with the following vector field (first order differential operator)

$$A_j = A_j(x) = \sum_{i=1}^d A_j^i(x)\partial_i = \langle A_j(x)\,, \partial\rangle\,.$$

Thus,

$$\begin{aligned} A_j A_k &= \sum_{i=1}^d A_j^i \partial_i \sum_{\ell=1}^d A_k^\ell \partial_\ell \\ &= \sum_{i,\ell=1}^d A_j^i \partial_i A_k^\ell \partial_\ell + \sum_{i,\ell=1}^d A_j^i A_k^\ell \partial_{i,\ell}^2\,. \end{aligned}$$

The Lie bracket of A_j and A_k is defined as

$$[A_j, A_k] = A_jA_k - A_kA_j = \sum_{i,\ell=1}^{d} \left[A_j^i \partial_i A_k^\ell - A_k^i \partial_i A_j^\ell\right] \partial_\ell$$

which is again vector field (first order differential operator). We may write the above identity as

$$[A_j, A_k] = [\partial A_k]\, A_j - [\partial A_j]\, A_k \,.$$

The following condition is called the Hörmander's condition.

(H) The vector space spanned by the vector fields

$$A_1, \cdots, A_m, [A_i, A_j], 0 \le i, j \le m\,, [A_i, [A_j, A_k]]\,, 0 \le i, j, k \le m\,, \cdots \tag{10.2.29}$$

at point x_0 is $\mathbb{R}^d$ (for any fixed t and x, each of them can be considered as an element in $\mathbb{R}^d$).

Theorem 10.7. *Let $X(t)$ be the unique solution to (10.2.27). Assume that the coefficients are smooth with bounded derivatives and satisfy the Hörmander's condition (H). Then, for any $t > 0$, the law of the random variable $X(t)$ has a smooth density with respect to the Lebesgue measure.*

As in the proofs of many other theorems on smoothness of density a critical component in the proof of this theorem is to show that a certain Malliavin covariance matrix has negative moments of all order. The first one who showed the existence of negative moments for the solution of a stochastic differential equation under Hörmander's condition seems to be S. Watanabe [Watanabe (1984)]. Following the idea of Meyer ([Meyer (1982b)]) and Stroock ([Stroock (1983)]), Norris gives a simplified proof in [Norris (1986)]. The proof is based on some semimartingale estimates which have their own interest. First, we prove a lemma, which is often called Norris lemma. We follow a version presented in [Nualart (2006)].

Lemma 10.4. *Let $\alpha, y \in \mathbb{R}$. Let $\beta(t)$, $\gamma(t) = (\gamma_1(t), \cdots, \gamma_m(t))$ and $u(t) = (u_1(t), \cdots, u_m(t))$ be adapted processes satisfying that there exist $t_0 > 0$ and $p \ge 2$ such that*

$$c := \mathbb{E}\left(\sup_{0\le t\le t_0} (|\beta(t)| + |\gamma(t)| + |u(t)|)^p\right) < \infty\,. \tag{10.2.30}$$

For $0 \le t \le t_0$, set

$$a(t) = \alpha + \int_0^t \beta(s)ds + \sum_{i=1}^m \int_0^t \gamma_i(s)dW_s^i\,,$$

$$Y_t = y + \int_0^t a(s)ds + \sum_{i=1}^m \int_0^t u_i(s)dW_s^i\,.$$

[It is easy to see that $\mathbb{E}\left(\sup_{0\le t\le t_0}|a(t)|^p\right)<\infty$ *and hence* $\int_0^t a(s)ds$ *is well-defined.] Then, for any* $q>8$ *and for any* $r,\nu>0$ *such that* $14r+9\nu<q-8$*, there exists* $\varepsilon_0=\varepsilon_0(t_0,q,r,\nu)$ *such that for all* $\varepsilon<\varepsilon_0$,

$$\mathbb{P}\left\{\int_0^{t_0} Y_t^2 dt<\varepsilon^q\,,\ \int_0^{t_0}\left(|a(t)|^2+|u(t)|^2\right)dt\ge\varepsilon\right\}\le c\varepsilon^{rp}+e^{-\varepsilon^{-\nu}}\,. \tag{10.2.31}$$

Proof. We shall divide the proof into several steps.

Step 1. Denote $\theta_t:=|\beta(t)|+|\gamma(t)|+|a(t)|+|u(t)|$ and define the stopping time

$$T=\inf\left\{s\ge 0:\ \sup_{0\le u\le s}\theta_u>\varepsilon^{-r}\right\}\wedge t_0\,.$$

Then, it is easy to see

$$\mathbb{P}\left\{\int_0^{t_0} Y_t^2 dt<\varepsilon^q\,,\ \int_0^{t_0}\left(|a(t)|^2+|u(t)|^2\right)dt\ge\varepsilon\right\}\le A_1+A_2\,, \tag{10.2.32}$$

where $A_1=\mathbb{P}(T<t_0)$ and

$$A_2=\mathbb{P}\left\{\int_0^{t_0} Y_t^2 dt<\varepsilon^q\,,\ \int_0^{t_0}\left(|a(t)|^2+|u(t)|^2\right)dt\ge\varepsilon\,,\ T=t_0\right\}.$$

By the definition of the stopping time T and Chebyshev inequality, we have

$$A_1\le\mathbb{P}(\sup_{0\le s\le t_0}|\theta_s|>\varepsilon^{-r})\le\varepsilon^{rp}\mathbb{E}\sup_{0\le s\le t_0}|\theta_s|^p\le c\varepsilon^{rp} \tag{10.2.33}$$

for some constant c (which depends only on the constant c in (10.2.30), m and p).

Step 2. The remaining objective is to bound A_2. To do this, we denote

$$\begin{cases} A_t=\displaystyle\int_0^t a(s)ds\,, & M_t=\displaystyle\sum_{i=1}^m\int_0^t u_i(s)dW_s^i\,,\\ N_t=\displaystyle\sum_{i=1}^m\int_0^t Y_s u_i(s)dW_s^i\,, & Q_t=\displaystyle\sum_{i=1}^m\int A_s\gamma_i(s)dW_s^i\,. \end{cases}$$

We also denote for any $\rho_i>0$ and $\delta_i>0$, $i=1,2,3$,

$$\begin{cases} B_1=\left\{\langle N\rangle_T<\rho_1\,,\ \sup_{0\le s\le T}|N_s|\ge\delta_1\right\},\\ B_2=\left\{\langle M\rangle_T<\rho_2\,,\ \sup_{0\le s\le T}|M_s|\ge\delta_2\right\},\\ B_3=\left\{\langle Q\rangle_T<\rho_3\,,\ \sup_{0\le s\le T}|Q_s|\ge\delta_3\right\}. \end{cases}$$

By the martingale inequality (A.41), we have

$$\mathbb{P}(B_i) \le 2\exp\left\{-\frac{\delta_i^2}{2\rho_i}\right\}, \qquad \forall\, i = 1, 2, 3\,. \tag{10.2.34}$$

Step 3. We are going to show

$$A_2 \subseteq B_1 \cup B_2 \cup B_3 \tag{10.2.35}$$

for the following choices of ρ_i and δ_i:

$$\begin{cases} \rho_1 = \varepsilon^{-2r+q}\,, & \delta_1 = \varepsilon^{q_1}\,, \quad q_1 = \dfrac{q}{2} - r - \dfrac{\nu'}{2}\,, \\ \rho_2 = 2\sqrt{2t_0^{3/2} + 2t_0 + 1}\,\varepsilon^{\frac{q_1}{2} - r}\,, & \delta_2 = \varepsilon^{q_2}\,, \quad q_2 = \dfrac{q}{8} - \dfrac{3r}{4} - \dfrac{5\nu'}{8}\,, \\ \rho_3 = 64 t_0 \varepsilon^{-2r+2q_2}\,, & \delta_3 = \varepsilon^{q_3}\,, \quad q_3 = \dfrac{q}{8} - \dfrac{7r}{4} - \dfrac{9\nu'}{8}\,, \end{cases}$$

where ν' is a number such that $\nu' < \nu$ and $14r + 9\nu' < q - 8$.

Suppose $\omega \notin B_1 \cup B_2 \cup B_3$, $T(\omega) = t_0$ and $\int_0^T Y_t^2 dt < \varepsilon^q$. Then, (10.2.35) is proved if we can show

$$\int_0^T (|a(t)|^2 + |u(t)|^2)dt < \varepsilon\,, \tag{10.2.36}$$

where and in what follows all the random variables are evaluated at ω and we omit the explicit dependence. In particular, we shall use $T(\omega) = t_0$ freely in the remaining part of the proof. By the definition of T, we see that $\sup_{0\le t\le T} |u(t)| \le \varepsilon^{-r}$. Then

$$\langle N\rangle_T = \int_0^T |Y_s|^2 |u(s)|^2 ds < \varepsilon^{-2r+q} = \rho_1\,.$$

Thus, since $\omega \notin B_1$, we see that

$$\sup_{0\le t\le T}\left|\sum_{i=1}^m \int_0^t Y_s u_i(s) dW_s^i\right| < \delta_1 = \varepsilon^{q_1}\,. \tag{10.2.37}$$

On the other hand we have $\sup_{0\le t\le T} |a(t)| \le \varepsilon^{-r}$. By Cauchy-Schwarz inequality, we have

$$\sup_{0\le t\le T}\left|\int_0^t Y_s a(s) ds\right| \le \left(t_0 \int_0^T Y_s^2 a^2(s)\right)^{1/2} \le t_0^{1/2}\varepsilon^{-r+\frac{q}{2}}\,. \tag{10.2.38}$$

From (10.2.37), (10.2.38) and the definition of Y_t, we obtain

$$\sup_{0\le t\le T}\left|\int_0^t Y_s dY_s\right| \le t_0^{1/2}\varepsilon^{-r+\frac{q}{2}} + \varepsilon^{q_1}\,. \tag{10.2.39}$$

By Itô formula we have $Y_t^2 = y^2 + 2\int_0^t Y_s dY_s + \langle M\rangle_t$. Hence,

$$\int_0^T \langle M\rangle_t dt = \int_0^T Y_t^2 dt - Ty^2 - 2\int_0^T \left(\int_0^t Y_s dY_s\right) dt$$
$$< \varepsilon^q + 2t_0 \left(t_0^{1/2}\varepsilon^{-r+\frac{q}{2}} + \varepsilon^{q_1}\right) < (2t_0^{3/2} + 2t_0 + 1)\varepsilon^{q_1}$$

since $q > q_1$ and $\frac{q}{2} - r > q_1$. Since $\langle M\rangle_t$ is increasing in t, we see that for any $0 < \gamma < T$,

$$\gamma\langle M\rangle_{T-\gamma} \le \int_{T-\gamma}^T \langle M\rangle_t dt < (2t_0^{3/2} + 2t_0 + 1)\varepsilon^{q_1}\,.$$

Using again $\sup_{0\le t\le T}|u(t)| \le \varepsilon^{-r}$, we have

$$\langle M\rangle_T = \langle M\rangle_{T-\gamma} + \int_{T-\gamma}^T |u(s)|^2 ds \le \gamma^{-1}(2t_0^{3/2} + 2t_0 + 1)\varepsilon^{q_1} + \gamma\varepsilon^{-2r}\,.$$

This expression attains its minimum value ρ_2 when

$$\gamma = \sqrt{(2t_0^{3/2} + 2t_0 + 1)\varepsilon^{\frac{q_1}{2}+r}}\,,$$

which is less than T when ε is sufficiently small. Hence, $\langle M\rangle_T < \rho_2$. Now since $\omega \notin B_2$, we see

$$\sup_{0\le t\le T} |M_t| < \delta_2 = \varepsilon^{q_2}\,. \tag{10.2.40}$$

Recalling that $Y_t = y + A_t + M_t$ we have

$$\left|\left\{t\in[0,T] : |y + A_t| \ge \varepsilon^{q/3} + \varepsilon^{q_2}\right\}\right|$$
$$\le \left|\left\{t\in[0,T] : |Y_t - M_t| \ge \varepsilon^{q/3} + \varepsilon^{q_2}\right\}\right|$$
$$\le \left|\left\{t\in[0,T] : |Y_t(\omega)| \ge \varepsilon^{q/3}\right\}\right|\,|$$
$$\le \frac{\int_0^T Y_t^2 dt}{\varepsilon^{\frac{2q}{3}}} \le \varepsilon^{q/3}\,, \tag{10.2.41}$$

where $|A|$ denotes the Lebesgue measure of A and where we used the Chebyshev inequality. Choose ε sufficiently small so that $\varepsilon^{q/4} < \frac{T}{4} = \frac{t_0}{4}$. Fix an arbitrary $t \in [0,T]$. Define

$$G_1 = \left\{s\in[0,T] \;:\; |s-t| < 2\varepsilon^{q/3}\right\}\,,$$
$$G_2 = \left\{s\in[0,T] \;:\; |y + A_s| < \varepsilon^{q/3} + \varepsilon^{q_2}\right\}\,.$$

Clearly, $|G_1| \ge 2\varepsilon^{q/3}$. If $G_1 \cap G_2 = \emptyset$, then $G_1 \subseteq G_2^c$ (the complement of G_2 in $[0,T]$). However, $|G_2| \le \varepsilon^{q/3}$ by (10.2.41). This is a contradiction.

Thus $G_1 \cap G_2 \neq \emptyset$. This means that for each fixed $t \in [0,T]$ there exists $s \in [0,T]$ such that $|s-t| \le 2\varepsilon^{q/3}$ and $|y+A_s| < \varepsilon^{q/3} + \varepsilon^{q_2}$. Consequently,

$$|y+A_t| \le |y+A_s| + \left|\int_s^t a_r dr\right| \le (1+2\varepsilon^{-r})\varepsilon^{q/3} + \varepsilon^{q_2} \quad \forall\, t \in [0,T].$$

When $t=0$, we have $|y| \le (1+2\varepsilon^{-r})\varepsilon^{q/3} + \varepsilon^{q_2}$. This again implies

$$|A_t| \le 2\left[(1+2\varepsilon^{-r})\varepsilon^{q/3} + \varepsilon^{q_2}\right] \le 8\varepsilon^{q_2}$$

since $q_2 < \frac{q}{3} - r$. This implies that

$$\langle Q\rangle_T = \int_0^T A_t^2 |\gamma(t)|^2 dt < 64 t_0 \varepsilon^{2q_2 - 2r} = \rho_3 .$$

Since $\omega \notin B_3$ the above inequality implies that

$$|Q_T| = \left|\sum_{i=1}^m \int_0^T A_t \gamma_i(t) dW_t^i\right| < \delta_3 = \varepsilon^{q_3} .$$

Now from Itô's formula it follows

$$\begin{aligned}\int_0^T (a^2(t) + |u(t)|^2)dt &= \int_0^T a_t dA_t + \langle M\rangle_T \\ &= a_T A_T - \int_0^T A_t \beta(t) dt - \sum_{i=1}^m \int_0^T A_t \gamma_i(t) dW_t^i + \langle M\rangle_T \\ &\le 8\varepsilon^{q_2 - r} + 8t_0\varepsilon^{q_2-r} + \varepsilon^{q_3} + 2\sqrt{(2t_0^{3/2} + 2t_0 + 1)}\varepsilon^{\frac{q_1}{2} - r} < \varepsilon\end{aligned}$$

for sufficiently small ε because $q_2 - r > q_3 > 1$ ($q_3 > 1$ is equivalent to $14r + 9\nu' < q - 8$) and $\frac{q_1}{2} - r > 1$. Thus, we prove (10.2.36) and hence (10.2.35) holds true.

Step 4. Now from (10.2.34) and (10.2.35) it follows

$$\begin{aligned}\mathbb{P}(A_2) &\le \sum_{i=1}^3 \mathbb{P}(B_i) \le 2\sum_{i=1}^3 \exp\left[-\frac{\delta_i^2}{2\rho_i}\right] \\ &\le 2\Bigg(\exp\left[-\frac{\varepsilon^{2q_1 - q + 2r}}{2}\right] + \exp\left[-\frac{\varepsilon^{2q_2 + r - \frac{q_1}{2}}}{4\sqrt{(2t_0^{3/2} + 2t_0 + 1)}}\right] \\ &\qquad + \exp\left[-\frac{\varepsilon^{2q_3 + 2r - 2q_2}}{128 t_0}\right]\Bigg) \\ &\le \exp\left[-\varepsilon^{-\nu}\right]\end{aligned}$$

since

$$2q_1 - q + 2r = 2q_2 + r - \frac{q_1}{2} = 2q_3 + 2r - 2q_2 = -\nu' < -\nu .$$

[In fact, the q_i defined earlier is obtained by solving the above three equations.] □

Lemma 10.5. *Let C be a symmetric non-negative $d \times d$ random matrix. Assume that $C \in L^p$ for all $p \geq 1$ (which means each entry of C has this property) and that for all $p \geq 2$, there is a $\varepsilon(p) > 0$ such that*

$$\sup_{|v|\leq 1} \mathbb{P}(v^T Cv \leq \varepsilon) \leq \varepsilon^p, \quad \forall\, \varepsilon \leq \varepsilon(p)\,. \tag{10.2.42}$$

Then, $\det(C)^{-1} \in L^p$ *for all* $p \geq 1$.

Proof. Denote by $\lambda = \inf_{|v|=1} v^T Cv$ the smallest eigenvalue of C. Then, $0 \leq \lambda^d \leq \det(C)$. Hence to show the lemma it suffices to show that $\mathbb{E}(\lambda^{-p}) < \infty$ for any $p \geq 2$. Let $|C|^2 = \sum_{i,j=1}^d |c_{ij}|^2$ denote the Hilbert-Schmidt norm of C. Then, we have

$$\begin{aligned}\mathbb{P}(\lambda < \varepsilon) &= \mathbb{P}\left(\inf_{|v|=1} v^T Cv < \varepsilon\right)\\ &\leq \mathbb{P}\left(\inf_{|v|=1} v^T Cv < \varepsilon\,,\ |C| \leq \frac{1}{\varepsilon}\right) + \mathbb{P}\left(|C| > \frac{1}{\varepsilon}\right).\end{aligned} \tag{10.2.43}$$

Using the Chebyshev inequality and the assumption that $C \in L^p$ for any $p \geq 1$, we see that

$$\mathbb{P}\left(|C| > \frac{1}{\varepsilon}\right) \leq \varepsilon^p \mathbb{E}(|C|^p) \leq \kappa_p \varepsilon^p \quad \text{for some constant } \kappa_p\,. \tag{10.2.44}$$

We want to show a similar inequality for the first term in (10.2.43). The difference between this term and the assumption is that now the 'inf' is inside the probability. To move this 'inf' from inside the probability sign to outside the probability sign, we shall use the covering technique. Fix a $\varepsilon > 0$. One can find a finite set $v_1, \cdots, v_N$ of unit vectors in $\mathbb{R}^d$ such that $\cup_{i=1}^N B(v_i, \varepsilon^2/2)$ contains the unit sphere S^{d-1} in $\mathbb{R}^d$, where $B(a, r)$ denotes the open ball of center a and radius r. It is well-known that one can choose N so that $N \leq \kappa\varepsilon^{-2d}$ for some constant κ independent of ε (see for example [Verger-Gaugry (2005)], [Böröczky (2004)] and the references therein for some further results). Now, we assume $|C| \leq 1/\varepsilon$ and $v_k^T Cv_k \geq 2\varepsilon$ for all $k = 1, 2, \cdots, N$. For any unit vector v, there is a v_k such that $|v - v_k| \leq \frac{\varepsilon^2}{2}$. Thus, we have

$$\begin{aligned}v^T Cv &\geq v_k^T Cv_k - |v^T Cv - v_k^T Cv_k|\\ &\geq 2\varepsilon - |v^T C(v - v_k)| - |(v - v_k)^T Cv_k|\\ &\geq 2\varepsilon - 2|C||v - v_k| \geq \varepsilon\,.\end{aligned}$$

As a consequence, we have

$$\mathbb{P}\left(\inf_{|v|=1} v^T C v < \varepsilon\,,\ |C| \le \frac{1}{\varepsilon}\right) \le \mathbb{P}\left(\cup_{k=1}^N \left\{v_k^T C v_k < 2\varepsilon\right\}\right)$$
$$\le N \max_k \mathbb{P}\left(v_k^T C v_k < 2\varepsilon\right) \le N(2\varepsilon)^p \le \kappa_p \varepsilon^{-2d+p}\,. \qquad (10.2.45)$$

Combining (10.2.44) and (10.2.45), we see that for any $q \ge 1$, there is a finite constant κ_q such that $P(\lambda < \varepsilon) \le \kappa_q \varepsilon^q$. This shows $\mathbb{E}(\lambda^{-p}) < \infty$ for any $p \ge 2$. The lemma is then proved. □

Proof of Theorem 10.7. We divide the proof into several steps.
Step 1. Let $\Phi(t)$ satisfy

$$\Phi(t) = I + \sum_{j=0}^m \int_0^t \partial A_j(X_s)\Phi(s)d^\circ W_s^j\,, \quad 0 \le t \le T\,. \qquad (10.2.46)$$

Then, as for (10.2.22)-(10.2.25) one can show the following.

(i) $\Phi(t)$ is invertible and

$$\phi^{-1}(t) = I - \sum_{j=0}^m \int_0^t \Phi^{-1}(s)\partial A_j(X(s))d^\circ W_s^j\,. \qquad (10.2.47)$$

(ii) For every $t \ge 0$, $X(t) \in \mathbb{D}_\infty$ and

$$D_r^{(\ell)} X(t) = \Phi(t)\Phi^{-1}(r)A_\ell(X(r))\,. \qquad (10.2.48)$$

(iii) The Malliavin covariance matrix

$$Q(t) = \left(\langle DX^i(t), DX^j(t)\rangle_{\mathbb{H}}\right)_{1\le i,j\le d}$$
$$= \left(\sum_{\ell=1}^m \int_0^t D_r^{(\ell)} X^i(t) D_r^{(\ell)} X^j(t) dr\right)_{1\le i,j\le d}$$

can be represented as

$$Q(t) = \Phi(t)C(t)\Phi(t)^T\,, \qquad (10.2.49)$$

where if we define $\Sigma(s) = \sum_{\ell=1}^m A_\ell(X(s))A_\ell^T(X(s))$, then

$$C(t) = \int_0^t \Phi^{-1}(s)\Sigma(s)\left(\Phi^{-1}(s)\right)^T ds\,. \qquad (10.2.50)$$

By Theorem 6.6, to show the theorem it suffices to show that $\det(Q(t))^{-1} \in L^p$. It is easy to see that $\Phi^{-1}(t) \in L^p$ for all $p \ge 1$ from equation (10.2.47). Since $Q^{-1}(t) = \left(\Phi^{-1}(t)\right)^T C^{-1}(t)\Phi(t)^{-1}$, to show $\det(Q(t))^{-1} \in L^p$ for all

$p \geq 1$, it suffices to show that $\det(C(t))^{-1} \in L^p$ for all $p \geq 1$, which is, by Lemma 10.4, implied by

$$\sup_{|v|=1} \mathbb{P}(v^T C(t) v \leq \varepsilon) \leq \varepsilon^p \quad \forall\, p \geq 2\,. \tag{10.2.51}$$

Step 2. Substituting the expression of $\Sigma(s)$ into (10.2.50) one can write for any $v \in \mathbb{R}^d$,

$$v^T C(t) v = \sum_{j=1}^{m} \int_0^t |v^T \Phi^{-1}(s) A_j(X(s))|^2 ds\,. \tag{10.2.52}$$

Introduce the following set of vector fields:

$$\begin{cases} \Xi_0 = \{A_1, \cdots, A_m\}\,, \\ \Xi_n = \{[A_k, V]\,, k = 1, 2, \cdots, m, V \in \Xi_{n-1}\}\,, n = 1, 2, \cdots \\ \Xi = \cup_{n=0}^{\infty} \Xi_n\,, \\ \Xi'_0 = \{A_1, \cdots, A_m\}\,, \\ \Xi'_n = \Big\{[A_k, V]\,, k = 1, 2, \cdots, m, V \in \Xi'_{n-1}\,, \\ \qquad [A_0, V] + \frac{1}{2}[A_j, [A_j, V]], V \in \Xi'_{n-1}\Big\}\,, n = 1, 2, \cdots \\ \Xi' = \cup_{n=0}^{\infty} \Xi'_n\,. \end{cases} \tag{10.2.53}$$

It is clear that the linear spans of $\Xi(x)$ and $\Xi'(x)$ are the same. From the Hörmander's condition, $\Xi'(x_0)$ spans $\mathbb{R}^d$. Thus, there are constants $R > 0, c > 0$ such that

$$\sum_{j=0}^{j_0} \sum_{V \in \Xi'_j} (v^T V(y))^2 \geq c\,, \quad \forall\, v \in \mathbb{R}^d \text{ with } |v| - 1 \text{ and } |y - x_0| \leq R.$$

For any $j = 0, 1, \cdots, j_0$ set $m(j) = 2^{4j}$ and

$$E_j = \left\{ \sum_{V \in \Xi'_j} \int_0^t (v^T \Phi^{-1}(s) V(X(s)))^2 ds < \varepsilon^{m(j)} \right\}\,.$$

Notice $E_0 = \{v^T C(t) v \leq \varepsilon\}$. We decompose

$$E_0 \subset (E_0 \cap E_1^c) \cup (E_1 \cap E_2^c) \cup \cdots \cup (E_{j_0-1} \cap E_{j_0}^c) \cup F\,,$$

where $F = E_0 \cap E_1 \cap \cdots \cap E_{j_0}$. Thus, for any unit vector $v \in \mathbb{R}^d$, we have

$$\mathbb{P}(v^T C(t) v \leq \varepsilon) = \mathbb{P}(E_0) \leq \mathbb{P}(F) + \sum_{j=0}^{j_0-1} \mathbb{P}(E_j \cap E_{j+1}^c)\,. \tag{10.2.54}$$

Step 3. In this step we estimate $\mathbb{P}(F)$. Consider the following stopping time

$$S = \inf\left\{r \geq 0 : \sup_{0\leq s\leq r} |X(s) - x_0| \geq R \text{ or } \sup_{0\leq s\leq r} |\Phi^{-1}(s) - I| \geq 1/2\right\} \wedge t\,.$$

We have

$$\mathbb{P}(F) \leq \mathbb{P}(F \cap \{S \geq \varepsilon^\beta\}) + \mathbb{P}(S < \varepsilon^\beta)\,.$$

We claim that for some choice of $\beta > 0$, $F \cap \{S \geq \varepsilon^\beta\} = \emptyset$ for sufficiently small ε. In fact, if $S \geq \varepsilon^\beta$, then

$$\begin{aligned}
&\sum_{j=0}^{j_0} \sum_{V\in\Xi'_j} \int_0^t (v^T\Phi^{-1}(s)V(X(s)))^2 ds \\
&\geq \sum_{j=0}^{j_0} \sum_{V\in\Xi'_j} \int_0^S \left(\zeta^T V(X(s))\right)^2 |v^T\Phi^{-1}(s)|^2 ds \geq \frac{c\varepsilon^\beta}{4}\,, \quad (10.2.55)
\end{aligned}$$

where $\zeta = \frac{v^T\Phi^{-1}(s)}{|v^T\Phi^{-1}(s)|}$. This is because when $s < S$, we have

$$|v^T\Phi^{-1}(s)| = |v^T - v^T(I - \Phi^{-1}(s))| \geq 1 - |I - \Phi^{-1}(s)| \geq 1/2\,.$$

On the other hand, the left hand side of (10.2.55) is bounded by $(j_0 + 1)\varepsilon^{m(j_0)}$ on the set $F \subseteq E_{j_0}$. Thus, if we choose $0 < \beta < m(j_0)$, then $F \cap \{S \geq \varepsilon^\beta\} = \emptyset$ for sufficiently small ε. Therefore, for sufficiently small $\varepsilon > 0$ we have

$$\begin{aligned}
\mathbb{P}(F) = \mathbb{P}(S < \varepsilon^\beta) &\leq \mathbb{P}\left(\sup_{0\leq s\leq \varepsilon^\beta} |X(s) - x| \geq R\right) \\
&\quad + \mathbb{P}\left(\sup_{0\leq s\leq \varepsilon^\beta} |\Phi^{-1}(s) - I| \geq 1/2\right) \\
&\leq R^{-q}\mathbb{E}\left(\sup_{0\leq s\leq \varepsilon^\beta} |X(s) - x|\right) + 2^q\mathbb{E}\left(\sup_{0\leq s\leq \varepsilon^\beta} |\Phi^{-1}(s) - I|\right)
\end{aligned}$$

for any $q \geq 2$. Using the equation satisfied by $X(s)$, $\Phi^{-1}(s)$ and Burkholder inequality, one can show $\mathbb{P}(F) \leq C\varepsilon^{q\beta/2}$ by a routine method, where $q \geq 2$ is arbitrary.

Step 4. An application of the Itô formula yields

$$\begin{aligned}
\Phi^{-1}(t)V(X(t)) &= V(x_0) + \sum_{k=1}^m \int_0^t \Phi^{-1}(s)[A_k, V](X(s))dW_s^k \\
&\quad + \int_0^t \Phi^{-1}(s)\left\{[A_0, V] + \frac{1}{2}\sum_{k=1}^m [A_k, [A_k, V]]\right\}(X(s))ds\,.
\end{aligned} \quad (10.2.56)$$

Denote $n(j) = \#\Xi'_j$, the cardinality of Ξ'_j and

$$\begin{aligned} Y_V(t) &= v^T\Phi^{-1}(t)V(X(t)), \\ u_{k,V}(t) &= v^T\Phi^{-1}(s)[A_k, V](X(s)), \quad k = 1, 2, \cdots, m, \\ a_V(t) &= v^T\Phi^{-1}(s)\left([A_0, V] + \frac{1}{2}\sum_{k=1}^m [A_k, [A_k, V]]\right)(X(s)). \end{aligned}$$

For $j = 0, 1, \cdots, j_0 - 1$, by definition of E_j, we have

$$\begin{aligned} &\mathbb{P}(E_j \cap E^c_{j+1}) \\ &= \mathbb{P}\left(\sum_{V\in\Xi'_j}\int_0^t Y_V^2(s)ds \le \varepsilon^{m(j)}, \sum_{V\in\Xi'_{j+1}}\int_0^t Y_V^2(s)ds > \varepsilon^{m(j+1)}\right) \\ &\le \sum_{V\in\Xi'_j}\mathbb{P}\left(\int_0^t Y_V^2(s)ds \le \varepsilon^{m(j)}, \int_0^t [\sum_{k=1}^m u^2_{k,V}(s) + a^2_V(s)]ds > \frac{\varepsilon^{m(j+1)}}{n(j)}\right). \end{aligned}$$

In the above inequality we have used the fact that a vector V in Ξ'_{j+1} can be defined through the vectors in Ξ'_j by (10.2.53). From (10.2.56), we can write

$$Y_V(t) = V(x_0) + \int_0^t a_V(s)ds + \sum_{k=1}^m \int_0^t u_{k,V}(s)dW^k_s.$$

Applying the Lemma 10.4 to the semimartingale $Y_V(t)$, we have

$$P(E_j \cap E^c_{j+1}) \le \kappa_p \varepsilon^p$$

due to the fact that $m(j)/m(j+1) > 8$. □

Remark 10.3. We refer to [Hairer (2011)] for a short, self-contained, elegant proof of the Hörmander theorem.

10.3 Exponential Integrability

In this section we are concerned with the exponential integrability of certain diffusion processes.

Let $\sigma_i(\cdot,\cdot)$, $i = 1, 2, \cdots, m$, $b(\cdot,\cdot)$ be continuous mappings from $\mathbb{R}_+ \times \mathbb{R}^d$ to $\mathbb{R}^d$ and let W^i_t, $i = 1, 2, \cdots, m$ be m independent Brownian motions on probability space $(\Omega, \mathcal{F}, P)$ with filtration $(\mathcal{F}_t)_{0\le t<\infty}$. Consider the following stochastic differential equation

$$dX_t = b(t, X_t)dt + \sum_{i=1}^m \sigma_i(t, X_t)dW^i_t, \quad 0 \le t \le T, \quad X_0 = x, \qquad (10.3.1)$$

where dW_t^i denotes Itô differential. The generator of the transition semigroup associated with this equation is

$$\mathrm{L}_t f(x) = \frac{1}{2}\sum_{i,j=1}^{d} a_{ij}(t,x)\frac{\partial^2 f}{\partial x_i \partial x_j}(x) + \sum_{i=1}^{d} b_i(t,x)\frac{\partial f}{\partial x_i}(x)\,, \qquad (10.3.2)$$

where

$$(a_{ij}(t,x))_{1\le i,j\le d} = A = \sum_{k=1}^{m} \sigma_k(t,x)\sigma_k^T(t,x)\,.$$

If A is bounded and uniformly elliptic, i.e. there are two constants $0 < \lambda < \mu < \infty$ such that

$$\lambda I \le A \le \mu I\,,$$

and if L is of divergence form, i.e. $\mathrm{L}f(x) = \frac{1}{2}\sum_{i,j=1}^{d} \frac{\partial}{\partial x_i}\left(a_{ij}(t,x)\frac{\partial f}{\partial x_j}(x)\right)$, then from [Stroock (1988)], we know that there is a constant M, dependent on λ and μ such that

$$P\left(\sup_{0\le s\le t} |X_s - x| \ge r\right) \le M e^{-\frac{r^2}{Mt}} \qquad (10.3.3)$$

for all $(t,x) \in \mathbb{R}^d$ and $r > 0$. This implies that there is an $\alpha_0 > 0$ such that for any $\alpha \in (0,\alpha_0)$,

$$\mathbb{E}\exp\left\{\alpha \sup_{0\le s\le t} |X_s - x|^2\right\} < \infty\,.$$

In what follows we deal with a different exponential integrability problem. We assume that the coefficients satisfy the following assumptions:

(I) There are $0 < \lambda_1\,,\gamma\,,\beta < \infty$ and $\phi(x)$ such that ϕ is of class C^2, $\phi(x) \ge 0$ and

$$\mathrm{L}_t\phi(x) \le \lambda_1 \phi^{1-\gamma}(x)\,, \quad \forall \quad x \in \mathbb{R}^d\,.$$

(II) There is $0 < \lambda_2 < \infty$ such that

$$|\Gamma_t(\phi)(x)| \le \lambda_2 \phi^{2-\gamma}(x)\,, \quad \forall \quad (t,x) \in \mathbb{R}_+ \times \mathbb{R}^d\,,$$

where $\Gamma_t(\phi)$ is the "carré du champ" operator:

$$\Gamma_t(\phi)(x) = \sum_{i,j=1}^{d} a_{ij}(t,x)\frac{\partial\phi(x)}{\partial x_i}\frac{\partial\phi(x)}{\partial x_j}\,.$$

Under the assumption (I), by an analogous argument to that of Theorem 10.2 one can show that the unique solution of (10.3.1) exists for all ≥ 0.

We say that X_T is exponentially integrable if there are $\beta, \alpha > 0$ such that

$$\mathbb{E}e^{\alpha|X_T|^\beta} < \infty\,.$$

We say that the solution X of (10.3.1) is uniformly exponentially integrable if there are $\beta, \alpha > 0$ such that

$$\mathbb{E}e^{\alpha \sup_{0\leq t\leq T}|X_t|^\beta} < \infty\,.$$

It is easy to see that uniformly exponential integrability implies exponential integrability.

The above assumption (II) will be used in the following way.

Lemma 10.6. *Let the coefficients satisfy the assumption (II). Then*

$$\sum_{k=1}^m |\langle \nabla\phi(x)\,, \sigma_k(x)\rangle|^2 \leq \lambda_2 \phi(x)^{2-\gamma}\,.$$

Proof. Let Tr denote the trace of a matrix and $\nabla\phi\times\nabla\phi$ denote the tensor product of $\nabla\phi$, i.e. a matrix with entries $\left(\frac{\partial\phi}{\partial x_i}\cdot\frac{\partial\phi}{\partial x_j}\right)$. Then

$$\begin{aligned}\sum_{k=1}^m |\langle \nabla\phi(x)\,, \sigma_k(x)\rangle|^2 &= \sum_{k=1}^m \mathrm{Tr}\left[\sigma_k(x)\sigma_k(x)^T\nabla\phi\times\nabla\phi\right]\\ &= \sum_{i,j=1}^d a_{ij}(t,x)\frac{\partial\phi}{\partial x_i}\frac{\partial\phi}{\partial x_j} \leq \lambda_2\phi^{2-\gamma}(x)\,.\end{aligned}$$

This completes the proof. □

The first result concerns with the exponential integrability of the solution. To this end the assumption (I) can be weakened as

(I)' There are $0 < \lambda_1 < \infty$ and a positive function $\phi(x)$ such that

$$\mathrm{L}_t\phi(x) \leq \lambda_1\phi(x)\,.$$

Theorem 10.8. *Let $X_t\,, 0\leq t\leq T$ be the solution to (10.3.1) and let the coefficients satisfy (I)' and (II).*

(i) If $\gamma > 1$, then for any real value α

$$\mathbb{E}e^{\alpha\phi(X_T)} < \infty\,.$$

(ii) If $\gamma = 1$, then there is an $\alpha_0 > 0$ such that

$$\mathbb{E}e^{\alpha\phi(X_T)} < \infty$$

for all $\alpha < \alpha_0$.

Proof. Using Itô formula, we have

$$\begin{aligned}\mathbb{E}\phi(X_t)^n &= \phi(X_0)^n + n\mathbb{E}\int_0^t \phi^{n-1}(X_s)\mathrm{L}_s\phi(X_s)ds \\ &\quad + n(n-1)\int_0^t \mathbb{E}\phi^{n-2}(X_s)\sum_{i,j=1}^d a_{ij}(s,X_s)\frac{\partial\phi}{\partial x_i}(X_s)\frac{\partial\phi}{\partial x_j}(X_s)ds \\ &\le \phi(X_0)^n + \lambda_1 n\mathbb{E}\int_0^t \phi^n(X_s)ds + \lambda_2 n^2\int_0^t \mathbb{E}\phi^{n-\gamma}(X_s)ds\,.\end{aligned}$$

Let $Z_s = \mathbb{E}\phi^n(X_s)$. Then

$$Z_t \le \phi^n(X_0) + \lambda_1 n\int_0^t Z_s ds + \lambda_2 n^2\int_0^t Z_s^{1-\frac{\gamma}{n}}ds\,.$$

Denote

$$g(t) = \lambda_1 nt + \lambda_2 n^2 t^{1-\frac{\gamma}{n}}\,.$$

The above inequality can be written as

$$Z_t = \phi(X_0)^n + \int_0^t g(Z_s)ds\,.$$

By an inequality of Bihari and Langenhop [Beckenbach and Bellman (1965)] (p.135, Theorem 3), we have

$$Z_t \le G_n^{-1}\left(G_n\left(\phi(X_0)^n\right) + t\right)\,, \tag{10.3.4}$$

where

$$G_n(u) = \int_1^u \frac{1}{g(t)}dt\,.$$

We are going to bound $G_n(u)$. When $u \ge 1$,

$$\begin{aligned}G_n(u) &= \int_1^u \frac{1}{\lambda_1 nt + \lambda_2 n^2 t^{1-\frac{\gamma}{n}}}dt \\ &\le \int_1^u \frac{1}{\lambda_1 nt}dt \le \frac{1}{\lambda_1 n}\log u\,.\end{aligned}$$

Thus

$$G_n\left(\phi(X_0)^n\right) \le \frac{\log\phi(X_0)}{\lambda_1}\,. \tag{10.3.5}$$

We can also bound $G_n(u)$ from below by a nondecreasing function as follows. When $u \geq n^{\frac{n}{\gamma}}$, we have

$$\begin{aligned} G_n(u) &= \int_1^u \frac{dt}{\lambda_1 nt + \lambda_2 n^2 t^{1-\frac{\gamma}{n}}} dt \geq \int_{n^{\frac{n}{\gamma}}}^u \frac{dt}{\lambda_1 nt + \lambda_2 n^2 t t^{-\frac{\gamma}{n}}} dt \\ &\geq \int_{n^{\frac{n}{\gamma}}}^u \frac{dt}{\lambda_1 nt + \lambda_2 n^2 t \left(n^{\frac{n}{\gamma}}\right)^{-\frac{\gamma}{n}}} dt = \int_{n^{\frac{n}{\gamma}}}^u \frac{dt}{Cnt} \\ &\geq \frac{1}{Cn}\left[\log u - \log n^{\frac{n}{\gamma}}\right] = \frac{1}{Cn} \log \frac{u}{n^{\frac{n}{\gamma}}} . \end{aligned}$$

Let $v = G_n(u)$. The above inequality says $v \geq \frac{1}{Cn} \log \frac{u}{n^{\frac{n}{\gamma}}}$ or in other word

$$G_n^{-1}(v) = u \leq e^{Cvn} n^{\frac{n}{\gamma}} .$$

This combined together with (10.3.4) and (10.3.5) gives

$$\begin{aligned} \mathbb{E}\phi^n(X_t) = \mathbb{E}(Z_T) &\leq G_n^{-1}\left(\frac{\log \phi(X_0)}{\lambda_1} + t\right) \\ &\leq n^{\frac{n}{\gamma}} e^{(c_0 + c_1 t)n} . \end{aligned} \tag{10.3.6}$$

By Stirling's formula $n! \sim n^n e^{-n}\sqrt{2\pi n}$, it is straightforward to check that if $\gamma > 1$, then

$$\mathbb{E} e^{\alpha\phi(X_T)} < \infty$$

for any real value α and if $\gamma = 1$, then

$$\mathbb{E} e^{\alpha\phi(X_T)} < \infty$$

for all real values $\alpha \leq \alpha_0$, where $\alpha_0 > 0$ is positive number, proving the theorem. □

Theorem 10.9. *Let $X_t, 0 \leq t \leq T$ be the solution of (10.3.1) and let the coefficients satisfy (I)' and (II).*

(i) If $\gamma > 1$, then for any $\alpha \in \mathbb{R}$,

$$\mathbb{E} e^{\alpha \sup_{0\leq t\leq T} \phi(X_t)} < \infty .$$

(ii) If $\gamma = 1$, then there is an $\alpha_0 > 0$ such that

$$\mathbb{E} e^{\alpha \sup_{0\leq t\leq T} \phi(X_t)} < \infty$$

for all real values $\alpha \leq \alpha_0$.

Proof. Following an application of Itô formula

$$f(X_t) = f(X_0) + \int_0^t \mathrm{L}_s f(X_s)ds + \sum_{k=1}^m \int_0^t \langle \nabla f(X_s), \sigma_k(s, X_s)\rangle dW_s^k$$

to $f(x) = \phi^n(x)$, we have

$$\begin{aligned}\phi^n(X_t) &= \phi^n(X_0) + n\int_0^t \phi^{n-1}(X_s)\mathrm{L}_s\phi(X_s)ds \\ &\quad + n(n-1)\int_0^t \phi^{n-2}(X_s)\sum_{i,j=1}^d a_{ij}(s, X_s)\frac{\partial \phi}{\partial x_i}(X_s)\frac{\partial \phi}{\partial x_j}(X_s)ds \\ &\quad + n\sum_{k=1}^m \int_0^t \phi^{n-1}(X_s)\langle \nabla\phi(X_s)\, , \sigma_k(s, X_s)\rangle dW_s^k\,.\end{aligned}$$

Thus,

$$\begin{aligned}&\sup_{0\le s\le t} \phi^{2n}(X_s) \\ &\le 9\Bigg\{\phi^{2n}(X_0) + n^2 \sup_{0\le s\le t}\left(\int_0^s \phi^{n-1}(X_\rho)\mathrm{L}_\rho\phi(X_\rho)d\rho\right)^2 \\ &\quad + n^4\left(\sup_{0\le s\le t}\int_0^s \phi^{n-2}(X_\rho)\sum_{i,j=1}^d a_{ij}(\rho, X_\rho)\frac{\partial\phi}{\partial x_i}(X_\rho)\frac{\partial\phi}{\partial x_j}(X_\rho)d\rho\right)^2 \\ &\quad + n^2\left(\sup_{0\le s\le t}\sum_{k=1}^m \int_0^s \phi^{n-1}(X_\rho)\langle\nabla\phi(X_\rho)\, , \sigma_k(\rho, X_\rho)\rangle dW_\rho^k\right)^2\Bigg\}.\end{aligned}$$

Consequently, we have

$$\begin{aligned}\mathbb{E}\sup_{0\le s\le t}\phi^{2n}(X_s) &\le C^n + Cn^2\mathbb{E}\left(\int_0^t \phi^{n-\gamma}(X_\rho)d\rho\right)^2 \\ &\quad + Cn^4\mathbb{E}\left(\int_0^t \phi^{n-\gamma}(X_\rho)d\rho\right)^2 \\ &\quad + n^2\sum_{k=1}^m\int_0^t \mathbb{E}\left|\phi^{n-1}(X_\rho)\langle\nabla\phi(X_\rho)\, , \sigma_k(\rho, X_\rho)\rangle\right|^2 d\rho \\ &\le C^n + Cn^4 t\int_0^t \left(\mathbb{E}\left[\phi^{2n}(X_s)\right]\right)^{(n-\gamma)/n} ds \\ &\quad + Cn^2\int_0^t \left(\mathbb{E}\left[\phi^{2n}(X_s)\right]\right)^{(n-\gamma)/n} ds\,. \end{aligned} \tag{10.3.7}$$

This inequality can be written as

$$\mathbb{E}\left[\sup_{0\le t\le T}|\phi(t)|^{2n}\right]\le C^n+C_Tn^4\int_0^T\left(\mathbb{E}\left[\phi^{2n}(X_s)\right]\right)^{(n-\gamma)/n}ds\,.$$

Combining the above inequality with (10.3.6) gives us

$$\begin{aligned}\mathbb{E}\left[\sup_{0\le t\le T}|\phi(t)|^{2n}\right]&\le C^n+C_Tn^4n^{\frac{2n}{\gamma}\cdot\frac{n-\gamma}{n}}\int_0^T e^{(c_0+c_1s)\cdot(2n)\cdot\frac{n-\gamma}{n}}\,ds\\&\le C^n+C_T^n n^{2n/\gamma}\,.\end{aligned}\tag{10.3.8}$$

This can be used to show the theorem. □

To explain the above two results, we consider their application to the solution of the following simple equations:

$$dX_t=X_t^\alpha dW_t\,,\quad 0\le t\le T\,,\quad X_0=x>0\,.\tag{10.3.9}$$

From [Yamada and Watanabe (1971)] it follows that when $\alpha\in[1/2,1]$, the above equation has a unique solution. It is easy to see that when $\alpha=1$ (i.e. this is the case of the so-called geometric Brownian motion), the solution is *not* exponentially integrable. But we can show that when $\alpha<1$, the solution is uniformly exponentially integrable.

In fact, in this case, the corresponding formal generator is

$$\mathrm{L}f(x)=\frac{1}{2}x^{2\alpha}f''(x)\quad\text{and}$$

$$\Gamma_t(f)(x)=x^{2\alpha}(f'(x))^2\,.$$

Let $\phi:\mathbb{R}\to\mathbb{R}_+$ be a C^3 function such that $|x|>1$, $\phi(x)=|x|^\beta$, where $\beta>0$.

By the choice of ϕ we know that when $|x|>1$

$$\phi'(x)=\beta\mathrm{sign}(x)|x|^{\beta-1}\quad\text{and}\quad\phi''(x)=\beta(\beta-1)|x|^{\beta-2}\,.$$

It is easy to see that when $|x|>1$

$$\mathrm{L}\phi(x)=\beta(\beta-1)\frac{1}{2}|x|^{2\alpha+\beta-2}\,.$$

The assumption (I)' is satisfied if and only if $\alpha\le1$. The assumption (II) is satisfied if

$$2\alpha+2\beta-2\le(2-\gamma)\beta$$

which is implied by also choosing

$$\beta=\frac{2-2\alpha}{\gamma}\quad\text{when}\quad\alpha<1\,.$$

Therefore, for any $\alpha<1$, we see that $\gamma>1$ is equivalent to $\beta<2-2\alpha$ and $\gamma=1$ is equivalent to $\beta=2-2\alpha$. Hence, we have the following corollary.

Corollary 10.1. *When $1/2\le\alpha<1$, the equation (10.3.9) has a unique solution X_t and the following statements hold true.*

(i) For any $c > 0$, $T > 0$, *and* $\beta < 2 - 2\alpha$, *we have*

$$\mathbb{E}\exp\left\{c \sup_{0\le t\le T} |X_t|^\beta\right\} < \infty .$$

(ii) There is a $c_0 > 0$ *(dependent on* T *and the data of the equation (10.3.9)), such that for all* $c < c_0$,

$$\mathbb{E}\exp\left\{c \sup_{0\le t\le T} |X_t|^\beta\right\} < \infty , \quad \text{where} \quad \beta = 2 - 2\alpha .$$

Remark 10.4. The above results are sharp in the following sense. When $\alpha = 0$, the solution X_t of (4.1) is the Brownian motion, i.e. $X_t = W_t$. The above corollary, part (ii) implies then

$$\mathbb{E}e^{C\sup_{0\le t\le T}|W_t|^2} < \infty$$

for some $C > 0$. This result is however well-known.

10.4 Itô-Wiener Chaos Expansion

For simplicity of notation, we first consider a one dimensional SDE:

$$dX_t = \sigma(X_t)dB_t + b(X_t)dt, \quad 0 \le t < \infty, \quad X_0 = x \in \mathbb{R}. \tag{10.4.1}$$

We assume that the coefficients $\sigma, b : \mathbb{R} \to \mathbb{R}$ are in $C_b^\infty(\mathbb{R}, \mathbb{R})$. Denote by $P_t(x, dy)$ the transition probability associated with the equation (10.4.1):

$$P_t f(x) = \mathbb{E}\int_{\mathbb{R}} f(y)P_t(x, dy) = \mathbb{E}\left[f(X_t)\right], \quad \forall f \in \mathcal{P}(\mathbb{R}),$$

where $\mathcal{P}(\mathbb{R})$ denotes the set of measurable functions of polynomial growth. $P_t(x, dy)$ then satisfies the following parabolic equation:

$$\frac{\partial P_t}{\partial t} = \frac{1}{2}\sigma^2\frac{\partial^2 P_t}{\partial x^2} + b\frac{\partial P_t}{\partial x} \tag{10.4.2}$$

with initial condition $P_0(x, dy) = \delta(dy - x)$.

Under the above assumption, (10.4.1) has a unique solution which belongs to L^p for any $p \ge 1$. Thus, it admits an Itô-Wiener's expansion. We will concern with the explicit form of the coefficients and the explicit residual for the expansion of $f(X_t)$ for some nice function f.

Denote $\nabla_\sigma f(x) = \sigma(x)\frac{d}{dx}f(x)$ and $P_t'(x, dy) = \frac{d}{dx}P_t(x, dy)$. In what follows P_t and ∇_σ are considered as two operators acting the function space C_b^∞. These two operators do not commute. The following examples will explain our notation

$$\nabla_\sigma\nabla_\sigma f(x) = \sigma(x)\sigma'(x)f'(x) + (\sigma(x))^2 f''(x);$$

$$\nabla_\sigma P_s \nabla_\sigma P_t f(x) = \sigma(x) \int_{\mathbb{R}^2} P_s'(x, dy_1)\sigma(y_1)P_t'(y_1, dy_2)f(y_2)\,.$$

The following theorem gives the finite chaos expansion with explicit form of the residual.

Theorem 10.10. *For any $f \in \mathcal{P}(\mathbb{R})$ and any fixed t,*

$$f(X_t) = \mathbb{E}f(X_t) + \sum_{k=1}^{n} J_n(f_n(t)) + J_{n+1}(g_{n+1}(t)), \tag{10.4.3}$$

where

$$g_n(t) = g_n(t; s_1, \cdots, s_n) = \nabla_\sigma P_{s_2 - s_1} \cdots \nabla_\sigma P_{t-s_n} f(X_{s_1}); \tag{10.4.4}$$

$$f_n(t) = \mathbb{E}g_n(t; s_1, \cdots, s_n) = P_{s_1}\nabla_\sigma P_{s_2-s_1} \cdots \nabla_\sigma P_{t-s_n} f(x)\,; \tag{10.4.5}$$

$$J_n(g_n(t)) = \int_{0<s_1<\cdots<s_n<t} g(t; s_1, \cdots, s_n) dW_{s_1} \cdots dW_{s_n}\,. \tag{10.4.6}$$

Proof. Applying the Itô formula to the process $P_{t-s}f(X_s)$, $0 \le s \le t$ and noticing $P_0 f = f$ and the equation (10.4.2), we obtain

$$f(X_t) = P_t f(x) + \int_0^t \nabla_\sigma P_{t-s} f(X_s) dB_s. \tag{10.4.7}$$

Consider the integrand of the above integral. Applying (10.4.7) to $\nabla_\sigma P_{t-s} f(X_s)$ (the function f is replaced by $\nabla_\sigma P_{t-s} f$) we obtain

$$\nabla_\sigma P_{t-s} f(X_s) = P_s \nabla_\sigma P_{t-s} f(x) + \int_0^s \nabla_\sigma P_{s-u} \nabla_\sigma P_{t-s} f(X_u) dB_u\,.$$

Hence, we obtain

$$f(X_t) = \mathbb{E}f(X_t) + \int_0^1 P_s \nabla_\sigma P_{t-s} f(x) dB_s + \int_0^1 \int_0^s \nabla_\sigma P_{s-u} \nabla_\sigma P_{t-s} f(X_u) dB_u dB_s\,.$$

Continuing this procedure proves the theorem. □

From the chaos expansion theorem, if $f \in \mathcal{P}(\mathbb{R})$, then $f(X_t)$ admits a chaos expansion $f(X_t) = \sum_{n=0}^\infty F_n$, F_n is in the n-th chaos. One can verify easily that $J_{n+1}(g_{n+1})$ is orthogonal to all Itô-Wiener chaos of order less or equal to n. Thus, by (10.4.3), $F_n = J_n(f_n)$. This yields the following chaos expansion for $f(X_t)$

Corollary 10.2. *Let $f \in \mathcal{P}(\mathbb{R})$. Then*

$$f(X_t) = \mathbb{E}f(X_t) + \sum_{n=1}^{\infty} J_n(f_n)\,, \tag{10.4.8}$$

where f_n is given by (10.4.5) *and the series is convergent in L^2.*

Remark 10.5. The formula (10.4.8) with (10.4.5) is called the Krylov-Veretennikov formula or the Isobe-Sato formula.

The high dimensional chaos expansion formula can be obtained in exactly the same way. We shall describe the formula. Now let $A_j : [0,T] \times \mathbb{R}^d \to \mathbb{R}^d$ be in $C_b^2(\mathbb{R} \times \mathbb{R}^d, \mathbb{R}^d)$ for all $j = 0, 1, \cdots, m$. Let us return to consider the equation (10.2.1). Denote

$$\begin{cases} b(t,x) = (b_i(t,x)) = A_0(t,x)\,, \\ a(t,x) = (a_{ij}(t,x)) = \displaystyle\sum_{k=1}^{m} A_k(t,x)A_k(t,x)^T\,. \end{cases}$$

Denote by $P_{s,t}(x,dy)$ the transition probability associated with X_t and $P_{s,t}f(x) = \int_{\mathbb{R}^d} f(y)P_{s,t}(x,dy)$. Namely,

$$\begin{aligned} \frac{\partial}{\partial t}P_{s,t}f(x) &= \frac{1}{2}\sum_{1\le i,j\le d} a_{ij}(t,x)\frac{\partial^2}{\partial x_i \partial x_j}P_{s,t}f(x) + \sum_{i=1}^{d} b_i(t,x)\frac{\partial}{\partial x_i}P_{s,t}f(x)\,, \\ P_{s,s}f(x) &= f(x)\,, \quad \forall\, t \ge s\,,\ x \in \mathbb{R}^d\,. \end{aligned} \tag{10.4.9}$$

We also use the notation

$$A_j(t)f(x) = \langle A_j(t)\,, \nabla f(x)\rangle = \sum_{i=1}^{d} A_j^i(t,x)\frac{\partial}{\partial x_i}f(x)\,.$$

If $A_j \in C_b^\infty$ and if $f \in \mathcal{P}(\mathbb{R}^d)$, then we have the following chaos expansion for the solution X_t (composed with f)

$$\begin{aligned} f(X_t) = \mathbb{E}\left[f(X_t)\right] &+ \sum_{n=1}^{\infty}\sum_{\ell_1,\cdots,\ell_n=1}^{m} \int_{0\le s_1<\cdots<s_n\le t} P_{0,s_1}A_{\ell_1}P_{s_1,s_2}A_{\ell_2} \\ &\cdots A_{\ell_n}P_{s_n,t}f(x)dW_{s_1}^{\ell_1}\cdots dW_{s_n}^{\ell_n}\,. \end{aligned} \tag{10.4.10}$$

10.5 FKG Inequality and Variance Inequality

In this section let us consider again the one dimensional stochastic differential equation

$$dX_t = \sigma(X_t)dW_t + b(X_t)dt\,, \quad X_0 = x\,, \tag{10.5.1}$$

where $\sigma, b : \mathbb{R} \to \mathbb{R}$ are continuously differentiable functions with bounded derivatives. The solution to the above equation exists uniquely and is denoted by $X_t = X_t(x)$. We need a simple lemma.

Lemma 10.7. *Let $P_t(x, dy)$ be the transition probability of* (10.5.1). *Let f be continuously differentiable such that $f' \geq 0$. Then,*

$$P_t' f(x) \equiv \frac{d}{dx} P_t f(x) \geq 0, \quad \forall\, x \in \mathbb{R}. \tag{10.5.2}$$

Proof. Denote the solution of (10.5.1) by $X_t(x)$. From the well-known formula $P_t f(x) = \mathbb{E} f(X_t(x))$, we have

$$\frac{d}{dx} P_t f(x) = \mathbb{E}\{f'(X_t(x)) \frac{d}{dx} X_t(x)\}. \tag{10.5.3}$$

Differentiating (10.5.1) with respect to x and letting $z_t = \frac{d}{dx} X_t(x)$, we obtain

$$\frac{d}{dx} X_t(x) \equiv z_t = 1 + \int_0^t \sigma'(X_s) z_s dW_s + \int_0^t b'(X_s) z_s ds.$$

The solution z_t of this equation has the following explicit form

$$z_t = \exp\{\int_0^t \sigma'(X_s) dW_s + \int_0^t [b'(X_s) - \frac{1}{2}|\sigma'(X_s)|^2] ds\},$$

which is almost surely positive. Combining this fact with (10.5.3), we prove the lemma. □

Theorem 10.11. *Let $\sigma, b : \mathbb{R} \to \mathbb{R}$ be continuously differentiable functions with bounded derivatives and let X_t be the solution of* (10.5.1). *Assume $f' \geq 0$ and $g' \geq 0$. Then,*

$$\mathbb{E}[f(X_t) g(X_t)] \geq \mathbb{E} f(X_t) \mathbb{E} g(X_t). \tag{10.5.4}$$

Proof. Taking $n = 1$ in formula (10.4.3), we have

$$f(X_t) = \mathbb{E} f(X_t) + \int_0^t \sigma(X_s) P_{t-s}' f(X_s) dW_s. \tag{10.5.5}$$

A similar formula holds for g. So

$$\begin{aligned}\mathbb{E} f(X_t) g(X_t) &= \mathbb{E} f(X_t) \mathbb{E} g(X_t) + \int_0^t \mathbb{E}\{|\sigma(X_t)|^2 P_{t-s}' f(X_s) P_{t-s}' g(X_s)\} ds \\ &\geq \mathbb{E} f(X_t) \mathbb{E} g(X_t),\end{aligned}$$

where the last inequality follows from Lemma 10.7. □

Denote by A the infinitesimal generator associated with (10.5.1):

$$Af(x) = \frac{1}{2}\sigma^2(x)\frac{\partial^2 f}{\partial x^2}(x) + b(x)\frac{\partial f}{\partial x}(x) = \frac{1}{2}\sigma^2 f'' + bf', \quad f \in C^2(\mathbb{R})$$

and $\mathbf{L}g = \frac{\partial g}{\partial s}g + Ag$ for $g \in C^{1,2}([0,1] \times \mathbb{R}; \mathbb{R})$. We will establish a variance inequality for the measure determined by X_t. Precisely, we have

Theorem 10.12. *If* $\sigma b' \le A\sigma$, *then*

$$\mathbb{E}|f(X_t) - \mathbb{E}f(X_t)|^2 \le \mathbb{E}|\sigma(X_t)f'(X_t)|^2 . \tag{10.5.6}$$

Proof. From (10.5.5), we obtain

$$\mathbb{E}|f(X_t) - \mathbb{E}f(X_t)|^2 = \int_0^t \mathbb{E}|\sigma(X_r)P'_{t-r}f(X_r)|^2 dr . \tag{10.5.7}$$

Set $g_r(x) = \sigma(x)P'_{t-r}f(x)$ and apply the Itô formula $\mathbb{E}(h_t(X_t)) = \mathbb{E}(h_r(X_r)) + \int_r^t \mathbb{E}\left[\mathbf{L}h_s(X_s)\right] ds$ to the process $h_s(X_s) = [g_s(X_s)]^2$, $r \le s \le t$. Then, we obtain

$$\mathbb{E}|\sigma(X_t)f'(X_t)|^2 = \mathbb{E}|g_r(X_r)|^2 + \int_r^t \mathbb{E}[\mathbf{L}(g_s^2)(X_s)]ds .$$

Using the simple identity $\mathbf{L}(fg) = f\mathbf{L}g + g\mathbf{L}f + \sigma^2 f'g'$, we have

$$\begin{aligned}\mathbb{E}|\sigma(X_t)f'(X_t)|^2 &= \mathbb{E}|g_r(X_r)|^2 + 2\int_r^t \mathbb{E}[g_s(X_s)\mathbf{L}(g_s)(X_s)]ds \\ &\quad + \int_r^t \mathbb{E}\left[\sigma^2(X_s)\left[g'_s(X_s)\right]^2\right] ds \\ &\ge \mathbb{E}|g_t(X_t)|^2 + 2\int_r^t \mathbb{E}[g_s(X_s)\mathbf{L}(g_s)(X_s)]ds . \end{aligned} \tag{10.5.8}$$

On the other hand, using again the simple identity $\mathbf{L}(fg) = f\mathbf{L}g + g\mathbf{L}f + \sigma^2 f'g'$, we have

$$\mathbf{L}(g_s) = (A\sigma)P'_{t-s}f + \sigma\mathbf{L}P'_{t-s}f + \sigma^2\sigma' P''_{t-s}f . \tag{10.5.9}$$

Differentiating the following equation

$$\frac{1}{2}\sigma^2 P''_{t-s}f + bP'_{t-s}f + \frac{\partial}{\partial s}P_{t-s}f = \mathbf{L}P_{t-s}f = 0$$

with respect to x, we obtain

$$\begin{aligned} 0 &= \sigma\sigma' P''_{t-s}f + b'P'_{t-s}f + \mathbf{L}P'_{t-s}f \\ &= \sigma\sigma' P''_{t-s}f + b'P'_{t-s}f . \end{aligned}$$

Or

$$\mathbf{L}P'_{t-s}f = -\sigma\sigma' P''_{t-s}f - b'P'_{t-s}f . \tag{10.5.10}$$

Combining (10.5.9) and (10.5.10) yields

$$\mathbf{L}g_s = [(A\sigma) - \sigma b']P'_{t-s}f = [\frac{A\sigma}{\sigma} - b']g_s\,. \tag{10.5.11}$$

Inserting (10.5.11) into (10.5.8), we have

$$\begin{aligned}&\mathbb{E}|\sigma(X_t)f'(X_t)|^2\\ &\ge \mathbb{E}|g_t(X_t)|^2 + 2\int_r^t \mathbb{E}\{\sigma(X_s)[(A\sigma) - \sigma b'](X_s)[P'_{t-s}f(X_s)]^2\}ds\,.\end{aligned}$$

When the condition of the theorem is satisfied, we have

$$\mathbb{E}|\sigma(X_t)f'(X_t)|^2 \ge \mathbb{E}|\sigma(X_t)P'_{t-r}f(X_t)|^2\,.$$

This together with (10.5.7) prove the theorem. □

10.6 Hypercontractivity, Spectral and Logarithmic Sobolev Inequality

We shall first consider a more general case of Markov semigroup and then we will apply the general theory to stochastic differential equations.

Let $(E, \mathcal{E}, \mu)$ be a measure space and let $L^2(\mu)$ be the set of all square integrable functions from E to $\mathbb{R}$. Denote $\mu(f) = \int_E f(x)\mu(dx)$. Let L be a Markov generator associated with the continuous Markov semigroup P_t in $L^2(\mu)$.

Assumptions: *We assume that μ is invariant with respect to L and $(P_t)_{t\ge 0}$. There is a nice algebra $\mathcal{A}$ of bounded functions on E which is dense in the L^2 domain of L, stable by L and P_t and by the action of composition with C^∞ real functions which is 0 at 0. P_t is ergodic such that for any f, $P_tf \to \mu(f)$ μ almost surely.*

Following P.-A. Meyer, we introduce the so-called "carré du champs" operator Γ as the symmetric bilinear form on $\mathcal{A}$ defined by

$$\Gamma(f,g) = \frac{1}{2}\{L(fg) - fLg - gLf\}\,, \quad \forall\, f,g\in\mathcal{A}\,. \tag{10.6.1}$$

Denote $\Gamma(f) = \Gamma(f,f)$. It is well-known that Γ is a positive-definite bilinear form on $\mathcal{A}$ if L generates a Markov semigroup. Following [Bakry and Emery (1985)], we define

Definition 10.1. L is called a diffusion if for every C^∞ function Ψ on $\mathbb{R}^n$, and for every finite family $f_1, \cdots, f_n \in \mathcal{A}$,

$$\begin{aligned}L\Psi(f_1,\cdots,f_n) = &\sum_{i=1}^n \frac{\partial\Psi}{\partial x_i}(f_1,\cdots,f_n)L(f_i)\\ &+ \sum_{i,j=1}^n \frac{\partial^2\Psi}{\partial x_i\partial x_j}(f_1,\cdots,f_n)\Gamma(f_i,f_j)\,.\end{aligned} \tag{10.6.2}$$

The iterated "carré du champs" operator is defined as

$$\Gamma_2(f,g) = \frac{1}{2}\left\{L\Gamma(f,g) - \Gamma(f,Lg) - \Gamma(g,Lf)\right\}, \quad \forall\, f,g \in \mathcal{A}. \tag{10.6.3}$$

We also denote $\Gamma_2(f) = \Gamma_2(f,f)$. From (10.6.1) and (10.6.2) it follows that for all C^∞ functions Φ on $\mathbb{R}^m$, Ψ on $\mathbb{R}^n$, and for all finite families $F = (f_1, \cdots, f_m)$, $G = (g_1, \cdots, g_n)$, where $f_1, \cdots, f_m, g_1, \cdots, g_n \in \mathcal{A}$,

$$\Gamma(\Phi(F), \Psi(G)) = \sum_{1\le i\le m, 1\le j\le n} \frac{\partial \Phi}{\partial x_i}(f_1, \cdots, f_m)\frac{\partial \Psi}{\partial y_j}(g_1, \cdots, g_n)\Gamma(f_i, g_j). \tag{10.6.4}$$

In the remaining part of this section, when the argument(s) are clear, we shall write ϕ for $\phi(f)$, ϕ', for $\phi'(f)$ etc. By (10.6.2) and (10.6.3), and (10.6.4) we have

$$\begin{aligned}
\Gamma_2(\phi(f)) &= \frac{1}{2}L\left[(\phi')^2\Gamma(f)\right] - \Gamma(\phi, \phi' Lf + \phi''\Gamma(f)) \\
&= \frac{1}{2}\phi'^2 L\Gamma(f) + \frac{1}{2}\Gamma(f)L(\phi'^2) + \Gamma((\phi')^2, \Gamma(f)) - \phi'\Gamma(f, \phi' Lf + \phi''\Gamma(f)) \\
&= \frac{1}{2}\phi'^2 L\Gamma(f) + \phi'\phi'' Lf\Gamma(f) + \left[\phi'\phi^{(3)} + \phi''^2\right]\Gamma(f)^2) + 2\phi'\phi''\Gamma(f,\Gamma(f)) \\
&\quad -(\phi')^2\Gamma(f,Lf) - \phi'\phi'' Lf\Gamma(f) - \phi'\phi''\Gamma(f,\Gamma(f)) - \phi'\phi^{(3)}\Gamma(f)^2 \\
&= \phi'^2\Gamma_2(f) + \phi'\phi''\Gamma(f,\Gamma(f)) + \phi''\Gamma(f)^2.
\end{aligned} \tag{10.6.5}$$

In fact, we have a more general formula which is also useful later: Let Φ be a C^∞ function on $\mathbb{R}^2$, let $f, g \in \mathcal{A}$. Then,

$$\Gamma_2(\Phi(f,g), \Phi(f,g)) = X^T M X, \tag{10.6.6}$$

where

$$X = \left(\frac{\partial \Phi(x,y)}{\partial x}, \frac{\partial \Phi(x,y)}{\partial y}, \frac{\partial^2 \Phi(x,y)}{\partial x^2}, \frac{\partial^2 \Phi(x,y)}{\partial x \partial y}, \frac{\partial^2 \Phi(x,y)}{\partial y^2}\right)$$

and if we denote $2(f,g,h) = \Gamma(g,\Gamma(f,h)) + \Gamma(h,\Gamma(f,g)) - \Gamma(f,\Gamma(g,h))$, then

$$M = \begin{pmatrix}
\Gamma_2(f) & \Gamma_2(f,g) & (f,f,f) & 2(f,f,g) & (f,g,g) \\
 & \Gamma_2(g) & (g,f,f) & 2(g,g,f) & (g,g,g) \\
 & & \Gamma(f)^2 & 2(\Gamma(f)\Gamma(g) & \Gamma(f,g)^2 \\
 & & & 2\Gamma(f,g) + 2\Gamma(f)\Gamma(g) & 2\Gamma(g)\Gamma(f,g) \\
\text{symmetry} & & & & \Gamma(g)^2
\end{pmatrix}. \tag{10.6.7}$$

To simplify computations, we denote $L_t := L - \frac{\partial}{\partial t}$. Let ϕ be a C^∞ function and let f be an element in $\mathcal{A}$. Then

$$L_t\phi(f_t) = \phi'(f_t)L_t f_t + \phi''(f_t)\Gamma(f_t). \tag{10.6.8}$$

Denote $F_t = P_t f$. By (10.6.8) and the fact that $L_t F_t = 0$.

$$L_t \phi(F_t) = \phi''(F_t)\Gamma(F_t)\,. \tag{10.6.9}$$

We have also if $G_t = P_t g$ for some $g \in \mathcal{A}$, then

$$\begin{aligned}\frac{d}{dt}\Gamma(F_t\,, G_t) &= \frac{1}{2}\frac{d}{dt}\{L(F_t G_t) - F_t L G_t - G_t L F_t\} \\ &= \frac{1}{2}\left\{L(G_t L F_t) + L(F_t L G_t) - 2LF_t LG_t - F_t L^2 G_t - G_t L^2 F_t\right\} \\ &= \Gamma(F_t, LG_t) + \Gamma(LF_t, G_t)\,.\end{aligned}$$

This formula combined with the definition of Γ_2 yields

$$L_t \Gamma(F_t, G_t) = 2\Gamma_2(F_t, G_t)\,. \tag{10.6.10}$$

The following lemma is an important ingredient in the proof of logarithmic Sobolev inequalities etc.

Lemma 10.8. *Assume that there is a $R \in \mathbb{R}$ such that*

$$\Gamma_2(f) \geq R\Gamma(f)\,, \quad \forall\, f \in \mathcal{A}\,. \tag{10.6.11}$$

Then, for every $T \geq 0$ and every $f \in \mathcal{A}$,

$$\sqrt{\Gamma(P_T f)} \leq e^{-RT} P_T \sqrt{\Gamma(f)})\,. \tag{10.6.12}$$

Proof. Let $\phi(x) = x^{1/2}$ and $\Xi_t = e^{-R(T-t)} P_{T-t}\phi(\Gamma(P_t f))$, $0 \leq t \leq T$. Denote $F_t = P_t f$. To simplify notation we write ϕ for $\phi(\Gamma(F_t))$, ϕ' for $\phi'(\Gamma(F_t))$, ϕ'' for $\phi''(\Gamma(F_t))$. Then, by (10.6.8), (10.6.10),

$$\begin{aligned}\Xi_t' &= -e^{-R(T-t)} P_{T-t}\{-R\phi + L_t(\phi)\} \\ &= -e^{-R(T-t)} P_{T-t}\{-R\phi + 2\phi'\Gamma_2(F_t) + \phi''\Gamma(\Gamma(F_t))\} \\ &= -\frac{1}{4} e^{-R(T-t)} P_{T-t}\left\{\phi^{-3}\left[-4R\Gamma(F_t)^2 + 4\Gamma_2(F_t)\Gamma(F_t) - \Gamma(\Gamma(F_t))\right]\right\}\,.\end{aligned} \tag{10.6.13}$$

Since $\Gamma_2(\Phi(f,g)) \geq R\Gamma(\Phi(f,g))$ for all Φ, $M - A$ is positive definite by (10.6.6), where A is a 5×5 matrix whose elements A_{ij} satisfy that $A_{11} = \Gamma(f)$, $A_{12} = A_{21} = \Gamma(f,g)$, $A_{22} = \Gamma(g)$ and all other elements are 0.

Taking the determinant of first and third rows and columns of $M - A$, we obtain that

$$(f,f,f)^2 \leq [\Gamma_2(f) - R\Gamma(f)]\,\Gamma(f)^2\,, \quad \text{or} \quad \Gamma(f,\Gamma(f))^2 \leq 4\Theta\Gamma(f)^2\,, \tag{10.6.14}$$

where $\Theta = \Gamma_2(f) - R\Gamma(f)$. Taking the determinant of first and forth rows and columns of $M - A$, we obtain that

$$\Gamma(g,\Gamma(f))^2 \leq 2\Theta\left\{\Gamma(f,g)^2 + \Gamma(f)\Gamma(g)\right\}\,. \tag{10.6.15}$$

Letting $g = \Gamma(f)$ in (10.6.15) and dominating $\Gamma(f, \Gamma(f))$ by (10.6.14), we obtain

$$\Gamma(g)^2 - 2\Theta\Gamma(f)\Gamma(g) \le 8\Gamma(f)^2\Theta^2 .$$

This is simplified as

$$[\Gamma(g) + 2\Gamma(f)]\,[\Gamma(g) - 4\Theta\Gamma(f)] \le 0 .$$

Thus, for any $f \in \mathcal{A}$,

$$\Gamma(\Gamma(f)) \le 4\Theta\Gamma(f) = 4\,[\Gamma_2(f) - R\Gamma(f)]\,\Gamma(f) .$$

This implies that the sum inside [ ] in (10.6.13) is negative. Thus, $\Xi_t' \le 0$, namely, $\Xi_T \le \Xi_0$, proving the lemma. □

Let $f, g \in \mathcal{A}$ and ϕ, ψ be C^∞ functions on $\mathbb{R}$. Denote $F_t = P_t f$ and $G_t = P_t g$.

Theorem 10.13. *If there is a $R \in \mathbb{R}$ such that $\Gamma_2(f) \ge R\Gamma(f)$, $\forall\ f \in \mathcal{A}$ and if ϕ and ψ are positive concave functions satisfying*

$$\phi \ge 0 , \quad \psi \ge 0 , \quad \phi'' \le 0 , \quad \psi'' \le 0 , \quad \left(e^{-R\tau}\phi'\psi'\right)^2 \le \phi\psi\phi''\psi'' , \tag{10.6.16}$$

then

$$\mu\left\{\psi(g)\ P_\tau(\phi(f))\right\} \le \phi(\mu(f))\psi(\mu(g)) . \tag{10.6.17}$$

Proof. For any fixed $\tau > 0$ and $T > 0$, set

$$h_t = P_{T-t}\left\{(P_\tau(\phi(F_t)))\psi(G_t)\right\} , \quad 0 \le t \le T .$$

Then, by (10.6.2)

$$\begin{aligned} \frac{dh_t}{dt} &= -P_{T-t}\left\{L_t\left[(P_\tau(\phi))\psi\right]\right\} \\ &= -P_{T-t}\left\{L_t\left(P_\tau(\phi)\right)\psi + P_\tau(\phi)L_t\left(\psi\right) + 2\Gamma(P_\tau(\phi)\,,\psi)\right\} \\ &= -P_{T-t}\left\{P_\tau\left[\phi''\Gamma(F_t)\right]\psi + P_\tau[\phi]\psi''\Gamma(G_t) + 2\Gamma(P_\tau(\phi)\,,\psi)\right\} . \end{aligned} \tag{10.6.18}$$

Since Γ is positive definite, we have

$$|\Gamma(P_\tau(\phi), \psi)|^2 \le \Gamma(P_\tau(\phi))\Gamma(\psi)$$

By (10.6.12), we have

$$\begin{aligned} |\Gamma(P_\tau(\phi), \psi)| &\le \sqrt{P_\tau[\Gamma(\phi)]}\sqrt{\Gamma(\psi)} \le e^{-R\tau}P_\tau\sqrt{\Gamma(\phi)}\sqrt{\Gamma(\psi)} \\ &= e^{-R\tau}P_\tau\left[|\phi'|\sqrt{\Gamma(F_t)}\right]|\psi'|\sqrt{\Gamma(G_t)} . \end{aligned}$$

Let us introduce the operator $\tilde{P}_\tau$ acting on $\mathcal{A} \times \mathcal{A}$ (a function of the form $\rho(x, y)$, x , $y \in E$ by $\tilde{P}_\tau(f \times g) = P_\tau(f)(x)g(x)$ (where $f \times g(x, y) = f(x)g(y)\, x, y \in E$). Since P_τ is positive preserving, the same is true for $\tilde{P}_\tau$. Then, we have

$$\frac{dh_t}{dt} \geq -P_{T-t}\left\{\tilde{P}_\tau\left[\phi''\Gamma(F_t) \times \psi + \phi \times \psi''\Gamma(G_t)\right.\right.$$
$$\left.\left. +2e^{-R\tau}|\phi'||\psi'|\sqrt{\Gamma(F_t)} \times \sqrt{\Gamma(G_t)}\right]\right\}. \qquad (10.6.19)$$

If (10.6.16) holds, then $\frac{dh_t}{dt} \geq 0$. Thus, $h_0 \leq h_T$. Computing h_T and h_0, we obtain the following inequality

$$P_T\left\{(P_\tau(\phi(f)))\psi(g)\right\} \leq P_\tau(\phi(P_T(f)))\psi(P_T g) \qquad (10.6.20)$$

Letting $T \to \infty$, the theorem is proven [the smoothness of ϕ and ψ poses no problem by a limiting argument]. □

This theorem is a generalization of the hypercontractivity theorem. To see this let us take $\phi(x) = |x|^{1/p}$ and $\psi(y) = |y|^{1/q'}$ with $1/q + 1/q' = 1$. (10.6.16) is true if and only if $e^{-R\tau} \leq (p-1)/(q-1)$. In this case (10.6.17) reads

$$\mu\left\{P_\tau(|f|^{1/p}))|g|^{1/q'}\right\} \leq (\mu(f))^{1/p}(\mu(g))^{1/q'} .$$

Replacing f by f^p and g by g^q, we have

$$\mu((P_\tau f)g) \leq (\mu(f^p))^{1/p}(\mu(g^{q'}))^{1/q'} .$$

Denote $\|f\|_p^p = \mu(f^p)$, The above inequality implies that $\|P_\tau f\|_q \leq \|f\|_p$. Thus, we have the following hypercontractivity theorem:

Corollary 10.3. *If there is a $R > 0$ such that $\Gamma_2(f) \geq R\Gamma(f)$, $\forall\ f \in \mathcal{A}$ and if $e^{-R\tau} \leq (p-1)/(q-1)$, then $\|P_\tau f\|_q \leq \|f\|_p$.*

Theorem 10.14. *Let ϕ be a C^∞ function satisfying*

$$\phi'' \geq 0, \quad \phi^{(4)} \geq 0, \quad 2(\phi^{(3)})^2 \leq \phi''\phi^{(4)} . \qquad (10.6.21)$$

Assume there is a $R > 0$ such that $\Gamma_2(f) \geq R\Gamma(f)$, $\forall\ f \in \mathcal{A}$. Then,

$$P_T(\phi(f)) \leq \phi(P_T(f)) + \frac{1-e^{-2RT}}{2R} P_T(\phi''(f)\Gamma(f)) . \qquad (10.6.22)$$

Letting $T \to \infty$, we have

$$\mu(\phi(f)) \leq \phi(\mu(f)) + \frac{1}{2R}\mu(\phi''(f)\Gamma(f)) . \qquad (10.6.23)$$

Remark 10.6. In (10.6.22) letting $\phi(x) = x^2$, one gets the Poincaré inequality and letting $\phi(x) = x \log x$, one gets the logarithmic Sobolev inequality

Proof. We continue to use the notation of last section. Let us take $\psi(x) \equiv 1$ and $\tau = 0$. Then, we see easily

$$\frac{dh_t}{dt} = P_{T-t}\{\phi''\Gamma(F)\}.$$

Differentiating this identity once more with respect to t and using formula (2.2) for $\Psi(x, y) = \phi''(x)y$ and $F_1 = F$ and $F_2 = \Gamma(F)$, we obtain

$$\begin{aligned}\frac{d^2h_t}{dt^2} &= P_{T-t}\{L_t[\phi''\Gamma(F)]\} \\ &= P_{T-t}\left\{\phi^{(4)}\Gamma(F)^2 + \phi''L_t\Gamma(F) + 2\phi^{(3)}\Gamma(F, \Gamma(F))\right\} \\ &= P_{T-t}\left\{\phi^{(4)}\Gamma(F)^2 + 2\phi''\Gamma_2(F) + 2\phi^{(3)}\Gamma(F, \Gamma(F))\right\} \\ &= P_{T-t}\{\mathrm{Tr}(AB) + 2R\phi''\Gamma(F)\},\end{aligned}$$

where $A = \begin{pmatrix} \phi''/2 & \phi^{(3)} \\ \phi^{(3)} & \phi^{(4)} \end{pmatrix}$ and $B = \begin{pmatrix} 4[\Gamma_2(F) - R\Gamma(F)] & \Gamma(F, \Gamma(F)) \\ \Gamma(F, \Gamma(F)) & \Gamma(F)^2 \end{pmatrix}$. B is positive definite by (10.6.14). By (10.6.21), A is also positive definite, we have

$$\frac{d^2h_t}{dt^2} \geq 2RP_{T-t}[\phi''(F)\Gamma(F)] = 2R\frac{dh_t}{dt}. \tag{10.6.24}$$

This implies that the derivative of $h_t - h'_t\frac{1-e^{-2Rt}}{2R}$ is positive. Thus,

$$h_T - h_0 \leq h'_T(1 - e^{-2RT})/(2R),$$

proving the theorem. □

Let us consider the hypercontractivity and logarithmic Sobolev inequality for the semirgroup generated by stochastic differential equation. For notational simplicity it seems more convenient for us to consider Stratonovich type stochastic differential equation.

We assume that $A_j : \mathbb{R}^d \to \mathbb{R}^d, j = 0, 1, \cdots, m$ are smooth and their derivatives of order greater than or equal to 1 are bounded by a constant independent of t (the constant may depend on the order). For our purpose in this section we shall consider only time independent Stratonovich type of stochastic differential equation:

$$X(t) = x + \sum_{j=0}^{m} \int_0^t A_j(X(s))d^\circ W_s^j, \quad 0 \leq t \leq T, \tag{10.6.25}$$

where d° denotes the Stratonovich integral. This equation can be written as Itô type as

$$X(t) = x + \int_0^t \tilde{A}_0(X(s)ds + \sum_{j=1}^m \int_0^t A_j(X(s))dW_s^j\,, \quad 0 \le t \le T \quad (10.6.26)$$

with $\tilde{A}_0(x) = A_0(x) + \frac{1}{2}\sum_{j=1}^m [\partial A_j(x)]\, A_j(x)$, where and in what follows $\partial A_j = \left(\partial_k A_j^i\right)_{1\le i,k\le d}$. We identify A_j with the following vector field

$$A_j = A_j(x) = \sum_{i=1}^d A_j^i(x)\partial_i = \langle A_j(x)\,, \partial\rangle\,.$$

Or we may also write the above as

$$A_j f(x) = \sum_{k=1}^d A_j^k(x)\frac{\partial}{\partial x_k} f(x)\,.$$

We see that A_j is an derivative operator, namely,

$$A_j(fg) = fA_j g + gA_j f\,, \quad \forall \;\; f, g \in C^1\,.$$

With these notations, the infinitesimal generator of the semigroup associated with (10.6.25) is given by

$$\mathcal{L}f = \frac{1}{2}\sum_{k=1}^m A_k^2 f + A_0 f\,. \quad (10.6.27)$$

Thus, using the fact that A_k is derivative operator, we have

$$\begin{aligned}\mathcal{L}(fg) &= \frac{1}{2}\sum_{k=1}^m \left[gA_k^2 f + fA_k^2 g + 2A_k f A_k g\right] + gA_0 f + fA_0 g \\ &= g\mathcal{L}(f) + f\mathcal{L}(g) + \sum_{k=1}^m A_k f\, A_k g\,. \quad (10.6.28)\end{aligned}$$

Thus,

$$\Gamma(f,g) = \frac{1}{2}\left[\mathcal{L}(fg) - f\mathcal{L}(g) - g\mathcal{L}(f)\right] = \frac{1}{2}\sum_{k=1}^m A_k f\, A_k g\,. \quad (10.6.29)$$

Applying (10.6.28) to $A_k f$ and $A_k g$, we have

$$\mathcal{L}(\Gamma(f,g)) = \frac{1}{2}\sum_{k=1}^m \left[A_k f \mathcal{L} A_k g + A_k g \mathcal{L} A_k f\right] + \frac{1}{2}\sum_{j,k=1}^m A_j A_k f A_j A_k g\,.$$

We have then

$$\Gamma_2(f,g) = \frac{1}{4}\sum_{k=1}^{m}\left[A_k f\,(\mathcal{L}A_k g - A_k\mathcal{L}g) + A_k g\,(\mathcal{L}A_k g - A_k\mathcal{L}g)\right]$$
$$+\frac{1}{4}\sum_{j,k=1}^{m} A_j A_k f A_j A_k g\,. \tag{10.6.30}$$

With these computations we can state the consequences of Theorem 10.13, Corollary 10.3 and Theorem 10.14 applied to Equation (10.6.25) as follows.

Theorem 10.15. *Let $\Gamma(f) = \Gamma(f,f)$ and $\Gamma_2(f) = \Gamma_2(f,f)$ be defined by* (10.6.29) *and* (10.6.30) *repesctively. Let the solution process X_t to* (10.6.25) *have the limiting probability measure μ. Assume*

$$\Gamma_2(f) \geq R\Gamma(f) \qquad \textit{for some positive number } R. \tag{10.6.31}$$

Then, we have the following statements.

(i) If ϕ and ψ are positive concave functions satisfying for some $T > 0$

$$\phi \geq 0\,, \quad \psi \geq 0\,, \quad \phi'' \leq 0\,, \quad \psi'' \leq 0\,, \quad \left(e^{-RT}\phi'\psi'\right)^2 \leq \phi\psi\phi''\psi''\,, \tag{10.6.32}$$

then

$$\mu\left\{\psi(g)\; P_T(\phi(f))\right\} \leq \phi(\mu(f))\psi(\mu(g))\,, \quad \forall\; f, g\,. \tag{10.6.33}$$

In particular, if $e^{-RT} \leq (p-1)/(q-1)$, then

$$\|P_T f\|_q \leq \|f\|_p\,, \quad \forall\; f. \tag{10.6.34}$$

(ii) Let ϕ be a C^∞ function satisfying

$$\phi'' \geq 0\,, \quad \phi^{(4)} \geq 0\,, \quad 2(\phi^{(3)})^2 \leq \phi''\phi^{(4)}. \tag{10.6.35}$$

Then,

$$P_T(\phi(f)) \leq \phi(P_T(f)) + \frac{1-e^{-2RT}}{2R} P_T(\phi''(f)\Gamma(f))\,, \quad \forall\; f. \tag{10.6.36}$$

Letting $T \to \infty$, we have ($R > 0$)

$$\mu(\phi(f)) \leq \phi(\mu(f)) + \frac{1}{2R}\mu(\phi''(f)\Gamma(f)). \tag{10.6.37}$$

Remark 10.7. If we take $\phi(x) = x^2$ or $\phi(x) = x\log x$ in (10.6.36) or in (10.6.37) we obtain the the Poincaré and logarithmic Sobolev inequality, respectively. We refer to [Bakry (1985)], [Bakry (1994)] [Bakry and Emery (1985)], [Bakry and Ledoux (1996)], [Hu (2000)] for more readings. See also [Driver and Hu (1996)] for a direct proof of (10.6.36).

10.7 Convergence to Density for Ergodic Diffusion

Let $(B_t, 0 \le t < \infty)$ be the d-dimensional standard Brownian motion on some probability space $(\Omega, \mathcal{F}, P)$ with filtration $(\mathcal{F}_t)_{\{0\le t<\infty\}}$. Let $u : \mathbb{R}^d \to \mathbb{R}$ be a nice function whose properties are given below. Denote

$$b(x) = -\frac{1}{2}\nabla u(x)\,. \tag{10.7.1}$$

Consider the stochastic differential equation

$$dx_t = b(x_t)dt + dB_t\,, 0 \le t < \infty\,, \quad x_0 = \xi\,, \tag{10.7.2}$$

where ξ has the probability density function ρ and is independent of the Brownian motion $(B_t\,, t \ge 0)$. To emphasize the dependence on ξ, we also denote the above solution by x_t^ξ. The infinitesimal generator associated with the above equation (10.7.2) is given by

$$Af(x) = \frac{1}{2}\Delta f(x) + b(x)\cdot\nabla f(x)\,, \quad f \in D_A\,,$$

where D_A is the domain of A. Assume that $b(x)$ is bounded on compact set and assume there are $r_0 > 0$ and $z \in \mathbb{R}^d$ such that

$$\int_{r_0}^\infty \exp\{-I_z(r)dr\}\,dr = \infty\,, \tag{10.7.3}$$

where

$$\beta_z(u) = \inf_{|x'|=u}\left(d - 1 + 2\sum_{i=1}^d x_i' b_i(x'+z)\right)\,, \quad I_z(r) = \int_{r_0}^r \frac{\beta_z(u)}{u}du\,. \tag{10.7.4}$$

Then from [Bhattacharya (1982)] we know that x_t is (stationary and) recurrent. If f is in the range of the operator A, then the ergodic theorem holds. This means

$$\frac{1}{T}\int_0^T f(x_r)dr \quad \text{converges to } \mu_\rho(f) \text{ almost surely as } T \to \infty\,, \tag{10.7.5}$$

where

$$\mu_\rho(f) = \int_{\mathbb{R}^d} f(x)\rho(x)dx\,, \quad \text{with} \quad \rho(x) = e^{-u(x)}\,. \tag{10.7.6}$$

Moreover, we also have the following functional central limit type theorem: As $n \to \infty$, the stochastic processes

$$\left(\sqrt{n}\left(\frac{1}{n}\int_0^{nT} f(x_r)dr - \mu_\rho(f)\right)\,, \; T \ge 0\right) \tag{10.7.7}$$

converges in law to a Brownian motion with variance

$$\sigma^2 = -2\int_{\mathbb{R}^d} f(x)g(x)\rho(x)dx\,, \tag{10.7.8}$$

where g is an element in the domain of A such that $Ag = f$. The strict positivity of the above σ^2 is also proved in [Bhattacharya (1982)].

Formally, if we take $f(x) = \delta(x-y)$ in (10.7.7), then we have the central limiting type theorem for convergence to the density

$$\left(\sqrt{n}\left(\frac{1}{n}\int_0^{nT}\delta(x_r - y)dr - \rho(y)\right),\ T\ge 0\right) \tag{10.7.9}$$

converges to a normal distribution. One of the objective of this section is to prove that. We shall only deal with one dimensional case.

Lemma 10.9. *If f is a continuous function with polynomial growth, namely, there are positive positive constants m and K such that*

$$|f(x)| \le K(1+|x|^m)\,. \tag{10.7.10}$$

Then, for all $0\le r<t<\infty$, we have

$$|\mathbb{E}\left[\delta(B_t - y)f(B_r + x)\right]| \le CG_t(y)\left(1 + r^{m/2} + \left|x + \frac{r}{t}y\right|^m\right), \tag{10.7.11}$$

where and throughout this section, we denote by C a generic constant whose value may vary from place to place.

Remark 10.8. In the above lemma, $\mathbb{E}\left[\delta(B_t-y)f(B_r+x)\right]$ is understood as the pairing between the generalized Wiener functional $\delta(B_t - y)$ and the test Wiener functional $f(B_r + x)$.

Proof. Let us consider the approximation of the Donsker delta functional $\delta(B_t - y)$ by the Gaussian kernel $G_\varepsilon(B_t - y)$, where $G_\varepsilon(x) = (2\pi\varepsilon)^{-1/2}\exp\{-|x|^2/(2\varepsilon)\}$. We use $\mathbb{E}\left[\delta(B_t-y)f(B_r+x)\right] = \lim_{\varepsilon\downarrow 0}\mathbb{E}\left[G_\varepsilon(B_t-y)f(B_r+x)\right]$. By the Markov property of the Brownian motion, we have

$$\begin{aligned}
&|\mathbb{E}\left[G_\varepsilon(B_t - y)f(B_r+x)\right]| = |\mathbb{E}\left\{f(B_r+x)\mathbb{E}\left[G_\varepsilon(B_t-y)|\mathcal{F}_r\right]\right\}| \\
&= \left|\int_{\mathbb{R}^2} G_{t-r}(z_1)G_r(z_2)G_\varepsilon(z_2+z_1-y)f(z_2+x)dz_1dz_2\right| \\
&= \left|\int_{\mathbb{R}} G_r(z_2)G_{t-r+\varepsilon}(z_2-y)f(z_2+x)dz_2\right|.
\end{aligned}$$

By the expression of the Gaussian kernel, we have

$$\begin{aligned}
&|\mathbb{E}\left[G_\varepsilon(B_t-y)f(B_r+x)\right]| \\
&= G_{t+\varepsilon}(y)\left|\int_{\mathbb{R}} G_{\frac{r(t-r+\varepsilon)}{t+\varepsilon}}(z)f(z+x+\frac{r}{t+\varepsilon}y)dz\right| \\
&= G_{t+\varepsilon}(y)\mathbb{E}\left|f\left(B_{\frac{r(t-r+\varepsilon)}{t+\varepsilon}}+x+\frac{r}{t+\varepsilon}y\right)\right| \\
&\le C^m K G_{t+\varepsilon}(y)\left(1+\mathbb{E}\left|B_{\frac{r(t-r+\varepsilon)}{t+\varepsilon}}\right|^m+\left|x+\frac{r}{t+\varepsilon}y\right|^m\right) \\
&\le C^m K G_{t+\varepsilon}(y)\left(1+\left|\frac{r(t-r+\varepsilon)}{t+\varepsilon}\right|^{m/2}+\left|x+\frac{r}{t+\varepsilon}y\right|^m\right).
\end{aligned}$$

The lemma follows easily from this estimate. □

Lemma 10.10. *Let $f_n : [0,T]^n \to \mathbb{R}$ be a symmetric (measurable) function of n-variable such that $\int_{[0,T]^n} |f_n(t_1,\cdots,t_n)|^2 dt_1\cdots dt_n < \infty$. Denote the multiple Itô-Wiener integral $I_n(f_n) = \int_{[0,T]^n} f_n(t_1,\cdots,t_n)dB_{t_1}\cdots dB_{t_n}$. Then*

$$\mathbb{E}\left[\delta(B_t-y)I_n(f_n)\right] = \frac{i^n}{2\pi t^{\frac{n+1}{2}}n!}e^{-\frac{y^2}{2t}}\int_{\mathbb{R}} e^{-\frac{|\xi|^2}{2}}\left(\xi - i\frac{y}{\sqrt{t}}\right)^n d\xi \int_{[0,t]^n} f_n(t_1,\cdots,t_n)dt_1\cdots dt_n\,. \quad (10.7.12)$$

Proof. First, let us consider the approximation of the Donsker delta functional.

$$\begin{aligned}
G_\varepsilon(B_t-y) &= \frac{1}{2\pi}\int_{\mathbb{R}} e^{i\xi(B_t-y)-\frac{\varepsilon|\xi|^2}{2}}d\xi \\
&= \frac{1}{2\pi}\int_{\mathbb{R}} e^{-i\xi y-\frac{(\varepsilon+t)|\xi|^2}{2}}e^{i\xi B_t+\frac{t\xi^2}{2}}d\xi \\
&= \sum_{n=0}^{\infty}\frac{1}{2\pi n!}\int_{\mathbb{R}} e^{-i\xi y-\frac{(\varepsilon+t)|\xi|^2}{2}}i^n\xi^n d\xi\; I_n(I_{[0,t]^{\otimes n}}),
\end{aligned}$$

where $I_{[0,t]^{\otimes n}}(t_1,\cdots,t_n) = I_{[0,t]}(t_1)\cdots I_{[0,t]}(t_n)$. Thus, we have obtained the Wiener chaos expansion of $G_\varepsilon(B_t-y)$. By the orthogonality of multiple Itô-Wiener integrals, we have

$$\begin{aligned}
\mathbb{E}\left[G_\varepsilon(B_t-y)I_n(f_n)\right] &= \frac{1}{2\pi n!}\int_{\mathbb{R}} e^{-i\xi y-\frac{(\varepsilon+t)|\xi|^2}{2}}i^n\xi^n d\xi\int_{[0,t]^n} f_n(t_1,\cdots,t_n)dt_1\cdots dt_n \\
&= \frac{i^n}{2\pi(t+\varepsilon)^{\frac{n+1}{2}}n!}e^{-\frac{y^2}{2(t+\varepsilon)}}\int_{\mathbb{R}} e^{-\frac{|\xi|^2}{2}}\left(\xi - i\frac{y}{\sqrt{t+\varepsilon}}\right)^n d\xi \\
&\qquad \int_{[0,t]^n} f_n(t_1,\cdots,t_n)dt_1\cdots dt_n\,.
\end{aligned}$$

Letting $\varepsilon \downarrow 0$, we obtain the lemma. □

Denote

$$F_T(y) = \frac{1}{T}\int_0^T \delta(x_r - y)dr\,. \tag{10.7.13}$$

In the remaining part of this section we shall use the Donsker delta functional and its derivative and so on freely. The justification of their use can follow from the above argument through Gaussian kernel approximation. For example, we may use $F_{T,\varepsilon}(y) = \frac{1}{T}\int_0^T G_\varepsilon(x_r - y)dr$ to approximate $F_T(y)$. To simplify notation, we also denote $F_T = F_T(y)$ (since we shall fix y).

From the Clark-Ocone formula and the chain rule of the derivative we have

$$\begin{aligned} F_T &= \mathbb{E}(F_T) + \int_0^T \mathbb{E}\left(D_t F_T|\mathcal{F}_t\right) dB_t \\ &= \frac{1}{T}\int_0^T p_t(x,y)dt + \frac{1}{T}\int_0^T \left(\int_t^T \mathbb{E}\left[\delta'(x_r - y)D_t x_r|\mathcal{F}_t\right] dr\right) dB_t\,. \end{aligned}$$

From the equation $x_r = x + \int_t^r b(x_u)du + B_r$, where $0 \le t < r < \infty$, we see that if $0 \le t < r < \infty$, then

$$D_t x_r = \int_t^r b'(x_u)D_t x_u du + 1\,.$$

This equation yields

$$D_t x_r = \exp\left\{\int_t^r b'(x_u)du\right\}\,.$$

Therefore we have

$$\begin{aligned} F_T &= \frac{1}{T}\int_0^T p_t(x,y)dt \\ &\quad + \frac{1}{T}\int_0^T \left(\int_t^T \mathbb{E}\left[\delta'(x_r - y)\exp\left\{\int_t^r b'(x_u)du\right\}\Big|\mathcal{F}_t\right] dr\right) dB_t\,. \end{aligned}$$

From the Markov property of the process x_t we have

Proposition 10.2. *Let $p_t(x,y)$ be the transition probability density function associated with x_t. Then*

$$F_T = \frac{1}{T}\int_0^T p_t(x,y)dt + \frac{1}{T}\int_0^T \left(\int_t^T q_{r-t}(x_t,y)dr\right) dB_t\,, \tag{10.7.14}$$

where

$$q_t(x,y) = \mathbb{E}\left[\delta'(x_t^x - y)\exp\left\{\int_0^t b'(x_u^x)du\right\}\right]\,. \tag{10.7.15}$$

Define the probability measure Q on $(\Omega, \mathcal{F}_t)$ by

$$\frac{dQ}{dP} = \exp\left\{-\int_0^t b(x_r)dB_r - \frac{1}{2}\int_0^t b^2(x_r)dr\right\},$$

where $(x_r\,, r \geq 0)$ is the solution of the equation (10.7.2) with initial condition $x_0 = x$. Then, under the probability measure Q, $(x_r, 0 \leq r \leq t)$ is a Brownian motion starting at x. The existence of such measure Q can be verified through Novikov or Kazamaki conditions. We can also write

$$\begin{aligned}\frac{dP}{dQ} &= \exp\left\{\int_0^t b(x_r)dB_r + \frac{1}{2}\int_0^t b^2(x_r)dr\right\}\\ &= \exp\left\{\int_0^t b(x_r)dx_r - \frac{1}{2}\int_0^t b^2(x_r)dr\right\}.\end{aligned}$$

From the Girsanov formula we have

$$\begin{aligned}q_t(x,y) &= \mathbb{E}\left[\delta'(x_t^x - y)\exp\left\{\int_0^t b'(x_u^x)du\right\}\right]\\ &= \mathbb{E}^Q\left[\delta'(x_t^x - y)\exp\left\{\int_0^t b'(x_u^x)du\right\}\exp\left\{\int_0^t b(x_r)dx_r - \frac{1}{2}\int_0^t b^2(x_r)dr\right\}\right]\\ &= \mathbb{E}\Bigg(\delta'(x + B_t - y)\exp\left\{\int_0^t b'(x + B_r)dr\right\}\\ &\quad \exp\left\{\int_0^t b(x + B_r)dB_r - \frac{1}{2}\int_0^t b^2(x + B_r)dr\right\}\Bigg).\end{aligned}$$

From the Itô formula and the equation (10.7.1) we have

$$-\frac{1}{2}u(x + B_t) = -\frac{1}{2}u(x) + \int_0^t b(x + B_r)dB_r - \frac{1}{4}\int_0^t u''(x + B_r)dr.$$

Namely,

$$\int_0^t b(x + B_r)dB_r = -\frac{1}{2}u(x + B_t) + \frac{1}{2}u(x) + \frac{1}{4}\int_0^t u''(x + B_r)dr.$$

Therefore,

$$\begin{aligned}q_t(x,y) &= \mathbb{E}\Bigg(\delta'(x + B_t - y)\exp\left\{-\frac{1}{2}\int_0^t u''(x + B_r)dr\right\}\\ &\quad \exp\left\{-\frac{1}{2}u(x+B_t)+\frac{1}{2}u(x)-\frac{1}{2}\int_0^t\left[\frac{1}{4}(u')^2-\frac{1}{2}u''\right](x+B_r)dr\right\}\Bigg)\\ &= \mathbb{E}\Bigg(\delta'(x + B_t - y)\exp\Bigg\{-\frac{1}{2}u(x + B_t)\\ &\quad +\frac{1}{2}u(x)-\frac{1}{2}\int_0^t\left[\frac{1}{4}(u')^2+\frac{1}{2}u''\right](x+B_r)dr\Bigg\}\Bigg).\end{aligned} \tag{10.7.16}$$

We shall use integration by parts formula to estimate (10.7.16). Denote $\langle f, g\rangle = \int_0^\infty f(r)g(r)dr$ and set for some $\alpha > 1/2$,

$$\begin{cases} \eta_t(r) = \alpha r^{\alpha-1} I_{[0,t]}(r) \quad \text{and} \\ \Theta_t = \exp\left\{-\frac{1}{2}u(x+B_t) - \frac{1}{2}\int_0^t \left[\frac{1}{4}(u')^2 + \frac{1}{2}u''\right](x+B_r)dr\right\}. \end{cases}$$

Then, the equation (10.7.16) becomes

$$\begin{aligned} q_t(x,y) &= e^{\frac{1}{2}u(x)}\mathbb{E}\left(\delta'(x+B_t-y)\Theta_t\right) \\ &= e^{\frac{1}{2}u(x)}\mathbb{E}\left(\langle D\delta(x+B_t-y), \eta_t\rangle\langle DB_t, \eta_t\rangle^{-1}\Theta_t\right) \\ &= e^{\frac{1}{2}u(x)}\langle DB_t, \eta_t\rangle^{-1}\mathbb{E}\left(\delta(x+B_t-y)\delta\left[\eta_t\Theta_t\right]\right) \\ &= e^{\frac{1}{2}u(x)}\langle DB_t, \eta_t\rangle^{-1}\mathbb{E}\left(\delta(x+B_t-y)\left[\delta(\eta_t)\Theta_t - \langle\eta_t, D\Theta_t\rangle\right]\right), \end{aligned} \tag{10.7.17}$$

where in the above last two expressions the first δ is the Dirac delta function and the second δ is the divergence operator (Skorohod integral). We use the same notation to be consistent with the literature and because there is no confusion here. Now

$$\begin{aligned} \langle DB_t, \eta_t\rangle &= \int_0^t \eta_t(r)dr = t^\alpha \\ \delta(\eta_t) &= \alpha\int_0^t r^{\alpha-1}dB_r \\ D_r\Theta_t &= -\Theta_t\left[\frac{1}{2}u'(x+B_t) + \int_r^t \tilde{u}(x+B_\rho)d\rho\right] \quad \forall\, 0 \le r \le t, \end{aligned}$$

where $\tilde{u} = \left[\frac{1}{4}u'u'' + \frac{1}{4}u'''\right]$. Therefore, we have

$$\begin{aligned} q_t(x,y) = e^{\frac{1}{2}u(x)}t^{-\alpha}\mathbb{E}\Big\{\delta(x+B_t-y)\Theta_t\Big[\alpha\int_0^t r^{\alpha-1}dB_r \\ +\frac{1}{2}t^\alpha u'(x+B_t) + \int_0^t \rho^\alpha\tilde{u}(x+B_\rho)d\rho\Big]\Big\}. \end{aligned} \tag{10.7.18}$$

Thus, we have

$$
\begin{aligned}
q_t^2(x,y) &\le e^{u(x)}t^{-2\alpha}\mathbb{E}\left\{\delta(x+B_t-y)\Theta_t^2\right\}\mathbb{E}\Bigg\{\delta(x+B_t-y) \\
&\qquad \left[\alpha\int_0^t r^{\alpha-1}dB_r+\frac{1}{2}t^\alpha u'(x+B_t)+\int_0^t \rho^\alpha\tilde{u}(x+B_\rho)d\rho\right]^2\Bigg\}. \\
&\le Ce^{u(x)}t^{-2\alpha}\mathbb{E}\left\{\delta(x+B_t-y)\Theta_t^2\right\} \\
&\qquad \Bigg\{\mathbb{E}\left(\delta(x+B_t-y)\left[\int_0^t r^{\alpha-1}dB_r\right]^2\right) \\
&\qquad +t^{2\alpha}\mathbb{E}\left[\delta(x+B_t-y)(u'(x+B_t))^2\right] \\
&\qquad +t^{\alpha+1}\mathbb{E}\left[\delta(x+B_t-y)\int_0^t \tilde{u}^2(x+B_r)dr\right]\Bigg\}. \qquad (10.7.19)
\end{aligned}
$$

Similar to Lemma 10.9 if u' satisfies (10.7.10), then we have

$$
\begin{aligned}
&\mathbb{E}\left[\delta(x+B_t-y)(u'(x+B_t))^2\right] \\
&\quad = G_t(x-y)|u'(y)|^2 \le CG_t(x-y)[1+|y|^{2m}]. \qquad (10.7.20)
\end{aligned}
$$

From Lemma 10.9, we see that

$$
\begin{aligned}
&\mathbb{E}\left[\delta(x+B_t-y)\int_0^t \tilde{u}^2(x+B_r)dr\right] \\
&\le CG_t(x-y)\int_0^t\left[1+r^m+|x|^{2m}+\left|\frac{r}{t}(y-x)\right|^{2m}\right]dr \\
&\le tC(t+|x|^2+|y-x|^2)^m G_t(x-y). \qquad (10.7.21)
\end{aligned}
$$

Itô formula yields

$$
\left[\int_0^t r^{\alpha-1}dB_r\right]^2 = I_2\left((r_1r_2)^{\alpha-1}I_{[0,t]}(r_1)I_{[0,t]}(r_2)\right)+\frac{1}{2\alpha-1}t^{2\alpha-1}.
$$

Thus

$$
\begin{aligned}
&\mathbb{E}\left\{\delta(x+B_t-y)\left[\int_0^t r^{\alpha-1}dB_r\right]^2\right\}\\
&=\mathbb{E}\left(\delta(x+B_t-y)I_2\left((r_1r_2)^{\alpha-1}I_{[0,t]}(r_1)I_{[0,t]}(r_2)\right)\right)\\
&\qquad+\frac{1}{2\alpha-1}t^{2\alpha-1}\mathbb{E}\left(\delta(x+B_t-y)\right)\\
&\le C\Bigg[t^{2\alpha-1}G_t(x-y)\\
&\qquad+t^{-3/2}e^{-\frac{y^2}{2t}}\int_{\mathbb{R}}e^{-\frac{|\xi|^2}{2}}\left(|\xi|+\frac{|y|}{\sqrt{t}}\right)^2d\xi\int_0^t\int_0^t(r_1r_2)^{\alpha-1}dr_1dr_2\Bigg]\\
&\le C\left[t^{2\alpha-1}G_t(x-y)+2t^{2\alpha-3/2}e^{-\frac{y^2}{2t}}\left(C+\frac{|y|^2}{t}\right)\right]\\
&\le Ct^{2\alpha-1}\left[G_t(x-y)+G_{2t}(y)\right] \qquad (10.7.22)
\end{aligned}
$$

since $\left(\frac{|y|}{\sqrt{t}}\right)^m G_{2t}(y)\le C_m$. Define

$$
\begin{aligned}
\tilde{q}_t(x,y)&:=\mathbb{E}\left\{\delta(x+B_t-y)\Theta_t^2\right\}\\
&=\mathbb{E}\Bigg\{\delta(x+B_t-y)\exp\Bigg[-u(x+B_t)\\
&\qquad-\int_0^t\left[(u')^2+\frac{1}{2}u''\right](x+B_r)dr\Bigg]\Bigg\}. \qquad (10.7.23)
\end{aligned}
$$

If $\left[(u')^2+\frac{1}{2}u''\right](x)\ge\kappa\,,\forall x\in\mathbb{R}$, for some positive constant κ, then

$$
\begin{aligned}
|\tilde{q}_t(x,y)|&\le Ce^{-\kappa t}\mathbb{E}\left\{\delta(x+B_t-y)\exp\left[-u(x+B_t)\right]\right\}\,.\\
&=Ce^{-\kappa t}G_t(y-x)e^{-u(y)}\,. \qquad (10.7.24)
\end{aligned}
$$

Combining the estimates in (10.7.19)-(10.7.24), we have

$$
\begin{aligned}
q_t^2(x,y)\le C^m e^{u(x)}t^{-2\alpha}e^{-\kappa t}G_t(x-y)e^{-u(y)}\cdot\Big\{&t^{2\alpha-1}\left[G_t(x-y)+G_{2t}(y)\right]\\
&t^{2\alpha}|y|^mG_t(x-y)+t^{\alpha+2}(|x|^2+t)^mG_t(x-y)\Big\}.
\end{aligned}
$$

Thus, we have the following estimate for $q_t(x,y)$

$$
\begin{aligned}
q_t(x,y)\le C^m e^{\frac{u(x)-u(y)}{2}}e^{-\kappa t/2}\cdot\Big\{&t^{-1/2}G_t(x-y)+t^{-1/2}\sqrt{G_t(x-y)G_{2t}(y)}\\
&+(|x|^2+|y|^2+1+t)^mG_t(x-y)\Big\}.
\end{aligned}
$$

Since $G_{2t}(x-y)\le Ct^{-1/2}$ and $\sqrt{G_t(x-y)}\le Ct^{1/4}G_{2t}(x-y)$, we have

Proposition 10.3. *Let u' and u'' satisfy the polynomial growth condition* (10.7.10) *for some $m > 0$ and let a positive constant $\kappa > 0$ exists such that*

$$(u'(x))^2 + 2u''(x) \geq \kappa\,, \forall\; x \in \mathbb{R}\,. \tag{10.7.25}$$

Let $q_t(x,y)$ be defined by (10.7.15)*. Then, there is a positive constant $C > 1$ such that*

$$q_t(x,y) \leq C^m e^{\frac{u(x)-u(y)}{2}} e^{-\kappa t/2}\left\{t^{-1/2}G_{2t}(x-y)+(|x|^2+|y|^2+1+t)^m G_t(x-y)\right\}. \tag{10.7.26}$$

Now we consider the rate of convergence of F_T to invariant measure μ_ρ.

We have from (10.7.14)

$$\begin{aligned}
&\sqrt{T}\,[F_T(y)-\rho(y)]\\
&= \frac{1}{\sqrt{T}}\int_0^T [p_t(x,y)-\rho(y)]\,dt + \frac{1}{T}\int_0^T\left(\int_0^{T-t} q_t(x_t,y)dr\right)dB_t\\
&= \frac{1}{\sqrt{T}}\int_0^T [p_t(x,y)-\rho(y)]\,dt + \frac{1}{\sqrt{T}}\int_0^T\left(\int_0^{\infty} q_t(x_t,y)dr\right)dB_t\\
&\qquad -\frac{1}{\sqrt{T}}\int_0^T\left(\int_{T-t}^{\infty} q_t(x_t,y)dr\right)dB_t\\
&=: I_1(T)+I_2(T)+I_3(T)\,.
\end{aligned}$$

From Theorem 1.1 of [Ganidis, Roynette and Simonot (1999)] it follows that if $u'(x) \sim |x|^\alpha$ when $x \to \infty$ for some $\alpha > 1$, then for any fixed x,

$$\left|\int_{\mathbb{R}} p_t(x,z)f(z)dz - \int_{\mathbb{R}}\rho(z)f(z)dz\right| \leq Ce^{-\sigma t}\int_{\mathbb{R}}|f(y)|dy \quad \forall f \in L^1(\mathbb{R},dx)\,.$$

Choose $f(z) = \frac{1}{2\varepsilon}I_{\{|z-y|<\varepsilon\}}(z)$ and let $\varepsilon \downarrow 0$. Then, we see that

$$|p_t(x,y)-\rho(y)| \leq Ce^{-\sigma t}\,.$$

This implies that $I_1(T)$ goes to 0 for every x and y fixed.

From the estimate (10.7.26), we use $G_t(x-y) \leq Ct^{-1/2}$. It is then elementary to see that

$$\int_{T-t}^\infty |q_r(x,y)|dr \leq Ce^{-\kappa_1(T-t)}\left[1+\log(T-t)+(|x|^2+|y|^2)^m\right]$$

for some positive constant $\kappa_1 > 0$. From this estimate we see that

$$\begin{aligned}
&\mathbb{E}\,|I_3(T)|^p\\
&\leq \frac{1}{\sqrt{T^p}}\left\{\int_0^T Ce^{-2\kappa_1(T-t)}\left[1+\log(T-t)+(|x_t|^2+|y|^2)^m\right]^2 dt\right\}^{p/2},
\end{aligned}$$

which converges to 0. From the estimate (10.7.26), we see that

$$\int_0^\infty |q_r(x,y)|dr \le C^m e^{\frac{u(x)-u(y)}{2}} \left\{ \left(1+|x|^2+|y|^2\right)^m + \log(|x-y|) \right\} .$$

This can be used to prove that

$$\mathbb{E}\left|\int_0^\infty q_r(x_t,y)dr\right|^p < \infty$$

for all $p \ge 1$. Thus, as a function T, $\int_0^T \left(\int_0^\infty q_t(x_t,y)dr\right) dB_t$ is a martingale. There is a Brownian motion $\tilde{B}_t$ such that

$$\int_0^T \left(\int_0^\infty q_r(x_t,y)dr\right) dB_t = \tilde{B}_{\tau_T} ,$$

where

$$\tau_T = \int_0^T \left(\int_0^\infty q_r(x_t,y)dr\right)^2 dt .$$

Therefore,

$$I_2(T) = \frac{1}{\sqrt{T}}\tilde{B}_{\tau_T} \overset{d}{=} \tilde{B}_{\tau_T/T} ,$$

where $X \overset{d}{=} Y$ means that X and Y have the same distribution.

But by the ergodic theorem we know that

$$\frac{\tau_T}{T} = \frac{1}{T}\int_0^T \left(\int_0^\infty q_r(x_t,y)dr\right)^2 dt$$

converges almost surely to

$$\sigma(y) := \int_{\mathbb{R}} \left(\int_0^\infty q_r(x,y)dr\right)^2 \rho(x)dx .$$

Thus, we have that $I_2(T)$ converges in distribution to a normal random variable with mean 0 and variance $\sigma(y)$.

Summarizing the above argument, we have

Theorem 10.16. *Let u' and u'' satisfy the polynomial growth condition* (10.7.10) *for some $m > 0$ and let a positive constant $\kappa > 0$ exist such that*

$$(u'(x))^2 + 2u''(x) \ge \kappa , \forall\, x \in \mathbb{R} . \tag{10.7.27}$$

Let $u'(x)$ have the same order as $|x|^\alpha$ when $x \to \infty$ for some $\alpha > 1$. Then

$$\sqrt{T}\left[\frac{1}{T}\int_0^T \delta(x_t - y)dt - \rho(y)\right] \overset{d}{\to} N(0,\sigma(y)) , \tag{10.7.28}$$

where

$$\sigma(y) = \int_{\mathbb{R}} \left(\int_0^\infty q_r(x,y)dr\right)^2 \rho(x)dx . \tag{10.7.29}$$

Remark 10.9. Other conditions other than (10.7.25) may also be used. For example one can assume that

$$\left[(u')^2 + 2u''\right](x) \geq \kappa |x|^\alpha \,, \forall x \in \mathbb{R}$$

for some positive constant κ and some $\alpha > 0$. Denote

$$k_t(x,y) := \mathbb{E}\left\{\delta(x + B_t - y)\exp\left[-\int_0^t \left[(u')^2 + \frac{1}{2}u''\right](x + B_r)dr\right]\right\}.$$

Then, from Corollary 1 of [Sikora (1997)], we see that

$$k_t(x,y) \leq \begin{cases} Ct^{-d/2}e^{-ct(|x|^{\alpha'}+|y|^{\alpha'})} & \text{when } t \leq 1 \\ \\ Ce^{-c\lambda_1 t}e^{-ct(|x|^{\alpha''}+|y|^{\alpha''})} & \text{when } t > 1\,, \end{cases}$$

where λ_1 is the smallest eigenvalue of $-\frac{1}{2}\Delta + |x|^\alpha$, $\alpha' = 2 \wedge \alpha$ and $\alpha'' = \alpha \wedge (1 + \alpha/2)$ and C and c are two positive constants.

Remark 10.10. Here are some formal computations for $q_t(x,y)$. If we denote

$$\bar{q}_t(x,y) = \mathbb{E}\left[\delta(x_t^x - y)\exp\left\{\int_0^t b'(x_u^x)du\right\}\right], \tag{10.7.30}$$

then $q_t(x,y) = \frac{\partial}{\partial y}\bar{q}_t(x,y)$. From the Feynman-Kac formula we know that $\bar{q}_t(x,y)$ is the solution of the following equation:

$$\frac{\partial}{\partial t}\bar{q}_t(x,y) = \frac{1}{2}\bar{q}_t''(x,y) + b(x)\bar{q}_t'(x,y) + b'(x)\bar{q}_t(x,y)\,,$$

where $\bar{q}_t'(x,y)$ and $\bar{q}_t''(x,y)$ denote the first and second derivatives with respect to x. Formally, if we denote

$$h(x,y) := \int_0^\infty \bar{q}_t(x,y)dt\,.$$

Then, $\int_0^\infty q_r(x,y)dr = \frac{\partial}{\partial y}h(x,y)$ and h satisfies

$$\frac{1}{2}\bar{h}''(x,y) + b(x)\bar{h}'(x,y) + b'(x)\bar{h}(x,y) = \delta(x-y)\,.$$

Chapter 11

Numerical Approximation of Stochastic Differential Eequation

In this section, we discuss two types of approximation of stochastic differential equations: time discretization schemes and Wong-Zakai approximation. For time discretization schemes there are two types of convergence: strong convergence $\mathbb{E}|X_t^\pi - X_t|^p$ or weak convergence $|\mathbb{E}(X_t^\pi) - \mathbb{E}(X_t)|$. If a rate is given, the number of terms to be included in the numerical scheme is different depending on that the rate is strong or weak. We shall present the proofs for both strong and weak convergence theorems for quite general schemes.

In statistics of stochastic differential equations, it is important to know the transition density to construct the maximum likelihood estimator. Since the explicit form of the transition density is usually not available, a natural way is to use numerical schemes. We shall also present a result on the rate of convergence of the density of X_t^π to the density of X_t.

Wong-Zakai approximation is another important approximation. These types of approximated processes are not adapted and they converge to a Stratonovich type equation (Itô equation with a correction term). The rate of convergence in L^p will be discussed.

11.1 L^p-Convergence Rate

Let $\left\{W_t = (W_t^1, \dots, W_t^m), 0 \le t \le T\right\}$ be m standard (real) independent Brownian motions on some time interval $[0,T]$ (bounded, and kept fixed below). Let $(\mathbb{B}, \mathbb{P})$ be the canonical Wiener space with the natural filtration $(\mathcal{F}_t)_{0\le t\le T}$. On $\mathbb{R}^d$ consider the following stochastic differential equation in Ito's sense

$$X_t = x + \sum_{j=0}^{m} \int_0^t A_j(s, X_s)\, dW_s^j \ , \ t \in [0,T] \ , \ x \in \mathbb{R}^d \ , \tag{11.1.1}$$

where $A_j = (A_j^1, \cdots, A_j^m) : [0,T] \times \mathbb{R}^d \to \mathbb{R}^d$ are some given smooth functions with bounded derivatives up to certain order specified later. We also use the convention $dW_s^0 = ds$ to simplify notation.

A time discretization method consists of the following.

(i) Divide the interval $[0,T]$ into smaller subintervals: $\pi : 0 = t_1 < t_2 < \cdots < t_{N-1} = t_N = T$.
(ii) Apply the Itô–Taylor formula (to be described later) on each subinterval and keep a given number of terms.
(iii) Piece together these approximations on each subinterval to get an approximate solution.

We then expect that these approximations will converge to the true solution when the subintervals become finer and finer.

To describe the Itô-Taylor formula, we introduce the following operators on functions $h : [0,T] \times \mathbb{R}^d \to \mathbb{R}$

$$
\begin{aligned}
L^j h(s,x) &= \sum_{k=1}^{d} A_j^k(s,x) \frac{\partial h}{\partial x^k}(s,x) , \quad j = 1, \cdots, m , \\
L^0 h(s,x) &= \frac{\partial h}{\partial s}(s,x) + \sum_{k=1}^{d} A_0^k(s,x) \frac{\partial h}{\partial x^k}(s,x) \\
&\quad + \frac{1}{2} \sum_{k,l=1}^{d} \sum_{j=1}^{m} A_j^k(s,x) A_j^l(s,x) \frac{\partial^2 h}{\partial x^k \partial x^l}(s,x),
\end{aligned}
$$

where A_j^k is the k-th component of the vector A_j $(k = 0, 1, \cdots, d)$. L^j is the operator A_j in section 10.2.

Consider a partition of the interval $[0,T]$, $\pi : 0 = t_0 < t_1 < \cdots < t_N = T$ and put $|\pi| = \sup_i (t_{i+1} - t_i)$, the *step* of the partition. On each subinterval $[t_n, t_{n+1}]$ we write (11.1.1) as

$$
X_t = X_{t_n} + \sum_{j=0}^{m} \int_{t_n}^{t} A_j(s, X_s) \, dW_s^j . \tag{11.1.2}
$$

For a sufficiently differentiable function $h : [0,T] \times \mathbb{R}^d \to \mathbb{R}$, an application of the Itô formula to $h(t, X_t)$ gives

$$
h(t, X_t) = h(t_n, X_{t_n}) + \sum_{j=0}^{m} \int_{t_n}^{t} L^j h(s, X_s) \, dW_s^j . \tag{11.1.3}
$$

This is the first order Itô-Taylor formula. To describe higher order Itô-Taylor formulas we introduce the following notation

$$\begin{aligned} \alpha &= (\alpha_1, \cdots, \alpha_l) \quad (0 \le \alpha_i \le m) \quad \text{(a multi-index)}, \\ l(\alpha) &= l, \quad n(\alpha) = \text{the number of zeroes among } \alpha_1, \ldots \alpha_l \,, \\ f_\alpha^k &= L^{\alpha_1} \cdots L^{\alpha_{l-1}} A_{\alpha_l}^k \quad (k = 1, \cdots, d), \quad f_\alpha = (f_\alpha^1, \ldots, f_\alpha^d), \end{aligned} \tag{11.1.4}$$

$$I_\alpha[g(\cdot)]_{s,t} = \int_{s<s_1<\cdots<s_l<t} g(s_1)\, dW_{s_1}^{\alpha_1} \cdots dW_{s_l}^{\alpha_l} \,, \tag{11.1.5}$$

where g is an adapted process. We put simply $I_{\alpha,s,t} = I_\alpha[1]_{s,t}$ in the case $g = 1$. These are the standard multiple integrals (including dt). They replace the monomials in the classical Taylor expansion. Notice that f_α^k is a reall valued function on $[0, T] \times \mathbb{R}^d$.

Now the generic scheme to obtain a general higher order Itô-Taylor formulas is as follows. In formula (11.1.2), we apply the Itô formula (11.1.3) to *some* processes $A_j(t, X_t)$ — usually to all, but the coefficient $A_0(t, X_t)$ may play a special role. Then, in the new formula we apply again (11.1.3) to *some* coefficients of the stochastic integrals, etc. This way, we obtain a general formula with the following structure of a main term plus a remainder

$$X_t = X_{t_n} + \sum_{\alpha \in \Gamma} f_\alpha(t_n, X_{t_n}) I_{\alpha,t_n,t} + \sum_{\alpha \in \Gamma'} I_\alpha[f_\alpha(\cdot, X_\cdot)]_{t_n,t} \,, \quad (t \in [t_n, t_{n+1}]) \,. \tag{11.1.6}$$

The "main term" is a sum over a finite set Γ of multi-indices, which has the following property: if $\alpha = (\alpha_1, \cdots, \alpha_l) \in \Gamma$, then $-\alpha := (\alpha_2, \cdots, \alpha_l) \in \Gamma$. On the other hand, Γ' is the set $\{\alpha : \alpha \notin \Gamma, -\alpha \in \Gamma\}$. This structure comes from the fact that each term is obtained by applying (11.1.3) to a preceding term.

Now a so called discretization scheme is obtained by discarding the remainder in (11.1.6). Since X_t is not known in the recursive computation, we replace it by its approximation $Y_t = Y_t^\pi$ (we omit its explicit dependence on partition π to simplify notation), and then we obtain the following time discretization approximation scheme, starting at $Y_0 = x$

$$Y_t = Y_{t_n} + \sum_{\alpha \in \Gamma} f_\alpha(t_n, Y_{t_n})\, I_{\alpha,t_n,t} \,, \quad (t \in [t_n, t_{n+1}]) \quad n = 0, 1, \cdots, N-1 \,. \tag{11.1.7}$$

This is a recursive formula. First, we know $Y_0 = x$. Then, we can use it to compute Y_t for all $t \in [t_0, t_1]$. Then, we can use this formula to compute Y_t for $t \in [t_1, t_2]$, and so on. This way we can compute Y_t for all

$t \in [0,T]$. In computer simulation it is common to use the above formula to compute the value of Y_t at the partition points t_k, $k = 1, 2, \cdots, N$ by the following recursive formula (restricted (11.1.7) at the partition points):

$$Y_{t_{n+1}} = Y_{t_n} + \sum_{\alpha \in \Gamma} f_\alpha(t_n, Y_{t_n})\, I_{\alpha, t_n, t_{n+1}}\,, \quad n = 0, 1, \cdots, N-1\,. \quad (11.1.8)$$

And then we use appropriate interpolation to find all values of Y_t.

In practice, we have to choose which terms to be included in Γ. The principle is to choose Γ which has the least number of elements (so that one needs to perform minimum computations) while it gives the desired rate of convergence. We shall consider two types of rates of convergence. One is the strong rate $\left(\mathbb{E}\left[|Y_t - X_t|^p\right]\right)^{1/p}$ and the other one is the weak rate: $|\mathbb{E}(Y_t) - \mathbb{E}(X_t)|$. For these two different rates, the choices of Γ will be different.

First, we describe the choice of Γ for the strong rate. We give ourselves a parameter γ which is an integer or half-integer. Our purpose is to have a norm estimate like (11.1.11) below. Then, the number of terms to take in the Itô-Taylor formula, namely, the choice of Γ, is

$$\Gamma = \mathcal{A}_\gamma = \left\{\alpha : l(\alpha) + n(\alpha) \le 2\gamma \quad \text{or} \quad l(\alpha) = n(\alpha) = \gamma + 1/2\right\}.$$

Let us recall that we use C to denote a generic constant whose precise value doesn't interest us, and may change from occurrence to occurrence. It may depend on several parameters (T, γ, ...) but never on the partition π.

Theorem 11.1. *Let γ be a positive integer or positive half integer. Let $\pi : 0 = t_0 < t_1 < \cdots < t_{N-1} < t_N = T$ be a partition of the interval $[0,T]$ with $|\pi| = \max_{j \le 0 \le N-1}(t_{j+1} - t_j)$. Let Y_t be defined by (11.1.7). Assume that for all $\alpha \in \mathcal{A}_\gamma$ the coefficients f_α defined by (11.1.4) satisfy the Lipschitz conditions*

$$|\, f_\alpha(t,x) - f_\alpha(t,y)\,| \le C\,|x - y| \quad (11.1.9)$$

and the linear growth condition

$$|\, f_\alpha(t,x)\,| \le C(1 + |x|) \quad (11.1.10)$$

for all $t \in [0,T]$ and $x, y \in \mathbb{R}^d$. Then, we have

$$\mathbb{E}\Big(\sup_{0 \le t \le T} |X_t - Y_t|^p\Big) \le C|\pi|^{p\gamma}\,. \quad (11.1.11)$$

We shall prove the theorem in the case $p = 2$ and general case is completely similar (see for example, Lemma 11.3). Before the proof we need two lemmas.

Lemma 11.1. *Let $g(s)$ be an adapted process.*

(i) When $l(\alpha) = n(\alpha), t \in [t_n, t_{n+1}]$,

$$\sup_{t_n \le s \le t} |I_\alpha[g(\cdot)]_{t_n,s}| \le C(t-t_n)^{l(\alpha)-1/2}\Big(\int_{t_n}^t |g(u)|^2 du\Big)^{1/2} \quad \text{a.e.} \tag{11.1.12}$$

(ii) When $l(\alpha) \ne n(\alpha)$,

$$\mathbb{E} \sup_{t_n \le s \le t} (I_\alpha[g(\cdot)]_{t_n,s})^2 \le C(t-t_n)^{l(\alpha)+n(\alpha)-1} \int_{t_n}^t \mathbb{E}|g(u)|^2 du \,. \tag{11.1.13}$$

Here and in what follows C *may depend on* α*, but since* α *ranges over a finite set* $\mathcal{A}_\gamma$ *this dependence is not important.*

Proof. (i) By the definition (11.1.5), $l(\alpha) = n(\alpha)$ means that there is no stochastic integral in $I_\alpha[g(\cdot)]_{t_n,s}$. Inequality (11.1.12) is obvious.
(ii) When $l(\alpha) \ne n(\alpha)$, we prove (11.1.13) by induction on the length $l(\alpha)$ of α. We define $\alpha- := (\alpha_1, \cdots, \alpha_l)$ if $\alpha = (\alpha_1, \cdots, \alpha_l, \alpha_{l+1})$. The case $l(\alpha) = 1$ is easy by discussing $n(\alpha) = 0$ and $n(\alpha) \ne 0$. For the passage from $\alpha- = (\alpha_1, \cdots, \alpha_l)$ to $\alpha = (\alpha_1, \cdots, \alpha_l, \alpha_{l+1})$ we also handle the cases $\alpha_{l+1} = 0$ and $\alpha_{l+1} \ne 0$ differently. When $\alpha_{l+1} = 0$, $I_\alpha[g(\cdot)]_{t_n,s} = \int_{t_n}^s I_{\alpha-}[g(\cdot)]_{t_n,u} du$. By Hölder's inequality

$$\begin{aligned}
\mathbb{E} \sup_{t_n \le s \le t} (I_\alpha[g(\cdot)]_{t_n,s})^2 &\le C(t-t_n) \int_{t_n}^t \mathbb{E}|I_{\alpha-}[g(\cdot)]_{t_n,u}|^2 du \\
&\overset{\text{induction}}{\le} C(t-t_n)^{l(\alpha-)+n(\alpha-)} \int_{t_n}^t \int_{t_n}^s \mathbb{E}|g(u)|^2 du ds \\
&\le C(t-t_n)^{l(\alpha-)+n(\alpha-)+1} \int_{t_n}^t \mathbb{E}|g(u)|^2 du \,.
\end{aligned}$$

But in this case $\alpha_{l+1} = 0$, $l(\alpha) = l(\alpha-)+1$, $n(\alpha) = n(\alpha-)+1$. So $l(\alpha-)+n(\alpha-)+1 = l(\alpha)+n(\alpha)-1$. This shows (11.1.13) in this case.

When $\alpha_{l+1} \ne 0$, $I_\alpha[g(\cdot)]_{t_n,s} = \int_{t_n}^s I_{\alpha-}[g(\cdot)]_{t_n,u} dW_s^{\alpha_{l+1}}$. By Doob's inequality, we have

$$\begin{aligned}
\mathbb{E} \sup_{t_n \le s \le t} (I_\alpha[g(\cdot)]_{t_n,s})^2 &\le C\mathbb{E}|I_\alpha[g(\cdot)]_{t_n,t}|^2 \le C \int_{t_n}^t \mathbb{E}|I_{\alpha-}[g(\cdot)]_{t_n,u}|^2 du \\
&\le C \int_{t_n}^t (u-t_n)^{l(\alpha-)+n(\alpha-)-1} \int_{t_n}^u \mathbb{E}|g(v)|^2 dv du \\
&\le C(t-t_n)^{l(\alpha-)+n(\alpha-)} \int_{t_n}^t \mathbb{E}|g(u)|^2 du \,.
\end{aligned} \tag{11.1.14}$$

But in this case $l(\alpha) = l(\alpha-)+1$ and $n(\alpha) = n(\alpha-)$. Thus $l(\alpha-)+n(\alpha-) = l(\alpha)+n(\alpha)-1$. This proves (11.1.13) in the case $\alpha_{l+1} \ne 0$. □

Lemma 11.2. *Let $g(s)$ be an adapted process. Denote $n_s = n$ if $t_n \le s < t_{n+1}$ and*

$$F_t^\alpha := \mathbb{E}\Big(\sup_{0\le s\le t} | \sum_{m=0}^{n_s-1} I_\alpha[g(\cdot)]_{t_m,t_{m+1}} + I_\alpha[g(\cdot)]_{t_{n_s},s}|^2\Big) . \tag{11.1.15}$$

Then,

$$F_t^\alpha \le \begin{cases} C|\pi|^{2(l(\alpha)-1)} \int_0^t R_{0,u} du & l(\alpha) = n(\alpha) \\ C|\pi|^{(l(\alpha)+n(\alpha)-1)} \int_0^t R_{0,u} du & l(\alpha) \ne n(\alpha), \end{cases} \tag{11.1.16}$$

where

$$R_{0,u} := \mathbb{E}\Big(\sup_{0\le s\le u} |g(s)|^2\Big) \le \infty . \tag{11.1.17}$$

Again C is independent of the subdivision but may depend on α.

Proof. The case $l(\alpha) = n(\alpha)$ is easy. We only need to discuss the case $l(\alpha) \ne n(\alpha)$. When the last index of α is not equal to 0, the sum of the the multiple integrals in (11.1.15) is a martingale as a process indexed by s. By Doob's inequality, we have that

$$\begin{aligned} F_t^\alpha &\le C\Big(\mathbb{E}\sum_{m=0}^{m(t)-1} I_\alpha[g(\cdot)]_{t_m,t_{m+1}} + I_\alpha[g(\cdot)]_{t_n(t),t}\Big)^2 \\ &= C\Big(\sum_{m=0}^{n(t)-1} \Big[\mathbb{E}\,|I_\alpha[g(\cdot)]_{t_m,t_{m+1}}|^2\Big] + \mathbb{E}|I_\alpha[g(\cdot)]_{t_{n(t)},t}|^2\Big) . \end{aligned}$$

Estimating the second moment for above each multiple integral on the corresponding interval by Lemma 1, inequality (11.1.13) and then taking the sum we will get the desired inequality (11.1.16).

If the last index of α is 0, then we have

$$| F_t^\alpha |^2 \le 2C\mathbb{E}\big(\sup_{0\le s\le t} \sum_{n=0}^{n_s-1} I_\alpha[g(\cdot)]_{t_n,t_{n+1}}\big)^2 + 2C\mathbb{E}\sup_{0\le s\le t} |I_\alpha[g(\cdot)]_{t_{n_s},s}|^2. \tag{11.1.18}$$

We estimate these two terms separately. The first one is bounded by

$$2C\mathbb{E}\big(\sup_{0\le r\le n(t)-1} \sum_{n=0}^{r} I_\alpha[g(\cdot)]_{t_n,t_{n+1}}\big)^2. \tag{11.1.19}$$

Now $\sum_{n=0}^r I_\alpha[g(\cdot)]_{t_n,t_{n+1}}$, $r = 0, \cdots, n(t)-1$ can be considered as a discrete martingale and we can then use Doob's inequality to complete the proof as in the case the last index is not equal to 0. The second term can be handled as follows:

$$\begin{aligned}\mathbb{E}\sup_{0\le s\le t}|I_\alpha[g(\cdot)]_{t_{n_s},s}|^2 &= \mathbb{E}\sup_{0\le s\le t}\int_{t_{n_s}}^s |I_{\alpha-}[g(\cdot)]_{t_{n_s},u}|^2du\\ &= \mathbb{E}\sup_{0\le s\le t}(s-n_s)\int_{t_{n_s}}^s |I_{\alpha-}[g(\cdot)]_{t_{n_s},u}|^2du\\ &\le |\pi|\int_0^t \mathbb{E}|I_{\alpha-}[g(\cdot)]_{t_{n_s},s}|^2ds.\end{aligned}$$

Application of Lemma 1 to bound $\mathbb{E}|\, I_{\alpha-}[g(\cdot)]_{t_{n_s},s}\,|^2$ yields the lemma. □

Proof of Theorem 11.1. We shall prove the theorem when $p=2$. General p case is similar. From (11.1.6) and (11.1.7) we have

$$Z(t)=\mathbb{E}\Big(\sup_{0\le s\le t}|X_s-Y_s|^2\Big)\le C\Big(\sum_{\alpha\in\mathcal{A}_\gamma}R_t^\alpha+\sum_{\alpha\in\mathcal{A}'_\gamma}U_t^\alpha\Big)\,, \tag{11.1.20}$$

where R_t^α and U_t^α are defined and estimated as follows:

$$\begin{aligned}R_t^\alpha \quad &:= \quad \mathbb{E}\Big(\sup_{0\le s\le t}\Big|\sum_{m=0}^{n_s-1}I_\alpha[f_\alpha(t_m,X_{t_m})-f_\alpha(t_m,Y_{t_m})]_{t_m,t_{m+1}}\\ &\qquad +I_\alpha[f_\alpha(t_{n_s},X_{n_s})-f_\alpha(t_{n_s},Y_{t_{n_s}})\,]_{t_{n_s},s}\,\Big|^2\Big)\\ &\overset{\text{Lemma 11.2}}{\le} \quad C\int_0^t \mathbb{E}\sup_{0\le s\le u}|\,f_\alpha(t_{n_s},X_{n_s})-f_\alpha(t_{n_s},Y_{t_{n_s}})\,|^2du\\ &\le \quad C\int_0^t Z(u)du\,.\end{aligned} \tag{11.1.21}$$

On the other hand, we have

$$\begin{aligned}U_t^\alpha \quad &:= \quad \mathbb{E}\Big(\sup_{0\le s\le t}\Big|\sum_{m=0}^{n_s-1}I_\alpha[f_\alpha(\cdot,X_\cdot)\,]_{t_m,t_{m+1}}+I_\alpha[f_\alpha(\cdot,X_\cdot)\,]_{t_{n_s},s}\,\Big|^2\Big)\\ &\overset{\text{Lemma 11.2}}{\le} C|\pi|^{\phi(\alpha)},\end{aligned} \tag{11.1.22}$$

where $\alpha\in\mathcal{A}'_\gamma$ and

$$\phi(\alpha)=\begin{cases}2(l(\alpha)-1) & : \quad l(\alpha)=n(\alpha)\\ l(\alpha)+n(\alpha)-1 & : \quad l(\alpha)\ne n(\alpha)\,.\end{cases} \tag{11.1.23}$$

Since $\alpha\in\mathcal{A}'_\gamma$ which implies

$$\begin{cases}2(l(\alpha)-1)\ge 2\gamma & \text{if } l(\alpha)=n(\alpha)\\ l(\alpha)+n(\alpha)-1\ge 2\gamma & \text{if } l(\alpha)\ne n(\alpha)\,,\end{cases}$$

we have

$$U_t^\alpha \le C|\pi|^{2\gamma}. \tag{11.1.24}$$

From (11.1.20), (11.1.21) and (11.1.24) we have

$$Z(t) \le C \int_0^t Z(u)\,du + C|\pi|^{2\gamma}\,. \tag{11.1.25}$$

Then, we deduce the theorem from Gronwall's inequality.

Remark 11.1. Let us return to Lemma 11.1, Formula (11.1.13). In the next section we will need the following extension to higher moments: If $\mathbb{E}\sup_{0\le t\le T}|g(t)|^p < \infty$, then

$$\mathbb{E}\sup_{t_n\le s\le t}(I_\alpha[g(\cdot)]_{t_n,s})^p \le C(t-t_n)^{p[l(\alpha)+n(\alpha)]/2}\,. \tag{11.1.26}$$

The easy proof is left to the reader.

Example 11.1. For the stochastic differential equation (11.1.1), the Euler-Maruyama scheme is when $\gamma = \frac{1}{2}$, namely,

$$\Gamma = \{\alpha = (\alpha), \alpha = 0, 1, \cdots, r\}$$

The recursive schemes is given by

$$X_t^e = X_{t_k}^e + \sum_{j=1}^m A_j(t_k, X_{t_k}^e)(W_t^j - W_{t_k}^j) + A_0(t_k, X_{t_k}^e)(t - t_k)\,, \quad t_k \le t \le t_{k+1}\,. \tag{11.1.27}$$

The Milstein Scheme is when $\gamma = 1$ with

$$\Gamma = \{\alpha = (\alpha), \alpha = 0, 1, \cdots, r;\ \text{or}\ \alpha = (\alpha_1, \alpha_2), 1 \le \alpha_1, \alpha_2 \le r\}\,.$$

The Milstein recursive scheme is

$$\begin{aligned} X_t^m &= \sum_{j=1}^m A_j(t_k, X_{t_k}^m)(W_t^j - W_{t_k}^j) + A_0(t_k, X_{t_k}^m)(t - t_k) \\ &\quad + \frac{1}{2}\sum_{\ell,j=1}^m A_\ell A_j(t_k, X_{t_k}^m)\int_{t_k}^{t_{k+1}} (W_s^\ell - W_{t_k}^\ell)dW_s^j\,, \quad t_k \le t \le t_{k+1}\,. \end{aligned} \tag{11.1.28}$$

For these two schemes the above general theorem becomes

Corollary 11.1. *For any $p \ge 1$, we have*

$$\mathbb{E}\left(\sup_{0\le t\le T}|X_t^e - X_t|^p\right) \le C_p|\pi|^{p/2}\,; \tag{11.1.29}$$

$$\mathbb{E}\left(\sup_{0\le t\le T}|X_t^m - X_t|^p\right) \le C_p|\pi|^p\,. \tag{11.1.30}$$

11.2 Convergence in $\mathbb{D}_{\beta,p}$ and Convergence in Density

We continue to use the notations in previous section. Let $x \in \mathbb{R}^d$ and $T > 0$ be fixed and set

$$F = F(\omega) = X_T(x); \qquad F_\pi = F_\pi(\omega) = X_T^\pi(x). \tag{11.2.1}$$

Then, F, $F_\pi \in \mathbb{D}_\infty[\mathbb{R}^d]$. To simplify notation, we denote $\|F\|_p = [\mathbb{E}(|F|^p)]^{1/p}$ and the norm of F in $\mathbb{D}_{\beta,p}$ by $\|F\|_{\beta,p}$. Theorem 11.1 says that

$$\|F_\pi - F\|_p = O(|\pi|^\gamma), \text{ as } |\pi| \to 0\,.$$

We are going to improve the above result to the convergence of F_π to F in $\mathbb{D}_{\beta,p}$. We need some estimates of multiple integrals of general kernel. Let

$$\mathcal{T} = \{(s_1, s_2, \ldots, s_l) \in \mathbb{R}^l \mid 0 \le s_1 < s_2, \ldots < s_l \le T\}$$

and let $L^2(\mathcal{T})$ be the usual L^2-space of real square-integrable functions on $\mathcal{T}$. Let $f(s_1, \ldots, s_l)$ be an $L^2(\mathcal{T})$-valued Wiener functional on W such that $f(s_1, \ldots, s_l)$ is $\mathcal{F}_{s_1}$-measurable for each fixed $s_1 < \ldots < s_l$. Define for a multi-index $\alpha = (\alpha_1, \cdots, \alpha_l)$ and $0 \le u \le v \le T$,

$$I_\alpha(f)_{u,v} = \int_{u<s_1<\cdots<s_l<v} f(s_1, \cdots, s_l)dW_{s_1}^{\alpha_1} \cdots dW_{s_l}^{\alpha_l}$$

by iterated Itô's stochastic integrals. In particular, if $f(s_1, \cdots, s_l) = f_\alpha(s_1, X_{s_1})$, then $I_\alpha(f)_{u,v}$ coincides with $I_\alpha\big(f_\alpha(\cdot, X_\cdot)\big)_{u,v}$ defined by (11.1.5). Set, for $0 \le u < s \le T$,

$$\|f\|_u(s) = \begin{cases} |f(s)| & \text{if} \quad l = 1 \\ \sup_{u<s_1<\ldots<s_{l-1}<s} |f(s_1, \ldots, s_{l-1}, s)| & \text{if} \quad l > 1. \end{cases}$$

Lemma 11.3.

(1) For $p \ge 1$ and $0 \le u < v \le T$,

$$\mathbb{E}\left[\sup_{u\le t\le v} |I_\alpha(f)_{u,t}|^{2p}\right] \le C(u-v)^{p[l(\alpha)+n(\alpha)]-1} \int_u^v \mathbb{E}[\|f\|_u(t)^{2p}]dt \tag{11.2.2}$$

(2) Let $\pi : 0 = t_0 < t_1 < \ldots < t_n = T$ be a partition of $[0,T]$. Set $|\pi| = \sup_i(t_{i+1} - t_i)$ and $m(s) = m$ if $t_m \le s < t_{m+1}$. Consider the following expectation for each $0 \le t \le T$:

$$F_t^\alpha = \mathbb{E}\left(\sup_{0\le s\le t} |\sum_{m=0}^{m(s)-1} I_\alpha(f)_{t_m,t_{m+1}} + I_\alpha(f)_{t_{m(s)},s}|^{2p}\right).$$

Then, for $p \geq 1$ and $0 \leq t \leq T$,

$$F_t^\alpha \leq \begin{cases} C|\pi|^{2p[l(\alpha)-1]} \int_0^t \mathbb{E}\left(\|f\|_{t_{m(s)}}(s)^{2p}\right) ds & \text{if } l(\alpha) = n(\alpha) \\ C|\pi|^{p[l(\alpha)+n(\alpha)-1]} \int_0^t \mathbb{E}\left(\|f\|_{t_{m(s)}}(s)^{2p}\right) ds & \text{if } l(\alpha) \neq n(\alpha). \end{cases} \tag{11.2.3}$$

Here, C is a positive constant depending on T, p and α which may vary from lines to lines.

Proof. Part (1) of the lemma can be proved by induction on the length $l(\alpha)$ of α. If $l(\alpha) > 1$,

$$I_\alpha(f)_{u,t} = \int_u^t I_{\alpha-}(f^s)_{u,s} dw_s^{\alpha_l},$$

where $f^s(s_1, \ldots, s_{l-1}) = f(s_1, \ldots, s_{l-1}, s)$. If $\alpha_l = 0$, then

$$\begin{aligned} \mathbb{E}[\sup_{u \leq t \leq v} |I_\alpha(f)_{u,t}|^{2p}] &\leq \mathbb{E}\left\{[\int_u^v |I_{\alpha-}(f^s)_{u,s}| ds]^{2p}\right\} \\ &\leq (v-u)^{2p-1} \int_u^v \mathbb{E}[|I_{\alpha-}(f^s)_{u,s}|^{2p}] ds \end{aligned}$$

by the Hölder inequality. If $\alpha_l \neq 0$, then

$$\begin{aligned} \mathbb{E}[\sup_{u \leq t \leq v} |I_\alpha(f)_{u,t}|^{2p}] &\leq C \cdot \mathbb{E}\left\{[\int_u^v |I_{\alpha-}(f^s)_{u,s}|^2 ds]^p\right\} \\ &\leq C(v-u)^{p-1} \int_u^v \mathbb{E}[|I_{\alpha-}(f^s)_{u,s}|^{2p}] ds \end{aligned}$$

by a standard martingale inequality of the Burkholder-Davis-Gundy type for stochastic integrals and the Hölder inequality. The inequality (11.2.2) for the case $l(\alpha) = 1$ can be obtained by the same estimates. Then, we can conclude the proof by induction if we notice the following facts: $l(\alpha-) = l(\alpha) - 1$, $n(\alpha-) = n(\alpha)$ or $n(\alpha) - 1$ accordingly as $\alpha_l \neq 0$ or $\alpha_l = 0$ and $\|f^s\|_u(t) \leq \|f\|_u(s)$ if $t \leq s$.

Next, we prove part (2). We notice that

$$\Xi(t) := \sum_{m=0}^{m(t)-1} I_\alpha(f)_{t_m, t_{m+1}} + I_\alpha(f)_{t_{m(t)}, t} = \int_0^t I_{\alpha-}(f^s)_{t_{m(s)}, s} dw_s^{\alpha_l}.$$

If $n(\alpha) = l(\alpha)$, then

$$\begin{aligned} \mathbb{E}[\sup_{0 \leq s \leq t} |\Xi(s)|^{2p}] &\leq \mathbb{E}\left\{\left[\int_0^t |I_{\alpha-}(f^s)_{t_{m(s)}, s}| ds\right]^{2p}\right\} \\ &\leq C \cdot \int_0^t \mathbb{E}\left\{|I_{\alpha-}(f^s)_{t_{m(s)}, s}|^{2p}\right\} ds \\ &\leq C \cdot \{\sum_{m=0}^{m(t)} \int_{t_m}^{t_{m+1} \wedge t} \mathbb{E}\left(|I_{\alpha-}(f^s)_{t_m, s}|^{2p}\right) ds. \end{aligned}$$

By estimate (11.2.2), this is dominated by

$$C|\pi|^{2pl(\alpha-)-1}\cdot\sum_{m=0}^{m(t)}\int_{t_m}^{t_{m+1}\wedge t}ds\int_{t_m}^{s}\mathbb{E}\left[\|f^s\|_{t_m}(\tau)^{2p}\right]d\tau$$

$$\leq C|\pi|^{2pl(\alpha-)}\int_0^t\mathbb{E}\left[\|f\|_{t_{m(s)}}(s)^{2p}\right]ds.$$

Since $l(\alpha-)=l(\alpha)-1$, (11.2.3) is obtained in this case.

If $n(\alpha)\neq l(\alpha)$, and $\alpha_l\neq 0$, then by the Burkholder-Davis-Gundy inequality applied to stochastic integral $\Xi(s)$, we have

$$\mathbb{E}[\sup_{0\leq s\leq t}|\Xi(s)|^{2p}]\leq C\cdot\mathbb{E}\left\{\left[\int_0^t|I_{\alpha-}(f^s)_{t_{m(s)},s}|^2ds\right]^p\right\}$$

$$\leq C\cdot\int_0^t\mathbb{E}\left\{|I_{\alpha-}(f^s)_{t_{m(s)},s}|^{2p}\right\}ds.$$

By the same estimate as above and by (11.2.2), this is further bounded by

$$C|\pi|^{p[l(\alpha-)+n(\alpha-)]}\int_0^t\mathbb{E}\left[\|f\|_{t_{m(s)}}(s)^{2p}\right]ds\,.$$

Since $\alpha_l\neq 0$, we have $l(\alpha-)+n(\alpha-)=l(\alpha)+n(\alpha)-1$ and hence (11.2.3) is obtained.

Finally, we consider the case $n(\alpha)\neq l(\alpha)$ and $\alpha_l=0$. We have

$$F_t^{\alpha}\leq C\cdot\mathbb{E}\left\{\sup_{0\leq s\leq t}|\sum_{m=0}^{m(s)-1}I_\alpha(f)_{t_m,t_{m+1}}|^{2p}\right\}$$

$$+C\cdot\mathbb{E}\left\{\sup_{0\leq s\leq t}|I_\alpha(f)_{t_{m(s)},s}|^{2p}\right\}$$

$$=:I_1+I_2\,.$$

We shall estimate these two terms separately. First, notice that

$$I_1=C\cdot\mathbb{E}\left\{\sup_{0\leq k\leq m(t)-1}|\sum_{m=0}^{k}I_\alpha(f)_{t_m,t_{m+1}}|^{2p}\right\}$$

and $S_k=\sum_{m=0}^{k}I_\alpha(f)_{t_m,t_{m+1}}$ forms a discrete time martingale. Then, we can apply the Burkholder-Davis-Gundy inequality for the discrete time martingale to obtain that

$$I_1\leq C\cdot\mathbb{E}\left\{\left[\sum_{m=0}^{m(t)-1}|I_\alpha(f)_{t_m,t_{m+1}}|^2\right]^p\right\}$$

$$=C\cdot\mathbb{E}\left\{\left[\sum_{m=0}^{m(t)-1}|\int_{t_m}^{t_{m+1}}I_{\alpha-}(f^s)_{t_m,s}ds|^2\right]^p\right\}.$$

Since

$$|\int_{t_m}^{t_{m+1}} I_{\alpha-}(f^s)_{t_m,s}ds|^2 \le (t_{m+1}-t_m)\cdot\int_{t_m}^{t_{m+1}} |I_{\alpha-}(f^s)_{t_m,s}|^2 ds,$$

this is further dominated by

$$\begin{aligned}
&C|\pi|^p\cdot\mathbb{E}\left\{\left[\sum_{m=0}^{m(t)-1}\int_{t_m}^{t_{m+1}} |I_{\alpha-}(f^s)_{t_m,s}|^2 ds\right]^p\right\}\\
&\le C|\pi|^p\cdot\mathbb{E}\left\{\left[\int_0^t |I_{\alpha-}(f^s)_{t_{m(s)},s}|^2 ds\right]^p\right\}\\
&\le C|\pi|^p\cdot\int_0^t \mathbb{E}\left\{|I_{\alpha-}(f^s)_{t_{m(s)},s}|^{2p}\right\} ds.
\end{aligned}$$

Then, by the same estimate as above and by (11.2.2), we deduce that the above is dominated by

$$C|\pi|^{p[l(\alpha-)+n(\alpha-)+1]}\int_0^t \mathbb{E}(\|f\|_{t_{m(s)}}(s)^{2p})ds.$$

Since $l(\alpha-)+n(\alpha-)+1 = l(\alpha)+n(\alpha)-1$, I_1 is now dominated as desired.

As for I_2, we have,

$$\begin{aligned}
&\mathbb{E}\left[\sup_{0\le s\le t}|I_\alpha(f)_{t_{m(s)},s}|^{2p}\right] = \mathbb{E}\left[\sup_{0\le s\le t}|\int_{m(s)}^s I_{\alpha-}(f^\tau)_{t_{m(s)},\tau}d\tau|^{2p}\right]\\
&\le \mathbb{E}\left\{\sup_{0\le s\le t}\left[(s-t_{m(s)})^{2p-1}\int_{m(s)}^s |I_{\alpha-}(f^\tau)_{t_{m(s)},\tau}|^{2p}d\tau\right]\right\}\\
&\le |\pi|^{2p-1}\mathbb{E}\left[\sum_{m=0}^{m(t)}\int_{t_m}^{t_{m+1}\wedge t} |I_{\alpha-}(f^s)_{t_m,s}|^{2p}ds\right].
\end{aligned}$$

By using (11.2.2), this can be dominated by

$$\begin{aligned}
&C|\pi|^{2p-1}\cdot|\pi|^{p[l(\alpha-)+n(\alpha-)]}\int_0^t \mathbb{E}(\|f\|_{t_{m(s)}}(s)^{2p})ds\\
&= C|\pi|^{p-1}\cdot|\pi|^{p[l(\alpha)+n(\alpha)-1]}\int_0^t \mathbb{E}(\|f\|_{t_{m(s)}}(s)^{2p})ds.
\end{aligned}$$

Since $p\ge 1$, we obtained the desired estimate for I_2 and the proof of (11.2.3) is now complete. □

In the remaining part of this section we use D^k to denote the derivative with respect to the k-th Brownian motion. If f is deterministic, i.e.,

$f \in L^2(\mathcal{T})$, then $I_\alpha(f)_{u,v} \in \mathbb{D}_\infty$ and we have, for each $s \in [0,T]$ and $k = 1, \ldots, m$

$$D_s^k I_\alpha(f)_{u,v}$$
$$= \begin{cases} \sum_{i:\alpha_i=k} \int_{u<s_1<\cdots<\hat{s}_i<\cdots<s_l<v} f(s_1,\cdots,s_l)|_{s_i=s} \\ \quad 1_{[s_{i-1},s_{i+1}]}(s) dW_{s_1}^{\alpha_1} \cdots d\hat{W}_{s_i}^{\alpha_i} \cdots dW_{s_l}^{\alpha_l}, & \text{if } l(\alpha) > 1 \\ 1_{[u,v]}(s) \cdot \delta_{k,\alpha} \cdot f(s), & \text{if } l(\alpha) = 1, \end{cases} \tag{11.2.4}$$

where the symbol ˆ means to discard the element which bears it.

Set, for fixed $1 \le i \le l$ and $s \in [0,T]$,

$$\alpha_{\setminus i} = (\alpha_1, \cdots, \hat{\alpha}_i, \cdots, \alpha_l) \tag{11.2.5}$$

and

$$f_{\setminus i}[s](s_1,\cdots,s_{l-1}) = f(s_1,\cdots s_{i-1}, s, s_i, \cdots, s_{l-1}) 1_{[s_{i-1},s_i]}(s) \quad \text{if} \quad l > 1. \tag{11.2.6}$$

Then, we may write (11.2.4) in the following form:

$$D_s^k I_\alpha(f)_{u,v} = \begin{cases} \sum_{i:\alpha_i=k} I_{\alpha \setminus i}(f_{\setminus i}[s])_{u,v}, & \text{if} \quad l(=l(\alpha)) > 1 \\ f(s) \cdot 1_{[u,v]}(s) \cdot \delta_{k,\alpha}, & \text{if} \quad l = 1. \end{cases} \tag{11.2.7}$$

Note that the right-hand side of (11.2.7) is well-defined for almost all s even if f is not deterministic and so we define generally, for each $s \in [0,T]$ and $k = 1, \ldots, r$,

$$\tilde{D}_s^k I_\alpha(f)_{u,v} = \begin{cases} \sum_{i:\alpha_i=k} I_{\alpha \setminus i}(f_{\setminus i}[s])_{u,v}, & \text{if} \quad l(=l(\alpha)) > 1 \\ f(s) \cdot 1_{[u,v]}(s) \cdot \delta_{k,\alpha}, & \text{if} \quad l = 1. \end{cases}$$

Then, we have generally

$$D_s^k I_\alpha(f)_{u,v} = \tilde{D}_s^k I_\alpha(f)_{u,v} + I_\alpha(e_s)_{u,v}, \tag{11.2.8}$$

where $e_s(s_1,\cdots,s_l) = D_s^k[f(s_1,\cdots,s_l)]$.

Consider the solution $X_t = X_t(x)$ of stochastic differential equation (11.1.6) and suppose that the coefficients satisfy the same assumptions as in Theorem 11.1.

Lemma 11.4. *Suppose that the coefficients $A_j, j = 0, 1, \cdots, m$ are in C^∞ with bounded derivatives and further $f_\alpha^k(s,x)$ for every multi-index α and $k = 1, \cdots, d$ have bounded derivatives. Then, we have*

$$\sup_{s\in[0,T]} \sup_{1\le k\le m} \int_0^T \mathbb{E}[\sup_{0\le u\le t} |D_s^k X_u|^p] dt < \infty \tag{11.2.9}$$

for all $p > 1$.

Proof. Taking D_s^k in both sides of equation (11.1.6) we obtain for $0 \le s \le t$,

$$D_s^k X_t = \sum_{j=0}^m \int_0^t \nabla A_j(u, X_u) D_s^k X_u dW_u^j + A_k(s, X_s).$$

So there are two constants C_1 and C_2 such that

$$\mathbb{E}\Big[\sup_{0\le u\le t} |D_s^k X_u|^p\Big] \le C_1 \sum_{j=0}^m \int_0^t \mathbb{E}\Big[\sup_{0\le u\le v} |D_s^k X_u|^p\Big] dv + C_2.$$

This shows the boundedness of (4.9) by the Gronwall lemma. □

Theorem 11.2. *Suppose that the coefficients $A_j, j = 0, 1, \cdots, m$ are in C^∞ with bounded derivatives and further $f_\alpha^k(s,x)$ have bounded derivatives for every multi-index α and $k = 1, \cdots, d$. Then, for every $p > 1$ and $k = 1, \cdots, m$*

$$\sup_{0\le s\le T} \mathbb{E}[\sup_{0\le u\le T} |D_s^k X_u^\pi - D_s^k X_u|^{2p}] = O(|\pi|^{2p\gamma}) \quad as \quad |\pi| \to 0. \tag{11.2.10}$$

Proof. Taking the derivative in both sides of equations (11.1.7) and (11.1.6), we have for $t \in [t_m, t_{m+1}]$,

$$\begin{aligned} D_s^k X_t^\pi = D_s^k X_{t_m}^\pi &+ \sum_{\alpha\in\Gamma} \nabla f_\alpha(t_m, X_{t_m}^\pi) D_s^k X_{t_m}^\pi I_{\alpha,t_m,t} \\ &+ \sum_{\alpha\in\Gamma} f_\alpha(t_m, X_{t_m}^\pi) D_s^k I_{\alpha,t_m,t} \end{aligned} \tag{11.2.11}$$

and

$$\begin{aligned} D_s^k X_t = D_s^k X_{t_m} &+ \sum_{\alpha\in\Gamma} \nabla f_\alpha(t_m, X_{t_m}) D_s^k X_{t_m} I_{\alpha,t_m,t} \\ &+ \sum_{\alpha\in\Gamma} f_\alpha(t_m, X_{t_m}) D_s^k I_{\alpha,t_m,t} \\ &+ \sum_{\alpha\in\mathcal{F}_\Gamma} I_\alpha[\nabla f_\alpha(\cdot, X_\cdot) D_s^k X_\cdot]_{t_m,t} \\ &+ \sum_{\alpha\in\mathcal{F}_\Gamma} \tilde{D}_s^k I_\alpha[f_\alpha(\cdot, X_\cdot)]_{t_m,t}\,, \end{aligned} \tag{11.2.12}$$

where $\nabla f(s,x) = (\frac{\partial f}{\partial x_j}(s,x))_{1\le j\le d}$.

Now subtract equation (11.2.11) from (11.2.12) to obtain

$$
\begin{aligned}
D_s^k X_t - D_s^k X_t^\pi \;&=\; D_s^k X_{t_m} \;-\; D_s^k X_{t_m}^\pi \qquad (11.2.13)\\
&+\sum_{\alpha\in\Gamma}\Big[\nabla f_\alpha(t_m, X_{t_m}) D_s^k X_{t_m} I_{\alpha,t_m,t} - \nabla f_\alpha(t_m, X_{t_m}^\pi) D_s^k X_{t_m}^\pi I_{\alpha,t_m,t}\Big]\\
&+\sum_{\alpha\in\Gamma}\Big[f_\alpha(t_m, X_{t_m}) - f_\alpha(t_m, X_{t_m}^\pi)\Big] D_s^k I_{\alpha,t_m,t}\\
&+\sum_{\alpha\in\mathcal{F}_\Gamma} I_\alpha[\nabla f_\alpha(\cdot, X.) D_s^k X.]_{t_m,t} + \sum_{\alpha\in\mathcal{F}_\Gamma} \tilde{D}_s^k I_\alpha[f_\alpha(\cdot, X.)]_{t_m,t}\\
&= D_s^k X_{t_m} - D_s^k X_{t_m}^\pi + \sum_{\alpha\in\Gamma} \nabla f_\alpha(t_m, X_{t_m}^\pi)\Big[D_s^k X_{t_m} - D_s^k X_{t_m}^\pi\Big] I_{\alpha,t_m,t}\\
&+R_{t_m,t},
\end{aligned}
$$

where we denote

$$
\begin{aligned}
R_{t_m,t}(s) \;&=\; \sum_{\alpha\in\Gamma}\Big[\nabla f_\alpha(t_m, X_{t_m}) - \nabla f_\alpha(t_m, X_{t_m}^\pi)\Big] D_s^k X_{t_m} I_{\alpha,t_m,t} \qquad (11.2.14)\\
&+\sum_{\alpha\in\Gamma}\Big[f_\alpha(t_m, X_{t_m}) - f_\alpha(t_m, X_{t_m}^\pi)\Big] D_s^k I_{\alpha,t_m,t}\\
&+\sum_{\alpha\in\mathcal{F}_\Gamma} I_\alpha[\nabla f_\alpha(\cdot, X.) D_s^k X.]_{t_m,t} + \sum_{\alpha\in\mathcal{F}_\Gamma} \tilde{D}_s^k I_\alpha[f_\alpha(\cdot, X.)]_{t_m,t}\\
&=: R_{t_m,t}^1(s) + R_{t_m,t}^2(s) + R_{t_m,t}^3(s) + R_{t_m,t}^4(s).
\end{aligned}
$$

Repeatedly using the formula (11.2.13) for $t = t_m$ $(m = m(t), m(t)-1, \cdots, 1)$, we have

$$
\begin{aligned}
D_s^k X_t - D_s^k X_t^\pi &= \sum_{\alpha\in\Gamma}\sum_{m=0}^{m(t)} \nabla f_\alpha(t_m, X_{t_m}^\pi)\Big[D_s^k X_{t_m} - D_s^k X_{t_m}^\pi\Big] I_{\alpha,t_m,t_{m+1}\wedge t}\\
&\quad + \sum_{m=0}^{m(t)} R_{t_m,t_{m+1}\wedge t}(s).
\end{aligned}
$$

Now fix s. By Lemma 11.3, part (2), we obtain (11.2.14)

$$
\begin{aligned}
\mathbb{E}\sup_{0\le v\le t} |D_s^k X_v - D_s^k X_v^\pi|^{2p} &\le C\sum_{\alpha\in\Gamma}\int_0^t \mathbb{E}\sup_{0\le u\le v} |D_s^k X_u - D_s^k X_u^\pi|^{2p} dv\\
&+\mathbb{E}\sup_{0\le u\le t} \Big|\sum_{m=0}^{m(u)} R_{t_m,t_{m+1}\wedge u}(s)\Big|^{2p}. \qquad (11.2.15)
\end{aligned}
$$

The last expectation of the above formula can be estimated as follows.

$$
\text{The last term of (11.2.15)} \le C(R_1(s)+R_2(s)+R_3(s)+R_4(s)), \qquad (11.2.16)
$$

where

$$R_i(s) = \mathbb{E} \sup_{0\le u\le t} |\sum_{m=0}^{m(u)} R^i_{t_m,t_{m+1}\wedge u}(s)|^{2p}. \tag{11.2.17}$$

Denoting $t_{m(u)}$ by ϕ_u for simplicity, $R_1(s)$ can be estimated by Lemma 11.3, part (2) as

$$\begin{aligned}
R_1(s) &\le \sum_{\alpha\in\Gamma} \int_0^t \mathbb{E}\left|\left[\nabla f_\alpha(\phi_u, X_{\phi_u}) - \nabla f_\alpha(\phi_u, X^\pi_{\phi_u})\right] D^k_s X_{\phi_u}\right|^{2p} du \\
&\le C\int_0^T \left\{\mathbb{E} \sup_{0\le u\le v} |D^k_s X_u|^{4p}\right\}^{1/2} \cdot \left\{\mathbb{E} \sup_{0\le u\le v} |X_u - X^\pi_u|^{4p}\right\}^{1/2} dv.
\end{aligned}$$

By Lemma 11.4 and (11.1.11), this is further dominated by

$$R_1(s) \le C|\pi|^{2p\gamma}. \tag{11.2.18}$$

Next, we estimate $R_2(s)$. It is easy to see that

$$D^k_s I_{\alpha,t_m,t_{m+1}} = 1_{[t_m,t_{m+1}]}(s)\cdot D^k_s I_{\alpha,t_m,t_{m+1}} = \delta_{m,m(s)}\cdot D^k_s I_{\alpha,t_{m(s)},t_{m(s)+1}}$$

and

$$\sup_\pi \sup_{0\le s\le T} \mathbb{E}|D^k_s I_{\alpha,t_{m(s)},t_{m(s)+1}}|^q < \infty$$

for every $q > 1$. Then,

$$\begin{aligned}
&R_2(s) \\
&\le C\sum_{\alpha\in\Gamma} \mathbb{E} \sup_{0\le u\le t} |\sum_{m=0}^{m(u)} \left[f_\alpha(t_m, X_{t_m}) - f_\alpha(t_m, X^\pi_{t_m})\right] D^k_s I_{\alpha,t_m,t_{m+1}\wedge u}|^{2p} \\
&\le C\sum_{\alpha\in\Gamma} \left\{\mathbb{E}|f_\alpha(t_{m(s)}, X_{t_{m(s)}}) - f_\alpha(t_{m(s)}, X^\pi_{t_{m(s)}})|^{4p}\right\}^{1/2} \\
&\quad \cdot \left\{\mathbb{E}|D^k_s I_{\alpha,t_{m(s)},t_{m(s)+1}\wedge t}|^{4p}\right\}^{1/2} \\
&\le C\left[\mathbb{E} \sup_{0\le u\le T} |X_u - X^\pi_u|^{4p}\right]^{1/2} \le C|\pi|^{2p\gamma}.
\end{aligned}$$

To estimate $R_3(s)$, we first notice that if $\alpha \in \mathcal{F}_\Gamma$, then

$$\begin{cases} 2l(\alpha) - 2 \ge 2\gamma & \text{when } l(\alpha) = n(\alpha) \\ l(\alpha) + n(\alpha) - 1 \ge 2\gamma & \text{when } l(\alpha) \ne n(\alpha). \end{cases}$$

Then, using Lemma 11.3, part (2), we obtain

$$
\begin{aligned}
R_3(s) &\le C|\pi|^{2\gamma p} \sum_{\alpha\in\mathcal{F}_\Gamma} \int_0^T \mathbb{E} \sup_{0\le u\le v} |\nabla f_\alpha(u, X_u) D_s^k X_u|^{2p} dv \\
&\le C|\pi|^{2\gamma p} \int_0^T \mathbb{E} \sup_{0\le u\le v} |D_s^k X_u|^{2p} dv \\
&\le C|\pi|^{2\gamma p}.
\end{aligned}
$$

Finally we estimate $R_4(s)$. Put $f(s_1, \cdots, s_l) = f_\alpha(s_1, X_{s_1})$. Since

$$
I_{\alpha\setminus i}(f_{\setminus i}[s])_{t_m,t_{m+1}} = 1_{[t_m,t_{m+1}]}(s) \cdot I_{\alpha\setminus i}(f_{\setminus i}[s])_{t_m,t_{m+1}},
$$

$$
\begin{aligned}
R_4(s) &= C \sum_{\alpha\in\mathcal{F}_\Gamma} \sum_{\alpha_i=k} \mathbb{E}\Big\{ \sup_{0\le u\le t} \Big| \sum_{m=0}^{m(u)} I_{\alpha\setminus i}[f_{\setminus i}[s]]_{t_m,t_{m+1}\wedge u} \Big|^{2p} \Big\} \\
&\le C \sum_{\alpha\in\mathcal{F}_\Gamma} \sum_{\alpha_i\ne 0} \mathbb{E}\{ \sup_{t_{m(s)}\le u\le t_{m(s)+1}\wedge t} |I_{\alpha\setminus i}[f_{\setminus i}[s]]_{t_{m(s)},u}|^{2p} \},
\end{aligned}
$$

and by Lemma 11.3, part (1), this is dominated by

$$
\begin{aligned}
&C \sum_{\alpha\in\mathcal{F}_\Gamma} \sum_{\alpha_i\ne 0} |\pi|^{p[l(\alpha_{\setminus i})+n(\alpha_{\setminus i})]-1} \int_{t_{m(s)}}^{t_{m(s)+1}\wedge t} \mathbb{E}\left[\|f_{\setminus i}[s]\|_{t_{m(s)}}(u)^{2p} \right] du \\
&\quad \le C|\pi|^{p[l(\alpha)+n(\alpha)-1]}.
\end{aligned}
$$

Since $l(\alpha) + n(\alpha) \ge 2\gamma + 1$ if $\alpha \in \mathcal{F}_\Gamma$, $l(\alpha) + n(\alpha) - 1 \ge 2\gamma$ and hence

$$
R_4(s) \le C|\pi|^{2\gamma p}.
$$

Thus we have shown

$$
\mathbb{E} \sup_{0\le u\le t} |D_s^k X_u - D_s^k X_u^\pi|^{2p} \le C_1 \int_0^t \mathbb{E} \sup_{0\le u\le v} |D_s^k X_u - D_s^k X_u^\pi|^{2p} dv + C_2|\pi|^{2\gamma p}
$$

with positive constants C_1 and C_2 independent of $s \in [0, T]$. Now (11.2.10) follows from the Gronwall lemma. □

To show the density convergence of X_t^π to X_t, we need to prove the nondegeneracy of X_t^π. This is given by the following more general lemma.

Lemma 11.5. *Let $H_n, H \in \mathbb{D}_{1,\infty-}$ satisfy the following:*

(i) There exists $\gamma > 0$ such that for any $1 < p < \infty$,

$$
\|H_n - H\|_{p,1} = O(n^{-\gamma}), \quad \text{as} \quad n \to \infty. \tag{11.2.19}
$$

(ii) $(\det\ \sigma_H)^{-1} \in L^{\infty-}$.

(iii) For any $1 \le p < \infty$, there is $\nu(p) > 0$ such that $\|(\det\ \sigma_{H_n})^{-1}\|_p = O(n^{\nu(p)})$ as $n \to \infty$.

Then, for any $1 \le p < \infty$, we have

$$\sup_n \|(\det\ \sigma_{H_n})^{-1}\|_p < \infty. \tag{11.2.20}$$

Proof. In the following, $C_1(p)$, $C_2(p)$, $\cdots$ are positive constants depending only on $1 < p < \infty$. Writing $\det\ \sigma_{H_n} = \tau_n$ and $\det\ \sigma_H = \tau$, for simplicity, we have for a given $1 < p < \infty$,

$$\begin{aligned}\mathbb{E}(|\tau_n|^{-p}) &= \mathbb{E}\left(|\tau_n|^{-p} : |\frac{\tau_n}{\tau} - 1| \le \frac{1}{2}\right) + \mathbb{E}\left(|\tau_n|^{-p} : |\frac{\tau_n}{\tau} - 1| > \frac{1}{2}\right) \\ &= : I_1 + I_2.\end{aligned}$$

If $|\frac{\tau_n}{\tau} - 1| \le \frac{1}{2}$, then $\tau_n \ge \frac{1}{2}\tau$ and hence $\tau_n^{-1} \le 2\tau^{-1}$. Consequently, by the assumption (ii),

$$\begin{aligned}I_1 &\le 2^p \mathbb{E}(|\tau|^{-p} : |\frac{\tau_n}{\tau} - 1| \le \frac{1}{2}) \\ &\le 2^p \mathbb{E}|\tau|^{-p} =: C_1(p) < \infty.\end{aligned}$$

As for I_2, we can estimate by the Schwartz inequality

$$I_2 \le \mathbb{E}(|\tau_n|^{-2p})^{1/2} \mathbb{P}(|\frac{\tau_n - \tau}{\tau}| > \frac{1}{2})^{1/2}.$$

By the assumption (iii),

$$\mathbb{E}(|\tau_n|^{-2p}) \le C_2(p) \cdot n^{\nu(2p) \cdot 2p}$$

and by (11.2.19) it is easy to deduce that

$$\mathbb{E}(|\tau_n - \tau|^{2k}) \le C_3(k) \cdot n^{-2k\gamma}$$

for every $k > 1$. Combining this with (ii),

$$E(|\frac{\tau_n - \tau}{\tau}|^{2k}) \le C_4(k) \cdot n^{-2k\gamma}$$

and hence

$$\begin{aligned}P(|\frac{\tau_n - \tau}{\tau}| > \frac{1}{2}) &\le 2^{2k} \mathbb{E}(|\frac{\tau_n - \tau}{\tau}|^{2k}) \\ &\le C_5(k) \cdot n^{-2k\gamma}.\end{aligned}$$

Consequently

$$I_2 \le \sqrt{C_2(p) \cdot C_5(k) \cdot n^{\nu(2p) \cdot 2p - 2k\gamma}}.$$

By taking k large enough so that $\nu(2p) \cdot 2p < 2k\gamma$ we can conclude (11.2.20). □

Remark 11.2. From the above proof, we see that the assumption (i) can be replaced by the following weaker assumption:
(i)' For every $1 < p < \infty$, there exists a $\mu(p) > 0$ such that $\mu(p) \uparrow \infty$ as $p \uparrow \infty$ and

$$\mathbb{E}(|H_n - H|^p + |DH_n - DH|^p) = O(n^{-\mu(p)}) \quad \text{as} \quad n \to \infty .$$

Let a d-dimensional Wiener functional $F : \mathbb{B} \to \mathbb{R}^d$ be given which is smooth and non-degenerate. Then, for every $\beta > 0$ and $y \in \mathbb{R}^d$, $(1 - \Delta)^{\beta/2}\delta_y \circ F$ is defined in $\mathbb{D}^{1+}_{-\infty}$. Suppose that a sequence $\{F_n : n = 1, 2, \cdots\}$ of d-dimensional smooth Wiener functionals converge to F as $n \to \infty$. We want to obtain an approximation of $(1 - \Delta)^{\beta/2}\delta_y \circ F$ in terms of F_n.

Theorem 11.3. *Let $F_n, n = 1, 2, \cdots$ and F be smooth d-dimensional Wiener functionals, i.e.,*

$$F_n, \quad F \in \mathbb{D}^{\infty-}_{\infty}[\mathbb{R}^d].$$

We suppose that F and F_n satisfy the following:

(i) F_n approximates F in $\mathbb{D}_\infty[\mathbb{R}^d]$ with order γ ($\gamma > 0$) in the sense that for every $1 < p < \infty$ and $\alpha > 0$,

$$\|F_n - F\|_{p,\alpha} = O(n^{-\gamma}) \quad \textit{as} \quad n \to \infty. \tag{11.2.21}$$

(ii) F is non-degenerate, i.e.,

$$(\det\ \sigma_F)^{-1} \in L^{\infty-}. \tag{11.2.22}$$

Then, for every $\alpha > 0$, $\beta \geq 0$, $\delta > 0$ and $1 < p < \infty$ such that $\alpha > \beta + d/q + 1$, $1/p + 1/q = 1$, it holds that

$$\sup_{y \in \mathbb{R}^d} \|[(1-\Delta)^{\beta/2}\phi_{n^{-\delta}}](F_n - y) - (1-\Delta)^{\beta/2}\delta_y \circ F\|_{p,-\alpha} = O(n^{-\gamma\wedge\delta}) \tag{11.2.23}$$

as $n \to \infty$, where

$$\phi_\rho(x) = (2\pi\rho^2)^{-d/2} e^{-\frac{|x|^2}{2\rho^2}}, \quad x \in \mathbb{R}^d, \quad \rho > 0. \tag{11.2.24}$$

Proof. We take a product Wiener space $(\bar{\mathbb{B}} = \mathbb{B} \times \mathbb{B}', \bar{\mathbb{P}} = \mathbb{P} \times \mathbb{P}')$, where $(\mathbb{B}', \mathbb{P}')$ is a classical d-dimensional standard Wiener space and consider on $(\bar{\mathbb{B}}, \bar{\mathbb{P}})$ the following Wiener functionals

$$H_n(\bar{w}) = F_n(w) + \frac{1}{n^\delta} w'(1), \quad H(\bar{w}) = F(w)$$

for $\bar{w} = (w, w') \in \mathbb{B} \times \mathbb{B}'$. Then, we can apply the above lemma 11.5 to obtain

$$\sup_n \|(\det\ \sigma_{H_n})^{-1}\|_p < \infty.$$

Set further $K_{n,t}(\bar{w}) = (1-t)F(w) + tH_n(\bar{w})$ for $t \in [0,1]$. Then,

$$K_{n,t}(\bar{w}) - F(w) = t(H_n(\bar{w}) - F(w))$$

and

$$\|(\det\ \sigma_{K_{n,t}})^{-1}\|_p = O(n^{2|\pi|d} t^{-2d}).$$

Applying Lemma 11.5 by fixing n, we deduce that, for each $p > 1$, $\mu > 0$ exists such that

$$\sup_{t\in[0,1]} \|(\det\ \sigma_{K_{n,t}})^{-1}\|_p = O(n^{\mu}).$$

Then, applying Lemma 11.5 again, we can conclude that, for each $p > 1$,

$$\sup_{t,n} \|(\det\ \sigma_{K_{n,t}})^{-1}\|_p < \infty.$$

Set $T = (1-\Delta)^{\beta/2}\delta_y$. Then

$$T \circ H_n - T \circ F = \int_0^1 \partial T \circ K_{n,t} \cdot (H_n - F)dt$$

and, if $\alpha > (\beta+1) + d/q, 1/p + 1/q = 1$, then by the fact that $(1-\Delta)^{\beta/2}\delta_y \circ F \in \mathbb{D}_{-\alpha,p}$ we can choose $\bar{p} > p$ such that

$$\sup_{t,n} \|\partial T \circ K_{n,t}\|_{-\alpha,\bar{p}} < \infty.$$

Since $\|H_n - F\|_{\bar{q},\alpha} = O(n^{-\gamma\wedge\delta})$ for $\bar{q} > 1$ such that $1/\bar{p} + 1/\bar{q} \leq 1/p$, we deduce that

$$\begin{aligned}
\|T \circ H_n - T \circ F\|_{p,-\alpha} &\leq \int_0^1 \|\partial T \circ K_{n,t} \cdot (H_n - F)\|_{p,-\alpha} dt \\
&\leq const \int_0^1 \|\partial T \circ K_{n,t}\|_{\bar{p},-\alpha} \|H_n - F\|_{\bar{q},\alpha} dt \\
&= O(n^{-\gamma\wedge\delta}).
\end{aligned}$$

From this, we can conclude (11.2.23) by noting

$$\int_{W'} [(1-\Delta)^{\beta/2}\delta_y](H_n(w,w'))P'(dw') = [(1-\Delta)^{\beta/2}\phi_{n^{-\delta}}](F_n(w) - y)$$

and a general fact that the map

$$\Phi(\bar{w}) = \Phi(w, w') \to \hat{\Phi}(w) = \int_{W'} \Phi(w, w')P'(dw')$$

is a contraction from $\mathbb{D}^p_{-\alpha}(\bar{\mathbb{B}})$ to $\mathbb{D}^p_{-\alpha}(\mathbb{B})$ for every $1 < p < \infty$ and $\alpha \in \mathbb{R}$. This follows at once from the following relation which can be easily verified by the Wiener chaos expansion:

$$[(1-\bar{L})^{-\alpha}\Phi]^\wedge = (1-L)^{-\alpha}\hat{\Phi},$$

with $\bar{L}$ and L being the Ornstein-Uhlenbeck operator on $\bar{W}$ and W, respectively. □

Remark 11.3. Obviously, $(1-\Delta)^{\beta/2}\phi_{n^{-\delta}} \in \mathcal{S}(\mathbb{R}^d)$ and hence $[(1-\Delta)^{\beta/2}\phi_{n^{-\delta}}](F_n - y) \in \mathbb{D}_\infty$ for each n and $y \in \mathbb{R}^d$.

Corollary 11.2. *Let $\beta \geq 0$, $1 < q < \infty$ and $\alpha > \beta + d/q + 1$ and G_n, $G \in \mathbb{D}^\alpha_q$, $n = 1, 2, \cdots$ such that*

$$\|G_n - G\|_{q,\alpha} = O(\frac{1}{n^\varepsilon}), \quad as \quad n \to \infty \tag{11.2.25}$$

for some $\varepsilon > 0$. Then, it holds that

$$\begin{aligned}\sup\nolimits_{y\in\mathbb{R}^d} &|(1-|\pi|)_y^{\beta/2}\mathbb{E}[G_n \cdot \phi_{n^{-\delta}}(F_n - y)] - (1-|\pi|)_y^{\beta/2}\mathbb{E}[G\cdot\delta_y(F)]| \\ &= O(n^{-\gamma\wedge\delta\wedge\varepsilon}) \quad as \quad n \to \infty.\end{aligned}$$

In particular, setting

$$u_n(y) = \mathbb{E}[G_n \cdot \phi_{n^{-\delta}}(F_n - y)] \quad and \quad u(y) = \mathbb{E}[G\cdot\delta_y(F)],$$

we have

$$\|u_n - u\|_\beta = O(n^{-\gamma\wedge\delta\wedge\varepsilon}) \quad as \quad n \to \infty. \tag{11.2.26}$$

We remark that we are not assuming that F_n is non-degenerate so that we do not even know that $\delta_y \circ F_n = \delta_0(F_n - y)$ can be defined.

Corollary 11.3. *If F_n in Theorem 11.3 is uniformly non-degenerate, i.e.*

$$\sup_n \|(\det\ \sigma_{F_n})^{-1}\|_p < \infty,$$

then, $[(1-\Delta)^{\beta/2}\phi_{n^{-\delta}}](F_n - y)$ and $O(n^{-\gamma\wedge\delta})$ in (11.2.23), and $\phi_{n^{-\delta}}(F_n - y)$ and $O(n^{-\gamma\wedge|\pi|\wedge\varepsilon})$ in (2.6), may be replaced by $(1-\Delta)^{\beta/2}\delta_y \circ F_n$ and $O(n^{-\gamma})$, and $\delta_y \circ F_n$ and $O(n^{-\gamma\wedge\varepsilon})$, respectively.

Indeed, if $T = (1-\Delta)^{\beta/2}\delta_y$, we can deduce by the same proof that

$$\|T(H_n) - T(F_n)\|_{p,-\alpha} = O(n^{-\delta}).$$

Also we may take δ as large as we want.

Theorem 11.4. *Let $F_n = X_T^{\pi_n}(x)$ and $F = X_T(x)$ for fixed $x \in \mathbb{R}^d$ and $T > 0$ and assume that equation (11.1.1) satisfy the non-degeneracy condition*

$$(\det \sigma_F)^{-1} \in L^{\infty-}. \tag{11.2.27}$$

For all $\beta > 0$, $1 < q < \infty$ and $\alpha > d/q + \beta + 1$, and for all G_n, $G \in \mathbb{D}_q^\alpha$ satisfying

$$\|G_n - G\|_{q,\alpha} = O(n^{-\varepsilon}) \quad \text{for some} \quad \varepsilon > 0 \quad \text{as} \quad n \to \infty,$$

we define

$$u_n(y) = \mathbb{E}[G_n \cdot \phi_{n^{-\delta}}(F_n - y)] = \mathbb{E}[G_n \cdot \phi_{n^{-\delta}}(X_T^{\pi_n}(x) - y)]$$

and

$$u(y) = \mathbb{E}[G \cdot \delta_y(F)] = \mathbb{E}[G \cdot \delta_y(X_T(x))] .$$

Then, we have

$$\|u_n - u\|_\beta = O(n^{-\gamma\wedge\delta\wedge\varepsilon}) \quad \text{as} \quad n \to \infty. \tag{11.2.28}$$

Note that

$$u(y) = P(T, x, y)\mathbb{E}[G|F = y]$$

where $P(T, x, y) = \mathbb{E}[\delta_y(X_T(x))]$ is the probability density of the solution $X_T(x)$. We know that $P(T, x, y)$ coincides with the heat kernel for the heat operator L^0.

Remark 11.4. If the non-degeneracy condition is also satisfied for F_n, i.e.,

$$\sup_n \|(\det\ \sigma_{F_n})^{-1}\|_p < \infty,$$

then, u_n in (11.2.28) may be replaced by

$$u_n(y) = \mathbb{E}[G_n \cdot \delta_y(F_n)] = P_n(T, x, y)\mathbb{E}[G_n|F_n = y],$$

where $P_n(T, x, y) = \mathbb{E}[\delta_y(X_T^{\pi_n}(x))]$ and it holds that

$$\|u_n - u\|_\beta = O(n^{-\gamma\wedge\varepsilon}), \quad \text{as} \quad n \to \infty.$$

In particular, for any $\beta > 0$,

$$\|P_n(T, x, \cdot) - P(T, x, \cdot)\|_\beta = O(n^{-\gamma}) \quad \text{as} \quad n \to \infty.$$

Example 11.2. Suppose that the coefficients of equation (11.1.1) satisfies the uniform ellipticity condition:

$$\inf_{(s,x)\in[0,T]\times\mathbb{R}^d} \det\,[a(s,x)] > 0,$$

where $a^{kl}(s,x) = \sum_{j=1}^m A_j^k(s,x)A_j^l(s,x), \quad k,l = 1,\cdots,d$ and consider the Euler-Maruyama scheme:

$$X_{t_{k+1}}^\pi = X_{t_k}^\pi + \sum_{j=1}^m A_j(t_k, X_{t_k}^\pi)[W_{t_{k+1}}^j - W_{t_k}^j] + A_0(t_k, X_{t_k}^\pi)(t_{k+1} - t_k).$$

Then, for any $\beta > 0$,

$$\|P^\pi(T,x,\cdot) - P(T,x,\cdot)\|_\beta = O(|\pi|^{1/2})$$

as $|\pi| = \max_k(t_{k+1}-t_k) \to 0$ and $P^\pi(T,x,y) = \mathbb{E}[\delta_y(X_T^\pi)]$ is given explicitly by

$$P^\pi(T,x,y) = \int_{\mathbb{R}^d}\cdots\int_{\mathbb{R}^d}\prod_{k=0}^{n-1}\Lambda(t_k,x_k;t_{k+1},x_{k+1})dx_1\cdots dx_{n-1},$$

where $\pi: \ 0 = t_0 < t_1 < \cdots < t_{n-1} < t_n = T$, $x_0 = x$, $x_n = y$ and, in general

$$\begin{aligned}
&\Lambda(\sigma,\xi;\tau,\eta)\\
&= \left[(2\pi(\tau-\sigma))^d \det[a(\sigma,\xi)]\right]^{-1/2}\\
&\times \exp\left[-\frac{1}{2}\langle a^{-1}(\sigma,\xi)(\eta-\xi-A_0(\sigma,\xi)(\tau-\sigma)), \eta-\xi-A_0(\sigma,\xi)(\tau-\sigma)\rangle\right],\\
&\qquad 0\le\sigma<\tau\le T, \quad \xi,\eta\in\mathbb{R}^d.
\end{aligned}$$

The above theorem says that the rate of convergence of $P^\pi(T,x,y)$ to $P(T,x,y)$ is $1/2$.

11.3 Weak Convergence Rate

Now we want to treat the *weak convergence rate problem*, that is to say, to estimate $|\mathbb{E}[h(X_T) - h(Y_T)]|$ for a continuous function h of polynomial growth, where $Y_T = Y_t^\pi$ is the time discretization approximation of X_T defined by (11.1.6) but with the choice of Γ defined by (11.3.1) below. Note that the absolute value sign is outside the expectation. We can also estimate $\sup_{0\le t\le T}|\,\mathbb{E}[\,h(X_t) - h(Y_t)\,]\,|$ without essential modifications except notational complexity. To get a weak convergence rate of order γ (here, an integer), we take from now on

$$\Gamma = \mathcal{B}_\gamma = \{\alpha, l(\alpha) \le \gamma\}\,. \tag{11.3.1}$$

Then, we put $\mathcal{B}'_\gamma = \{\alpha : \alpha \notin \mathcal{B}_\gamma, -\alpha \in \mathcal{B}_\gamma\} = \{\alpha : l(\alpha) = \gamma + 1\}$.

We have the following theorem.

Theorem 11.5. *Let γ be a positive integer. Assume that all coefficients A_j $(j = 0, 1, \dots, m)$ are Lipschitz continuous and their components belong to $C_p^{2(\gamma+1)}$ (the space of functions from $\mathbb{R}^d$ to $\mathbb{R}$ whose derivatives of order $\le 2(\gamma+1)$ are continuous and are of polynomial growth). Assume that for all $\alpha \in \Gamma$, $f_\alpha = L^{\alpha_1} \cdots L^{\alpha_{l-1}} A_{\alpha_l}$, defined by (11.1.4), is of linear growth:*

$$| f_\alpha(x) | \le C(1 + |x|) \ . \tag{11.3.2}$$

Then, for each $h \in C_p^{2(\gamma+1)}$ there is a constant C_h independent of $|\pi|$ such that

$$| \mathbb{E}[h(X_T) - h(Y_T)] | \le C_h \ |\pi|^\gamma \ . \tag{11.3.3}$$

We need two lemmas to prove this theorem. We introduce the following notation: $X_\cdot^{s,x}$ is the solution of the s.d.e.

$$X_t^{s,x} = x + \sum_{j=0}^m \int_s^t A_j(X_r^{s,x}) \, dW_r^j \ , \quad s \le t \le T \tag{11.3.4}$$

and for notational simplicity we denote $\tilde{X}_s = X_s^{t_n, Y_{t_n}}$ for $t_n \le s \le t_{n+1}$.

Lemma 11.6. *Let $g(\cdot)$ be adapted and $\sup_{0 \le s \le T} \mathbb{E}|g(s)|^p < \infty$ for any $1 \le p < \infty$. Then, for $t_n \le t \le t_{n+1}$,*

$$|\mathbb{E}\{I_\alpha[g(\cdot)]_{t_n,t} | \mathcal{F}_{t_n}\}| \le C(\omega)(t - t_n)^{l(\alpha)}, \tag{11.3.5}$$

where and in what follows $C(\omega)$ denotes a positive generic random constant independent of partition (which may vary from line to line) such that $\mathbb{E}C(\omega)^p < \infty$ for any $1 \le p < \infty$.

Proof. It is a straightforward consequence of our assumptions. □

Lemma 11.7. *Under the assumptions of Lemma 11.6, we have for $t_n \le t \le t_{n+1}$,*

$$|M_\alpha| := |\mathbb{E}\{[h(\tilde{X}_t) - h(Y_{t_n})] I_\alpha[g(\cdot)]_{t_n,t} | \mathcal{F}_{t_n}\}| \le C(\omega)(t - t_n)^{l(\alpha)}. \tag{11.3.6}$$

Proof. We shall prove this lemma by induction on $l(\alpha)$. It is easy to see that (11.3.6) is true when $l(\alpha) = 1$. Let (11.3.6) be true for $l(\alpha) \le k$. We are going to prove that it is true for $l(\alpha) = k + 1$.

Let $l(\alpha) = k + 1$. Applying the Itô formula (11.1.3), we have

$$h(\tilde{X}_t) - h(Y_{t_n}) = \sum_{j=0}^m \int_{t_n}^t L^j h(\tilde{X}_s) dW_s^j, \quad t_n \le t \le t_{n+1}. \tag{11.3.7}$$

When $\alpha_l = 0$ ($l(\alpha) = k+1$), by the Itô formula (11.1.3), Lemma 11.6 and the induction assumption we have

$$
\begin{aligned}
|M_\alpha| \le & \int_{t_n}^t |\mathbb{E}\{[h(\tilde{X}_s) - h(Y_{t_n})]I_{\alpha-}[g(\cdot)]_{t_n,s}|\mathcal{F}_{t_n}\}|ds \\
& + \int_{t_n}^t |\mathbb{E}\{[L^0h(\tilde{X}_s) - L^0h(Y_{t_n})]I_\alpha[g(\cdot)]_{t_n,s}|\mathcal{F}_{t_n}\}|ds \\
& + \int_{t_n}^t \mathbb{E}|L^0h(Y_{t_n})||\mathbb{E}\{I_\alpha[g(\cdot)]_{t_n,s}|\mathcal{F}_{t_n}\}|ds \\
\le & \, C(\omega)(t-t_n)^{k+1} + \int_{t_n}^t |\mathbb{E}\{[L^0h(\tilde{X}_s) - L^0h(Y_{t_n}]I_\alpha[g(\cdot)]_{t_n,s}|\mathcal{F}_{t_n}\}|ds.
\end{aligned}
\tag{11.3.8}
$$

Applying (11.3.8) repeatedly, we obtain

$$
\begin{aligned}
|M_\alpha| \le C(\omega)(t-t_n)^{k+1} + \int_{t_n \le s_1 < \cdots < s_{k+1} < t} & |\mathbb{E}\{[(L^0)^{k+1}h(\tilde{X}_{s_1}) \\
& -(L^0)^{k+1}h(Y_{t_n})]I_\alpha[g(\cdot)]_{t_n,s_1}|\mathcal{F}_{t_n}\}|ds_1 \cdots ds_{k+1}.
\end{aligned}
$$

Now it is easy to see that the conditional expectation inside the above multiple integral is in L^p for any $1 \le p < \infty$. This proves (11.3.6) for $\alpha_l = 0$. In the same way we can prove (11.3.6) for $\alpha_l \ne 0$. □

Proof of Theorem 11.5. Set $u(s,x) = \mathbb{E}[h(X_T^{s,x})]$ (see (11.3.4)). Then, for $h \in C_p^{(2\gamma+1)}$ we have $u(s,\cdot) \in C_p^{(2\gamma+1)}$ (this can be shown easily by Malliavin calculus for example). We have $\mathbb{E}[h(X_T)] = u(0, X_0)$ and

$$
\mathbb{E}u(t_n, X_{t_n}^{t_{n-1}, Y_{t_{n-1}}}) = \mathbb{E}u(t_{n-1}, Y_{t_{n-1}}), \quad n \ge 1\,. \tag{11.3.9}
$$

We compute the following expectation

$$
\begin{aligned}
|\,\mathbb{E}[\,h(Y_T) - h(X_T)\,]\,| &= |\,\mathbb{E}[\,u(T, Y_T) - u(0, X_0)\,]\,| \\
&= |\,\mathbb{E}[\,u(T, Y_T) - u(0, Y_0)\,]\,| \\
&\le |\,\mathbb{E}\sum_{n=1}^N \big[u(t_n, Y_{t_n}) - u(t_{n-1}, Y_{t_{n-1}})\,\big]| \\
&= |\,\mathbb{E}\sum_{n=1}^N \big[u(t_n, Y_{t_n}) - u(t_n, X_{t_n}^{t_{n-1}Y_{n-1}})\,\big]| \\
&= |\,\mathbb{E}\sum_{n=1}^N u'(t_n, X_{t_n}^{t_{n-1}Y_{n-1}})\big(\,Y_{t_n} - X_{t_n}^{t_{n-1}Y_{n-1}}\,\big) \\
&\quad + \frac{1}{2}\sum_{n=1}^N \mathbb{E}u''(t_n, Z_{\theta,n})\big(\,X_{t_n}^{t_{n-1}Y_{n-1}} - Y_{t_n}\big)^2\,|,
\end{aligned}
\tag{11.3.10}
$$

where u' and u'' are derivatives of $u(s,x)$ w.r.t. x, $Z_{\theta,n} := X_{t_n}^{t_{n-1}Y_{n-1}} + \theta(Y_{t_n} - X_{t_n}^{t_{n-1}Y_{n-1}})$ and $0 \le \theta \le 1$. By (11.1.26) we know that the last term is dominated by the sum of

$$\begin{aligned} C\big(\mathbb{E}|u''(Z_{\theta,n})|^2\big)^{1/2}&\big(\mathbb{E}(X_{t_n}^{t_{n-1}Y_{n-1}} - Y_{t_n})^4\big)^{1/2} \\ &\le C\sum_{\alpha\in\mathcal{B}'_\gamma}\big(\mathbb{E}|\,I_\alpha[f_\alpha(\cdot,X.)]_{t_n,t_{n+1}}\,|^4\big)^{1/2} \\ &\le C\sum_{\alpha\in\mathcal{B}'_\gamma}\Big((t_{n+1}-t_n)^{2(l(\alpha)+n(\alpha))}\Big)^{1/2} \le C\sum_{\alpha\in\mathcal{B}'_\gamma}(t_{n+1}-t_n)^{l(\alpha)}. \end{aligned}$$

But when $\alpha\in\mathcal{B}'_\gamma$ we have $l(\alpha)\ge\gamma+1$, so the last term of (11.3.10) is at most $C(t_{n+1})-t_n)^\gamma$.

As for the first term of (11.3.10), first we notice that by the assumptions of Theorem 11.5, we have $\sup_{0\le t\le T}|f_\alpha(t,X_t)|^p<\infty$ for any $1\le p<\infty$ and $\alpha\in\mathbb{B}'_\gamma$. We have

$$\begin{aligned} \Big|\,\mathbb{E}\sum_{n=1}^N &u'(t_n,X_{t_n}^{t_{n-1}Y_{n-1}})\big(Y_{t_n}-X_{t_n}^{t_{n-1}Y_{n-1}}\big)\,\Big| \\ &\le \Big|\,\mathbb{E}\sum_{n=1}^N u'(t_n,Y_{t_{n-1}})\,\big(Y_{t_n}-X_{t_n}^{t_{n-1}Y_{n-1}}\big)\,\Big| \\ &\quad+\Big|\,\mathbb{E}\sum_{n=1}^N\big[u'(t_n,X_{t_n}^{t_{n-1}Y_{n-1}})-u'(t_n,Y_{t_{n-1}})\big]\,\big(Y_{t_n}-X_{t_n}^{t_{n-1}Y_{n-1}}\big)\,\Big|. \end{aligned} \tag{11.3.11}$$

The second term of (11.3.11) is dominated by

$$\begin{aligned} \sum_{\alpha\in\mathbb{B}'_\gamma}\sum_{n=1}^N\Big|\mathbb{E}\Big\{\mathbb{E}\big[&u'(t_n,X_{t_n}^{t_{n-1},Y_{n-1}})-u'(t_n,Y_{n-1})\big] \\ &I_\alpha[f_\alpha(\cdot,X.)]_{t_{n-1},t_n}|\mathcal{F}_{t_n-1}\big]\Big\}\Big| \\ &\le \sum_{\alpha\in\mathbb{B}'_\gamma}\sum_{n=1}^N\mathbb{E}[C(\omega)](t_n-t_{n-1})^{l(\alpha)} \\ &\le C\sum_{n=1}^N(t_n-t_{n-1})^{\gamma+1}\le C\delta^\gamma. \end{aligned}$$

This gives the necessary estimate for the second term of (11.3.11). It is easy to see that the first term of (11.3.11) is dominated by $C\delta^\gamma$. This proves the theorem. □

Remark 11.5. For more discussion about numerical solution, see [Kloeden and Platen (1992)].

11.4 Wong-Zakai Approximation

The following result is an slight improvement of a result in [Hu, Kallianpur and Xiong (2002)]

Lemma 11.8. *Let $(W_t, t \geq 0)$ be a Brownian motion on a probability space $(\mathbb{B}, \mathbb{H}, \mathbb{P})$. Let $\pi : 0 = t_0 < t_1 < \cdots < t_{n-1} < t_n = T$ be a partition of interval $[0, T]$ and let*

$$W_t^\pi = W_{t_k} + \frac{W_{t_{k+1}} - W_{t_k}}{t_{k+1} - t_k}(t - t_k), \quad t_k \leq t \leq t_{k+1}, \tag{11.4.1}$$

be the polygonal approximation of the Brownian motion W. Then, there is a constant $C_{p,T}$ (independent of π) such that (for all π such that $\log|\pi| < -2$)

$$\left(\mathbb{E}\left[\sup_{0\leq t\leq T} |W_t^\pi - W_t|^p\right]\right)^{1/p} \leq C_{p,T}(\log|\pi|)^2 |\pi|^{\frac{1}{2}+\frac{1}{\log|\pi|}}. \tag{11.4.2}$$

Remark 11.6. An obvious consequence is that for any $\varepsilon > 0$, there is a constant $C_{p,T,\varepsilon}$ such that

$$\left(\mathbb{E}\left[\sup_{0\leq t\leq T} |W_t^\pi - W_t|^p\right]\right)^{1/p} \leq C_{p,T,\varepsilon}|\pi|^{\frac{1}{2}-\varepsilon}. \tag{11.4.3}$$

Proof. For any $r \geq p$, we have

$$\begin{aligned}
&\left\{\mathbb{E}\sup_{0\leq t\leq 1}|W_t^\pi - W_t|^p\right\}^{\frac{r}{p}} \leq \mathbb{E}\sup_{0\leq t\leq 1}\{|W_t^\pi - W_t|^r\} \\
&\leq \mathbb{E}\sup_{1\leq k\leq n}\sup_{t_{k-1}\leq t\leq t_k}\left\{|W_t - W_{t_{k-1}}| + \left|\frac{W_{t_k} - W_{t_{k-1}}}{t_k - t_{k-1}}(t - t_{k-1})\right|\right\}^r \\
&\leq 2^r\mathbb{E}\sup_{1\leq k\leq n}\sup_{t_{k-1}\leq t\leq t_k}|W_t - W_{t_{k-1}}|^r \\
&\qquad +\mathbb{E}\sup_{1\leq k\leq n}\left|W_{t_k} - W_{t_{k-1}}\right|^r \sup_{t_{k-1}\leq t\leq t_k}\frac{(t - t_{k-1})^r}{(t_k - t_{k-1})^r} \\
&\leq 2^{r+1}\mathbb{E}\sup_{1\leq k\leq n}\left\{\sup_{t_{k-1}\leq t\leq t_k}|W_t - W_{t_{k-1}}|^r\right\} \\
&\leq 2^{r+1}\sum_{k=1}^n \mathbb{E}\left\{\sup_{t_{k-1}\leq t\leq t_k}|W_t - W_{t_{k-1}}|^r\right\}.
\end{aligned}$$

By Burkholder-Davis-Gundy inequality (A.61), there is a constant C_1 independent of r such that

$$\mathbb{E}\sup_{t_{k-1}\le t\le t_k}|W_t - W_{t_{k-1}}|^r \le C_1^r r^{r+1}|t_k - t_{k-1}|^{r/2}.$$

Thus

$$\begin{aligned}\left\{\mathbb{E}\sup_{0\le t\le T}|W_t^\pi - W_t|^p\right\}^{\frac{r}{p}} &\le 2^{r+1}\sum_{k=1}^n C_1^r r^{r+1}|t_k - t_{k-1}|^{r/2}\\ &\le 2^{r+1}C_1^r r^{r+1}|\pi|^{\frac{r}{2}-1}\sum_{k=1}^n |t_k - t_{k-1}|\\ &= 2^{r+1}C_1^r r^{r+1}|\pi|^{\frac{r}{2}-1}T.\end{aligned}$$

Therefore, we have

$$\mathbb{E}\sup_{0\le t\le T}|W_t^\pi - W_t|^p \le 2^{p+p/r}C_1^p r^{p+\frac{p}{r}}T^{\frac{p}{r}}|\pi|^{\frac{p}{2}-\frac{p}{r}}.$$

Or

$$\left(\mathbb{E}\sup_{0\le t\le T}|W_t^\pi - W_t|^p\right)^{1/p} \le 2^{1+1/r}C_1 r^{1+\frac{1}{r}}T^{\frac{1}{r}}|\pi|^{\frac{1}{2}-\frac{1}{r}}.$$

Taking $r = -\log|\pi|$, the lemma follows from the above inequality. □

Let us now consider the Wong-Zakai approximation. To simplify notation we consider the homogeneous equation (similar to (10.2.28) with a different notation for A_0)

$$X_t = x + \sum_{j=0}^m \int_0^t A_j(X_s)dW_s^j, \quad 0 \le t \le T. \tag{11.4.4}$$

Let $\pi : 0 = t_0 < t_1 < \cdots < t_n = T$ be a partition and we consider the polygonal approximation (11.4.1) for each Brownian motion W_t^j to obtain $W_t^{j,\pi}$. Denote $\Delta_k = t_{k+1} - t_k$ and $\Delta_k W^j = W_{t_{k+1}}^j - W_{t_k}^j$. The Wong-Zakai approximation of the solution is

$$X_t^\pi = x + \sum_{j=0}^m \int_0^t A_j(X_s^\pi)\dot{W}_s^{j,\pi}ds, \quad 0 \le t \le T. \tag{11.4.5}$$

It is an interesting remark by Wong and Zakai ([Wong and Zakai (1965)]) that contradictory to the common sense, the processes X_t^π do not converge to X_t. Instead, they converge to a process Y_t, which is different from

X_t. More precisely, the limiting process Y_t satisfies the following stochastic differential equation:

$$Y_t = x + \sum_{j=0}^{m} \int_0^t A_j(Y_s) dW_s^j + \frac{1}{2} \sum_{j=1}^{m} \int_0^t A_j^2(Y_s) ds \,, \quad 0 \le t \le T \,, \quad (11.4.6)$$

where $A_j^2 = A_j A_j$ and $A_j A_k = \langle A_j, \nabla A_k \rangle$. We state the following

Theorem 11.6. *Let $A_j : \mathbb{R}^d \to \mathbb{R}^d$, $j = 0, 1, \cdots, m$, be three times continuously differentiable. Let A_j together with all the derivatives up to order three be bounded. Then, there is a constant $C_{p,T}$ independent of the partition π (may depend on the bound of the derivatives of A_j) such that*

$$\left(\mathbb{E} \left[\sup_{0 \le t \le T} |X_t^\pi - Y_t|^p \right] \right)^{1/p} \le C_{p,T} (\log |\pi|)^2 |\pi|^{\frac{1}{2} + \frac{1}{\log |\pi|}} \,. \quad (11.4.7)$$

Proof. The proof is similar to that of Theorem 11.1 but in its simplest case of Euler-Marauyama scheme. We shall outline the idea. We also point out where we use Lemma 11.8.

Equation (11.4.5) can be written on each sub-interval as

$$X_t^\pi = X_{t_k}^\pi + \sum_{j=1}^{m} \int_{t_k}^t A_j(X_s^\pi) \frac{\Delta_k W^j}{\Delta_k} ds \,, \quad t_k \le t \le t_{k+1} \,. \quad (11.4.8)$$

This identity shows that X_t^π is differentiable on $[t_k, t_{k+1}]$ with an explicit derivative. Using the usual chain rule and the fact that A_j has bounded derivatives up to order three, we have

$$\begin{aligned} A_j(X_s^\pi) = A_j(X_{t_k}^\pi) &+ \sum_{\ell=1}^{m} A_\ell A_j(X_{t_k}^\pi) \frac{\Delta_k W^\ell}{\Delta_k} (s - t_k) \\ &+ A_0 A_j(X_{t_k}^\pi)(s - t_k) + O(\Delta_k) \,, \end{aligned}$$

where $O(\Delta_k)$ means that $|O(\Delta_k)/\Delta_k| \le C < \infty$. Substituting this into equation (11.4.8), we obtain

$$\begin{aligned} X_t^\pi = X_{t_k}^\pi &+ A_0(X_{t_k}^\pi)(t - t_k) + \sum_{j=1}^{m} A_j(X_{t_k}^\pi) \frac{\Delta_k W^j}{\Delta_k} (t - t_k) \\ &+ \frac{1}{2} \sum_{\ell,j=1}^{m} A_\ell A_j(X_{t_k}^\pi) \frac{\Delta_k W^\ell \Delta_k W^j}{\Delta_k^2} (t - t_k)^2 + O(|\Delta_k|^{3/2}) \,. \end{aligned}$$

(11.4.9)

Consider the Euler-Maruyama scheme for equation (11.4.6):

$$
\begin{aligned}
Y_t^\pi = Y_{t_k}^\pi &+ \sum_{j=1}^m A_j(Y_{t_k}^\pi)(W_t^j - W_{t_k}^j) + A_0(Y_{t_k}^\pi)(t-t_k) \\
&+ \frac{1}{2}\sum_{j=1}^m A_j^2(Y_{t_k}^\pi)(t-t_k)\,, \quad t_k \le t \le t_{k+1}\,.
\end{aligned} \tag{11.4.10}
$$

By (11.1.29), we have

$$
\mathbb{E}\left(\sup_{0\le t\le T} |Y_t^\pi - X_t|^p\right) \le C_{p,T}|\pi|^{1/2}\,.
$$

Thus, to show the theorem, we need only to show the estimate for $\mathbb{E}\left(\sup_{0\le t\le T} |X_t^\pi - Y_t^\pi|^p\right)$ (one can also show $\mathbb{E}\left(\sup_{0\le t\le T} |X_t^\pi - X_t|^p\right)$ directly with a little more effort).

Subtracting Equation (11.4.10) from (11.4.9) we have when $t \in [t_k, t_{k+1}]$,

$$
\begin{aligned}
X_t^\pi - Y_t^\pi &= X_{t_k}^\pi - Y_{t_k}^\pi + \left[A_0(X_{t_k}^\pi) - A_0(Y_{t_k}^\pi)\right](t-t_k) \\
&+ \frac{1}{2}\sum_{j=1}^m \left[A_j^2(X_{t_k}) - A_j^2(Y_{t_k})\right](t-t_k) \\
&+ \sum_{j=1}^m \left[A_j(X_{t_k}^\pi) - A_j(Y_{t_k}^\pi)\right](W_t^j - W_{t_k}^j) \\
&+ \sum_{j=1}^m A_j(X_{t_k}^\pi)\left[\frac{\Delta_k W^j}{\Delta_k}(t-t_k) - W_t^j + W_{t_k}^j\right] \\
&+ \frac{1}{2}\sum_{j=1}^m A_j^2(X_{t_k}^\pi)\left[\frac{(\Delta_k W^j)^2}{\Delta_k^2}(t-t_k)^2 - (t-t_k)\right] \\
&+ \frac{1}{2}\sum_{\ell\ne j}^m A_\ell A_j(X_{t_k}^\pi)\frac{\Delta_k W^\ell \Delta_k W^j}{\Delta_k^2}(t-t_k)^2 + O(|\Delta_k|^{3/2})\,.
\end{aligned} \tag{11.4.11}
$$

Denote $n_s = \max\{k, t_k \le s\}$ as before. Set $Z_t = X_t^\pi - Y_t^\pi$. Then, we have after we recursively use (11.4.11)

$$
Z_t = R_1(t) + R_2(t) + R_3(t) + R_4(t) + O(|\Delta_k|^{3/2})\,. \tag{11.4.12}
$$

where

$$R_1(t) = \int_0^t \left[A_0(X^\pi_{n_s}) - A_0(Y^\pi_{n_s})\right] ds + \frac{1}{2}\sum_{j=1}^m \int_0^t \left[A_j^2(X_{n_s}) - A_j^2(Y_{t_{n_s}})\right] ds$$

$$+\sum_{j=1}^m \int_0^t \left[A_j(X^\pi_{t_{n_s}}) - A_j(Y^\pi_{t_{n_s}})\right] dW_s^j\,;$$

$$R_2(t) = \sum_{j=1}^m A_j(X^\pi_{t_{n_s}}) \left[\frac{\Delta_{n_t} W^j}{\Delta_k}(t - t_{n_t}) - W_t^j + W^j_{t_{n_t}}\right];$$

$$R_3(t) = \frac{1}{2}\sum_{k=0}^{n_t-1}\sum_{j=1}^m A_j^2(X^\pi_{t_k})\left[(\Delta_k W^j)^2 - \Delta_k\right]$$

$$+\frac{1}{2}\sum_{j=1}^m A_j^2(X^\pi_{t_{n_t}})\left[\frac{(\Delta_{n_t}W^j)^2}{\Delta^2_{n_t}}(t-t_{n_t})^2 - (t-t_{n_t})\right];$$

$$R_4(t) = \frac{1}{2}\sum_{k=0}^{n_t-1}\sum_{\ell\neq j}^m A_\ell A_j(X^\pi_{t_k})\Delta_k W^\ell \Delta_k W^j$$

$$+\frac{1}{2}\sum_{\ell\neq j}^m A_\ell A_j(X^\pi_{t_{n_t}})\frac{\Delta_{n_t}W^\ell\Delta_{n_t}W^j}{\Delta_k^2}(t-t_{n_t})^2\,.$$

In the same way as for (11.1.21), we see that

$$\mathbb{E}\sup_{0\le s\le t}|R_1(s)|^p \le C\int_0^t \mathbb{E}\sup_{0\le r\le s}|Z_r|^p dr\,.$$

Since A_j is bounded, we can use Lemma 11.8 to bound R_2:

$$\mathbb{E}\sup_{0\le s\le t}|R_2(s)|^p \le C_{p,T}(\log|\pi|)^2|\pi|^{\frac{1}{2}+\frac{1}{\log|\pi|}}\,.$$

Denote the first and the second sums of R_3 by R_{31} and R_{32}. R_{31} is a discrete time martingale (it is constant on any subinterval). We can use Doob's maximal inequality to obtain $\mathbb{E}\sup_{0\le t\le T}|R_{31}(t)|^p \le |\pi|^{p/2}$. There is no summation over k in R_{32}. This term can be obviously bounded as a constant multiple of

$$\mathbb{E}\sup_{0\le t\le T}|R_{32}(t)|^p \le |\pi|^p + \mathbb{E}\sup_{0\le t\le T}(W_t - W_{t_{n_t}})^2 \le |\pi|^p$$

$$+\mathbb{E}\sup_k \sup_{t_k\le t\le t_{k+1}}(W_t - W_{t_k})^{2p}$$

which can be bounded again by Lemma 11.8. $R_4(t)$ can be dealt with in the same as $R_3(t)$. Thus we have

$$\mathbb{E}\sup_{0\le s\le t}|Z_s|^p \le C\int_0^t \mathbb{E}\sup_{0\le r\le s}|Z_r|^p ds + O(|\pi|^{1/2}) + C_{p,T}(\log|\pi|)^2|\pi|^{\frac{1}{2}+\frac{1}{\log|\pi|}}\,.$$

An application of the Gronwall lemma yields the theorem. □

Remark 11.7. In [Hu and Nualart (2009)] it is proved that for the Wong-Zakai approximation we have the following almost sure convergence in the Hölder norm. For any $\beta \in (1/3, 1/2)$, there is a random constant $C_{T,\beta} \in (0, \infty)$ such that

$$\|Y - X^\pi\|_\beta \le C_{T,\beta} |\pi|^{1/2-\beta} \sqrt{|\log|\pi||}\,, \tag{11.4.13}$$

where $\|f\|_\beta = \sup_{0\le s<t\le T} \frac{|f(t)-f(s)|}{|t-s|^\beta}$ is the β Hölder norm of f.

Appendix A

Appendix

A.1 Some Elementary Results from Analysis

In this section, we will present some results from analysis which may not be found in a single source. We also give proofs and hope it will facilitate the reading of this manuscript.

Let $a : \mathbb{R}_+ \to \mathbb{R}_+$ be a right continuous and non-decreasing function. The function

$$a^{(-1)}(r) := \inf \{s \geq 0 \,, a(s) > r\} \,, \quad r \in \mathbb{R}_+ \tag{A.1}$$

is called the right inverse of a (we define the inf of an empty set as ∞: namely, $\inf \emptyset = \infty$). It is easy to see that

Proposition A.1. *$a^{(-1)}(t)$ is also a non-negative, right continuous and non-decreasing function and we have*

$$a(a^{(-1)}(r)) \geq r \,, \quad \forall\, r \geq 0 \,; \tag{A.2}$$

and

$$a(r) = \sup\{t, a^{(-1)}(t) \leq r\} \,, \quad \forall\, r \geq 0 \,. \tag{A.3}$$

Proof. From the definition of (A.1) we see that for any $\varepsilon > 0$, there is an s_ε such that

$$a(s_\varepsilon) > r \quad \text{and} \quad s_\varepsilon < a^{(-1)}(r) + \varepsilon \,.$$

This implies $r < a(s_\varepsilon) \leq a(a^{(-1)}(r) + \varepsilon)$. Letting $\varepsilon \to 0$ yields (A.2).

Denote $t_r = \sup\{t, a^{(-1)}(t) \leq r\}$. For any $t \in \{t, a^{(-1)}(t) \leq r\}$, by (A.2) we have $t \leq a(a^{(-1)}(t)) \leq a(r)$. Thus, $t_r \leq a(r)$. We want to show that $t_r \geq a(r)$. If this is not the case, then there is a $\delta > 0$ such that $t_r < a(r) - \delta$. Namely, $t_r + \delta < a(r)$. This can be written as $r \in \{s \geq 0 \,, a(s) > t_r + \delta\}$. By (A.1), $c(t_r + \delta) \leq r$. This contradicts the definition of t_r. The identity (A.3) is then proved. □

Lemma A.1. *If $a : \mathbb{R}_+ \to \mathbb{R}_+$ is a right continuous and non-decreasing function and if $f : \mathbb{R}_+ \to \mathbb{R}_+$ is bounded and Borel measurable, then*

$$\int_0^\infty f(r)da(r) = \int_0^\infty f(a^{(-1)}(r))I_{\{a^{(-1)}(r)<\infty\}}dr\,. \tag{A.4}$$

Proof. By linearity and limiting argument, it suffices to show the lemma for the simple functions of the form $f(r) = I_{[0,u]}(r)$, where $u \in [0,\infty)$ is a given number. In this case, we have from (A.3)

$$\begin{aligned}\int_0^\infty f(r)da(r) &= a(u) = \sup\{r, a^{(-1)}(r) \le u\}\\ &= \int_0^\infty I_{\{r,a^{(-1)}(r)\le u\}}dr = \int_0^\infty f(a^{(-1)}(r))I_{\{r,a^{(-1)}(r)<\infty\}}dr\,.\end{aligned}$$

This proves the lemma. □

Lemma A.2. *Let $f, g : \mathbb{R}_+ \to \mathbb{R}$ be two right continuous functions with bounded variations. Then, for any $0 \le a < b < \infty$, we have the following integration by parts formula:*

$$f(b)g(b) - f(a) - f(b) = \int_a^b f(s)dg(s) + \int_a^b g(s-)df(s)\,, \tag{A.5}$$

where $g(s-) = \lim_{t\to s-} g(t)$ is the left limit of g at s.

Proof. We have

$$\begin{aligned}(f(b) - f(a))(g(b) - g(a)) &= \int_a^b\int_a^b df(s)dg(t)\\ &= \int_{a<s\le t\le b} df(s)dg(t) + \int_{a<t<s\le b} df(s)dg(t)\\ &= \int_a^b dg(t)\int_a^t df(s) + \int_a^b df(s)\int_a^{s-} dg(t)\\ &= \int_a^b [f(t) - f(a)]dg(t) + \int_a^b [g(s-) - g(a)]df(s)\\ &= \int_a^b f(t)dg(t) + \int_a^b g(s-)df(s)\\ &\quad -f(a)[g(b) - g(a)] - g(a)[f(b) - f(a)]\,,\end{aligned}$$

proving the lemma. □

Definition A.1. Let $\Phi : \mathbb{R}_+ \to \mathbb{R}_+$ be a non-decreasing convex function with $\Phi(0) = 0$. Then, there is a non-negative, right continuous, non-decreasing function $\phi : \mathbb{R}_+ \to \mathbb{R}_+$ such that $\Phi(t) = \int_0^t \phi(r)dr$. The right inverse $\psi(t) = \phi^{(-1)}(t) = \inf\{s \geq 0\,, a(s) > r\}$ is also a non-negative, right continuous, non-decreasing function. The function $\Psi(t) = \int_0^t \psi(r)dr$ is also non-negative, right-continuous, and nondecreasing and it is called the conjugate convex function of Φ.

Example A.1. If $\Phi(t) = \dfrac{t^p}{p}\,, t \geq 0$ and $p > 1$, then $\phi(t) = t^{p-1}$. We have $\psi(t) = t^{1/(p-1)}$ and $\Psi(t) = \dfrac{t^q}{q}$, where $q = \dfrac{p}{p-1}$ is the conjugate number of p (namely, $1/p + 1/q = 1$).

Lemma A.3. *Let $\Phi : \mathbb{R}_+ \to \mathbb{R}_+$ be any non-decreasing convex function with $\Phi(0) = 0$ and let Ψ be the conjugate convex function of Φ. Then for any $u, v > 0$, the following Young inequality holds*

$$uv \leq \Phi(u) + \Psi(v)\,. \tag{A.6}$$

Proof. If $\phi(u) > v$, then by the definition of ψ we have $\psi(v) \leq u$. Thus, by the integration by parts formula and then by (A.4), we have

$$\begin{aligned}
\Phi(u) + \Psi(v) &= \int_0^u \phi(r)dr + \int_0^v \psi(r)dr \\
&= u\phi(u) - \int_0^u r d\phi(r) + \int_0^v \psi(r)dr \\
&= u\phi(u) - \int_0^\infty r I_{[0,u]}(r)d\phi(r) + \int_0^\infty I_{\{r\leq v\}}\psi(r)dr \\
&= u\phi(u) - \int_0^\infty \psi(r)I_{\{\psi(r)\leq u\}}dr + \int_0^\infty I_{\{r\leq v\}}\psi(r)dr \\
&= u\phi(u) - \int_{\{r:\psi(r)\leq u, r>v\}} \psi(r)dr\,,
\end{aligned} \tag{A.7}$$

where the last equality follows from

$$\{r \leq v\} \subseteq \{\psi(r) \leq \psi(v)\} \subseteq \{r : \psi(r) \leq u\}\,.$$

We also have

$$\begin{aligned}
\int_{\{r:\psi(r)\leq u, r>v\}} \psi(r)dr &\leq \int_{\{\psi(v)\leq\psi(r)\leq u\}} \psi(r)dr = \int_{\psi(v)}^u r d\phi(r) \\
&\leq u\int_{\psi(v)}^u d\phi(r) = u\left[\phi(u) - \phi(\psi(v))\right] \leq u\left[\phi(u) - v\right]
\end{aligned} \tag{A.8}$$

by (A.2). Substituting (A.8) into (A.7), we have

$$\Phi(u) + \Psi(v) \geq u\phi(u) - u\left[\phi(u) - v\right] = uv\,.$$

This proves the lemma in the case $\phi(u) > v$. The case $\phi(u) \leq v$ can be proved in a similar way. □

Definition A.2. Let $\Phi : \mathbb{R}_+ \to \mathbb{R}_+$ be a nondecreasing and convex function such that $\Phi(0) = 0$. We call Φ moderate if there is a constant $c > 0$

$$\Phi(2t) \le c\Phi(t)\,, \quad \forall t \ge 0\,.$$

Lemma A.4. *Let $\Phi : \mathbb{R}_+ \to \mathbb{R}_+$ be a moderate and convex function such that $\Phi(0) = 0$. Let $\Psi(t)$ be its conjugate function. Let ϕ and ψ be the right derivative of Φ and Ψ, namely, $\Phi(t) = \int_0^t \phi(r)dr$ and $\Psi(t) = \int_0^t \psi(r)dr$ for all $t \ge 0$ with ψ being the right inverse of ϕ. Define*

$$c = \sup_{t>0} \frac{\Phi(2t)}{\Phi(t)} \quad \text{and} \quad \rho = \sup_{t>0} \frac{t\phi(t)}{\Phi(t)}\,. \tag{A.9}$$

Then

(i) $c \ge 2$ and $1 \le \rho \le c - 1$.
(ii) For all $t \ge 1$ and $u \ge 0$, we have $\Phi(tu) \le t^\rho \Phi(u)$.
(iii) For all $u \ge 0$, $\Psi(\phi(u)) \le (\rho - 1)\Phi(u)$.

Proof. We assume that Φ is not identically 0. For any $u \ge 0$, we have

$$\begin{aligned}\Phi(u) &= \int_0^u \phi(r)dr \le u\phi(u) \le \int_u^{2u} \phi(s)ds \\ &= \Phi(2u) - \Phi(u) \le (c-1)\Phi(u)\,.\end{aligned}$$

This implies $c \ge 2$. The inequality $\Phi(u) \le u\phi(u)$ yields $\rho \ge 1$ and the inequality $u\phi(u) \le (c-1)\Phi(u)$ yields $c - 1 \ge \rho$. This proves (i).

From the definition ρ, we have for any $s > 0$ and $u \ge 0$, the following is true:

$$\frac{su\phi(su)}{\Phi(su)} \le \rho\,.$$

Thus, we have

$$\log \frac{\Phi(tu)}{\Phi(u)} = \int_1^t \frac{\frac{d}{ds}\Phi(su)}{\Phi(su)} ds = \int_1^t \frac{u\phi(su)}{\Phi(su)} ds \le \int_1^t \frac{\rho}{s} ds = \rho \log t = \log t^\rho\,.$$

This means $\Phi(tu) \le t^\rho \Phi(u)$. Hence, (ii) is proved.

Now we prove (iii). Notice $\sup\{s\,, \psi(s) \le u\} = \phi(u)$. By the identity (A.4) we have

$$\int_0^u r d\phi(r) = \int_{\{s:\psi(s)\le u\}} \psi(s)ds = \int_0^{\phi(u)} \psi(s)ds = \Psi(\phi(u))\,.$$

Thus

$$\begin{aligned}\Psi(\phi(u)) + \Phi(u) &= \Psi(\phi(u)) + \int_0^u \phi(r)dr \\ &= \int_0^u r d\phi(r) + u\phi(u) - \int_0^u r d\phi(r) = u\phi(u) \le \rho\Phi(u)\,.\end{aligned}$$

This is (iii). □

Example A.2. If $\Phi(t) = t^p$, then $\phi(t) = pt^{p-1}$. In this case we have $\rho = p$.

Lemma A.5. *If $A \subseteq \mathbb{R}$ is measurable set of positive Lebesgue measure, then $A - A = \{a - a' \, ; \; a, a' \in A\}$ contains an open interval containing the origin.*

Proof. We shall use $|A|$ to denote the Lebesgue measure of A. Without loss of generality we can assume that $0 < |A| < \infty$. Since A is measurable there is an open set O containing A such that $|A| \geq \frac{3}{4}|O|$. Since O is the union of countably many disjoint intervals, there must be an interval I such that $|A \cap I| \geq \frac{3}{4}|I|$. We want to show that $A - A$ contains $(-\delta, \delta)$, where $\delta = \frac{1}{2}|I|$. If $x \in (-\delta, \delta)$, then $(x + I) \cup I$ is an interval of length less than $\frac{3}{2}|I|$ which contains both $A \cap I$ and $x + A \cap I$. These two sets must intersect since $|x + A \cap I| = |A \cap I| > \frac{3}{4}|I|$. This means that there are $a, a' \in A \cap I$ such that $a = x + a'$. This implies $x = a - a' \in A - A$. □

Lemma A.6. *If A and B are two measurable sets with positive Lebesgue measures., then $D = A - B = \{a - b; \; a \in A \, , b \in B\}$ contains a nonempty open interval.*

Proof. As in the previous proof, we may find two bounded intervals I and J such that $|A \cap I| > \frac{3}{4}|I|$ and $B \cap J| > \frac{3}{4}|J|$. For any positive integer n we divide I into n subintervals of equal length. There must be one such subinterval I' such that $|A \cap I'| > \frac{3}{4}|I'|$. For any integer m we can do the same thing for J to find a subinterval J' of length $|J|/m$ such that $|B \cap J'| > \frac{3}{4}|J'|$. Appropriately choosing m and n we can assume $\frac{1}{2}|I'| \leq |J'| < |I'|$. For such chosen I' and J', one can find an open interval Δ of length $|I'| - |J'| > 0$ such that

$$x + J' \subseteq I' \quad \text{and} \quad x + (B \cap J') \subseteq I'$$

for all $x \in \Delta$. For any $x \in \Delta$,

$$|x + (B \cap J')| = |(B \cap J')| > \frac{3}{4}|J'| \geq \frac{3}{8}|I'| \, .$$

Now since both $x + (B \cap J')$ and $A \cap I'$ are contained in I' and the sum of their measures is greater than $\frac{3}{4}|I'| + \frac{3}{8}|I'| = \frac{9}{8}|I'|$ we see that $x + (B \cap J')$ and $A \cap I'$ intersect. This implies that there are $a \in A$ and $b \in B$ such that $x + b = a$, proving the lemma. □

Lemma A.7. *If S is a measurable set of $\mathbb{R}$ with positive Lebesgue measure, then $S + S$ contains a nonempty open interval.*

Proof. Take $B = -S$ in Lemma A.6. $\square$

Definition A.3. We say a set $S \subseteq \mathbb{R}$ is an additive semigroup if $S+S \subseteq S$.

Lemma A.8. *If a set $S \subseteq (0, \infty)$ is an additive semigroup and S contains a set of positive Lebesgue measure, then there exists a real number $b \in (0, \infty)$ such that S contains the interval (b, ∞).*

Proof. By Lemma A.7 the set $S + S$ contains some interval (α, β) with $0 < \alpha < \beta < \infty$. Thus, the semigroup contain all intervals $(n\alpha, n\beta)$. For a positive integer k such that $k > \alpha/(\beta - \alpha)$, each set $(n\alpha, n\beta)$ overlap with the next. Hence, $\bigcup_{n=k}^{\infty}(n\alpha, n\beta) = (k\alpha, \infty)$, which is contained in S. $\square$

Let $f : \mathbb{R}_+ \to \mathbb{R}_+$ be measurable and we define for $\lambda > 0$

$$f^*(\lambda) = \limsup_{x\to\infty} \frac{f(\lambda x)}{f(x)} \quad \text{and} \quad f_*(\lambda) = \liminf_{x\to\infty} \frac{f(\lambda x)}{f(x)}$$

if the limits exist.

Theorem A.1. *If $f : \mathbb{R}_+ \to \mathbb{R}_+$ is measurable and positive and if there is a measurable set $\Lambda \subseteq \mathbb{R}_+$ of positive Lebesgue measure such that $f^*(\lambda) < \infty$ for every $\lambda \in \Lambda$, then there is a real number $a_0 \geq 1$ such that $f^*(\lambda) < \infty$ for all $\lambda \geq a_0$, and for every a and b with $a_0^2 < a < b$,*

$$\limsup_{x\to\infty} \sup_{\lambda\in[a,b]} \frac{f(\lambda x)}{f(x)} < \infty\,. \tag{A.10}$$

Proof. Denote $h(x) = \log f(e^x)$ and

$$h^*(u) = \limsup_{x\to\infty} \{h(x+u) - h(x)\}\,.$$

From

$$h(x+u+v) - h(x) = h(x+u+v) - h(x+v) + h(x+v) - h(x)$$

we see immediately that

$$h^*(u+v) \leq h^*(u) + h^*(v)\,. \tag{A.11}$$

Let $S = \{u > 0\,,\ h^*(u) < \infty\}$. Then, by (A.11) we know that S is an additive semigroup of $\mathbb{R}$. The assumption says that S has a positive Lebesgue measure. Then, by Lemma A.8, there is a real number $a_0 \geq 1$ such that S contains $[\log a_0, \infty)$. Namely, $h^*(u) < \infty$ for all $u \geq \log a_0$.

Now we prove (A.10). Denote $A_0 = \log a_0$, $A = \log a$ and $B = \log b$. The inequality (A.10) is equivalent to

$$\limsup_{x\to\infty} \sup_{u\in[A,B]} \{h(x+u) - h(x)\} < \infty\,. \tag{A.12}$$

We will show (A.12) by contradiction. Suppose the contrary. Then, there exists a sequence $x_n \to \infty$ and $u_n \in [A, B]$ such that

$$h(x_n + u_n) - h(x_n) > n \quad \text{for } n = 1, 2, \cdots .$$

Passing to a subsequence, we may assume that $u_n \to u \in [A, B]$. Since $A > 2A_0$, the interval $I = [A_0, u - A_0]$ has a positive length. By $h^*(\lambda) < \infty$ for all $\lambda \geq A_0$ it follows that for every $y \in I$ there is an integer n_y such that $h(x_n + y) - h(x_n) \leq \frac{1}{2}n$ for all $n \geq n_y$. Hence $h(x_n + u_n) - h(x_n + y) > \frac{1}{2}n$ for all $n \geq n_y$. This means that $I = \cup_{k=1}^{\infty} I_k$, where

$$I_k := I \cap \left\{ y : \ \forall n \geq k \, , \ h(x_n + u_n) - h(x_n + y) > \frac{1}{2}n \right\} .$$

Since h is measurable, each I_k is measurable so that there is a K such that $|I_K| > 0$. Let $Z_n := u_n - I_K$. Then, $|Z_n| = |I_K| > 0$ for all n. Since any element in I_K is in I we see that all the Z_n are subsets of the bounded interval $[A - B + A_0, B - A_0]$. Therefore, the set $\cap_{j=1}^{\infty} \cup_{n=j}^{\infty} Z_n$ is a subset of the same interval. This set has measure equal to $\lim_{j\to\infty} |\cup_{n=j}^{\infty} Z_n| \geq |I_K|$ and is then nonempty. Thus, there exists z belonging to infinitely many Z_n. Hence $h(x_n + u_n) - h(x_n + u_n - z) > \frac{1}{2}n$ for an infinite set of n. This contradicts $h^*(z) < \infty$, proving the proposition. □

Definition A.4. A measurable function $f : (0, \infty) \to (0, \infty)$ is called regularly varying at ∞ with index $\rho \in (-\infty, \infty)$ (or ρ-varying) if for any $\lambda > 0$,

$$\lim_{x\to\infty} \frac{f(\lambda x)}{f(x)} = \lambda^{\rho} . \tag{A.13}$$

ρ is called the exponent of variation. If $\rho = 0$, then it is called slowly varying at infinity.

It is easy to see that if f is regularly varying with exponent ρ, then $L(x) = x^{-\alpha} f(x)$ is slowly varying.

Example A.3. A typical ρ-varying function is $f(x) = x^{\rho}$. The functions $\log(1 + x)$, $\log\log(e + x)$, and $\exp[(\log x)^{\alpha}]$ with $0 < \alpha < 1$ are slowly varying functions.

Theorem A.2 (Uniform Convergence Theorem or UCT). *If ℓ is slowly varying, then*

$$\frac{\ell(\lambda x)}{\ell(x)} \to 1 \quad (\textit{as } x \to \infty \textit{ uniformly on } \lambda \textit{ on each compact set}). \tag{A.14}$$

Proof. We follow the idea as in the proof of Proposition A.1. Define $h(x) = \log \ell(e^x)$. Then, $\ell(x)$ is slowly varying if and only if

$$h(x+u) - h(x) \to 0 \quad (x \to \infty), \quad \forall\, u \in \mathbb{R}. \tag{A.15}$$

It suffices to show that the convergence in (A.15) is uniform in u on any compact set of $\mathbb{R}$.

Suppose the contrary. Then, there exist $\varepsilon > 0$, $x_n \to \infty$, and a bounded sequence u_n such that

$$|h(x_n + u_n) - h(x_n)| > 2\varepsilon, \quad n = 1, 2, \cdots$$

passing to a suitable sequence, we may assume that $u_n \to u_0$ for some point $u_0 \in \mathbb{R}$ and $|u_n - u_0| \le 1$ for all n. For every y, we know that $|h(x_n + y) - h(x_n)| \le \varepsilon$ for all large n, hence $|h(x_n + u_n) - h(x_n + y)| > \varepsilon$ for all large n. Thus, $[-1, 1] = \cup_{k=1}^{\infty} I_k$, where

$$I_k := [-1, 1] \cap \{y\,,\ \forall n \ge k\,,\ |h(x_n + u_n) - h(x_n + y)| > \varepsilon\}\,.$$

Each I_k is measurable so that there exists a K such that I_K has a positive Lebesgue measure. Denote

$$Z_n := u_n - I_K = \{u_n - y\,,\ y \in I_K\} \quad (n = 1, 2, \cdots),$$

and $Z := \cap_{j=1}^{\infty} \cup_{n=j}^{\infty} Z_n$. It is easy to see that all the Z_n have measure $|I_K|$, and all the Z_n are subset of the fixed interval $[u_0 - 2, u_0 + 2]$. Thus, Z is a subset of the same interval having measure

$$|Z| = \lim_{j \to \infty} |\cup_{n=j}^{\infty} Z_n| \ge |I_K| > 0\,.$$

This means that Z is not empty. So, there is a z such that $z \in Z$, hence z belongs to infinitely many Z_n. This means that

$$|h(x_n + u_n) - h(x_n + u_n - z)| > \varepsilon$$

for infinitely many n, contradicting (A.15). This proves the theorem. □

Theorem A.3. *A function ℓ is slowly varying if and only if it can be represented in the form*

$$\ell(x) = c(x) \exp\left\{\int_a^x \frac{\varepsilon(u)}{u} du\right\}, \quad x \ge a \tag{A.16}$$

for some $a > 0$, where $c(x)$ and $\varepsilon(x)$ are measurable and

$$\lim_{x \to \infty} c(x) = c \in (0, \infty)\,, \quad \lim_{x \to \infty} \varepsilon(x) = 0\,. \tag{A.17}$$

Proof. The "if" part. If (A.16) holds true, then

$$I(\lambda, x) := \frac{\ell(\lambda x)}{\ell(x)} = \frac{c(\lambda x)}{c(x)} \exp\left\{\int_x^{\lambda x} \frac{\varepsilon(u)}{u} du\right\} .$$

Fix $\lambda > 0$. For any $\varepsilon > 0$, by the assumptions on $c(x)$ and $\varepsilon(x)$, there is a x_0 such that when $x \geq x_0$,

$$1 - \varepsilon \exp(-\varepsilon|\log\lambda|) \leq I(\lambda, x) \leq 1 - \varepsilon \exp(-\varepsilon|\log\lambda|) .$$

This implies that $\lim_{x\to\infty} I(\lambda, x) = 1$.

The "only if" part. Define $h(x) = \log \ell(e^x)$. It suffices to show that

$$h(x) = d(x) + \int_b^x e(t)dt \tag{A.18}$$

for some constant b and $d(x) \to d \in \mathbb{R}$ and $e(x) \to 0$ as $x \to \infty$.

First, by the uniform convergence theorem (Theorem A.2) we see that for A sufficiently large

$$|h(x + u) - h(x)| \leq 1 , \quad \forall\ x \geq A \quad \text{and} \quad u \in [0, 1] .$$

This implies $|h(x)| \leq 1 + |h(A)|$, whenever $x \in [A, A+1]$. Or by induction, we have $|h(x)| \leq n + |h(A)|$, whenever $x \in [A, A + n]$ for $n = 1, 2, \cdots$. Hence $h(x)$ is integrable on any finite interval far enough to the right.

When A is sufficiently large, we may write

$$h(x) = \int_x^{x+1} [h(x) - h(t)]\, dt + \int_A^x [h(t + 1) - h(t)]\, dt + \int_A^{A+1} h(t)dt \qquad (x \geq A) .$$

The last term on the right hand side is a constant, c say. Denote $e(x) = h(x + 1) - h(x)$. Then, it is easy to check that $e(x) \to 0$ as $x \to \infty$ by the definition of $h(x) = \log \ell(e^x)$ and the slowly varying property of ℓ. The first term on the right is $\int_0^1 [h(x) - h(x + u)]du$, which goes to 0 as $x \to \infty$ by Theorem A.2. Thus, (A.18) holds with $d(x) = c + \int_0^1 [h(x) - h(x + u)]du$. The proof of the theorem is completed. □

Corollary A.1. *If ℓ is slowly varying, then for any constant $A > 1$, $\delta > 0$, there is a $B(A, \delta)$ such that*

$$\ell(y)/\ell(x) \leq A \max\left\{(y/x)^{\delta}, (y/x)^{-\delta}\right\} , \quad (\forall\ x, y \geq B(A, \delta)) . \tag{A.19}$$

Proof. By (A.16) we have that there is a $B(A,\delta)$ such that for all $x, y \geq B(A,\delta)$)

$$\frac{\ell(y)}{\ell(x)} = \frac{c(y)}{c(x)} \exp\left\{\int_x^y \frac{\varepsilon(u)}{u} du\right\} \leq A \exp\left\{\delta\left|\int_x^y \frac{1}{u} du\right|\right\}.$$

The corollary is proved easily. □

Proposition A.2. *let J be a slowly varying function at infinity such that $J(x^\rho) \approx J(x)$ as $x \to \infty$ for each $\rho > 0$. Then*

$$\sup_{a \leq \rho \leq b} J(x^\rho) \preccurlyeq J(x) \quad \textit{for all } 0 < a < b < \infty. \tag{A.20}$$

Proof. Define $f(x) = J(e^x)$ or $J(x) = f(\log x)$. Then $J(x^\rho)/J(x) = f(\rho \log x)/f(\log x) = f(\rho y)/f(y)$, where $y = \log x$. Using the assumption on the function J and applying Theorem A.1 to f we have

$$\limsup_{x\to\infty} \sup_{a \leq \rho \leq < b} J(x^\rho)/J(x) < \infty$$

for all $1 < a < b < \infty$. If $0 < a < b < \infty$, then we have

$$\limsup_{x\to\infty} \sup_{2 \leq \rho \leq 2b/a} J(x^{a\rho/2})/J(x^{a/2}) = \limsup_{y\to\infty} \sup_{2 \leq \rho \leq 2b/a} J(y^\rho)/J(y) < \infty.$$

By using again $J(x^{a/2}) \preccurlyeq J(x)$, we have

$$\sup_{a \leq \rho \leq b} J(x^\rho) \preccurlyeq J(x) \quad \text{for all } 0 < a < b < \infty, \tag{A.21}$$

which is the proposition. □

Definition A.5.

(i) Let E be a Borel subset of $\mathbb{R}^d$. The diameter of E is $\text{dia}(E) := \sup\{|x-y|,\ x, y \in E\}$.

(ii) Let $\delta > 0$ be a positive number and let $E \in \mathcal{B}(\mathbb{R}^d)$. A δ-cover of E is a family of Borel subsets $\{V_\ell, \ell \in S\} \subseteq \mathcal{B}(\mathbb{R}^d)$, where S is a set of indices, such that $\text{dia}(V_\ell) \leq \delta$ and $E \subseteq \cup_{\ell \in S} V_\ell$.

(iii) For any given $s, \delta > 0$, we define a function from $\mathcal{B}(\mathbb{R}^d) \to [0, \infty]$

$$\mathcal{H}_\delta^s(E) := \inf\left\{\sum_{\ell \in S} (\text{dia}(V_\ell))^s\ ,\ \{V_\ell\,, \ell \in S\} \text{ is a } \delta\text{-cover of } E\right\}. \tag{A.22}$$

It is easy to see that for fixed s and E, $\mathcal{H}^s_\delta(E)$ is a decreasing function of δ. One can prove that for any fixed $s > 0$,

$$\mathcal{H}^s(E) := \lim_{\delta \downarrow 0} \mathcal{H}^s_\delta(E) = \sup_{\delta > 0} \mathcal{H}^s_\delta(E) \,, \quad \forall\, E \in \mathcal{B}(\mathbb{R}^d) \tag{A.23}$$

is an outer measure on $\mathcal{B}(\mathbb{R}^d)$.

Definition A.6. This set function $\mathcal{H}^s$ is called the s-dimensional Hausdorff measure on $\mathbb{R}^d$.

Definition A.7. Let $\phi : \mathbb{R} \to \mathbb{R}$ be a measurable function. It is called *approximately differentiable* at $a \in \mathbb{R}$ with an approximate derivative $\mathrm{ap}\,\phi'(a) \in \mathbb{R}$ if

$$\lim_{\eta \to 0} \frac{1}{\eta} \lambda \left\{ x \in [a-\eta, a+\eta] : |\phi(x) - \phi(a) - (x-a)\mathrm{ap}\,\phi'(a)| > \varepsilon |x-a| \right\} = 0$$

for any $\varepsilon > 0$, where λ denotes the Lebesgue measure.

It is obvious that if $\phi : \mathbb{R} \to \mathbb{R}$ is continuously differentiable, then ϕ is approximately differentiable and $\mathrm{ap}\,\phi'(a) = \phi'(a)$.

If $\phi : \mathbb{R}^d \to \mathbb{R}$ is measurable, then we denote $\mathrm{ap}\,\partial_i \phi$ (if it exists) the approximate partial derivative of ϕ with respect to the i-th variable while others are kept fixed.

If $\phi = (\phi_1, \cdots, \phi_d) : \mathbb{R}^m \to \mathbb{R}^d$, where $m \geq d$, is measurable such that $\mathrm{ap}\,\partial_j \phi_i$ exists for all $i = 1, \cdots, d$ and $j = 1, \cdots, m$, then we denote

$$J_d\phi(x) = \left\{ \sum_{\lambda \in \Lambda(d,m)} \left[\det \left(\mathrm{ap}\,\partial_{\lambda(j)} \phi_i(x) \right)_{1 \leq i,j \leq d} \right]^2 \right\}^{1/2} \,, \tag{A.24}$$

where $\Lambda(d, m)$ denotes the set of all strictly increasing bijections from $\{1, \cdots, d\}$ to $\{1, \cdots, m\}$.

The following theorem is from [Federer (1969), Theorem 3.2.11], [Evans and Gariepy (2015), p. 319] (see also [Bouleau and Hirsch (1986)]).

Theorem A.4. *Assume $m \geq d$. If $\phi, f : \mathbb{R}^m \to \mathbb{R}^d$ are two measurable functions such that $\mathrm{ap}\,\partial_j \phi_i$ exists almost surely for all $i = 1, \cdots, d$ and $j = 1, \cdots, m$. Then,*

$$\int_{\mathbb{R}^m} f(x) J_d\phi(x) dx = \int_{\mathbb{R}^d} \int_{\phi^{-1}(y)} f(x) \mathcal{H}^{m-d}(dx) dy \,, \tag{A.25}$$

where dx and dy are the Lebesgue measures on $\mathbb{R}^m$ and $\mathbb{R}^d$ respectively. If f is taken to be an indicator function $f(x) = I_B(x)$ for some $B \in \mathcal{B}(\mathbb{R}^m)$, then

$$\int_B J_d\phi(x) dx = \int_{\mathbb{R}^d} \mathcal{H}^{m-d}(B \cap \phi^{-1}(y)) dy \,. \tag{A.26}$$

An immediate consequence of this theorem (see [Bouleau and Hirsch (1986)]) is

Corollary A.2. *Assume $m \ge d$. If $\phi : \mathbb{R}^m \to \mathbb{R}^d$ is measurable and* $\operatorname{ap}\partial_j\phi_i$ *exist almost surely for all $i = 1, \cdots, d$ and $j = 1, \cdots, m$. Then, for any $B \in \mathcal{B}(\mathbb{R}^d)$ with Lebesgue measure 0, we have*

$$\int_{\phi^{-1}(B)} \det\left(\langle \operatorname{ap}\partial\phi_i\,, \operatorname{ap}\partial\phi_j\rangle_{1\le i,j\le d}\right) dx = 0\,. \tag{A.27}$$

Definition A.8. Let $(\Omega, \mathcal{F}, \mathbb{P})$ be a measure space and let $\mathcal{U}$ be a subset of $L^1(\Omega, \mathcal{F}, \mathbb{P})$. We say the $\mathcal{U}$ is uniformly integrable if

$$\lim_{c\to+\infty} \sup_{f\in\mathcal{U}} \int_{\{|f|\ge c\}} |f(\omega)|\,\mathbb{P}(d\omega) = 0\,. \tag{A.28}$$

Theorem A.5. *Let $\mathcal{U}$ be a subset of $L^1(\Omega, \mathcal{F}, \mathbb{P})$. The following two statements are equivalent.*

(i) $\mathcal{U}$ is uniformly integrable.

(ii) There is a function $G : \mathbb{R}_+ \to \mathbb{R}_+$ such that $\lim_{t\to\infty} \frac{G(t)}{t} = \infty$ and

$$\sup_{f\in\mathcal{U}} \mathbb{E}\left[G(f))\right] < \infty\,. \tag{A.29}$$

Proof. First, we show "(ii) ⇒ (i)". Denote $M = \sup_{f\in\mathcal{U}} \mathbb{E}\left[G(f))\right]$. For any $\varepsilon > 0$ let $a = M/\varepsilon$. From the property of G there is a c such that $G(t)/t \ge a$ (or $G(t)/a \ge t$) for all $t \ge c$. This means that on $\{|f| \ge c\}$, $|f| \le \frac{G(|f|)}{a}$. Therefore, we have for any $\kappa \ge c$,

$$\int_{\{|f|\ge\kappa\}} |f(\omega)|\mathbb{P}(d\omega) \le \frac{1}{a}\int_{\{|f|\ge\kappa\}} G(|f(\omega)|)\mathbb{P}(d\omega) \le \frac{1}{a}\mathbb{E}(G(|f|)) \le \frac{M}{a} = \varepsilon\,.$$

Now we show "(i) ⇒ (ii)". Denote $a_n(f) = \mathbb{P}\left(|f| > n\right)$, $n = 1, 2, \cdots$. Then, there is an increasing sequence c_n, such that

$$\sup_{f\in\mathcal{U}} \int_{\{|f|\ge c_n\}} |f|d\mathbb{P} \le 2^{-n}\,.$$

Thus, we have

$$\begin{aligned} 2^{-n} &\ge \int_{\{|f|\ge c_n\}} |f|d\mathbb{P} \ge \sum_{k=c_n}^{\infty} k\mathbb{P}(k < |f| \le k+1) \\ &= \sum_{m=c_n}^{\infty} \mathbb{P}(|f| > m) = \sum_{m=c_n}^{\infty} a_m(f)\,. \end{aligned}$$

This implies that

$$\sup_{f\in\mathcal{U}}\sum_{n}\sum_{m=c_n}^{\infty} a_m(f) < \infty\,.$$

Define g_m to be the number of integers n such that $c_n \le m$. Namely, $g_m = \#\{n \ge 1\,,: c_n \le m\}$. Now we define $g(t) = g_m\,,\ m \le t < m+1$ and $G(t) = \int_0^t g(r)dr$. Then

$$\begin{aligned}\mathbb{E}\left[G(|f|)\right] &\le g_1\mathbb{P}(1 < |f| \le 2) + (g_1+g_2)\mathbb{P}(2 < |f| \le 3) + \cdots \\ &= \sum_{n=1}^{\infty} g_m a_m(f) = \sum_{n}\sum_{m=c_n}^{\infty} a_m(f) < \infty\end{aligned}$$

uniformly for $f \in \mathcal{U}$. This proves the theorem. □

Remark A.1. We refer to [Dellacherie and Meyer (1975)] for more about the uniform integrability.

Lemma A.9 (Paley-Zygmund Inequality). *If $X \ge 0$ is a random variable with finite variance, and if $0 < \theta < 1$, then*

$$\mathrm{P}(X \ge \theta\,\mathrm{E}[X]) \ge (1-\theta)^2\frac{(\mathrm{E}[X])^2}{\mathrm{E}[X^2]}\,. \tag{A.30}$$

Proof. First, we have

$$\mathrm{E}[X] = \mathrm{E}[X\,\mathbf{1}_{\{X<\theta\,\mathrm{E}[X]\}}] + \mathrm{E}[X\,\mathbf{1}_{\{X\ge\theta\,\mathrm{E}[X]\}}]\,.$$

The first addend is at most $\theta\,\mathrm{E}[X]$. From the Cauchy-Schwarz inequality it follows that the second addend is bounded by $\left[\mathrm{E}(X^2)\,\mathrm{P}(X\ge\theta\,\mathrm{E}[X])\right]^{1/2}$. This yields $(1-\theta)\,\mathrm{E}[X] \le \left[\mathrm{E}(X^2)\,\mathrm{P}(X\ge\theta\,\mathrm{E}[X])\right]^{1/2}$. The desired inequality then follows by taking the square. □

A.2 Martingales

Let T be an interval of $\mathbb{N}$ or $\mathbb{R}_+$. Let $(\Omega,\mathcal{F},P)$ be a probability space with an increasing right continuous filtration $\{\mathcal{F}_t\}_{\{t\in T\}}$ satisfying the usual conditions.

Definition A.9. A real valued stochastic process $X = \{X_t,\ t \in T\}$, adapted to $\{\mathcal{F}_t\}_{\{t\in T\}}$, is called a martingale (supermartingale, submartingale respectively) with respect to the filtration $\{\mathcal{F}_t\}_{\{t\in T\}}$ if

(i) X_t is integrable for each $t \in T$.

(ii) $X = \{X_t,\ t \in T\}$ is adapted to $\{\mathcal{F}_t\}_{\{t\in T\}}$. Namely, for any $t \in T$, $X_t \in \mathcal{F}_t$.

(iii) $\mathbb{E}(X_t|\mathcal{F}_s) = X_s$ $(\mathbb{E}(X_t|\mathcal{F}_s) \le X_s\,,\ \mathbb{E}(X_t|\mathcal{F}_s) \ge X_s)$ for all $s, t \in T$ and $s \le t$.

There are countless examples of martingales.

Example A.4. Let $(\Omega, \mathcal{F}, P)$ be a probability space with an increasing right continuous filtration $\{\mathcal{F}_t\}_{\{t\in T\}}$. Let Y be an integrable real valued random variable. Then, process

$$Y_t := \mathbb{E}\left(Y|\mathcal{F}_t\right)\,, t \in T$$

is a martingale with respect to $\{\mathcal{F}_t\}_{\{t\in T\}}$.

Example A.5. Let $X_1\,, \cdots, X_n\,, \cdots$ be independent random variables of mean zero. Assume they are integrable. Then

$$S_n = X_1 + \cdots + X_n\,, \quad n = 1, 2, \cdots$$

is a martingale with respect to the filtration $\mathcal{F}_n = \sigma(X_1, \cdots, X_n)$.

Example A.6. Let $B = \{B_t\,, t \ge\}$ be a d-dimensional Brownian motion on on some probability space with in increasing and right continuous filtration $(\Omega, \mathcal{F}, \{\mathcal{F}_t\}_{\{t\in T\}}, P)$.

(i) If $h : \mathbb{R}^d \to \mathbb{R}$ is harmonic, namely, $\Delta h(x) := \sum_{i=1}^d \frac{\partial^2 h}{\partial x_i^2}(x) = 0$ for all $x \in \mathbb{R}^d$, then $h(B_t)$ is a martingale.

(ii) If $h : \mathbb{R}_+ \times \mathbb{R}^d$ satisfies $\left[\partial_t + \frac{1}{2}\Delta\right] h(t, x) = 0$ for all $t > 0$ and all $x \in \mathbb{R}^d$, then $h(t, B_t)$ is a martingale.

In particular, B_t, $B_t^2 - t$, $\exp\left\{\sigma B_t - \frac{\sigma^2 t}{2}\right\}$ $(\forall\ \sigma \in \mathbb{R})$ are martingales.

Definition A.10. Let $(\Omega, \mathcal{F}, \{\mathcal{F}_t\}_{\{t\in T\}}, P)$ be a probability space with a right continuous filtration. A mapping $\tau : \Omega \to \mathbb{R}_+$ is called a stopping time if for any $t \ge 0$, $\{\omega \in \Omega\,,\ \tau \le t\} \in \mathcal{F}_t$.

Example A.7. If $X = \{X_t\,, t \ge 0\}$ is a right continuous, $\mathcal{F}_t$-adapted $\mathbb{R}^d$-valued process and E is Borel set in $\mathbb{R}^d$, then

$$\tau_E(\omega) = \inf\{t\,,\ \ X_t(\omega) \in E\}$$

is a stopping time. It is called the hitting time of E or the exit time of E^c.

Definition A.11. For any stopping time τ, we define

$$\mathcal{F}_\tau = \{A \in \mathcal{F}\,, \ \text{ such that for any } t \in T \ A \cap \{\tau \leq t\} \in \mathcal{F}_t\}\,. \tag{A.31}$$

Proposition A.3. *Let $\{X_n, n \geq 0\}$ be a (sub)martingale with respect to the filtration $(\mathcal{F}_n)$ and let $\{H_n, n \geq 0\}$ be a positive, bounded process such that $H_n \in \mathcal{F}_{n-1}$ for all $n \geq 1$. Then, the process $Y = H \cdot X$ defined by*

$$Y_0 = X_0\,, \quad Y_{n+1} = Y_n + H_n(X_{n+1} - X_n)\,, \quad n = 0, 1, \cdots,$$

is a (sub)martingale (with respect to the filtration $(\mathcal{F}_n)$). In particular, if T is a stopping time, then the stopped process $X^T = \{X_{n\wedge T}, n \geq 0\}$ is a (sub)martingale.

Proof. The first statement is straightforward from the definition of (sub)martingale. The process $Y = H \cdot X$ is the (discrete time analogue of) stochastic integral of H with respect to X. The second statement follows from the fact that $X^T = (H \cdot X)$ with $H_n = I_{\{n \leq T\}} = 1 - I_{\{T \leq n-1\}} \in \mathcal{F}_{n-1}$. □

Proposition A.4. *If X is a martingale and if $S \leq T$ are two bounded stopping times, namely there is a constant M such that $S \leq T \leq M$ a.s., then*

$$\mathbb{E}\left[X_T \middle| \mathcal{F}_S\right] = X_S\,. \tag{A.32}$$

Proof. Denote $H_n = I_{\{n \leq T\}} - I_{\{n \leq S\}}$. When $n > M$, $(H \cdot X)_n - X_0 = X_T - X_S$. Taking the expectation, we have

$$\mathbb{E}(X_T - X_S) = \mathbb{E}((H \cdot X)_n - X_0) = \mathbb{E}((H \cdot X)_n) - \mathbb{E}(X_0) = 0$$

since $(H \cdot X)$ is a martingale. This implies $\mathbb{E}(X_T) = \mathbb{E}(X_S)$.

For any set $B \in \mathcal{F}_S$, set $S^B = SI_B + MI_{B^c}$ and $T^B = TI_B + MI_{B^c}$. Then, S^B and T^B are also two stopping times. Thus, $\mathbb{E}(X_{T^B}) = \mathbb{E}(X_{S^B})$. This means

$$\mathbb{E}\left[X_T I_B + MI_{B^c}\right] = \mathbb{E}\left[X_S I_B + MI_{B^c}\right]\,.$$

Or we have $\mathbb{E}\left[X_T I_B\right] = \mathbb{E}\left[X_S I_B\right]$. This is (A.32). □

The proof of the above proposition also yields

Proposition A.5. *If X is an adapted and integrable process such that for any bounded two stopping times S and T, $\mathbb{E}(X_T) = \mathbb{E}(X_S)$, then X is a martingale. If X is a submartingale and if $S \leq T$ are two bounded stopping times, then $\mathbb{E}\left[X_T \middle| \mathcal{F}_S\right] \geq X_S$.*

For a stochastic process $\{M_t, t \geq 0\}$, we define its maximal process by

$$M_t^* := \sup_{0 \leq s \leq t} M_s \,.$$

Proposition A.6. *If $\{X_n, n = 0, 1, \cdots, N\}$ is an integrable submartingale, then for any $\lambda > 0$,*

$$\lambda \mathbb{P}\left[\sup_n X_n \geq \lambda\right] \leq \mathbb{E}\left[X_N I_{\{\sup_n X_n \geq \lambda\}}\right] \leq \mathbb{E}\left[|X_N| I_{\{\sup_n X_n \geq \lambda\}}\right] \,. \quad \text{(A.33)}$$

As a consequence, we have for any $p > 1$,

$$\mathbb{E}\left[|X_N|^p\right] \leq \mathbb{E}\left[\sup_n |X_n|^p\right] \leq \left(\frac{p}{p-1}\right) \mathbb{E}\left[|X_N|^p\right] \,. \quad \text{(A.34)}$$

Proof. Introduce

$$T = \begin{cases} \inf\{n\,;\ X_n \geq \lambda\} & \text{if this set is not empty} \\ N & \text{if this set is empty}\,. \end{cases}$$

This is a stopping time. By the second part of the previous proposition, we have

$$\begin{aligned} \mathbb{E}\left[X_N\right] \geq \mathbb{E}\left[X_T\right] &= \mathbb{E}\left[X_T I_{\{\sup_n X_n \geq \lambda\}}\right] + \mathbb{E}\left[X_T I_{\{\sup_n X_n < \lambda\}}\right] \\ &\geq \lambda \mathbb{P}\left(\sup_n X_n \geq \lambda\right) + \mathbb{E}\left[X_N I_{\{\sup_n X_n < \lambda\}}\right] \,. \end{aligned}$$

This is the first inequality of (A.33). The second inequality of (A.33) is obvious.

The first inequality in (A.34) is trivial. Now, we show the second inequality. Denote $X^* = \sup_n |X_n|$. Then, for any $p > 1$,

$$\begin{aligned} \mathbb{E}\left[(X^*)^p\right] &= \mathbb{E}\left[\int_0^\infty p\lambda^{p-1} I_{\{X^* > \lambda\}} d\lambda\right] = \int_0^\infty p\lambda^{p-1}\, \mathbb{P}\left(X^* > \lambda\right) d\lambda \\ &\leq \int_0^\infty p\lambda^{p-2}\, \mathbb{E}\left[|X_N| I_{\{X^* > \lambda\}}\right] d\lambda \\ &= \mathbb{E}\left[|X_N| \int_0^\infty p\lambda^{p-2}\, I_{\{X^* > \lambda\}} d\lambda\right] \\ &= \frac{p}{p-1}\mathbb{E}\left[|X_N|(X^*)^{p-1}\right] \\ &\leq \frac{p}{p-1}\left(\mathbb{E}\left[|X_N|^p\right]\right)^{1/p} \left(\mathbb{E}\left[(X^*)^p\right]\right)^{(p-1)/p} \,. \end{aligned} \quad \text{(A.35)}$$

This yields $\left(\mathbb{E}\left[(X^*)^p\right]\right)^{1/p} \leq \frac{p}{p-1}\left(\mathbb{E}\left[|X_N|^p\right]\right)^{1/p}$, proving (A.34). □

Corollary A.3. *If $\{X_n, n = 0, 1, \cdots, N\}$ is a martingale or a positive submartingale, then for any $p \geq 1$ and $\lambda > 0$,*

$$\lambda^p \mathbb{P}\left[\sup_n |X_n| \geq \lambda\right] \leq \mathbb{E}\left[|X_N|^p\right]. \tag{A.36}$$

Proof. If $\mathbb{E}[|X_N|^p] < \infty$, then $|X|^p$ is a submartinagle. Thus, (A.36) is a direct consequence of (A.33) applied to $|X|^p$. □

We have also the continuous time analogies of the above results.

Theorem A.6 (Doob's inequality). *Let $\{X_t, t \in T\}$ be a right continuous martingale (or positive submartingale) on some interval T of $\mathbb{R}$. Denote $X^* = \sup_{t \in T} |X_t|$. Then, for any $p \geq 1$,*

$$\lambda^p \mathbb{P}(X^* \geq \lambda) \leq \sup_{t \in T} \mathbb{E}\left[|X_t|^p\right], \quad \forall\, \lambda > 0. \tag{A.37}$$

For any $p > 1$,

$$\mathbb{E}\left[(X^*)^p\right] \leq \left(\frac{p}{p-1}\right)^p \sum_{t \in T} \mathbb{E}\left[|X_t|^p\right]. \tag{A.38}$$

Definition A.12. An adapted, continuous process X is called an $(\mathcal{F}_t)$ local martingale if there exist stopping times $T_n, n \geq 1$ such that

(i) the sequence $\{T_n\}$ is increasing and $\lim_{n \to \infty} T_n = \infty$ a.s.;
(ii) for every n, the process $\left\{X_{t \wedge T_n} I_{\{T_n > 0\}}\right\}$ is a uniformly integrable martingale.

Definition A.13.

(i) Let M be a continuous local martingale. Then, there is a unique process $\langle M \rangle$ vanishing at 0 such that $\left\{M_t^2 - \langle M \rangle_t\right\}$ is a continuous local martingale. This process $\langle M \rangle$ is called the quadratic variation process of M.
(ii) If M and N are two continuous local martingales, then we define their bracket as

$$\langle M, N \rangle = \frac{1}{4}\left[\langle M + N \rangle - \langle M - N \rangle\right].$$

(iii) A continuous semimartingale X is a process of the form $X = M + A$, where M is a continuous local martingale and A is a continuous adapted process of finite variation. Its quadratic variation is defined as $\langle X \rangle = \langle M \rangle$. We can also define the bracket of two semimartingales as the bracket of their martingales parts.

The following theorems are ones of the most important theorems in stochastic analysis, whose proofs will not be given in this book. We refer to [Dellacherie and Meyer (1982)], [He, Wang and Yan (1992)], [Revuz and Yor (1999)] and [Protter (2004)] for references. There are also many other books containing this theorem.

Theorem A.7. *Let M be a local martingale. For an adapted process H satisfying $\mathbb{P}\left(\int_0^T H_s^2 d\langle M\rangle_s < \infty\right) = 1$, one can define the stochastic integral $\int_0^T H_s dM_s$. And $\left\{N_t = \int_0^t H_s dM_s\,, 0 \le t \le T\right\}$ is also a local martingale with quadratic variation $\langle N\rangle_t = \int_0^t H_s^2 d\langle M\rangle_s$.*

Theorem A.8 (Itô formula). *Let $(X = (X^1, \cdots, X^d)$ be a continuous d-dimensional semimartingale. Let $F \in C^2(\mathbb{R}^d, \mathbb{R})$. Then, $F(X)$ is a semimartingale and*

$$F(X_t) = F(X_0) + \sum_{i=1}^d \int_0^t \partial_i F(X_s) dX_s^i + \frac{1}{2}\sum_{i,j=1}^d \int_0^t \partial_{ij}^2 F(X_s) d\langle X^i, X^j\rangle_s\,. \tag{A.39}$$

As a consequence of this theorem one can prove

Theorem A.9. *If $\{M_t, 0 \le t \le T\}$ is a local martingale, then*

$$\left\{N_t^\sigma := \exp\left\{\sigma M_t - \frac{\sigma^2}{2}\langle M\rangle_t\right\}, 0 \le t \le T\right\} \tag{A.40}$$

is also a local martingale.

Proof. By Itô formula we have $N_t^\sigma = 1 + \sigma\int_0^t N_s^\sigma dM_s$. □

As an application of the Doob's inequality, we have the following exponential inequality.

Proposition A.7. *If $\{M_t, t \in [0, T]\}$ is a continuous martingale such that $M_0 = 0$ and $\mathbb{E}\exp\{\lambda\langle M\rangle_T\} < \infty$ for any $\lambda > 0$, then*

$$\mathbb{P}\left(\sup_{0\le t\le T} |M_t| \ge a\,, \langle M\rangle_T \le b\right) \le 2\exp\left\{-\frac{a^2}{2b}\right\}, \quad \forall\, a, b > 0\,. \tag{A.41}$$

Proof. Under the above condition for any $\sigma \in \mathbb{R}$, N^σ defined by (A.40) is also a martingale. If $\sup_{0\le t\le T} M_t \ge a$ and $\langle M\rangle_T \le b$, then $\sup_{0\le t\le T} N_t^\sigma \ge$

$\exp\left\{\sigma a - \frac{\sigma^2 b}{2}\right\}$. Thus, we have

$$\begin{aligned} I_{a,b}(M) &:= \mathbb{P}\left(\sup_{0\le t\le T} M_t \ge a\,, \langle M\rangle_T \le b\right) \\ &\le \mathbb{P}\left(\sup_{0\le t\le T} N_t^\sigma \ge \exp\left\{\sigma a - \frac{\sigma^2 b}{2}\right\}\right) \\ &\le \exp\left\{-\sigma a + \frac{\sigma^2 b}{2}\right\} \sup_{0\le t\le T} \mathbb{E}\left[N_t^\sigma\right] = \exp\left\{-\sigma a + \frac{\sigma^2 b}{2}\right\}, \end{aligned}$$

where in the above second inequality we use (A.37) with $p = 1$. Since σ is arbitrary we can take $\sigma = b/a$ to obtain $I_{a,b}(M) \le \exp\left(-\frac{a^2}{2b}\right)$. Finally since

$$\mathbb{P}\left(\sup_{0\le t\le T} |M_t| \ge a\,, \langle M\rangle_T \le b\right) \le I_{a,b}(M) + I_{a,b}(-M)$$

the theorem is proved. □

Lemma A.10. *Let M be a continuous local martingale such that $M_0 = 0$.*

(i) For $p \ge 2$, there is a constant C_p such that

$$\mathbb{E}\left[(M_\infty^*)^p\right] \le C_p \mathbb{E}\left[\langle M\rangle_\infty^{p/2}\right]. \tag{A.42}$$

(ii) For $p \ge 4$, there is a constant c_p such that

$$\mathbb{E}\left[\langle M\rangle_\infty^{p/2}\right] \le c_p \mathbb{E}\left[(M_\infty^*)^p\right]. \tag{A.43}$$

Proof. (i). By stopping, we may assume that M is bounded. Applying the Itô formula we have

$$|M_\infty|^p = p\int_0^\infty |M_s|^{p-1}\mathrm{sign}(M_s)dM_s + \frac{p(p-1)}{2}\int_0^\infty |M_s|^{p-2}d\langle M\rangle_s\,.$$

Thus,

$$\begin{aligned} \mathbb{E}\left[|M_\infty|^p\right] &= \frac{p(p-1)}{2}\mathbb{E}\left[\int_0^\infty |M_s|^{p-2}d\langle M\rangle_s\right] \\ &\le \frac{p(p-1)}{2}\mathbb{E}\left[(M_\infty^*)^{p-2}\langle M\rangle_\infty\right] \\ &\le \frac{p(p-1)}{2}\left\{\mathbb{E}\left[(M_\infty^*)^p\right]\right\}^{\frac{p-2}{p}}\left\{\mathbb{E}\left[\langle M\rangle_\infty^{p/2}\right]\right\}^{2/p} \\ &\le \frac{p(p-1)}{2}\left(\frac{p}{p-1}\right)^{p-2}\left\{\mathbb{E}\left[|M_\infty|^p\right]\right\}^{\frac{p-2}{p}}\left\{\mathbb{E}\left[\langle M\rangle_\infty^{p/2}\right]\right\}^{2/p}, \end{aligned}$$

where the last inequality follows from Theorem A.6. This proves part (i) of the lemma.

(ii). From $M_t^2 = 2\int_0^t M_s dM_s + \langle M\rangle_t$, we have

$$\begin{aligned}
\mathbb{E}\left[\langle M\rangle_\infty^{p/2}\right] &\le c_p\left\{\mathbb{E}\left[(M_\infty^*)^p\right] + \mathbb{E}\left[\left|\int_0^\infty M_s dM_s\right|^{p/2}\right]\right\} \\
&\le c_p\left\{\mathbb{E}\left[(M_\infty^*)^p\right] + \mathbb{E}\left[\left(\int_0^\infty M_s^2 d\langle M\rangle_s\right)^{p/4}\right]\right\} \\
&\le c_p\left\{\mathbb{E}\left[(M_\infty^*)^p\right] + \mathbb{E}\left[(M_\infty^*)^{p/2}\left(\langle M\rangle_\infty\right)^{p/4}\right]\right\} \\
&\le c_p\left\{\mathbb{E}\left[(M_\infty^*)^p\right] + \left(\mathbb{E}\left[(M_\infty^*)^p\right]\right)^{1/2}\left(\mathbb{E}\left[\langle M\rangle_\infty^{p/2}\right]\right)^{1/2}\right\} \\
&\le c_p\mathbb{E}\left[(M_\infty^*)^p\right] + \frac{1}{2}\mathbb{E}\left[\langle M\rangle_\infty^{p/2}\right], \qquad \text{(A.44)}
\end{aligned}$$

which implies (A.43). Here in the above first inequality, we have used $(a+b)^p \le c_p(a^p + b^p)$, $\forall\, a, b \ge 0$; in the above second inequality, we have applied (A.42) to the martingale $\int_0^t M_s dM_s$ whose quadratic variation is $\int_0^t M_s^2 d\langle M\rangle_s$; and in the above last inequality, we have used $ab \le \varepsilon a^2 + \frac{1}{\varepsilon}b^2$, $\forall\, a, b \ge 0, \varepsilon > 0$. □

Definition A.14. A positive, adapted and right continuous process X is called dominated by an increasing process A if

$$\mathbb{E}\left[X_T|\mathcal{F}_0\right] \le \mathbb{E}\left[A_T|\mathcal{F}_0\right], \quad \forall \text{ any bounded stopping time } T. \qquad \text{(A.45)}$$

Lemma A.11. *Let X be dominated by a continuous increasing process A.*

(i) For any $x, y > 0$,

$$\mathbb{P}\left(X_\infty^* > x;\ A_\infty \le y\right) \le \frac{1}{x}\mathbb{E}\left[A_\infty \wedge y\right], \qquad \text{(A.46)}$$

where $X_\infty^* = \sup_s X_s$.

(ii) For any $k \in (0,1)$,

$$\mathbb{E}\left[(X_\infty^*)^k\right] \le \frac{2-k}{1-k}\mathbb{E}\left[A_\infty^k\right]. \qquad \text{(A.47)}$$

Remark A.2. If X is positive, $y = \infty$, and $A_t = \sup_{0\le s\le t} X_s$, then inequality (A.46) is the usual Chebyshev inequality.

Proof. (i). It suffices to show the inequality in the case $P(A_0 \le y) > 0$ or even $P(A_0 \le y) = 1$. In fact we can replace $\mathbb{P}$ by $\mathbb{P}' = \mathbb{P}(\cdot|A_0 \le 1)$.

Moreover, by Fatou's lemma, it suffices to show

$$\mathbb{P}(X_n^* > x\,;\ A_n \le y) \le \frac{1}{x}\mathbb{E}[A_n \wedge y] \le \frac{1}{x}\mathbb{E}[A_\infty \wedge y]\,. \tag{A.48}$$

Define

$$R_n = \begin{cases} \inf\{t\,:\ A_t > y\} & \text{not empty} \\ n & \text{empty set}\,; \end{cases}$$

and

$$S_n = \begin{cases} \inf\{t\,:\ X_t > x\} & \text{not empty} \\ n & \text{empty set}\,. \end{cases}$$

Then, $\{A_n \le y\} = \{R_n = n\}$. Therefore, we have

$$\begin{aligned} \mathbb{P}(X_n^* > x\,;\ A_n \le y) &= \mathbb{P}(X_n^* > x\,;\ R_n = n) \\ &\le \mathbb{P}(X_{S_n} > x\,; S_n < n\,;\ A_n \le y) \\ &\le \mathbb{P}(X_{S_n \wedge R_n} > x) \le \frac{1}{x}\mathbb{E}[X_{S_n \wedge R_n}] \\ &\le \frac{1}{x}\mathbb{E}[A_{S_n \wedge R_n}] \le \frac{1}{x}\mathbb{E}[A_n \wedge y]\,. \end{aligned}$$

This is (A.48). The part (i) of the lemma is proved by Fatou's lemma. (ii). Let F be a continuous increasing function from $\mathbb{R}_+$ to $\mathbb{R}_+$ with $F(0) = 0$. Then, from (A.45) it follows

$$\begin{aligned} \mathbb{E}[F(X_\infty^*)] &= \mathbb{E}\left[\int_0^\infty I_{\{X_\infty^* > x\}} dF(x)\right] \\ &\le \int_0^\infty [\mathbb{P}(X_\infty^* > x\,;\ A_\infty \le x) + \mathbb{P}(A_\infty > x)]\, dF(x) \\ &\le \int_0^\infty \left[\frac{1}{x}\mathbb{E}(A_\infty \wedge x) + \mathbb{P}(A_\infty > x)\right] dF(x) \\ &\le \int_0^\infty \left[\frac{1}{x}\mathbb{E}\left(A_\infty I_{\{A_\infty \le x\}}\right) + 2\mathbb{P}(A_\infty > x)\right] dF(x) \\ &= 2\mathbb{E}[F(A_\infty)] + \mathbb{E}\left[A_\infty \int_{A_\infty}^\infty \frac{dF(x)}{x}\right] = \mathbb{E}\left[\tilde{F}(A_\infty)\right], \end{aligned}$$

where $\tilde{F}(x) = 2F(x) + x\int_x^\infty \frac{dF(u)}{u}$. Inequality (A.47) follows by taking $F(x) = x^k$ and in this case $\tilde{F}(x) = \frac{2-k}{1-k}x^k$. □

Theorem A.10. *For every $p \in (0, \infty)$, there are two constants c_p and C_p such that for any continuous local martingale M vanishing at 0*

$$c_p\mathbb{E}\left[\langle M\rangle_\infty^{p/2}\right] \le \mathbb{E}[(M_\infty^*)^p] \le C_p\mathbb{E}\left[\langle M\rangle_\infty^{p/2}\right]. \tag{A.49}$$

The above inequality is called Burkholder-Davis-Gundy inequality.

Proof. Let us first prove the second inequality in (A.49). When $p \geq 2$, the inequality is (A.42). When $0 < p < 2$, (A.49) follows from an application of (A.47) to $X = (M^*)^2$ and $A = C\langle M\rangle$ (the domination is a consequence of (A.42) applied to stopped martingale).

Now we consider the first inequality in (A.49). When $p \geq 4$, the inequality follows from (A.43). If $0 < p < 4$, then the inequality follows from an application of (A.47) to $X = \langle M\rangle^2$ and $A = C\,(M^*)^4$. The necessary domination follows (A.43) applied to stopping martingale. □

The Burkholder-Davis-Gundy inequality has different forms. One is the explicit constant c_p and C_p appeared in the inequality. Another is the replacement of the function x^p by more general moderate convex functions. We will carry out some brief discussion. For details we refer to [Dellacherie and Meyer (1982)], [He, Wang and Yan (1992)], [Revuz and Yor (1999)] and [Protter (2004)].

If we let $p = 1$ in (A.49), we obtain $\mathbb{E}\,[M_\infty^*] \leq c\mathbb{E}\left[\langle M\rangle_\infty^{1/2}\right]$ for some constant c. In fact, with a careful analysis, this constant c can be chosen as $2\sqrt{2}$ (see [Revuz and Yor (1999), p.167]). Moreover, we can also replace the probability $\mathbb{P}$ by $\mathbb{P}(\cdot|\mathcal{F}_0)$. Therefore, we have

$$\mathbb{E}\,[M_\infty^*|\mathcal{F}_0] \leq 2\sqrt{2}\mathbb{E}\left[\langle M\rangle_\infty^{1/2}|\mathcal{F}_0\right]. \tag{A.50}$$

Definition A.15. Let M be a continuous local martingale. We denote

$$\mathcal{H}^1 = \Big\{M\,,\ M \text{ is continuous local martingale}$$
$$\text{and } \|M\|_{\mathcal{H}^1} := \mathbb{E}\left(\langle M\rangle\right) < \infty\Big\}.$$

Theorem A.11. *Let $M \in \mathcal{H}^1$ be a continuous integrable martingale. Then, for any stopping time T, we have*

$$\mathbb{E}\left[M_\infty^* - M_T^* I_{\{T>0\}}|\mathcal{F}_T\right] \leq 2\sqrt{2}\mathbb{E}\left[\sqrt{\langle M\rangle_\infty - \langle M\rangle_T I_{\{T>0\}}}\ |\mathcal{F}_T\right]. \tag{A.51}$$

Proof. Denote $\bar{M}_t = \left(M_{T+t} - M_T I_{\{T>0\}}\right) I_{\{T<\infty\}}$ and $\mathcal{G}_t = \mathcal{F}_{T+t}$, $t \geq 0$. Then, it is easy to verify that $\bar{M}$ is a $(\mathcal{G}_t)$ continuous local martingale and $\langle \bar{M}\rangle_t = \langle M\rangle_{T+t} - \langle M\rangle_T I_{\{T>0\}}$. Therefore, $\mathbb{E}\left\{\sqrt{\langle \bar{M}\rangle_\infty}\right\} < \infty$. This means that $\bar{M} \in \mathcal{H}^1$. By (A.50), we have

$$\mathbb{E}\left[\bar{M}_\infty^*|\mathcal{G}_0\right] \leq 2\sqrt{2}\mathbb{E}\left\{\sqrt{\langle M\rangle_\infty - \langle M\rangle_T I_{\{T>0\}}}|\mathcal{G}_0\right\}. \tag{A.52}$$

On the other hand, it is obvious from our definition of $\bar{M}$ that $M^*_\infty \le \bar{M}^*_\infty + M^*_T I_{\{T>0\}}$. Thus, (A.52) implies

$$\mathbb{E}\left[M^*_\infty - M^*_T I_{\{T>0\}}|\mathcal{G}_0\right] \le 2\sqrt{2}\mathbb{E}\left\{\sqrt{\langle M\rangle_\infty - \langle M\rangle_T I_{\{T>0\}}}|\mathcal{G}_0\right\}. \quad (A.53)$$

Since $\mathcal{G}_0 = \mathcal{F}_T$, we have (A.51). □

Lemma A.12. *Let $\Phi : \mathbb{R}_+ \to \mathbb{R}_+$ be a moderate convex function and let φ be its right derivative. Assume that ξ and η are two non-negative random variables such that*

$$\mathbb{E}\Phi(\xi) < \infty \quad \text{and} \quad \mathbb{E}\Phi(\xi) \le \mathbb{E}\left[\eta\varphi(\xi)\right]. \quad (A.54)$$

Then

$$\mathbb{E}\Phi(\xi) \le \rho^{\rho+1}\mathbb{E}\Phi(\eta). \quad (A.55)$$

Proof. Let Ψ be the conjugate convex function of Φ and let ψ be the right derivative of Ψ. Since Ψ is convex and $\rho \ge 1$, we see

$$\begin{aligned}\Psi\left(\frac{\varphi(\xi)}{\rho}\right) &= \Psi\left(\frac{\varphi(\xi)}{\rho} + \frac{\rho-1}{\rho}\times 0\right)\\ &\le \frac{1}{\rho}\Psi(\varphi(\xi)) + \frac{\rho-1}{\rho}\times\Psi(0) = \frac{1}{\rho}\Psi(\varphi(\xi)).\end{aligned}$$

From the Young inequality (A.6), Lemma (A.4), and the above inequality we have

$$\begin{aligned}\eta\varphi(\xi) = \rho\eta\frac{\varphi(\xi)}{\rho} &\le \Phi(\rho\eta) + \Psi\left(\frac{\varphi(\xi)}{\rho}\right)\\ &\le \rho^\rho\Phi(\eta) + \frac{1}{\rho}\Psi(\varphi(\xi)) \le \rho^\rho\Phi(\eta) + \frac{\rho-1}{\rho}\Phi(\xi).\end{aligned}$$

From the assumption (A.54), we have

$$\mathbb{E}\Phi(\xi) \le \mathbb{E}\left[\eta\varphi(\xi)\right] \le \rho^\rho\mathbb{E}\Phi(\eta) + \frac{\rho-1}{\rho}\mathbb{E}\Phi(\xi).$$

This yields the lemma. □

The following lemma (called Garsia lemma) combined with Lemma A.12 can transform a martingale inequality for $\Phi(x) = x^p$ to general moderate convex function Φ.

Lemma A.13. *Let $(A_t, t \ge 0)$ be an $\mathcal{F}_t$-adapted, non-decreasing process and let ξ and η be two non-negative integrable random variables. Assume that $\xi \ge A_\infty$ a.s. and assume one of the following conditions holds:*

(i) $\xi \in \mathcal{F}_\infty$ *and for any stopping time* T,

$$\mathbb{E}\left[\xi|\mathcal{F}_T\right] - A_T I_{\{T>0\}} \le \mathbb{E}\left[\eta|\mathcal{F}_T\right] \quad \textit{almost surely.} \tag{A.56}$$

(ii) $\xi \in \mathcal{F}_\infty$; A *is predictable with* $A_0 = 0$ *and for any predictable time* T

$$\mathbb{E}\left[\xi|\mathcal{F}_T\right] - A_T \le \mathbb{E}\left[\eta|\mathcal{F}_T\right] \quad \textit{almost surely.} \tag{A.57}$$

Then, for all $\lambda > 0$, *we have*

$$\mathbb{E}\left[(\xi - \lambda) I_{\{\xi \ge \lambda\}}\right] \le \mathbb{E}\left[\eta I_{\{\xi\ge\lambda\}}\right] . \tag{A.58}$$

Moreover, for any nondecreasing and convex function $\Phi : \mathbb{R}_+ \to \mathbb{R}_+$ *with* $\Phi(0) = 0$, *we have*

$$\mathbb{E}\Phi(\xi) \le \mathbb{E}\left[\eta\varphi(\xi)\right] . \tag{A.59}$$

Proof. We start by showing (A.58). First, we assume that (i) holds. Define

$$T = \inf\{t : A_t > \lambda\} .$$

Then, $A_T I_{\{t>0\}} \le \lambda$. Since $\{\xi \ge \lambda\} = \{T < \infty\} \cup \{T = \infty, \xi \ge \lambda\} \in \mathcal{F}_T$, it follows from (A.56) that

$$\mathbb{E}\left\{(\xi - \lambda) I_{\{\xi\ge\lambda\}}\right\} \le \mathbb{E}\left\{\left(\xi - A_T I_{\{t>0\}}\right) I_{\{\xi\ge\lambda\}}\right\} \le \mathbb{E}\eta I_{\{\xi\ge\lambda\}}$$

which is (A.58). Now, we assume (ii). Since T is predictable, $T := \inf\{t, A_t \ge \lambda\}$ is also predictable. Hence, there is a sequence of predictable times T_n such that $T_n \uparrow T$. From (A.57) it follows that

$$\mathbb{E}\left[\xi|\mathcal{F}_{T_n}\right] - A_{T_n} \le \mathbb{E}(\eta|\mathcal{F}_{T_n}) \quad \text{almost surely.}$$

Letting $n \to \infty$ yields

$$\mathbb{E}\left[\xi|\mathcal{F}_T\right] - A_T \le \mathbb{E}(\eta|\mathcal{F}_T) \quad \text{almost surely.}$$

Since $\{\xi \ge \lambda\} = \{T < \infty\} \cup \{T = \infty, \xi \ge \lambda\} \in \mathcal{F}_T$ and $A_T \le \lambda$, we have

$$\begin{aligned}\mathbb{E}\left\{\left(\mathbb{E}\left[\xi|\mathcal{F}_T\right] - \lambda\right) I_{\{\xi\ge\lambda\}}\right\} &\le \mathbb{E}\left\{\left(\mathbb{E}\ \left[\xi|\mathcal{F}_T\right] - A_T\right) I_{\{\xi\ge\lambda\}}\right\} \\ &\le \mathbb{E}\left\{\left[\mathbb{E}(\eta|\mathcal{F}_T)\right] I_{\{\xi\ge\lambda\}}\right\} \quad \text{almost surely.}\end{aligned}$$

This proves (A.58).

Integrating (A.58) from 0 to ∞ with respect to $d\varphi(\lambda)$, where φ is the right derivative of Φ, we obtain (A.59). □

Theorem A.12 (Burkholder-David-Gundy inequality). *Let* $\{M_t,\ t \ge 0\}$ *be a continuous local martingale and let* Φ *be a moderate convex function on* $\mathbb{R}_+$ *such that* $\Phi(M_\infty^*)$ *and* $\Phi(\sqrt{\langle M\rangle_\infty})$ *are integrable. Then*

$$\begin{aligned}\rho^{-(\rho+1)} 6^{-\rho} \mathbb{E}\Phi(\sqrt{\langle M\rangle_\infty}) &\le \mathbb{E}\Phi(M_\infty^*) \\ &\le \rho^{\rho+1}(2\sqrt{2})^\rho \mathbb{E}\Phi(\sqrt{\langle M\rangle_\infty})\end{aligned} \tag{A.60}$$

where ρ *is defined by (A.9).*

Proof. Denote $A_t = \sup_{0\le s\le t} M_s$, $\xi = A_\infty = M^*$, and $\eta = 2\sqrt{2}\sqrt{\langle M\rangle_\infty}$. Then, it follows from Theorem A.11 and Lemma A.13 that $\mathbb{E}\left[\Phi(\xi)\right] \le \mathbb{E}\left[\eta\varphi(\xi)\right]$. By Lemma A.12 we have

$$\mathbb{E}\left[\Phi(\xi)\right] \le \rho^{\rho+1}\mathbb{E}\left[\Phi(\eta)\right] .$$

By the property of the moderate convex function (Lemma A.4, part (ii)), we have $\Phi(\eta) \le (2\sqrt{2})^\rho\Phi(\sqrt{\langle M\rangle_\infty})$. This proves the second inequality in (A.60).

To show the first inequality, we first bound $\mathbb{E}\left(\langle M\rangle_\infty\right)$ by a multiple of $\mathbb{E}(M^*_\infty)$. This again can be completed as follows. (i) using (A.50) and then using (A.47) to the case $k = 1/2$, to obtain $\mathbb{E}\left[(N^*_\infty)^{1/2}\right] \le 3\cdot 2^{9/2}\mathbb{E}\left[\langle N\rangle_\infty^{1/4}\right]$, where $N_t = \int_0^t M_s dM_s$. (ii) using the idea of (A.44) to bound $\mathbb{E}\left[\langle M\rangle_\infty^{1/2}\right]$ by a constant multiple of $\mathbb{E}\left[M^*_\infty\right]$. But this way the constant is not the best known. From [Dellacherie and Meyer (1982), p. 286, Equation (91.3)], we know that

$$\mathbb{E}\left[\langle M\rangle_\infty^{1/2}\right] \le 6\mathbb{E}\left[M^*_\infty\right] .$$

Or more generally we have

$$\mathbb{E}\left[\sqrt{\langle M\rangle_\infty} - \sqrt{\langle M\rangle_T}\mathcal{F}_T\right] \le 6\mathbb{E}\left[M^*_\infty|\mathcal{F}_T\right] , \quad \forall \text{ any stopping time } T .$$

This can be used to prove the first inequality of (A.60). □

From Example A.2, we see that $\rho = p$ if $\Phi(x) = x^p$, $x > 0$. Thus, we have the following Burkholder inequality.

Corollary A.4. *Let $\{M_t\,,\ t\ge 0\}$ be a continuous local martingale. Then, for any $p > 1$,*

$$\begin{aligned} p^{-(p+1)}6^{-p}\mathbb{E}\left[\langle M\rangle_\infty^{p/2}\right] &\le \mathbb{E}\left[(M^*_\infty)^p\right] \\ &\le p^{p+1}(2\sqrt{2})^p\mathbb{E}\left[\langle M\rangle_\infty^{p/2}\right] . \end{aligned} \tag{A.61}$$

Remark A.3. For the discussion of the best constants in the Burkholder-David-Gundy inequality, see [Burkholder (2002)], [Burkholder (1997)] and related works.

Bibliography

Adler, R. J. An introduction to continuity, extrema, and related topics for general Gaussian processes. Institute of Mathematical Statistics Lecture Notes - Monograph Series, 12. Institute of Mathematical Statistics, Hayward, CA, 1990.

Adler, R. J. and Taylor, J. E. Random fields and geometry. Springer Monographs in Mathematics. Springer, New York, 2007.

Albeverio, S.; Hu,Y. and and Zhou, X.Y. A remark on non smoothness of self-intersection local time of planar Brownian motion, Statistics and Probability Letters 32 (1997), 57-65.

Bhattacharya, R. N. On the functional central limit theorem and the law of the iterated logarithm for Markov processes. Z. Wahrsch. Verw. Gebiete 60 (1982), 185-201.

Ball, K. An elementary introduction to modern convex geometry. Flavors of geometry, 1-58, Math. Sci. Res. Inst. Publ., 31, Cambridge Univ. Press, Cambridge, 1997.

Bargmann, V. On a Hilbert space of analytic functions and an associated integral transform. Communications on Pure and Applied Mathematics 14 (1961), 187-214.

Bakry, D. Transformations de Riesz pour les semigroupe symetriques seconde partie: Etude sous la condition $\Gamma_2 \geq 0$, Séminaire de probabilités, XIX, 1983/84, 145-174, Lecture Notes in Math., 1123, Springer, Berlin-New York, 1985.

Bakry, D. L'hypercontractivité et son utilisation en théorie des semigroupes. Lectures on probability theory (Saint-Flour, 1992), 1–114, Lecture Notes in Math., 1581, Springer, Berlin, 1994.

Bakry, D. and Emery, M. Diffusions hypercontractives, Sem. Prob. XIV, Lecture Notes in Mathematics 1123, Springer, 1985, 177-206.

Bakry, D. and Ledoux, M. Lévy-Gromov's isoperimetric inequality for an infinite-dimensional diffusion generator. Invent. Math. 123 (1996), no. 2, 259-281.

Bakry, D. and Michel, D. Sur les inégalités FKG, Sem. Prob. XXVI, 170-188, Lecture Notes in Mathematics 1526, Springer, 1992.

Basse-O'Connor, A. Integrability of seminorms. Electron. J. Probab. 16 (2011),

216-229.
Barthe, F. and Huet, N. On Gaussian Brunn-Minkowski inequalities. Studia Math. 191 (2009), no. 3, 283-304.
Bass, R. F. and Chen, X. Self-intersection local time: Critical exponent, large deviations and laws of the iterated logarithm. Ann. Probab. 32 (2004), 3221-3247.
Beckenbach E.F. and Bellman R. *Inequalities.* Springer, 1965.
Beckner, W. Inequalities in Fourier analysis. Ann. of Math. (2) 102 (1975), no. 1, 159-182.
Beran, J. *Statistics for Long-Memory Processes.* Chapman and Hall. 1984.
Bingham, N. H.; Goldie, C. M. and Teugels, J. L. Regular variation. Encyclopedia of Mathematics and its Applications, 27. Cambridge University Press, Cambridge, 1989.
Biagini, F.; Hu, Y.; Øksendal, B. and Zhang, T.S. Stochastic calculus for fractional Brownian motion and applications. Probability and its Applications (New York). Springer-Verlag London, Ltd., London, 2008.
Bismut, J.-M. The Atiyah-Singer theorems: a probabilistic approach. I. The index theorem. J. Funct. Anal. 57 (1984), no. 1, 5699.
Bismut, J.-M. The Atiyah-Singer theorems: a probabilistic approach. II. The Lefschetz fixed point formulas. J. Funct. Anal. 57 (1984), no. 3, 329348.
Bismut, J.-M. Hypoelliptic Laplacian and Bott-Chern cohomology. A theorem of Riemann-Roch-Grothendieck in complex geometry. Progress in Mathematics, 305. Birkhäuser/Springer, Cham, 2013.
Bogachev, V. I. Gaussian measures. Mathematical Surveys and Monographs, 62. American Mathematical Society, Providence, RI, 1998.
Bogachev, V. I. ; Röckner, M. and Schmuland, B. Generalized Mehler semigroups and applications, Probability Theory and Related Fields 105 (1996), 193-225.
Bonami, A. Ensembles $\Lambda(p)$ dans le dual de D^∞. Ann. Inst. Fourier, 18(2) (1968), 193-204 .
Bonami, A. Étude des coefficients de Fourier des fonctions de $L_p(G)$. Ann. Inst. Fourier, 20 (1970), 335-402.
Borell, C. Tail probabilities in Gauss space. Vector space measures and applications. 73-82, Lecture Notes in Math., 644, Springer, 1978.
Borell, C. The Ehrhard inequality, C. R. Math. Acad. Sci. Paris 337 (2003), 663-666.
Böröczky, K. Jr. Finite packing and covering. Cambridge Tracts in Mathematics, 154. Cambridge University Press, Cambridge, 2004.
Borodin, A. N. Brownian local time. *Russian Math. Surveys* 44 (1989), 1-51.
Borodin, A. N. and Salminen, P. Handbook of Brownian motion—facts and formulae. Second edition. Probability and its Applications. Birkhäuser Verlag, Basel, 2002.
Bouleau, N. and Hirsch, F. Propriétés d'absolue continuité dans les espaces de Dirichlet et application aux équations différentielles stochastiques. Séminaire de Probabilités, XX, 1984/85, 131-161, Lecture Notes in Math., 1204, Springer, Berlin, 1986.

Bouleau, N. and Lépingle, D. Numerical methods for stochastic processes, John Wiley & Sons Inc. 1994.

Borovkov, A. A. Ergodicity and stability of stochastic processes. Wiley Series in Probability and Statistics: Probability and Statistics. John Wiley & Sons, Ltd., Chichester, 1998.

Brockwell, P. and Davis, R. *Time Series: Theory and Methods*, 2nd ed. Springer, 1991.

Buckdahn, R. Anticipative Girsanov transformations and Skorohod stochastic differential equations. Mem. Amer. Math. Soc. 111 (1994), no. 533, viii+88 pp.

Budhiraja, A. and Kallianpur, G. The generalized Hu-Meyer formula for random kernels. Appl. Math. Optim. 35 (1997), 177-202.

Burago, Yu. D. and Zalgaller, V. A. Geometric inequalities. Grundlehren der Mathematischen Wissenschaften, 285. Springer Series in Soviet Mathematics. Springer, 1988.

Burkholder, D. L. Sharp norm comparison of martingale maximal functions and stochastic integrals. Proceedings of the Norbert Wiener Centenary Congress, 1994 (East Lansing, MI, 1994), 343358, Proc. Sympos. Appl. Math., 52, Amer. Math. Soc., Providence, RI, 1997.

Burkholder, D. L. The best constant in the Davis inequality for the expectation of the martingale square function. Trans. Amer. Math. Soc. 354 (2002), no. 1, 91-105.

Carbery, A. and Wright, J. Distributional and L^q norm inequalities for polynomials over convex bodies in $\mathbb{R}^n$. *Math. Research Lett.* 8 (2001). 233-248.

Carlen, E. A. Superadditivity of Fisher's information and logarithmic Sobolev inequalities. J. Funct. Anal. 101 (1991), no. 1, 194-211.

Carmona, R. A. and Nualart, D. Traces of random variables on Wiener space and the Onsager-Machlup functional. J. Funct. Anal. 107 (1992), no. 2, 402-438.

Chen, L. H. Y.; Goldstein, L. and Shao, Q.-M. Normal approximation by Stein's method. Probability and its Applications (New York). Springer, Heidelberg, 2011.

Chatterjee, S. and Meckes, E. Multivariate normal approximation using exchangeable pairs. ALEA Lat. Am. J. Probab. Math. Stat. 4 (2008), 257-283.

Chen, X. Random walk intersections. Large deviations and related topics. Mathematical Surveys and Monographs, 157. American Mathematical Society, Providence, RI, 2010.

Chen, X., Li, W., Marcus, M. B. and Rosen, J. Chen, Xia; Li, Wenbo V.; Marcus, Michael B.; Rosen, Jay A CLT for the L2 modulus of continuity of Brownian local time. Ann. Probab. 38 (2010), no. 1, 396-438.

Chung, K. L. A course in probability theory. Third edition. Academic Press, Inc., San Diego, CA, 2001.

Chung, K. L. and Zhao, Z. X. From Brownian motion to Schrödinger's equation. Grundlehren der Mathematischen Wissenschaften, 312. Springer-Verlag, Berlin, 1995.

Decreusefond, L.; Hu, Y. and Üstünel, A. S. Une inégalité d'interpolation sur l'espace de Wiener. C. R. Acad. Sci. Paris Sér. I Math. 317 (1993), no. 11,

1065-1067.

Delgado, R. and Jolis, M. A Hu-Meyer formula for generalized multiple anticipating integrals. Stochastics Stochastics Rep. 58 (1996), 115-138.

Delgado, R. and Sanz, M. The Hu-Meyer formula for nondeterministic kernels. Stochastics Stochastics Rep. 38 (1992), 149-158.

Dellacherie, C. and Meyer, P.-A. Probabilités et potentiel. Hermann, Paris, 1975.

Dellacherie, C. and Meyer, P.-A. Probabilities and potential. B. Theory of martingales. Translated from the French by J. P. Wilson. North-Holland Mathematics Studies, 72. North-Holland Publishing Co., Amsterdam, 1982.

Dellacherie, C. and Meyer, P.-A. Probabilités et potentiel. Chapitres XII-XVI. Théorie du potentiel associée à une résolvante. Théorie des processus de Markov. Second edition. Publications de l'Institut de Math matiques de l'Université de Strasbourg, Hermann, Paris, 1987.

Dellacherie, C.; Maisonneuve, B. and Meyer, P.-A. Probabilités et potentiel. Chapitres XVII à XXIV. Processus de Markov (fin). Compléments de calcul stochastique. Publications de l'Institut de Mathmatiques de l'Université de Strasbourg XX. Actualités Scientifiques et Industrielles, 1410. Hermann, Paris, 1992.

Diestel, J.; Jarchow, H. and Tonge, A. Absolutely Summing Operators, Cambridge Studies in Advanced Mathematics, vol. 43, Cambridge University Press, Cambridge, 1995.

Dimitrov, D. K. Higher order Turán inequalities. Proc. Amer. Math. Soc. 126 (1998), 2033-2037.

Dominici, D. Asymptotic analysis of the Hermite polynomials from their differential-difference equation. J. Difference Equ. Appl., 13 (12) (2007), pp. 1115-1128.

Donoghue, W. F., Jr. Monotone matrix functions and analytic continuation. Die Grundlehren der mathematischen Wissenschaften, Band 207. Springer, New York-Heidelberg, 1974.

Duncan, T. E.; Hu, Y. and Pasik-Duncan, B. Stochastic calculus for fractional Brownian motion. I. Theory. SIAM J. Control Optim. 38 (2000), no. 2, 582-612.

Driver, B.K. and Hu, Y. On heat kernel logarithmic Sobolev inequalities, Stochastic Analysis and Applications (Powys, 1995), World Sci. Publishing, River Edge, NJ, 1996, 189-200.

Ehrhard, A. Symétrisation dans l'espace de Gauss, Math. Scand. 53 (1983), 281-301.

Epperson, J. B. The hypercontractive approach to exactly bounding an operator with complex Gaussian kernel. J. Funct. Anal. 87 (1989), no. 1, 1-30.

Eisenbaum, N. Characterization of positively correlated squared Gaussian processes. Ann. Probab. 42 (2014), 559-575.

Elworthy, K. D.; Le Jan, Y. and Li, X. M. On the geometry of diffusion operators and stochastic flows. Lecture Notes in Mathematics, 1720. Springer-Verlag, Berlin, 1999.

Evans, L. C. and Gariepy, R. F. Measure theory and fine properties of functions. Revised edition. Textbooks in Mathematics. CRC Press, Boca Raton, FL,

2015.

Fang, L. and Zhang, X. Slepian's inequality with respect to majorization. Linear Algebra Appl. 434 (2011), 1107-1118.

Fang, S. Non-dégénérescence des pseudo-normes de Sobolev sur l'espace de Wiener. (French) [Nondegeneracy of Sobolev pseudonorms on Wiener space] Bull. Sci. Math. 115 (1991), no. 2, 223–234.

Farré, M.; Jolis, M. and Utzet, F. Multiple Stratonovich integral and Hu-Meyer formula for Lévy processes. Ann. Probab. 38 (2010), 2136-2169.

Federer, H. Geometric measure theory. Die Grundlehren der mathematischen Wissenschaften, Band 153 Springer-Verlag New York Inc., New York, 1969.

Fernique, M. X. Intégrabilité des Vecteurs Gaussiens. Academie des Sciences, Paris, Comptes Rendus, 270 Séries A (1970), 1698-1699.

Feynman, R. P. and Hibbs, A. R. Quantum mechanics and path integrals. Emended edition. Emended and with a preface by Daniel F. Styer. Dover Publications, Inc., Mineola, NY, 2010.

GANIDIS, H.; ROYNETTE, B. and SIMONOT, F. Convergence rate of some semi-groups to their invariant probability. Stochastic Process. Appl. 79 (1999), 243-263.

Gardner, R. J. The Brunn Minkowski inequality. Bull. Amer. Math. Soc. (N.S.) 39 (3) (2002), 355-405.

Gardner, R. J. The Brunn-Minkowski inequality: A survey with proofs. available at http://www.ac.wwu.edu/~gardner.

Gardner, R. J. and Zvavitch, A. Gaussian Brunn-Minkowski inequalities. Trans. Amer. Math. Soc. 362 (2010), no. 10, 5333-5353.

Garsia, A. M.; Rodemich, E.; and Rumsey, H., Jr. A real variable lemma and the continuity of paths of some Gaussian processes. Indiana Univ. Math. J. 20 (1970/1971), 565-578.

Glimm, J. and Jaffe, A. Quantum Physics, A Functional Integral Point of View, 2nd edition, Springer, 1987.

Gohberg, I.; Goldberg, S.; and Krupnik, N. Traces and determinants of linear operators. Operator Theory: Advances and Applications, 116. Birkhäuser Verlag, Basel, 2000.

Gohberg, I.; Goldberg, S.; and Krupnik, N. Hilbert-Carleman and regularized determinants for linear operators. Integral Equations Operator Theory 27 (1997), no. 1, 1047.

Grafakos, L. Classical Fourier analysis. Second edition. Graduate Texts in Mathematics, 249. Springer, 2008.

Gradshteyn, I. S. and Ryzhik, I. M. Table of integrals, series, and products. Academic Press, Inc., San Diego, CA, 2000.

Granger, C.W.J. and Joyeux, R. An introduction to long-memory time series models and fractional differencing. *J. Time Ser. Anal.* 1 (1980). 15-29.

Gross, L. Abstract Wiener spaces. 1967 Proc. Fifth Berkeley Sympos. Math. Statist. and Probability (Berkeley, Calif., 1965/66), Vol. II: Contributions to Probability Theory, Part 1 pp. 3142 Univ. California Press, Berkeley, Calif.

Grothendieck, A. Résumé de la Théorie Métrique des Produit Tensoriels

Topologiques, Bol. Mat. Sao Paulo, 8 (1956), 1-79.
Gross, L. Logarithmic Sobolev inequalities. Amer. J. Math. 97 (1975), 1061-1083.
Gubner, J.A. Theorems and fallacies in the theory of long-range-dependent processes. *IEEE Transactions on Information Theory* 51 (2005). 1234-1239.
Haagerup, U. The best constants in the Khintchine inequality, Studia Math. 70 (1981), no. 3, 231-283 (1982).
Hairer, M. On Malliavin's proof of Hörmander's theorem. Bull. Sci. Math. 135 (2011), no. 6-7, 650666.
Hall, B. C. Lie groups, Lie algebras, and representations. An elementary introduction. Graduate Texts in Mathematics, 222. Springer-Verlag, New York, 2003.
Hall, B. C. Holomorphic methods in analysis and mathematical physics, in First Summer School in Analysis and Mathematical Physics (S. Prez-Esteva and C. Villegas-Blas, Eds.), 1-59, Contemporary Mathematics 260, Amer. Math. Soc. 2000.
Hargé, G. A particular case of correlation inequality for the Gaussian measure. Ann. Probab. 27 (1999), no. 4, 1939-1951.
Hargé, G. Characterization of equality in the correlation inequality for convex functions, the U-conjecture. Ann. Inst. H. Poincaré Probab. Statist. 41 (2005), no. 4, 753-765.
He, S. W.; Wang, J. G. and Yan, J. A. Semimartingale theory and stochastic calculus. Kexue Chubanshe (Science Press), Beijing; CRC Press, Boca Raton, FL, 1992.
Hille, E. and Phillips, R. S. Functional analysis and semi-groups. Third printing of the revised edition of 1957. American Mathematical Society Colloquium Publications, Vol. XXXI. American Mathematical Society, Providence, R. I., 1974.
Hoffmann-Jøgensen, J.; Shepp, L. A. and Dudley, R. M. On the lower tail of Gaussian seminorms. Ann. Probab. 7 (1979), no. 2, 319-342.
Hosking, J. R. M. Fractional differencing. *Biometrika* 68 (1981). 165-176.
Houdré, C. and Kagan, A. Variance inequalities for functions of Gaussian variables. J. Theoret. Probab. 8 (1995), no. 1, 23-30.
Hu, Y. Calculs formels sur les EDS de Stratonovitch. Séminaire de Probabilités, XXIV, 1988/89, 453-460, Lecture Notes in Math., 1426, Springer, 1990.
Hu, Y. Sur un travail de R. Carmona et D. Nualart. Séinaire de Probabilités, XXVI, 587-594, Lecture Notes in Math., 1526, Springer, Berlin, 1992.
Hu, Y. Some operator inequalities, Séminaire de Probabilités XXVIII, Ed. by J. Azema, P.A. Meyer and M. Yor, Lecture Notes in Mathematics 1583, Springer-Verlag, 1994, 316-333.
Hu, Y. On the differentiability of functions of an operator, Séminaire de Probabilités XXIX, Ed. J. Azema, P.A. Meyer and M. Yor, Lecture Notes in Mathematics 1613, Springer-Verlag, 1995, 218-219. 3.
Hu, Y. Wiener-Itô chaos expansion with exact residual and correlation, variance inequalities, Journal of Theoretical Probability, 10 (1997), 835-848.
Hu, Y. A unified approach to several inequalities for Gaussian and diffusion measures. Séminaire de Probabilités, XXXIV, 329-335. Lecture Notes in Math.,

1729, Springer, 2000.

Hu, Y. Integral transformations and anticipative calculus for fractional Brownian motions. Mem. Amer. Math. Soc. 175 (2005), no. 825.

Hu, Y. Stochastic quantization and ergodic theorem for diffusion processes. Sci. China Math. 55 (2012), no. 11, 2285-2296.

Hu, Y.; Jolis M. and Tindel S. On Stratonovich and Skorohod stochastic calculus for Gaussian processes. Annals of Probability 41 (2013), no. 3A, 1656-1693.

Hu, Y. and Kallianpur, G. Exponential integrability and application to stochastic quantization. Appl. Math. Optim. 37 (1998), no. 3, 295–353.

Hu, Y.; Kallianpur, G. and Xiong, J. An approximation for the Zakai equation. Appl. Math. Optim. 45 (2002), no. 1, 23-44.

Hu, Y. and Le, K. A multiparameter Garsia-Rodemich-Rumsey inequality and some applications. Stochastic Process. Appl. 123 (2013), no. 9, 3359-3377.

Hu, Y. and Le, K. Nonlinear Young integrals and differential systems in Hölder media. To appear in Transaction of American Mathematical Society.

Hu, Y., Lu, F. and Nualart, D. Convergence of densities of some functionals of Gaussian processes. *J. Funct. Anal.* 266 (2014), 814-875.

Hu, Y. and Meyer, P.-A. Chaos de Wiener et intégrale de Feynman. Séminaire de Probabilités, XXII, 51–71, Lecture Notes in Math., 1321, Springer, 1988.

Hu, Y. and Meyer, P.-A. Sur les intégrales multiples de Stratonovitch. Séminaire de Probabilités, XXII, 72–81, Lecture Notes in Math., 1321, Springer, 1988.

Hu, Y. and Meyer, P. A. On the approximation of multiple Stratonovich integrals. Stochastic processes, 141-147, Springer, New York, 1993.

Hu, Y. and Nualart, D. Renormalized self-intersection local time for fractional Brownian motion. Ann. Probab. 33 (2005), no. 3, 948-983.

Hu, Y. and Nualart, D. Parameter estimation for fractional Ornstein-Uhlenbeck processes. Statist. Probab. Lett. 80 (2010), no. 11-12, 1030–1038. MR2638974.

Hu, Y. and Nualart, D. Stochastic integral representation of the L^2 modulus of Brownian local time and a central limit theorem. Electron. Commun. Probab. 14 (2009), 529-539.

Hu, Y. and Nualart, D. Rough path analysis via fractional calculus. Trans. Amer. Math. Soc. 361 (2009), no. 5, 2689-2718.

Hu, Y., Nualart, D.: Central limit theorem for the third moment in space of the Brownian local time increments. Elect. Commun. Probab. 15 (2010), 396-410.

Hu, Y. and Øksendal, B. Wick approximation of quasilinear stochastic differential equations. Stochastic analysis and related topics, V (Silivri, 1994), 203–231, Progr. Probab., 38, Birkhäuser Boston, 1996.

Hu, Y. and Watanabe, S. Donsker's delta functions and approximation of heat kernels by time discretization methods, J. Math. Kyoto University, 36 (1996), 499-518.

Hunt, B. R. and Sauer, T. and Yorke, J. A. Prevalence: a translation-invariant "almost every" on infinite-dimensional spaces. Bull. Amer. Math. Soc. (N.S.) 27 (1992). 217-238.

Isobe, E. and Sato, S. Wiener-Hermite expansion of a process generated by an It stochastic differential equation. J. Appl. Probab. 20 (1983), no. 4, 754-765.

Itô, K. Multiple Wiener integral. J. Math. Soc. Japan 3 (1951), 157-169.

Itô, K. Complex multiple Wiener integral. Jap. J. Math. 22 (1952), 63-86.

Janson, S. On complex hypercontractivity. J. Funct. Anal. 151 (1997), no. 1, 270-280.

Janson, S. Gaussian Hilbert Spaces. Cambridge University Press, 1997.

Joag-Dev, K.; Perlman, M. D. and Pitt, L. D. Association of normal random variables and Slepian's inequality. Ann. Probab. 11 (1983), 451-455.

Johnson, G. W. and Kallianpur, G. Some remarks on Y. Z. Hu and P.-A. Meyer's paper and infinite-dimensional calculus on finitely additive canonical Hilbert space: Teor. Veroyatnost. i Primenen. 34 (1989), 742-752; translation in Theory Probab. Appl. 34 (1989), 679-689 (1990).

Kallenberg, O. On an independence criterion for multiple Wiener integrals. Ann. Probab. 19 (1991), no. 2, 483-485.

Kazamaki, N. The equivalence of two conditions on weighted norm inequalities for martingales, in Proc. Intern. Symp. on Stochastic Differential Equations, Kyoto, 1976, ed. K. Itô, John Wiley and sons, 1978.

Krivine, J.L. Constants de Grothendieck et Fonctions de Type Positif sur les Sphères, Adv. Math. 31 (1979), 16-30.

Kloeden, P. E. and Platen, E. Numerical Solution of Stochastic Differential Equations, Springer, 1992.

Kuelbs, J. and Li, W. V. Metric entropy and the small ball problem for Gaussian measures. J. Funct. Anal. 116 (1993), no. 1, 133-157.

Kuo, H. H. Gaussian measures in Banach spaces. Lecture Notes in Mathematics, Vol. 463. Springer-Verlag, Berlin-New York, 1975.

Kusuoka, S. and D. Stroock, D. Applications of the Malliavin calculus. II. J. Fac. Sci. Univ. Tokyo Sect. IA Math. 32, no. 1, (1985), 1-76.

Lang, S. Real and functional analysis. Third edition. Graduate Texts in Mathematics, 142. Springer-Verlag, New York, 1993.

Latała, R. and Matlak, D. Royen's proof of the Gaussian correlation inequality, 2015. arXiv:1512.08776.

Lebedev,N. N. Special Functions & Their Applications, translated by Richard R. Silverman, Dover Books on Mathematics, 1972.

Lifshits, M. Lectures on Gaussian processes. Springer Briefs in Mathematics. Springer, Heidelberg, 2012.

M. Ledoux, L'algèbre de Lie des gradients itérés d'un générateur Markovien, C.R. 317 (1993), 1049-1052.

Ledoux, M. and Talagrand, M. Probability in Banach spaces. Isoperimetry and processes. Reprint of the 1991 edition. Classics in Mathematics. Springer, 2011.

Li, W. V. and Linde, W. Approximation, metric entropy and small ball estimates for Gaussian measures. Ann. Probab. 27 (1999), no. 3, 1556-1578.

Li, W. V. and Shao, Q.-M. Gaussian processes: inequalities, small ball probabilities and applications. Stochastic processes: theory and methods, 533–597, Handbook of Statist., 19, North-Holland, Amsterdam, 2001. MR1861734.

Lim, A. P. C. and Luo, D. A note on Gaussian correlation inequalities for non-symmetric sets. Statist. Probab. Lett. 82 (2012), no. 1, 196-202.

Malliavin, P. Stochastic calculus of variation and hypoelliptic operators. Proceedings of the International Symposium on Stochastic Differential Equations (Res. Inst. Math. Sci., Kyoto Univ., Kyoto, 1976), pp. 195263, Wiley, New York-Chichester-Brisbane, 1978.

Meckes, E. An infinitesimal version of Stein's method of exchangeable pairs. Ph.D. thesis, Stanford University (2006).

Meyer, P. A. Fonctionelles multiplicatives et additives de Markov. Ann. Inst. Fourier (Grenoble) 12 (1962), 125-230.

Meyer, P. A. Démonstration probabiliste de certaines inégalités de Littlewood-Paley. I-IV. Les inégalités classiques. Séminaire de Probabilités X, pp. 125-183. Lecture Notes in Math., Vol. 511, Springer, 1976.

Meyer, P. A. Retour sur la théorie de Littlewood-Paley. Seminar on Probability, XV, pp. 151-166, Lecture Notes in Math., 850, Springer, 1981.

Meyer, P. A. Note sur les processus d'Ornstein-Uhlenbeck. Seminar on Probability, XVI, pp. 95-133, Lecture Notes in Math., 920, Springer, 1982.

Meyer, P. A. Malliavin calculus, and some pedagogy. Unpublished note.

Meyer, P. A. Quantum probability for probabilists. Lecture Notes in Mathematics, 1538. Springer-Verlag, Berlin, 1993.

Nazarov, F. and Podkorytov, A. Ball, Haagerup, and distribution functions, Complex analysis, operators, and related topics, 247-267, Oper. Theory Adv. Appl., 113, Birkhäuser, Basel, 2000.

Nelson, E. A quartic interaction in two dimensions. In Mathematical Theory of Elementary Particles, 69-73, MIT Press, 1966.

Nelson, E. Construction of quantum fields from Markoff fields. J. Functional Analysis 12 (1973) , 97-112.

Neveu, J. Sur l'espérance conditionnelle par rapport à un mouvement brownien. Ann. Inst. H. Poincaré Sect. B (N.S.) 12 (1976), no. 2, 105-109.

Norris, J. Simplified Malliavin calculus. Séminaire de Probabilités, XX, 1984/85, vol. 1204 of *Lecture Notes in Math.* , 101-130. Springer, Berlin, 1986.

Nourdin, I. and Peccati, G. Stein's method on Wiener chaos. Probab. Theory Related Fields 145 (2009), no. 1-2, 75-118.

Nourdin I. and Peccati G. Normal approximations with Malliavin calculus. From Stein's method to universality. Cambridge Tracts in Mathematics. Cambridge University Press. 2012.

Nourdin, I. and Poly, G. (2013). Convergence in total variation on Wiener chaos. *Stoch. Process. Appl.* 123 651-674.

Nualart, D. The Malliavin calculus and related topics, 2nd edition. Springer-Verlag 2006.

Nualart, D. and Ortiz-Latorre, S. Central limit theorems for multiple stochastic integrals and Malliavin calculus. Stochastic Process. Appl. 118 (2008), no. 4, 614-628. MR2394845.

Nualart, D. and Peccati, G. Central limit theorems for sequences of multiple stochastic integrals. Ann. Probab. 33 (2005), no. 1, 177-93.

Nualart, D.; Üstïnel, A. S. and Zakai, M. Some remarks on independence and conditioning on Wiener space. Stochastic analysis and related topics, II

(Silivri, 1988), 122-127, Lecture Notes in Math., 1444, Springer, Berlin, 1990.

Palma, W. *Long-Memory Time Series: Theory and Methods.* Wiley Series in Probability and Statistics. Wiley-Interscience [John Wiley & Sons], Hoboken, NJ., 2007.

Olafsson, G. The SegalBargmann Transform on Euclidean Space and Generalizations. World Scientific. 2014.

Peccati, G. and Tudor, C. A. Gaussian limits for vector-valued multiple stochastic integrals. Séminaire de Probabilités, XXXVIII, 247-262, Lecture Notes in Math., 1857, Springer, Berlin, 2005. MR2126978.

Pitt, L. D. A Gaussian correlation inequality for symmetric convex sets, Ann. Prob. 5 (1977), 470-474.

Pitt, L. D. Positively correlated normal variables are associated. Ann. Probab. 10 (1982), 496-499.

Pisier, G. Factorization of Linear Operators and Geometry of Banach Spaces, CBMS Regional Conf. Series in Math. 60, AMS, 1986.

Pisier, G. Riesz transforms: a simpler analytic proof of P. A. Meyer's inequality. In Séminaire de Probabilités, XXII, 485-501. Lecture Notes in Mathematics 1321. Springer, 1988.

Prékopa, A. Logarithmic concave measures with application to stochastic programming. Acta Sci. Math. 32 (1971), 301-316.

Prékopa, A. On logarithmic concave measures and functions. Acta Sci. Math. 34 (1973), 335-343.

Protter, P. E. Stochastic integration and differential equations. Second edition. Applications of Mathematics (New York), 21. Stochastic Modelling and Applied Probability. Springer-Verlag, Berlin, 2004.

Raič, M. A multivariate CLT for decomposable random vectors with finite second moments. J. Theoret. Probab. 17 (2004), no. 3, 573-603.

Ral'chenko, K. V. The two-parameter Garsia-Rodemich-Rumsey inequality and its application to fractional Brownian fields. Theory Probab. Math. Statist. No. 75 (2007), 167-178.

Revuz, D. and Yor, M. Continuous martingales and Brownian motion. Third edition. Grundlehren der Mathematischen Wissenschaften, 293. Springer-Verlag, Berlin, 1999.

Rosen, J. A stochastic calculus proof of the CLT for the L^2 modulus of continuity of local time. Séminaire de Probabilités XLIII, 95-104, Lecture Notes in Math., 2006, Springer, Berlin, 2011.

Rosen, J. A CLT for the third integrated moment of Brownian local time increments. (English summary) Stoch. Dyn. 11 (2011), no. 1, 5-48.

Royen, T. A simple proof of the Gaussian correlation conjecture extended to multivariate gamma distributions. 2014, arXiv:1408.1028.

Russo, F. and Vallois, P. Forward, backward and symmetric stochastic integration. Probab. Theory Related Fields 97 (1993), no. 3, 403-421.

Segal, I. E. Mathematical problems of relativistic physics, in Kac, M., Proceedings of the Summer Seminar, Boulder, Colorado, 1960, Vol. II, Lectures in Applied Mathematics, American Mathematical Society, Chap. VI.

Schneider, R. Convex bodies: the BrunnMinkowski theory, Cambridge University Press, Cambridge, 1993.

Simon, B. The $P(\phi)_2$ Euclidean (quantum) field theory. Princeton Series in Physics. Princeton University Press, Princeton, N.J., 1974.

Sikora, A. On-diagonal estimates on Schrödinger semigroup kernels and reduced heat kernels, Commun. Math. Phys. 188 (1997), 233-249.

Simon, B. Functional integration and quantum physics. Second edition. AMS Chelsea Publishing, Providence, RI, 2005.

Simon, B. Trace ideals and their applications. Second edition. Mathematical Surveys and Monographs, 120. American Mathematical Society, Providence, RI, 2005.

Solé, J. L. and Utzet, F. The Hu-Meyer formula. A view of the problem with special attention to the Poisson case. Chaos expansions, multiple Wiener-Itô integrals and their applications (Guanajuato, 1992), 35-45, Probab. Stochastics Ser., CRC, Boca Raton, FL, 1994.

Stein, C. A bound for the error in the normal approximation to the distribution of a sum of dependent random variables. Proceedings of the Sixth Berkeley Symposium on Mathematical Statistics and Probability (1972). 583–602.

Schilling, R. L.; Song, R. and Vondraček, Z. Bernstein functions. Theory and applications. Second edition. de Gruyter Studies in Mathematics, 37. Walter de Gruyter & Co., Berlin, 2012.

Slepian, D. The one-sided barrier problem for Gaussian noise. Bell System Tech. J. 41 (1962), 463-501.

Stroock D. W. *Diffusion semigroups corresponding to uniformly elliptic divergence form operators.* Séminaire de Probabilités, XXII, 316-347, Lecture Notes in Math., 1321, Springer, Berlin-New York, 1988.

Szegö. G. Orthogonal Polynomials, Revised ed., American Mathematical Society Colloquium Publications, vol. 23, American Mathematical Society, Rhode Island, 1959.

Szegö. G. On an inequality of P. Turán concerning Legendre polynomials, American Mathematical Society Bulletin 54 (1948), 401-405.

Skorokhod, A. V. Asymptotic methods in the theory of stochastic differential equations. Translated from the Russian by H. H. McFaden. Translations of Mathematical Monographs, 78. American Mathematical Society, Providence, RI, 1989.

Stein, E. M. Singular integrals and differentiability properties of functions. Princeton Mathematical Series, No. 30. Princeton University Press, Princeton, N.J. 1970

Stein, E. M. Topics in harmonic analysis related to the Littlewood-Paley theory. Annals of Mathematics Studies, No. 63, Princeton University Press, Princeton, N.J.; University of Tokyo Press, Tokyo 1970.

Stroock, D.W. Some applications of stochastic calculus to partial differential equations. Ecole dEté de Probabilités de Saint Flour, Lecture Notes in Math. 976 (1983), 267-382.

Sugita, H. Hu-Meyer's multiple Stratonovich integral and essential continuity of multiple Wiener integral. Bull. Sci. Math. 113 (1989), 463-474.

Svante, J. On complex hypercontractivity. J. Funct. Anal. 151 (1997), no. 1, 270-280.

Thangavelu, S. Lectures on Hermite and Laguerre expansions. Mathematical Notes, 42. Princeton University Press, Princeton, NJ, 1993.

Toh, T.-L. and Chew, T.-S. Henstock's multiple Wiener integral and Henstock's version of Hu-Meyer theorem. Math. Comput. Modelling 42 (2005), 139-149.

Tuynman, G. M. The derivation of the exponential map of matrices. Amer. Math. Monthly 102 (1995), no. 9, 818-820.

Üstünel, A. S. and Zakai, M. On independence and conditioning on Wiener space. Ann. Probab. 17 (1989), no. 4, 1441-1453.

Üstünel, A. S. and Zakai, M. Transformation of measure on Wiener space. Springer Monographs in Mathematics. Springer, Berlin, 2000.

Vakhania, N. N.; Tarieladze, V. I. and Chobanyan, S. A. Probability distributions on Banach spaces. Mathematics and its Applications (Soviet Series), 14. D. Reidel Publishing Co., Dordrecht, 1987.

Varadarajan, V. S. Lie groups, Lie algebras, and their representations. Prentice-Hall Series in Modern Analysis. Prentice-Hall, Inc., Englewood Cliffs, N.J., 1974.

Verger-Gaugry, J.-L. Covering a ball with smaller equal balls in R^n. Discrete Comput. Geom. 33 (2005), no. 1, 143-155.

Wang, B. and Hu, T. A new proof of fractional Hu-Meyer formula and its applications. J. Inequal. Appl. 2012, 12 pp.

Watanabe, S. Stochastic differential equation and Malliavin calculus. Tata institute of fundamental research, Springer, 1984.

Watanabe, S. Fractional order Sobolev spaces on Wiener space, Probab. Theory Relat. Fields 95 (1993), 175-198.

Weissler, F. B. Two-point inequalities, the Hermite semigroup, and the Gauss-Weierstrass semigroup. J. Funct. Anal. 32 (1979), no. 1, 102-121.

Widder, D. V. The Laplace Transform. Princeton Mathematical Series, v. 6. Princeton University Press, Princeton, N. J., 1941.

Wong, E. and Zakai, M. On the relation between ordinary and stochastic differential equations. Internat. J. Engrg. Sci. 3 (1965), 213-229.

Yamada, T. and Watanabe, S. On the uniqueness of solutions of stochastic differential equations. Journal of Mathematics of Kyoto University 11 (1971), 155-167.

Yan, J.-A. Notes on the Wiener semigroup and renormalization. Séminaire de Probabilités, XXV, 79–94, Lecture Notes in Math., 1485, Springer, Berlin, 1991.

Yan, J.-A. Products and transforms of white-noise functionals (in general setting). Appl. Math. Optim. 31 (1995), no. 2, 137–153.

Yosida, K. Functional analysis, Springer,1980.

Zeidler, E. Nonlinear functional analysis and its applications. I. Fixed-point theorems. Translated from the German by Peter R. Wadsack. Springer-Verlag, New York, 1986.

Zeidler, E. Nonlinear functional analysis and its applications. II/A. Linear monotone operators. Translated from the German by the author and Leo F. Boron. Springer-Verlag, New York, 1990.

Nonlinear functional analysis and its applications. II/B. Nonlinear monotone operators. Translated from the German by the author and Leo F. Boron. Springer-Verlag, New York, 1990.

Nonlinear functional analysis and its applications. III. Variational methods and optimization. Translated from the German by Leo F. Boron. Springer-Verlag, New York, 1985.

Nonlinear functional analysis and its applications. IV. Applications to mathematical physics. Translated from the German and with a preface by Juergen Quandt. Springer-Verlag, New York, 1988.

Index